Microbial Exopolysaccharides

This book offers a complete coverage of microbial refinery for exopolysaccharides (EPS) production, including genetic improvements, production techniques, biotechnological applications in food, cosmetics, health and environment sector, economic perspectives, and commercialization of EPS products. It focuses on exopolysaccharides production from an organism perspective to offer a complete picture from isolation of microbes to commercialization of EPS-based products. It covers strategies for EPS production and future perspectives and the potential of microbial refinery.

Features:

- Provides a concise introduction to the science, biology, technology, and application of exopolysaccharides (EPS).
- Details upstream and downstream steps in EPS production from microbial resources.
- Acts as a complete guide from production to commercialization.
- Explores the potential application of EPS for socioeconomical benefits.
- Discusses the EPS applications in food, cosmetics, health, and bioremediation approach for clean and sustainable development.

This book is aimed at researchers and graduate students in fermentation technology, biochemical engineering, and biotechnology.

Microbial Exopolysaccharides

Production and Applications

Edited by
Shashi Kant Bhatia, Parmjit Singh Panesar
and Sanjeet Mehariya

CRC Press
Taylor & Francis Group
Boca Raton London New York

CRC Press is an imprint of the
Taylor & Francis Group, an **informa** business

Designed cover image: © Shutterstock

First edition published 2024
by CRC Press
2385 NW Executive Center Drive, Suite 320, Boca Raton FL 33431

and by CRC Press
4 Park Square, Milton Park, Abingdon, Oxon, OX14 4RN

CRC Press is an imprint of Taylor & Francis Group, LLC

ISBN: 978-1-032-37941-8 (hbk)
ISBN: 978-1-032-37942-5 (pbk)
ISBN: 978-1-003-34268-7 (ebk)

DOI: 10.1201/9781003342687

Typeset in Times
by Apex CoVantage LLC

Contents

Chapter 11 Biomedical applications of exopolysaccharides 250

Thaaranni Bashkeran, Shinji Sakai, Retno Wahyu Nurhayati,
Minh Hong Nguyen, Wildan Mubarok, Ryota Goto,
Dinda Shezaria Hardy Lubis, Auzan Luthfi,
and Masrina Mohd Nadzir

Chapter 12 Exopolysaccharides for bioremediation 275

Ramesh Sharma, Pinku Chandra Nath, Shubhankar Debnath,
Amiya Ojha, Biswajit Sarkar, Tarun Kanti Bandyopadhyay,
and Biswanath Bhunia

Preface

As we delve deeper into the world of microorganisms, we begin to uncover the many hidden wonders that lie within. Among these wonders are "exopolysaccharides" that are produced by certain microbes. Exopolysaccharides (EPS) are complex polymers consisting of long chains of sugar molecules, produced by many microorganisms, including bacteria, fungi, and algae. These are secreted into the extracellular environment and form a protective layer around the cells, providing them with mechanical strength, shielding them from environmental stresses, and facilitating interactions with other cells and surfaces. EPS is known to exhibit a wide range of physical and chemical properties, including viscosity, elasticity, adhesiveness, and water-holding capacity, which make them useful in a variety of industrial applications. For example, some EPS have been used as thickeners and stabilizers in food and beverage industries, while others have been used in the production of biodegradable plastics, as well as in the formulation of drugs and cosmetics. In addition to their industrial applications, EPS also plays important roles in natural ecosystems. These can act as carbon and nutrient sinks and contribute to the formation of biofilms, which are complex microbial communities that play crucial roles in many environmental processes, such as bioremediation, wastewater treatment, and soil formation. EPS production and composition can vary widely depending on the microorganism, environmental conditions, and growth phase. Researchers are actively studying the biosynthesis and regulation of EPS production in order to optimize their production and tailor their properties for specific applications. Overall, the study of exopolysaccharides represents a fascinating intersection between microbiology, chemistry, and materials science, with broad applications in industry, biotechnology, and environmental science.

So join us on this journey as we unlock the secrets of microbial exopolysaccharides and discover the many ways in which they are changing the world. Whether you are a scientist, a student, or simply curious about the hidden wonders of the microbial world, this book is sure to captivate and inspire. The content of this book is organized in 14 chapters, and all the chapters are contributed by global experts from microbial fermentation, metabolic engineering, food technology, environmental technology, and biomedical science area. Chapter 1 covers the introduction and provides an overview of microbial exopolysaccharides. Chapter 2 is about the research guide that will help students and researchers to plan their experiments starting from microbial isolation, identification of EPS-producing microbes, EPS production and extraction methods, techniques in purification, composition analysis of EPS, and methodologies for characterization of EPS for various biomedical and environmental applications. Chapter 3 is focused on information related to various microbes (bacteria, fungi, yeast, and algae) involved in EPS production and fermentation strategy for increased production with various factors which affect yield. Chapter 4 discusses molecular aspects to design strategies for the metabolic engineering of microbial hosts for improved production of EPS. Chapter 5 covers the distinct bioreactor designs and optimization of bioprocess parameters for improved EPS production. Chapter 6 is about the use of various agro-industry wastes by-products (like molasses, cheese whey, corn steep liquor, etc.) as a feedstock for microbial fermentation to produce a variety of exopolysaccharides like xanthan, pullulan, gellan, levan, etc. Chapter 7 summarizes the production of exopolysaccharides from different extremophiles and their potential application in health and bioremediation of the environment. Chapter 8 emphasizes marine microorganism's exploration for exopolysaccharides production and their application in food, medical as well as environmental fields. Chapters 9–12 are about the application of EPS in various sectors such as food, cosmetics, biomedical, and bioremediation. Chapter 13 discusses recent research in the area of EPS composites and blends and their applications in different fields especially healthcare and tissue engineering. Chapter 14 is about advances in microalgal EPS production, strategies to stimulate microalgal EPS biosynthesis, and potential biotechnological applications in various areas.

In closing, the world of microbial exopolysaccharides is a captivating realm full of untold wonders waiting to be discovered. From their awe-inspiring role in protecting microorganisms to their seemingly limitless potential in industry, biotechnology, and environmental science, these complex polymers are a true marvel of nature. This book serves as a comprehensive guide to the field, offering an entrancing exploration of the intricate processes involved in their production, the diverse range of organisms that produce them, and the cutting-edge applications that are transforming the world around us. We hope that this book ignites your curiosity and imagination, inspiring you to delve deeper into the captivating world of microbial exopolysaccharides, and unlocking new insights and discoveries along the way. Thank you to all of the contributors for generously sharing your knowledge and expertise to make this book possible. Your contributions have been invaluable and will undoubtedly benefit countless readers for years to come. We are also thankful to the expert reviewers for providing their useful comments and suggestions, which helped to shape the chapter organization and improve the scientific discussions and overall quality of the chapters. We sincerely thank the CRC Press team comprising Dr. Gagandeep Singh and Aditi Mittal for their support and hard work in publishing this book.

Editors: Shashi Kant Bhatia,
Parmjit Singh Panesar, Sanjeet Mehariya

About the editors

Shashi Kant Bhatia is an associate professor in the Department of Biological Engineering, Konkuk University, Seoul, South Korea, and has more than ten years' experience in the area of biowastes valorization into bioenergy, biochemicals, and biomaterials. He earned an MSc and a PhD in biotechnology at Himachal Pradesh University (India). He has qualified JNU combined biotechnology entrance, DBT-JRF, GATE, CSIR exams. Dr. Bhatia has worked as a Brain Pool Post Doc Fellow at Konkuk University (2014–2016), South Korea. Dr. Bhatia has contributed extensively to the industrial press and served as an editorial board member of *Carbohydrate Polymer Technologies and Applications, Bioprocess and Biosystems Engineering, Sustainability and Energies* journal and associate editor of *Biotechnology for Biofuel and Bioproducts, Frontier of Microbiology, Microbial Cell Factories, Biomass Conversion and Biorefinery, 3 Biotech, International Journal of Biomaterials and PLOS One* journals. He has published more than 200 research and review articles on industrial biotechnology, bioenergy production, biomaterial, biotransformation, microbial fermentation, and enzyme technology in international scientific peer-reviewed journals and holds 15 international patents. He has also edited four books on microbial biotechnology. Dr. Bhatia has more than 9,000 citations with h-index >50 and i10 index >160 (As per Google scholar). He was selected to the Top 2% Scientist List of Stanford University.

Parmjit Singh Panesar is working as a Professor and Dean (Planning and Development), Sant Longowal Institute of Engineering and Technology Longowal, Punjab, India. Prof. Panesar has also worked as Dean (Research and Consultancy) and Head, Department of Food Engineering and Technology, SLIET Longowal, India. His research is focused in the area of food biotechnology, especially bioprocessing of food industry by-products, food enzymes, prebiotics (lactulose, GOS), biopigments, etc. Prof. Panesar has more than 7,000 citations and h-index of 46 (as per Google scholar). In 2005, he has been awarded BOYSCAST (Better Opportunities for Young Scientists in Chosen Areas of Science and Technology) fellowship by Department of Science and Technology (DST), Govt. of India, to carry out advance research at Chembiotech labs, University of Birmingham Research Park, UK. Prof. Panesar has successfully completed seven research projects funded by DBT, CSIR, MHRD, AICTE, New Delhi, and two projects are in progress. He has published more than 180 international/national scientific papers, 50 book reviews in peer-reviewed journals, 50 chapters, and has authored/edited eight books. He has guided 15 PhD students and eight students are under progress. He is a member of the editorial advisory boards of national/international journals, including *International Journal of Biological Macromolecules, Journal of Food Science & Technology, Carbohydrate Polymer Technologies & Applications.*

In recognition of his work, Prof. Panesar was awarded "Fellow Award 2018" by The Biotech Research Society of India (BRSI) and "Fellow Award 2019" by National Academy of Dairy Science India (NADSI). He has been selected for the prestigious "INSA Teachers Award (2020)" by Indian National Science Academy (INSA). Recently, Prof. Panesar was also listed in the most coveted list of "World Ranking of Top 2% Scientists" published by Stanford University, USA (2020, 2021, 2022, 2023). He has also served as the national vice president of the Association of Food Scientists & Technologists of India (AFSTI), during 2019. He has also visited several countries, like UK, USA, Canada, Switzerland, New Zealand, Australia, Germany, France, Singapore, Malaysia, China, Iran, Thailand etc.

Sanjeet Mehariya is a postdoctoral researcher at the Department of Chemistry, Umeå University, Umeå, Sweden. He earned a PhD in engineering at the University of Campania "Luigi Vanvitelli," Italy. Dr. Mehariya's research expertise includes the development of blue-bioeconomy through the algal biorefinery, the production of value-added chemical and the formulation of functional food from algal biomass, and agro-industrial waste-based resource recovery. He has collaborated with the national and international stakeholders, including policymakers, industries, and prominent RTD institutes in the field of biobased economy. He has worked at CSIR-Institute of Genomics and Integrative Biology, Delhi, India; Konkuk University, Seoul, South Korea; ENEA-Italian National Agency for New Technologies, Energy, and Sustainable Economic Development, Rome, Italy; Hong Kong Baptist University, Hong Kong; University of Campania "Luigi Vanvitelli," Italy; and Sapienza—University of Rome, Italy.

Contributors

Abedin, Md Minhajul
Institute of Bioresources and
  Sustainable Development
Regional Centre, Tadong 737102
Sikkim, India.

Ahuja, Vishal
University Institute of Biotechnology
Chandigarh University (Punjab) India
  and University Centre for Research
  Development
Chandigarh University (Punjab) India.

Bandyopadhyay, Tarun Kanti
Department of Chemical Engineering
National Institute of Technology
Agartala, India-799046.

Bashkeran, Thaaranni
School of Chemical Engineering
Engineering Campus
Universiti Sains Malaysia, 14300
  Nibong Tebal
Pulau Pinang, Malaysia.

Behl, Manya
Department of Biotechnology
Himachal Pradesh University
Shimla, India.

Bhatia, Ravi Kant
Department of Biotechnology
Himachal Pradesh University
Shimla.

Bhatia, Shashi Kant
Department of Biological Engineering
Konkuk University, Seoul-05029
South Korea and Institute for
  Ubiquitous Information Technology
  and Application
Konkuk University
Seoul-05029, Republic of Korea.

Bhatt, Arvind Kumar
Department of Biotechnology
Himachal Pradesh University, Shimla.

Bhunia, Biswanath
Department of Bio Engineering
National Institute of Technology
Agartala, India-799046.

Chandel, Neha
School of Medical and Allied
  Sciences
GD Goenka University
Gurugram-122103, Haryana, India.

Chaudhuri, Surabhi
Department of Biotechnology
National Institute of Technology
Durgapur, Durgapur 713209, India.

Chourasia, Rounak
Institute of Bioresources and
  Sustainable Development
Regional Centre, Tadong 737102
Sikkim, India.

Costa, Jorge Alberto Vieira
College of Chemistry and Food
  Engineering
Federal University of Rio Grande
Rio Grande, RS, Brazil.

de Carvalho, Matheus Pereira
College of Chemistry and Food
  Engineering
Federal University of Rio Grande
Rio Grande, RS, Brazil.

de Morais, Michele Greque
College of Chemistry and Food
  Engineering
Federal University of Rio Grande
Rio Grande, RS, Brazil.

Debnath, Shubhankar
Department of Bio Engineering
National Institute of Technology
Agartala, India-799046.

Devi, Mamta
Department of Biotechnology
Himachal Pradesh University,
    Shimla.

Goto, Ryota
Department of Materials Engineering
    Science
Graduate School of Engineering Science
Osaka University, Toyonaka
Osaka, 560-8531, Japan.

Gupta, Saurabh
PG Department of Microbiology
Mata Gujri College
Fatehgarh Sahib, 140406
Punjab, India.

Gurav, Ranjit
Ingram School of Engineering
Texas State University
United States of America.

Hazari, Meheria
Department of Biotechnology
National Institute of Technology
    Durgapur
Durgapur 713209, India.

Jha, Harit
Department of Biotechnology
Guru Ghasidas Vishwavidyalaya
Bilaspur, Chhattisgarh 495009,
    India.

Kaur, Brahmeet
Department of Food Engineering and
    Technology
Sant Longowal Institute of Engineering
    and Technology
Longowal 148106, Punjab, India.

Kaur, Harinderjeet
Department of Food Processing
    Technology
Sri Guru Granth Sahib World
    University
Fatehgarh Sahib, 140406, Punjab, India.

Kumar, Bikash
Department of Biosciences and
    Bioengineering
Indian Institute of Technology
Guwahati, Assam, India, 781039, and
    Department of Biosciences and
    Biomedical Engineering
Indian Institute of Technology,
    Indore.

Kumari, Deeksha
Department of Biotechnology
Himachal Pradesh university,
    Shimla.

Le, Tuan
School of Biotechnology and Food
    Technology
Hanoi University of Science and
    Technology
No. 1 Daicoviet, Haibatrung, Hanoi.

Lopes, Danielle Rubim
College of Chemistry and Food
    Engineering
Federal University of Rio Grande
Rio Grande, RS, Brazil.

Lubis, Dinda Shezaria Hardy
University of Bern
Hochschulstrasse 6, 3012 Bern
Switzerland, and Universitas Indonesia
Kampus Baru UI, Depok
West Java, Indonesia.

Luthfi, Auzan
Universitas Indonesia
Kampus Baru UI
Depok, West Java, Indonesia.

Mehariya, Sanjeet
Department of Chemistry
Umeå University
Umeå, Sweden 901 87.

Moreira, Juliana Botelho
College of Chemistry and Food
  Engineering
Federal University of Rio Grande
Rio Grande, RS, Brazil.

Mubarok, Wildan
Department of Materials Engineering
  Science
Graduate School of Engineering Science
Osaka University, Toyonaka
Osaka, 560-8531, Japan.

Nadzir, Masrina Mohd
School of Chemical Engineering
Engineering Campus
Universiti Sains Malaysia
14300 Nibong Tebal
Pulau Pinang, Malaysia.

Namdeo, Neha
Department of Biotechnology
Guru Ghasidas Vishwavidyalaya
Bilaspur, Chhattisgarh 495009, India.

Nath, Pinku Chandra
Department of Bio Engineering
National Institute of Technology
Agartala, India-799046.

Nguyen, Lan-Huong
School of Biotechnology and Food
  Technology
Hanoi University of Science and
  Technology
No. 1 Daicoviet, Haibatrung, Hanoi.

Nguyen, Minh Hong
Bioresource Research Center
Phenikaa University
Hanoi 12116, Vietnam.

Nguyen, Tien-Thanh
School of Biotechnology and Food
  Technology
Hanoi University of Science and
  Technology
No. 1 Daicoviet, Haibatrung, Hanoi.

Nurhayati, Retno Wahyu
Universitas Indonesia
Kampus Baru UI
Depok, West Java, Indonesia.

Ojha, Amiya
Department of Bio Engineering
National Institute of Technology
Agartala, India-799046.

Panesar, Gaurav
Department of Food Engineering and
  Technology
Tezpur University, Tezpur 784028
Assam, India.

Panesar, Parmjit Singh
Department of Food Engineering and
  Technology
Sant Longowal Institute of Engineering
  and Technology
Longowal, 148106
Sangrur, Punjab, India.

Pham, Tuan-Anh
School of Biotechnology and Food
  Technology
Hanoi University of Science and
  Technology
No. 1 Daicoviet, Haibatrung, Hanoi.

Phukon, Loreni Chiring
Institute of Bioresources and
  Sustainable Development
Regional Centre
Tadong 737102, Sikkim, India.

Rai, Amit Kumar
Food and Nutritional Biotechnology

National Agri-Food Biotechnology
    Institute
SAS Nagar, Mohali, 160071
Punjab, India.

Randhawa, Mansimran Kaur
GSSDGS Khalsa College
Patiala, 147001, Punjab, India.

Rathour, Ranju Kumari
Department of Biotechnology
Himachal Pradesh university
Shimla.

Saha, Sourav
Department of Biotechnology
National Institute of Technology
    Durgapur
Durgapur 713209, India.

Sakai, Shinji
Department of Materials Engineering
    Science
Graduate School of Engineering
    Science
Osaka University, Toyonaka
Osaka, 560-8531, Japan.

Santos, Thaisa Duarte
College of Chemistry and Food
    Engineering
Federal University of
    Rio Grande
Rio Grande, RS, Brazil

Sarkar, Biswajit
Department of Bio Engineering
National Institute of Technology
Agartala, India-799046.

Sharma, Ramesh
Department of Bio Engineering
National Institute of Technology
Agartala, India-799046.

Sharma, Swati
School of Skill Building
SRM University
Sikkim 737102, India.

Singh, Amrit
PG Department of Microbiology
Mata Gujri College
Fatehgarh Sahib, 140406
Punjab, India.

Singh, Harpal
Department of Food Processing
    Technology
Sri Guru Granth Sahib World
    University
Fatehgarh Sahib, 140406
Punjab, India.

Singh, Rupinder Pal
Department of Food Processing
    Technology
Sri Guru Granth Sahib World University
Fatehgarh Sahib, 140406, Punjab, India.

Singh, Sudhir P.
Center of Innovative and Applied
    Bioprocessing
Sector-81 (Knowledge City)
SAS Nagar, Mohali-140306, India.

Singh, Yadvinder
Department of Botany and
    Environmental Sciences
Sri Guru Granth Sahib World
    University
Fatehgarh Sahib, 140406, Punjab, India.

Son, Chu-Ky
School of Biotechnology and Food
    Technology
Hanoi University of Science and
    Technology
No. 1 Daicoviet, Haibatrung, Hanoi.

Thakur, Kalpana
Department of Biotechnology
Himachal Pradesh university, Shimla.

Thakur, Monika
Center of Innovative and Applied
  Bioprocessing
Sector-81 (Knowledge City)
SAS Nagar, Mohali-140306, India

Yang, Yung-Hun
Department of Biological
  Engineering
College of Engineering
Konkuk University
Seoul-05029
Republic of Korea.

1 Introduction to microbial exopolysaccharides

Shashi Kant Bhatia, Parmjit Singh Panesar, and Sanjeet Mehariya

An exopolysaccharide (EPS) is a polymer secreted by microbes in response to environmental stress conditions such as pH, temperature, and osmotic pressures (Feng et al., 2023; Nguyen et al., 2020). Various microbes including bacteria (*Streptococcus, Lactococcus, Pediococcus, Lactobacillus, Leuconostoc*, etc.), fungi (*Agaricus blazi, Cordyceps* sp., *Ganoderma lucidum, Grifola frondose, Lentinus edodes*, etc.), and algae (*Chlorella zofingiensis, Scenedesmus* sp. *Chlorella vulgaris, Porphyridium cruentum, Chlorella pyrenoidosa*, etc.) are able to produce EPS (Morais et al., 2022; Nguyen et al., 2020; Wan Mohtar et al., 2016). Exopolysaccharides are mainly composed of various sugars, proteins, fatty acids, and nucleic acids. Naturally, it helps microbes in surface attachment and protects against phagocytosis and improves the osmotic environment of cells to reduce the damage caused by the passage of toxic molecules or ions (Angelin and Kavitha, 2020). The production of EPS generally occurs intracellularly, and various pathways (active and passive) are involved to secret EPS (Dueholm et al., 2023). Exopolysaccharides are composed of several different sugar building blocks, and their properties vary with composition (Bhatia et al., 2022). Different microbes can produce various types of EPS, and it's important to search for and explore new microbes for the production of EPS that have unique properties (Sathiyanarayanan et al., 2016). This research area is gaining interest around the globe, and the number of publications is increasing each year. According to the Scopus database, around 2,199 research and review articles have been published in the last few years (2019–2023). The top five countries involved in the microbial biopolymer research area include China (565), India (289), the United States (190), Italy (120), and Brazil (104) (Figure 1.1).

Screening and isolation of EPS producing microbes is an essential step in research and has resulted in significant developments with the involvement of genetics, metabolic engineering, and metagenomics. Solid surface culture and liquid culture methodology are the two main approaches used for the screening and isolation of EPS-producing microbes (Rühmann et al., 2015). In the solid-surface culture process, polysaccharide secreting organisms produce an easily visible polysaccharide-containing film on the surface of the culture medium. In liquid culture methodology, polysaccharide-producing microorganisms are cultured in liquid broth, and on completion of growth, the supernatant is analyzed for EPS production by solvent precipitation method (Gurav et al., 2022). After successful screening and selection of microbes, optimization of nutrients and cultivation conditions are required to

DOI: 10.1201/9781003342687-1

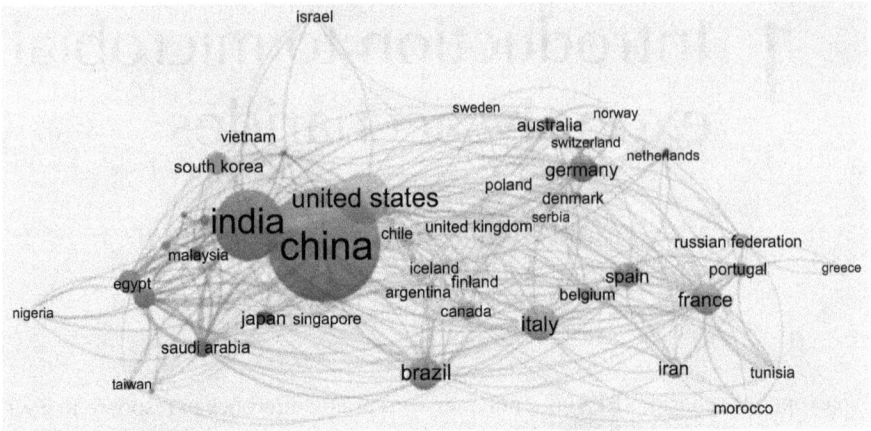

FIGURE 1.1 Co-occurrence mapping of publications related to exopolysaccharides (min. number of occurrence 10)

increase EPS production. The media components should be chosen according to the ability of microbes to utilize various carbon and nitrogen sources. The EPS production is highly dependent upon the type of fermenting substrate used, temperatures, and pH levels (Sørensen et al., 2022). For EPS production mainly two methods are used such as solid-state fermentation (SSF) and submerged fermentation (SmF). In solid-state fermentation (SSF), the production process does not involve a fermentation broth but rather a solid soaking material, such as soil and wood. These substrates usually contain organic substrates such as cellulose and starch, along with other nutrients in high quantities. This approach is beneficial as it results in a greater yield of exopolysaccharide production compared to submerged fermentation (SmF) (Seesuriyachan et al., 2012). In SmF, the microbial cells are normally suspended in a nutrient-containing broth and grown submerged (Yang et al., 2021). For large-scale EPS production bioreactors are used and various factors (pH, temperature, mass transfer rates, oxygen transfer rates, and agitation) may influence the reactor's performance (Freitas et al., 2017).

Separation of EPS is required from fermented broth, and the solvent precipitation method is the most widely used method. Ethanol and methanol are the most common solvents used for EPS precipitation (Ziadi et al., 2018). Protein content can be removed using trichloroacetic acid and EPS can be further purified using size exclusion chromatography (Wu et al., 2021). Exopolysaccharides can be characterized for their composition using various techniques such as gas chromatography (GC), gas chromatography-mass spectrometry (GC-MS), high performance liquid chromatography (HPLC), and size exclusion chromatography (Banerjee et al., 2022). The surface structure of EPS is very important and can be studied using instrumentation such as Scanning Electron Microscopes (SEMs), Field-Emission Scanning Electron Microscopes (FESEMs), and Atomic Force Microscopes (AFMs) (Bhatia et al., 2021). SEMs enable the visualization of the EPS surface by capturing the surface topography. FESEMs enable users to study the microstructure of EPS in detail and understand the chemical composition using different detector settings. AFMs enable the mapping of the nanoscale surfaces with atomic resolution, allowing for the observation of complex EPS structures at the nanolevel (Bhatia et al., 2021).

Exopolysaccharides are highly hydrated and able to absorb water, cross-linked, and form complex scaffolds. All these properties make them suitable candidates for diverse applications in food, cosmetics, drug delivery, tissue engineering, and environment-related applications (Bagnol et al., 2022). In food sectors EPS help to maintain texture and improve shelf life and taste. Xanthan gum is one of the highly used EPS as a thickening agent in processed foods (Bhat et al., 2022). In the cosmetic industry, EPS is used as fillers, thickeners, and moisturizers. Hyaluronic acid is found in many moisturizers and antiaging creams. In biomedical applications, EPS is used as a drug carrier to protect and release drugs at the right place and the right time (Saha and Datta, 2022). It is also used in wound healing products, scaffolds in regenerative medicine to help repair damaged or diseased tissue, as they provide a biocompatible matrix that encourages the growth of healthy cells (Mohd Nadzir et al., 2021). Bioremediation is a process of using biological systems to clean up environmental contamination and restore the environment. Exopolysaccharides improve remediation efforts in several ways. They can help increase the solubility of pollutants in the environment, making them easier to remove. They can also bind with pollutants, separating them from the environment and preventing them from being released back into the air, soil, or water (Gurav et al., 2022).

The properties of EPS can be further improved by preparing their composite material with natural polymers, synthetic polymers, and inorganic materials (Concórdio-Reis et al., 2023). Composite materials create a strong, porous material that has a number of appealing properties. In conclusion, exopolysaccharides composite materials are an attractive choice in the fields of biomedicine, electronics, and energy thanks to their low cost, low toxicity, and low weight combined with their strength, flexibility, and thermal insulation (Hussain et al., 2017). Additionally, their biodegradable nature adds yet another attractive property to this versatile material.

Exopolysaccharides can also be used to improve soil health and plant yields. The addition of exopolysaccharides to soils has been seen to lead to improvements in the physical and chemical properties of the soil, leading to greater water retention, decreased erosion, and improved fertility (Morcillo and Manzanera, 2021). Furthermore, it has been seen to increase the levels of sterilized soil additives, providing greater protection against pathogenic microorganisms and plants (Abdalla et al., 2021). Additionally, exopolysaccharides can improve plant growth by increasing their uptake of key nutrients.

Overall, microbial exopolysaccharides have been seen to have significant potential in the food industry, drug delivery, tissue engineering, and bioremediation. This book will cover topics related to recent advances in microbial exopolysaccharide production and their biotechnological applications.

ACKNOWLEDGMENTS

The authors acknowledge the KU Research Professor Program of Konkuk University, Seoul, South Korea.

REFERENCES

Abdalla, A.K., Ayyash, M.M., Olaimat, A.N., Osaili, T.M., Al-Nabulsi, A.A., Shah, N.P., Holley, R., 2021. Exopolysaccharides as antimicrobial agents: Mechanism and spectrum of activity. Front. Microbiol. 12.

Angelin, J., Kavitha, M., 2020. Exopolysaccharides from probiotic bacteria and their health potential. Int. J. Biol. Macromol. 162, 853–865.

Bagnol, R., Grijpma, D., Eglin, D., Moriarty, T.F., 2022. The production and application of bacterial exopolysaccharides as biomaterials for bone regeneration. Carbohydr. Polym. 291, 119550.

Banerjee, A., Mohammed Breig, S.J., Gómez, A., Sánchez-Arévalo, I., González-Faune, P., Sarkar, S., Bandopadhyay, R., Vuree, S., Cornejo, J., Tapia, J., Bravo, G., Cabrera-Barjas, G., 2022. Optimization and characterization of a novel exopolysaccharide from *Bacillus haynesii* CamB6 for food applications. Biomolecules. 12(6).

Bhat, I.M., Wani, S.M., Mir, S.A., Masoodi, F.A., 2022. Advances in xanthan gum production, modifications and its applications. Biocatal. Agric. Biotechnol. 42, 102328.

Bhatia, S.K., Gurav, R., Choi, Y.-K., Choi, T.-R., Kim, H.-J., Song, H.-S., Mi Lee, S., Lee Park, S., Soo Lee, H., Kim, Y.-G., Ahn, J., Yang, Y.-H., 2021. Bioprospecting of exopolysaccharide from marine Sphingobium yanoikuyae BBL01: Production, characterization, and metal chelation activity. Bioresour. Technol. 324, 124674.

Bhatia, S.K., Gurav, R., Kim, B., Kim, S., Cho, D.-H., Jung, H., Kim, Y.-G., Kim, J.-S., Yang, Y.-H., 2022. Coproduction of exopolysaccharide and polyhydroxyalkanoates from *Sphingobium yanoikuyae* BBL01 using biochar pretreated plant biomass hydrolysate. Bioresour. Technol. 361, 127753.

Concórdio-Reis, P., Ramos, K., Macedo, A.C., Serra, A.T., Moppert, X., Guézennec, J., Sevrin, C., Grandfils, C., Reis, M.A.M., Freitas, F., 2023. Bioactive exopolysaccharide-composites based on gold and silver nanoparticles tailored for wound healing. Mater. Today. Commun. 34, 105351.

Dueholm, M.K.D., Besteman, M., Zeuner, E.J., Riisgaard-Jensen, M., Nielsen, M.E., Vestergaard, S.Z., Heidelbach, S., Bekker, N.S., Nielsen, P.H., 2023. Genetic potential for exopolysaccharide synthesis in activated sludge bacteria uncovered by genome-resolved metagenomics. Water Res. 229, 119485.

Feng, L., Qian, T., Yang, G., Mu, J., 2023. Characteristics of exopolysaccharides produced by isolates from natural bioflocculant of *Ruditapes philippinarum* conglutination mud. Front. Microbiol. 13.

Freitas, F., Torres, C.A.V., Reis, M.A.M., 2017. Engineering aspects of microbial exopolysaccharide production. Bioresour. Technol. 245, 1674–1683.

Gurav, R., Bhatia, S.K., Choi, T.-R., Hyun Cho, D., Chan Kim, B., Hyun Kim, S., Ju Jung, H., Joong Kim, H., Jeon, J.-M., Yoon, J.-J., Yun, J., Yang, Y.-H., 2022. Lignocellulosic hydrolysate based biorefinery for marine exopolysaccharide production and application of the produced biopolymer in environmental clean-up. Bioresour. Technol. 359, 127499.

Hussain, A., Tabasum, S., Noreen, A., Ali, M., Iqbal, R., Zuber, M., 2017. Blends and composites of exopolysaccharides; properties and applications: A review. Int. J. Biol. Macromol. 94, 10–27.

Mohd Nadzir, M., Nurhayati, R.W., Idris, F.N., Nguyen, M.H., 2021. Biomedical applications of bacterial exopolysaccharides: A review. Polymers (Basel). 13(4).

Morais, M.G., Santos, T.D., Moraes, L., Vaz, B.S., Morais, E.G., Costa, J.A.V., 2022. Exopolysaccharides from microalgae: Production in a biorefinery framework and potential applications. Bioresour. Technol. Rep. 18, 101006.

Morcillo, R.J.L., Manzanera, M., 2021. The effects of plant-associated bacterial exopolysac-charides on plant abiotic stress tolerance. Metabolites. 11(6).

Nguyen, P.-T., Nguyen, T.-T., Bui, D.-C., Hong, P.-T., Hoang, Q.-K., Nguyen, H.-T., 2020. Exopolysaccharide production by lactic acid bacteria: The manipulation of environ-mental stresses for industrial applications. AIMS Microbiol. 6, 451–469.

Rühmann, B., Schmid, J., Sieber, V., 2015. Methods to identify the unexplored diversity of microbial exopolysaccharides. Front. Microbiol. 6.

Saha, I., Datta, S., 2022. Bacterial exopolysaccharides in drug delivery applications. J. Drug Deliv. Sci. Technol. 74, 103557.

Sathiyanarayanan, G., Bhatia, S.K., Kim, H.J., Kim, J.-H., Jeon, J.-M., Kim, Y.-G., Park, S.-H., Lee, S.H., Lee, Y.K., Yang, Y.-H., 2016. Metal removal and reduction potential of an exopolysaccharide produced by Arctic psychrotrophic bacterium *Pseudomonas* sp. PAMC 28620. RSC Adv. 6, 96870–96881.

Seesuriyachan, P., Kuntiya, A., Hanmoungjai, P., Techapun, C., Chaiyaso, T., Leksawasdi, N., 2012. Optimization of exopolysaccharide overproduction by *Lactobacillus confusus* in solid state fermentation under high salinity stress. Biosci. Biotechnol. Biochem. 76(5), 912–917.

Sørensen, H.M., Rochfort, K.D., Maye, S., MacLeod, G., Brabazon, D., Loscher, C., Freeland, B., 2022. Exopolysaccharides of lactic acid bacteria: Production, purification and health benefits towards functional food. Nutrients. 14(14), 2938.

Wan Mohtar, W.A.A.Q.I., Ab. Latif, N., Harvey, L.M., McNeil, B., 2016. Production of exo-polysaccharide by *Ganoderma lucidum* in a repeated-batch fermentation. Biocatal. Agric. Biotechnol. 6, 91–101.

Wu, J., Yan, D., Liu, Y., Luo, X., Li, Y., Cao, C., Li, M., Han, Q., Wang, C., Wu, R., Zhang, L., 2021. Purification, structural characteristics, and biological activities of exopolysac-charide isolated from leuconostoc mesenteroides SN-8. front. Microbiol. 12.

Yang, X., Yang, Y., Zhang, Y., He, J., Xie, Y., 2021. Enhanced exopolysaccharide produc-tion in submerged fermentation of *Ganoderma lucidum* by Tween 80 supplementation. Bioprocess. Biosyst. Eng. 44, 47–56.

Ziadi, M., Bouzaiene, T., M'Hir, S., Zaafouri, K., Mokhtar, F., Hamdi, M., Boisset-Helbert, C., 2018. Evaluation of the efficiency of ethanol precipitation and ultrafiltration on the purification and characteristics of exopolysaccharides produced by three lactic acid bacteria. BioMed Res. Int. 2018, 1896240.

2 A research guide for EPS production and characterization

Neha Chandel, Vishal Ahuja,
Yung-Hun Yang, and Shashi Kant Bhatia

2.1 INTRODUCTION

Carbohydrates are cosmopolitan natural biopolymers and act as structural as well as storage media in all forms of life (Zhang et al., 2021). In carbohydrates, smaller monomers are interconnected with acetal linkage to form large branched and unbranched oligosaccharides and polysaccharides molecules to perform different functions. Types of structure and bonding pattern determine the physical and chemical properties of polysaccharides and ultimately industrial fate in bakery, cosmetics, food-beverages, pharma, and health care, for example, alginate, arabic gum, agar, carrageenan, gaur gum, and starch, are some of the commonly used polysaccharides (Mohd Nadzir et al., 2021).

Microbial polysaccharides can be categorized into structural/cell wall polysaccharides (participate in formation and maintenance of cell membrane and wall), intracellular polysaccharides (localized within cell compartment), and extracellular polysaccharides (EPS). Cells release EPS outside the cellular compartment into the growth medium/surrounding (Bhatia et al., 2021). Several microorganisms, including bacteria, fungi, and algae, etc., like *Pseudomonas* sp. (Sathiyanarayanan et al., 2016), *Sphingobium yanoikuyae* BBL01 (Bhatia et al., 2022), *Echinicola sediminis* BBL-M-12 (Gurav et al., 2022), *Phanerochaete chrysosporium* (Cao et al., 2018) have been reported for extracellular polysaccharides production. Thermodynamically, EPS biosynthesis in microorganism is an energy-expensive process, thus it is only possible when EPS biosynthesis offer some exceptional advantage to the producer like defense, stress tolerance, pathogenesis, and drug resistance (Maeda et al., 2021).

Depending upon the microbes, the need of cells, and surrounding growth conditions, the composition of EPS vary (Ibrahim et al., 2022). Various microbes are able to produce different types of EPS, so it's important to search for new microbes for their EPS production potential. The properties of EPS depend upon its composition, and various techniques can be used to analyze it such as TLC, HPLC, GC-MS, FTIR, and NMR, etc. The properties of EPS also depend on the composition and can be used in different sectors, like cosmetics, food, nutraceuticals, biomedical, oil recovery, bioremediation, etc. This book chapter will provide complete research

DOI: 10.1201/9781003342687-2

guidance to researchers related to the isolation of EPS-producing microbes, composition analysis, structural analysis, and protocol to study various applications.

2.2 BIOSYNTHETIC PATHWAY

EPS are high molecular weight microbial metabolites, which are mainly composed of carbohydrates (sugar residues) along with protein and non-proteins fractions including DNA, phospholipids, acetate, carboxylate, glycerol, phosphates, pyruvate, sulfate, and succinate (Angelin and Kavitha, 2020). Mostly EPSs are synthesized intracellularly but in some cases like levans and dextrans, EPS are synthesized and polymerized outside the cells (Figure 2.1). Biosynthesis is a multistep process governed by numerous enzymes, located at different sites in the cellular compartment. Based on action and localization, enzyme groups have been divided into four categories: (i) intracellular enzymes involved in metabolic pathways that generate monosaccharides residues; (ii) intracellular enzymes, including hexokinases that phosphorylates sugar residues; (iii) enzymes bounded to periplasmic membrane, that is, glycosyltransferases catalyze the transfer of sugar residues; and (iv) enzymes located outside the cell membrane and catalyze the polymerization (Kumar et al., 2007a).

Biosynthetic pathways begin with cellular uptake of the substrate by active/passive diffusion, its metabolism in the central metabolite pathway followed by

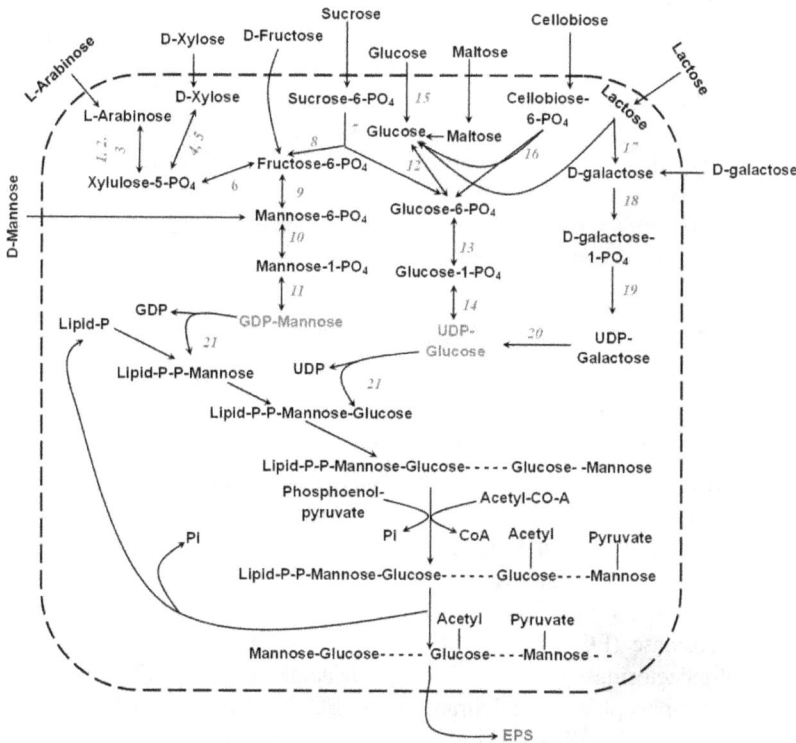

FIGURE 2.1 Biosynthesis mechanism of exopolysaccharides

polysaccharide synthesis (Freitas et al., 2011). In general, the EPS biosynthesis process is completed in three following steps:

- *Synthesis of nucleotide precursor:* Uridine diphosphate glucose (UDG) and thymidine diphosphate glucose (TDG) are the activated sugar precursors (ASPs) act as repeating units. These ASPs are produced from glycolytic intermediates, including glucose-6-phosphate (G6P) and fructose-6-phosphate (F-6-P) in multiple step process (Nguyen et al., 2020).
- *EPS chain synthesis:* Isoprenoid lipid carrier, undecaprenyl phosphate (UDeP) present at the cytoplasmic membrane is the core and initiation point for EPS synthesis. Priming glycosyltransferase catalyzes the attachment of sugar nucleotide with UDeP. Glycosyltransferases regulate the sequential addition of sugar nucleotides to it and govern chain elongation from repeating units (De Vuyst and Degeest, 1999).
- *Polymerization:* In the final stage, the repeating units are exported from the inner core to the outer part of the cell membrane. The whole export and polymerization process is catalyzed by three different proteins encoded by EPS gene cluster.
  - Flippase/translocase (encoded by wzx/cpsJ gene) facilitates the transportation of lipid carrier-repeating unit complex to the periplasmic site from the inner cytoplasmic surface.
  - Polymerase (encoded by wzy/cpsH) couples the repeating units (Laws et al., 2001).
  - Chain length determination protein that detects the chain length, stops polymerization after completion, and separates lipid carrier-repeating unit complex followed by export of final EPSs (Laws et al., 2001).

Exopolysaccharides are of two types: homo-polysaccharides and hetero-polysaccharides depending on composition. Homopolysaccharide synthesis involves two major components, that is, specific glycosyltransferase (glucansucrase or fructansucrase) and an extracellular sugar donor.

1. L-arabinose isomerase (EC 5.3.1.4); 2. Ribulokinase (EC 2.7.1.16); 3. L-ribulose-5-phosphate 4-epimerase (EC 5.1.3.4); 4. Xylose isomerase (EC 5.3.1.5); 5. Xylulose kinase (EC 2.7.1.17); 6. Transketolase (EC 2.2.1.1); 7. Fructokinase (EC 2.7.1.4); 8. Sucrose-6-phosphate hydrolase (EC 3.2.1.26); 9. Mannose-6-phosphate isomerase (EC 5.3.1.8); 10. Phosphomannomutase (EC 5.4.2.8); 11. Mannose-1-phosphate guanylyltransferase (EC 2.7.7.13); 12. Glucokinase (EC 2.7.1.2); 13. Phosphoglucomutase (EC 5.4.2.2); 14. UTP-glucose-1-phosphate uridylyltransferase (EC 2.7.7.9); 15. α-glucosidase (EC 3.2.1.20); 16. 6-phospho-β-glucosidase (EC 3.2.1.86); 17. β-galactosidase (EC 3.2.1.23); 18. Galactokinase (EC 2.7.1.6); 19. Galactose-1-phosphate uridylyltransferase (EC 2.7.7.10); 20. UDP-glucose-4-epimerase (EC 5.1.3.2); 21. Glucose-6-phosphate isomerase (EC 5.3.1.9)

For the synthesis of glucan, sucrose acts as an extracellular sugar donor, while fructans utilize fructose containing oligosaccharides. In the case of heteropolysaccharide, the synthesis process becomes a little bit complex due to the involvement of multiple factors and numerous gene products (enzymes). Based on the functionality, genes take part in EPS synthesis, which can be classified into four groups:

- Enzymes like priming glycosyltranferases, flippase, polymerase, and EpsA that act as polysaccharide assembly machinery
- Phosphoregulatory system including EpsB, EpsC, and EpsD that regulate the activity of polysaccharide assembly
- Sugar nucleotide biosynthetic pathways that supply sugar precursors
- Glycosyltranferases that initiate and elongate the glycan chains by transferring activated sugar residues

Besides, some other genes also encode accessories enzyme that decorates EPS by adding acetyl-, pyruvate, and other functional groups (Nguyen et al., 2020; Zeidan et al., 2017).

2.3 MICROORGANISMS PRODUCE EPS

For commercial production, microbial strains having high growth rates and EPS production capability are required parameters, and screening and selection of a potent microbe is an important step. Soil, water, and even fecal sample can be used as a source for the isolation of EPS producing microorganisms. Serial dilution and spreading are the common way to culture microorganisms on glucose-containing agar medium. Incubation conditions may vary with respect to fungi, and bacteria, etc., that is, 27±2°C for fungi and 35±2°C for bacteria for 5–7 d (Balasubramanian et al., 2019; Saleh Amer et al., 2020). Grown colonies observed for the appearance as EPS producing colonies appear to be highly mucoid (Jin et al., 2019). Based on the appearance, colonies can be further cultured and screened for quantitative and qualitative characterization of EPS.

Microbial biomass can be collected from the suspension culture via centrifugation or filtration through filter paper. Collected biomass should be dried at 60°C and treated with 10% trichloroacetic acid (TCA) for 20–30 min to degrade protein. To remove the cells, debris samples are centrifuged to remove the pellet and recovered supernatant mixed with icy-chilled absolute alcohol (3–4 times of sample) and kept still at 4°C for 24 h to precipitate the EPS. Precipitated EPS is collected by centrifugation and dried overnight at 40°C (Balasubramanian et al., 2019; Saleh Amer et al., 2020).

Several microbial strains have been identified for the production of different EPS, including *Bacillus subtilis* (Cerminati et al., 2021), *Escherichia coli* (Woo et al., 2019), *Streptococcus zooepidemicus* (Zakeri et al., 2017) for hyaluronic acid; *Tanticharoenia sakaeratensis* (Aramsangtienchai et al., 2020), *Neoasaia chiangmaiensis*, *Kozakia baliensis*, and *Gluconobacter cerinus* (Anguluri et al., 2022) for levan; *Aureobasidium melanogenum* for pullulan (Chen et al., 2019); *Leuconostoc mesenteroides* (Sarwat et al., 2008) and *Lactobacillus* (Jumma Kareem and Abdul Sattar Salman, 2019) for dextran. Microbial production might be a profitable method for EPS production due to its high growth rate and ability to utilize a wide range of feedstocks. Therefore, extensive efforts are needed to improve the feasibility and commercial acceptability. Table 2.1 summarizes the efforts of EPS production under varied environments.

TABLE 2.1

Microorganisms cultivation conditions and nutrient requirement for exopolysaccharide production

Microorganism	Samples used for isolation	Carbon source	Culture conditions	Productivity	Reference
Pseudomonas aeruginosa MTCC 1688	Outer ear infection (Culture collection)	–	Nutrient broth, 37°C, pH 7, 96 h	0.52 g/L	(Chug et al., 2021)
Echinicola sediminis BBL-M-12	Marine	Miscanthus hydrolysate	20 g/L glucose	6.18 g/L	(Gurav et al., 2022)
Gluconacetobacter sp.	-	Glycerol	-	25.4 ± 2.4 g/L 0.46 ± 0.04 g/L.h	(Rath et al., 2022)
Rhodotorula mucilaginosa sp. GUMS16	Leaf debris of Deylaman jungle, Guilan, Iran	Sucrose	0.01 g/mL sucrose, 0.14 g/mL ammonium sulfate, pH 5 and 25°C	130 g/L	(Okoro et al., 2021)
Xanthomonas campestris 1182	Phytobacteria Culture of the Biological Institute Brazil	Shrimp shell	0.01 g/L urea, KH_2PO_4 0.1 g/L, 10% shrimp shell extract, pH 7, 28°C, 102 h	4.64 g/L	(de Sousa Costa et al., 2014)
X. campestris PTCC 1473	Persian Type Collection Culture, Tehran, Iran	Alkali-pretreated rice straw	0.8 g/L NH_4NO_3, 2.0 g/L yeast extract, 2.5 g/L Na_2HPO_4, 2.5 g/L KH_2PO_4, pH 7.5, 72 h	9.7 g/100 biomass	(Jazini et al., 2017)
X. campestris	Leaves of infected plants	Date juice	20 g/L sucrose, 0.1 g/L urea, and 1 g/L K_2HPO_4	18.9 g/L	(Al-Roomi and Al-Sahlany, 2022)
Aspergillus sp.	Soil samples from agricultural field	Maltose	1% w/w maltose, 5.36 mg/g gelatin, 4.9 mg/g Cu^{2+}, 3.5 mg/g Ca^{2+}, 2.9 mg/g Zn^{2+}, 3.4 mg/g Mn^{2+} and 1.8 mg/g Mg^{2+}, pH 6, 30°C	22.2 mg/g	(Balasubramanian et al., 2019)
A. pullulans		Hazelnut husk	7.2 g/L $(NH_4)_2SO_4$, 2.5 mL H_2SO_4, 20 g ground hazelnut husk	74.39 g/L	(Akdeniz Oktay et al., 2022)
Fomitopsis meliae AGDP-2	Fruiting bodies of mushrooms cultivation sites of Anand	Maltose	37.5 g/L maltose, 6 g/L peptone and 0.9 g/L $MgSO_4$.	10.4 g/L	(Prajapati et al., 2022)

Organism	Source	Substrate	Conditions	Yield	Reference
Bacillus cereus SZ1	Stems of *A. annua*	Peptone, yeast extract	10 g peptone, 5 g yeast extract 10 g NaCl, pH 7.0 30°C, 180 rpm, 10 h	46 mg/L	(Zheng et al., 2016)
Pestalotiopsis sp. BC55	*Andrographis paniculata* (Burm.f.) Nees	Potato infusion, dextrose, glucose	30% potato dextrose broth, with 7.66% glucose, 0.29% urea, 0.05% $CaCl_2$, pH 6.93, 24°C, 3.76 days	4.32 g/L	(Mahapatra and Banerjee, 2016)
Bionectria ochroleuca M21	*Psidium guajava*	Glucose, yeast extract	55.7 g/L glucose, 6.04 g/L yeast extract, 0.25 g/L $MgSO_4$, 0.1% Tween 80, pH 6.5, 25°C, 150 rpm, 4 days	2.65 g/L	(Li et al., 2016b)
Porphyridium sordidum	Mahdia seawater in central eastern Tunisia	-	41.62 g/L NaCl, 0.63 g/L $NaNO_3$ and 7.2 g/L $MgSO_4$, 0.9 g/L of K_2HPO_4, 1.55 g/L of $CaCl_2$, 1 mL/L trace metal solution and vitamin solution; pH 7.5, 20°C, luminous intensity of 150 µmol photons/m²/s, 100 rpm	2.45 g/L	(Drira et al., 2021)
E. coli	Korean Culture Center of Microorganisms Korea	Glucose and galactose	5 g/L yeast extract, 10 g/L peptone, 10 g/L NaCl, 50 µg/mL ampicillin, 25 µg/mL kanamycin, and 17.5 µg/mL chloramphenicol, 37°C, 200 rpm	29.98 mg/L	(Woo et al., 2019)
Tanticharoenia sakaeratensis	Bioresource Research Center Thailand	Sucrose	20% (w/v) sucrose, peptone, 5 g/L NaCl, 1.5 g/L meat extract, 1.5 g/L yeast extract, 37°C, 250 rpm, 35 h	24.7 g/L	(Aramsangtienchai et al., 2020)
Aureobasidium melanogenum	Soil	Maltose	303 g/L maltose; initial pH 7; 8 g/L Tween 80, 30°C, 180rpm	122.34 g/L	(Chen et al., 2019)

2.4   EPS DETECTION AND QUANTIFICATION TECHNIQUES

Basic composition of EPS, that is, sugars, proteins, etc., can be determined spectrophotometrically using various standard protocols (Ozturk Urek and Ilgin, 2019). The molecular weight of EPS can be determined by gel permeation chromatography. Ethanol precipitated dried EPS is reconstituted in 0.2 M NaCl buffer and loaded onto Sepharose CL-6B column and eluted with buffer (flow rate 0.6 mL/min). Dextran standard of different molecular weight (25000, 80000, 270000, 670000, and 1100000 Da) is used as standards. In addition, total sugar and reducing sugar can be determined with the phenol sulfuric method (Masuko et al., 2005) and dinitrosalicylic acid (Miller, 1959) respectively. A glucose solution of 1 mg/mL can be used as a standard.

In the phenol sulfuric method, solid EPS powder (100 mg) is mixed with 5 mL of 2.5 N HCl and boiled for 3 h in a water bath to hydrolyze the polymer carbohydrate. Sodium carbonate is added to neutralize the acid, and the sample is diluted to 100 mL with water: 50–100 μL hydrolyzed EPS solution mixed with 1 mL 5% phenol reagent and 5 mL concentrated sulfuric acid. Sample mixed and incubated at 25°C for 10 min, and absorbance is recorded at 490 nm. Sugars are dehydrated in presence of sulfuric acid to hydroxyl methyl furfural, which conjugated with phenol to form orange-colored stable compound (Masuko et al., 2005). Similar kind of procedure can be adopted for DNS method as well. The 3, 5-Dinitrosalicylic acid (DNSA) is used extensively in biochemistry for the estimation of reducing sugars. Free carbonyl (C=O) group in reducing sugars get oxidized to aldehyde functional group (–CHO) in aldose and ketone functional group (–C=O) in keto sugars and simultaneously 3, 5-Dinitrosalicylic acid reduces to 3-amino-5-nitrosalicylic acid (ANSA). Reduced product converted to a reddish-brown-colored complex under alkaline conditions which can be read at 540 nm (Miller, 1959).

Protein content of EPS in dried powder can be determined by Bradford (1976) using bovine serum albumin as standard. In Bradford reagent, Coomassie Brilliant Blue G-250 binds to protein by hydrophobic as well as ionic interaction which causes a shift in absorbance from 465 nm to 595 nm. The color of the solution also changes from blue to dark blue in presence of protein. For protein estimation 50–100 μL sample is mixed with 200 μL Bradford reagent, and samples are incubated at 25°C for 10 min. Absorbance is recorded at 595 nm and compared with the standard solution (Kruger, 2009).

2.5   EPS EXTRACTION AND PURIFICATION

EPS comprised of biomolecules of different chemical and physical characteristics. The same properties would aid in the extraction of EPS from the cultivation system. The methods used for EPS extraction can be categorized into chemical and physical methods. In order to overcome the drawbacks of physical and chemical methods, some methods are used in combinations. The selection of EPS extraction methods greatly relies on chemical composition, surface characteristics, and native state. Preservation of the native state, physiochemical characteristics, and final yield are necessary parameters to assess extraction efficiency. There is no simple method for qualitatively and quantitatively extracting all microbial EPS (Donot et al., 2012).

Physical extraction of EPS: Physical methods exploit the physical characteristics of EPS like molecular weight, size, etc. and impose external force to separate it from the medium/producing cells. Ultra/high-speed centrifugation, heat/thermal extraction, and ultrasonic-assisted extraction are the common methods used for the physical extraction of EPS. In addition to separation, an external force like heat and ultrasound may also disrupt cells and intracellular impurities might be added to extracted EPS. High temperature also deactivates enzymes and redox-active substances. In both cases, the quality of the end product will be compromised, hence its treatment conditions has to be standardized carefully. Centrifugation-assisted extraction and separation are one of the most common methods used in microbiology, especially for the separation of cells from the broth. The difference in density is the main principle that aid in separation. Cell separation is induced under the influence of external forces (Galib et al., 2021). For EPS extraction, microbial culture/wet granules are centrifuged at 4000–6000 × g (depending upon the sample and difference in density) for 15–20 min. During the operation, samples are kept at 4°C to prevent any thermal damage. Post separation pellet is discarded, and the supernatant is collected to recover EPS. For the precipitation of EPS from the supernatant, twice the volume of ice-cold ethanol is added which modulated the solubility of EPS and precipitate it (Bhatia et al., 2021). The precipitate can be collected by filtration or centrifugation, dried and further used for characterization.

Thermal/heat is a mode of energy used for the extraction of EPS. For the optimum extraction of EPS, the treatment parameters must be optimized. For thermal extraction, a bacterial culture is heated for 8–10 min at 80°C and then centrifuged at 11,180 × g. The content is filtered with 0.45 µM acetate cellulose membranes to collect the supernatant. Excess salts and other impurities can be removed by membrane dialysis (3500 Da) for 24 h at 4°C (Senthilkumar et al., 2021). Another aspect is the associated cost and poor heat transfer in conventional heating, thus advanced methods can be adopted. Ultrasonication is a more efficient mode of heat transfer that can be used for the extraction and separation of EPS. Ultrasound waves create cavitation, the creation of microbubbles and cavities followed by an explosion. Microbial culture sonicated at 37 W for 1 min and centrifuged at 4000–6000 × g for 15–20 min, and EPS is collected by precipitation (Senthilkumar et al., 2021). The explosive burst of bubbles creates a pulse of a very high temperature of up to 5000 K and a pressure of around 1000 bar. Under such high temperatures and pressure, loosely bound EPS is separated from cells (Cheaburu-Yilmaz et al., 2019).

Cell lysis and cross-contamination with cell debris and other intracellular inclusions compromised the quality of EPS extracted (Duan et al., 2020). In general, it has been observed that the yield of physical extraction methods is lower due to lower EPS recovery. In addition, the high temperatures might disrupt the protein fraction and affect the native structure of EPS.

Chemical methods: Physical extraction offered lower product yield hence chemical reagents/solvents might improve the yield. Based on the composition, EPS exhibit different physicochemical behavior like hydrophobicity and available functional groups interacting with the surrounding environment. Chemical composition also plays a crucial role in the type and interaction of EPS with the cell wall. Based on the type of bonding and strength, EPS can be classified into soluble EPS (S-EPS), bound-EPS, which is further categorized into loosely bound EPS (L-EPS), and tightly bound EPS (T-EPS) (Guangyin and Youcai, 2017). In terms of composition, proteins, polysaccharides, and humic substances are the major components in EPS. It has been observed that protein fraction increased in T-EPS, which is mainly attributed to enzymes (Fernando et al., 2020). Alkali, alkali/formaldehyde, salts like NaCl, and EDTA; solvents including ethanol, ether, and cationic resins are commonly used extractants. Chemical extractants weaken the interaction forces but also affect the cell wall and may disrupt it. Hence, the operating condition must be selected carefully to prevent contamination with cellular metabolites and debris (Wang et al., 2022). Microbial culture mixed with 30 mL EDTA solution (2% w/v) in deionized water (DI) and kept for 3 h at 4°C. The sample is centrifuged at 11,180 × g for 20 min and filtered through 0.45 μM acetate cellulose membranes. The sample is dialyzed through 3500 Da membrane for 24 h at 4°C (Senthilkumar et al., 2021).

Alkali, that is, NaOH treatment can also be used for EPS extraction. Microbial biofilm suspension is mixed with 12 mL NaOH (1 N) and kept for 3 h at 4°C. The suspension is centrifuged at 11,180 × g for 20 min filter through 0.45 μM acetate cellulose membranes. The supernatant is collected and repurified via membrane dialysis (3500 Da) for 24 h at 4°C (Senthilkumar et al., 2021). To improve the extraction, alkali may be coupled with other chemical extractants methods. Alkali-formaldehyde treatment is a sequential process in which the pellet collected after the first centrifugation is kept in formaldehyde solution under refrigeration for 1 h and then mixed with 1 N NaOH and further kept at 4°C for 2 h. EDTA extraction also followed the same procedure as collected pellets is resuspended in 2% (m/v) EDTA solution under refrigeration and kept for 2–3 h. Posttreatment samples are again centrifuged at 10,000 rpm for 15–20 min, and the supernatant is collected, which is further purified by filtration with 0.2 μm syringe filter (Galib et al., 2021). Chemicals used in EPS extraction might contaminate the final product or further increase the requirement for purification. Dialysis and successive extraction are a method for EPS purification. The recovered/reconstituted sample dialyzed through 3500 Da membrane with cold buffer for 24 h. Buffer can be replaced after the saturation point to make the process effective (Senthilkumar et al., 2021).

Bacillus BU-4 was cultivated in pretreated liquids of steam-exploded quinoa stalks and enzymatic hydrolysates of Curupaú sawdust at 30°C for 48 h. Post-cultivation, broth was centrifuged to separate cells followed by ethanol/trichloroacetic acid treatment, followed by dialysis (Chambi et al., 2021). Chromatography is another approach for EPS extraction utilizing the physical properties (gel permeation) and chemical properties like charges (ion exchange). Molecular size of EPS was determined by gel permeation chromatography (GPC) using a combination of Sephacryl S-200 and Sephacryl S-500 columns equipped with a refractive index detector and fraction collector. For the analysis, the EPS sample was precipitated in deionized water at 50°C for 10–12 h followed by centrifugation at 2000 × g for 5 min. The fractionation was done with distilled water pumped at a flow rate of 0.45 mL/min (Buksa et al., 2021).

2.6 EPS CHARACTERIZATION

EPS are secreted from prokaryotic and eukaryotic organisms varying from different genera of bacteria, green and red microalgae, diatoms, yeasts, and molds (Xiao and Zheng, 2016). The composition of EPS varies significantly between different microbial species. They are mainly composed of carbohydrates and some non-carbohydrate moieties (proteins and nucleic acids) attached to them. Carbohydrates can either be homo (single repeating subunits) or heteropolysaccharides (2–8 residues repeating subunits) interlinked with glycosidic linkages. Biochemically they can either be polyanionic, polycationic, or neutral. Homopolymeric exopolysaccharides are produced either by synthase-dependent pathways or by extracellular enzymes. Heteropolysaccharides like xanthan, succinoglycan, and sphingans are produced by complex pathways like ABC transporter pathways and Wzx/Wzy pathways (Schmid, 2018). Different microorganisms produce structurally different EPS. Their composition, properties, and applications are summarized in Table 2.2.

Characterization of EPS includes studies on its composition, molecular mass, thermal properties, crystallinity, surface analysis, and conformation. There is no single standard technique to study any particular character. Researchers have used different techniques to study each character.

2.6.1 EPS COMPONENT ANALYSIS

Traditionally colorimetric methods were used to determine the biochemical composition of EPS. These methods were used to determine the total amount of polysaccharides and proteins in a given sample. For example, carbohydrate content can be measured using the anthrone method, where glucose can be used as a standard. Other dyes like Alcian blue, Aniline blue, Congo red, Calcofluor white can also be used. Proteins can be quantified using Coomassie Blue method, and humic substances can be measured using the Lowry method while DNA is quantified using the diphenylamine colorimetric method (Chen et al., 2013). Purification and structure elucidation can be done by electrophoresis followed by silver staining and Alcian blue staining.

2.6.2 MOLECULAR WEIGHT DETERMINATION

Molecular weight of exopolysaccharides may vary as it depends on different factors like the type of microbial strain used, composition, nature, pH of medium and substrate used, time of harvesting, etc. The carbon source, nitrogen source, the concentration of NaCl and phosphate, and viscosity of the medium also affect the molecular weight and polydispersity index of EPS like pullulans (Sugumaran and Ponnusami, 2017). For example, the molecular weight of EPS VITP14 isolated from *Virgibacillus dokdonensis* is ~555 kDa, while the gellan EPS isolated from *Pseudomonas* sp. and *Sphingomonas* sp. is ~500 kDa, and the anionic EPS isolated from *Alteromonas* sp. has a molecular weight of 780 kDa (Andrew and Jayaraman, 2022).

Techniques used for the determination of average molecular weight include sedimentation, viscometry, osmometry, chromatographic techniques like HPLC, high performance gel permeation chromatography (HPGPC), and size exclusion

TABLE 2.2

Different types of EPS with their compositions, properties, and potential applications

EPS type	Composition	Properties	Applications	References
Cellulose	Fructose/Glucose	Water holding capacity, tensile strength	Nondigestible fiber, used in the food industry	(Fang and Catchmark, 2015)
Alginate	B-D-Mannuronic acid and α-L-guluronic acid	Water holding capacity, gel forming, viscosifier, stabilizer	Food industry, alginate dressings	(Hay et al., 2010)
Dextran	Glucose	Moisturizers, thickeners, microcarriers	Food industry, tissue culture	(Moradi et al., 2021; Nacher-Vazquez et al., 2015; Robyt et al., 2008)
Curdlan	Glucose	Viscoelasticity, alkaline soluble, water insoluble,	Used in food industry, can be used as gelling agent and immobilization matrix	(Chien et al., 2017; Liu et al., 2015)
Hyaluronic acid	Glucuronic acid, N-acetylglucosamine	Film forming and gelling properties	Food and cosmetic industry	(Izawa et al., 2009; Moradi et al., 2021)
Xanthan	Two d-glucoses, two d-mannoses and a d-glucuronic acid	Thickener, stabilizer, viscosifier, thermally stable, biocompatible	Food and cosmetic industry, oil recovery, drug delivery, tissue engineering	(Bhat et al., 2022)
Levan	Fructose	Viscosifier, stabilizer, water holding capacity, prebiotic	Food industry	(Kekez et al., 2017; Moradi et al., 2021)
Mannans	Mannose	Thermally stable, antioxidant activity, antibiofilm activity	Food processing industry	(Lakra et al., 2020)
Gellan	Glucopyranose, glucuronic acid, mannopyranose and rhamnopyranose	Stabilizer, thickening agent, gelling agent, encapsulation matrix	Food additive, food industry	(Bajaj et al., 2007; Gansbiller et al., 2020; Li et al., 2016a)
Welan	Glucopyranose, glucuronic acid, mannopyranose and rhamnopyranose with L-Rhap or L-Manp, linked to first glucose	Rheological properties	Food industry, oil recovery, personal care	(Gansbiller et al., 2020)

Name	Composition	Properties	Applications	References
Diutan	Glucopyranose, glucuronic acid, mannopyranose and rhamnopyranose, with L-Rhap (1→4)-α-Rhap-dimer α (1→3), linked to first glucose	Rheological properties	Food industry, oil recovery, personal care	(Gansbiller et al., 2020; Li et al., 2016a)
Lubcan	Glucopyranose, glucuronic acid, mannopyranose, and rhamnopyranose	Rheological properties	Food industry, oil recovery, personal care	(Gansbiller et al., 2020)
Sphingan S-88	Glucopyranose, glucuronic acid, mannopyranose, and rhamnopyranose	Rheological properties	Food industry, oil recovery, personal care	(Gansbiller et al., 2020)
Succinoglycan	Fucose, galactose, glucose, and glucuronic acid	Thickener, emulsifier, stabilizing agents, antitumor, antibacterial activity	Food emulsions, cosmetic, nanoparticle stabilization, biomedical industry	(Jeong et al., 2022)
Colanic acid	L-fucose, D-glucose, D-galactose, D-glucuronic acid	Antitumor, anti-inflammatory, skin whitening	Food, cosmetics, and pharmaceutical industry	(Yun et al., 2021)
Pullulans	Isomaltose residues	Water soluble, oil resistant, transparent, oxygen impermeable films, adhesive	Food preservation, cosmetics industry	(Sugumaran and Ponnusami, 2017)
Emulsan	Glucose	Potent emulsifier	Oil recovery, cosmetics	(Choi et al., 1996)
Xanthan gum	Glucose, glucuronic acid, pyruvic acid	Viscosifier, stabilizer, emulsifier, suspending agent	Food industry, oil recovery, textile industry, pharmaceutical, cosmetic and agricultural industry	(Kim et al., 2017; Wang et al., 2017)
Kefiran	D-glucose and D-galactose	Antibacterial, antioxidant, antifungal, gelling agent	Food industry, pharmaceutical, drug delivery	(Elsayed et al., 2017; Moradi et al., 2021)

chromatography with multi-angle laser light scattering. Molecular weight can also be determined using gel electrophoresis, low-pressure size exclusion chromatography, and high-pressure size exclusion chromatography (HPSEC) (Xiao and Zheng, 2016).

2.6.3 COMPOSITIONAL ANALYSIS

Compositional analysis helps to understand the primary conformation of polysaccharides, and various techniques are used for this purpose.

TLC: Thin layer chromatography is used to study the monomeric subunits of EPS. For the analysis, take 10 mg of freeze dried EPS sample. Treat it with 2 mL of (2 mol/L) trifluoroacetic acid at 100°C for 5 h. Load EPS hydrolysate on TLC silica gel along with standard sugars like glucose, galactose, mannose, rhamonose, xylose, glucuronic acid, etc. The solvent system consists of a mixture of chloroform (3 parts), acetic acid (3.5 parts), and water (0.5 parts). Blots are developed by spraying aniline-diphenylamine in acetone-orthophophoric acid (10:1). The chromatograms are then analyzed based on their retention force values and compared with standards. This technique has been used to check the monomeric composition of EPS isolated from *Rhodobacter johrii* CDR-SL Cii (Sran et al., 2019). These monomeric compositions can be further confirmed using techniques like HPLC.

HPLC: High performance liquid chromatography is also used to determine the monomeric composition of EPS. For HPLC analysis, the EPS samples are first hydrolyzed by treating them with 2 mL (2 mol/L) of trifluoroacetic acid at 100°C for 5 h. This hydrolysate and standard sugars are then injected into HPLC incorporated with RI detector. 0.1 mM calcium EDTA dissolved in HPLC grade water is used as a mobile phase (Sran et al., 2019). HPLC chromatogram is then analyzed for the presence of peaks at different retention times, and these results are compared with the standard.

GC-MS: GC-MS analysis is also done to investigate the monosaccharide composition of EPS. GC analysis requires derivatization of sugars; hence extracted EPS is hydrolyzed with acid that recovers monomers from the polymeric units followed by derivatization to trimethylsilyl/alditol acetates derivatives. Obtained derivatives can be analyzed through GC, GC-MS. For the analysis GC system must be equipped with flame-ionization detector (Freitas et al., 2009). Some recent modifications use high pressure anion exchange and amperometric detection without any derivatization like required in GC. The sample preparation is done as follows:

take 10 mg of EPS sample and hydrolyze it with 1M trifluoroacetic acid at 100°C for 8 h. Dissolve 10 mg of $NaBH_4$ in 1 mL of 1M ammonium hydroxide. Add 100 μL of it to hydrolyze EPS sample and left it for 16 h at room temperature. This will reduce the hydrolyzed EPS sample. Add N, O-bis (trimethylsilyl) trifluoroacetamide in pyridine to the reduced sample at 60°C for 30 min. The obtained methylsilanes are then analyzed using GC-MS (Gupta and Thakur, 2016).

HPAEC-PAD: High Performance Anion-Exchange Chromatography with Pulsed Amperometric detection (HPAEC-PAD) is another technique of monosaccharide composition analysis. EPS can be hydrolyzed using 2 N trifluoroacetic acid at 100°C for 1 h. The Carbopac guard column and Carbopac PA1 analytical column are used at 1 mL/min at 30°C. Monosaccharides can be analyzed using the multiple step gradient method. This method uses isocratic 15 mM NaOH for 10 min followed by a linear NaOAc gradient with fixed 15 mM NaOH for 30 min (Zhang et al., 2012). Different carbohydrates like glucose, galactose, mannose, arabinose, glucosamine, galactosamine, N-acetylneuraminic acid, and N-glycolylneuraminic acid are used as standard. The retention time for each monosaccharide standard can be measured individually. This method has been used to detect the monosaccharide composition of EPS isolated from the psychrotrophic bacterium *Pseudomonas* sp. BGI-2 (Ali et al., 2020). Raman Spectroscopy along with 3D-EEM fluorescence spectroscopy has been used to analyze the composition of EPS fraction from suspended sludge and biofilm samples in sequencing batch biofilm reactors (Zhang et al., 2014). These studies will provide information about sugar residues.

FTIR: FTIR spectroscopy revealed the bonds, functional groups, and atoms involved in bond formation. Certain functional groups and linkages, present in the sample, can absorb infrared energy at a particular wavelength (Liu et al., 2017). It gives information about C=O bonds and O-H bonds. The protocol for preparation of the sample is as follows: 2 mg of EPS sample is mixed with 200 mg of KBr. The FITR spectra and bonds are detected in the range of 4000–400 cm^{-1} while sulfate content is observed in the range of 1250–1050 cm^{-1} (Caruso et al., 2019; Mancuso Nichols et al., 2004). The technique has been used to analyze the structure of pullulan (Sugumaran and Ponnusami, 2017). For example in FITR analysis of pullulans, peaks are observed due to O-H stretching, C-H stretching, O-C-O starching, C-O-H bending, α (1→6) glycosidic linkage, and α (1→4) glycosidic linkage (Sugumaran et al., 2013; Sugumaran and Ponnusami, 2017).

NMR: NMR is one of the most powerful tools for the structural characterization of EPS. In NMR available protons generate signals that aid in qualitative as well as quantitative characterization. Samples are analyzed for ^{1}H and ^{13}C-NMR spectra at 500 Mhz and 125 MHz respectively at 55°C, and 30 mg of EPS sample is dissolved in 500 μL of D$_2$O. For ^{1}H, chemical shifts are reported in parts per million (ppm) relative to sodium 2, 2, 3, 3-d 4-(trimethylsilyl) propanoate while for ^{13}C shifts are reported relative to CDCl$_3$ (Caruso et al., 2019; Mancuso Nichols et al., 2004). Several research groups have used NMR spectral analysis, where signals for pullulan are observed between 4.5 and 5.6 ppm. Anomeric carbon α (1→6) peak is seen at 100.1 ppm (Singh et al., 2009; Sugumaran and Ponnusami, 2017). Branched oligosaccharides like fucoidan have been characterized using NMR (Clement et al., 2010).

2.6.4 GLYCOSYL COMPOSITION AND LINKAGE ANALYSIS

For analyzing the glycosyl composition, a combined GC-MS analysis is done. The EPS samples are mixed with 20 µg inositol, which is an internal standard. Add 1M methanolic HCl in a screw capped tube and heat it for 18 h at 80°C. Let it cool down and remove the solvent using nitrogen. Treat samples with methanol, pyridine, and acetic anhydride for 30 min. When samples have evaporated, use Tri-Sil to derivatize the sample at 80°C for 30 min. The samples are then subjected to GC-MS analysis and data analyzed for glycosyl composition (Padmanabhan and Shah, 2020; Santander et al., 2013).

For linkage analysis, EPS samples (0.5 mg) are dissolved in dimethyl sulfoxide in presence of NaOH. Samples are permethylated with methyl iodide. The sample is hydrolyzed using 2 M trifluoroacetic acid for 2 h at 121°C. This will give rise to partially methylated monosaccharide samples. Reduce it with Deuterated sodium borohydride and acetylate with acetic anhydride and TFA. These methylated alditol acetates are then subjected to GC-MS analysis (Padmanabhan and Shah, 2020). These techniques have been used for the characterization of EPS isolated from *Streptococcus thermophilus* ASCC 1275.

2.7 STRUCTURAL AND PHYSICAL PROPERTIES ANALYSIS

Various techniques are used for EPS surface structure (SEM, CLSM, and AFM) and physical properties (TGA and XRD) analysis as discussed here:

SEM: The most pioneer and basic technique to analyze the structure of biofilm is microscopy. With the evolution of microscopy, new techniques like confocal laser scanning microscopy (CLSM), electron microscopy, scanning transmission X-ray microscopy (STXM), atomic force microscopy, etc. have been used to study the detailed three-dimensional structure of EPS.

The morphology and bio-architecture of EPS can be studied using SEM. For SEM analysis, lyophilized EPS samples are used by fixing them to SEM stubs, followed by gold coating. Samples are then observed at an accelerating voltage of 5.0 kV, providing magnification of 250 to 2500X (Ispirli et al., 2019). It provides excellent details on EPS and cell distribution. But the major disadvantage remains that live biofilms cannot be monitored. These techniques can be coupled with energy dispersive spectroscopy (EDS) to understand the distribution of metals, their elemental composition, crystallinity, and localization. However, during sample preparation for electron microscopy, sections can shrink and create artifacts. The technique doesn't provide information on metal speciation and its quantification. STXM uses synchrotron light to understand the distribution of metals within the live biofilms.

CLSM: CLSM is a nondestructive, non-invasive method used to understand the three-dimensional architecture of biofilms. Sample preparation for CLSM analysis is done by growing biofilm on steel coupons. Biofilm coated coupons can be removed at desired time intervals. The coupons are rinsed with phosphate buffer pH −7.4. This is followed by staining using stains like 0.01% Acridine orange for 20 min or SYTO-9. Rinse the biofilms with phosphate buffer pH −7.4 for 10 min and observe under a microscope (Zhang and Fang, 2001). The microscope TCS-NT is equipped with Argon-Krypton laser and has a water immersion lens. Fluorescence is measured using different excitation and emission wavelengths. The technique uses multiple fluorescent channels to detect the presence of different biofilm components using specific markers. The cell number, morphotypes, and image analysis is done by using software like COMSTAT, ISA, and PHLIP. EPS are visible as amorphous organic material or as bright spots if they are treated and labeled with lectin concanavaline A (ConA) (Mueller et al., 2006). EPS can also be visualized and characterized using fluorescently labeled wheat germ agglutinin (WGA) and ConA in combination with Enzyme-linked lectino sorbent assay in *Pseudomonas* biofilms (Strathmann et al., 2002).

AFM: Atomic Force Microscopy is a powerful tool to probe the surface structure and properties at nanometer resolution for EPS. It can be used in contact mode, non-contact mode, and tapping mode. For sample preparation EPS (1 mg/mL) is dissolved in deionized water. 500 uL samples uniformly distributed on a mica sheet and dried at room temperature to form a uniform thin film. AFM images can be captured using a scanning probe microscope. The microscope uses a tip present on the end of the cantilever to scan the changes in height of the sample surface. The deflections are amplified, and a map of the surface is generated. The tip can also be used to probe interaction forces between the tip and EPS (Hansma et al., 2000). Characterization is done using 2 μm and 100 μm scanners at a temperature of 23°C and 45% humidity. For contact mode, pyramidal silicon tips attached to the cantilever are used with a spring constant of 0.032 N/m, while for non-contact mode, silicon tips and cantilevers with a spring constant of 42 N/m and resonant frequency of 320 kHz are used (Pham et al., 2003). Bhatia et al. (2021) performed AMF analysis of EPS produced by *Sphingobium yanoikuyae*. EPS surface showed a round lump of 8 nm height with irregular and rough surface.

TGA: TGA analysis is done to check the thermal properties of EPS, its melting point, and enthalpy changes. This helps to predict the thermostability of the compound for potential applications in industries. For thermogravimetric analyses (TGA), the freeze dried sample (~15 mg) is taken and scanned in a temperature range of 30–1000°C, in a dinitrogen environment, applying the heating rate of 10°C/min (Shen et al., 2013; Solmaz et al., 2018). TGA curve is then plotted, and different stages of weight loss are analyzed. The first stage of weight loss is mainly due to the breakdown of carboxyl groups and moisture loss. The second stage is usually due to the pyrolysis of EPS (Solmaz et al., 2018). EPS isolated from *Lactobacillus plantarum* HY has shown a melting point of 320°C (Liu et al., 2019). EPS isolated from *Pleurotus pulmonarius* has shown a degradation temperature of 217°C (Shen et al., 2013). The EPS isolated from *Streptococcus thermophilus* has shown a degradation temperature of 110.84°C (Kanamarlapudi and Muddada, 2017).

XRD: This technique is used to study the physical characteristics of lyophilized EPS, using X-Ray diffractometer. Exopolysaccharides pellets are used for analysis using X-ray diffractometer. Diffractometer has Ni-filtered CuKα radiation, and running conditions are 40 kV, 40 mA, scan speed 0.005°, time/scan of 0.1 sec, and 0.2 mm slit. The technique gives information about the crystallinity of the EPS. The X-ray diffraction scan is done in different ranges of two theta angles 5–80°. Diffraction is calculated using Bragg's law:

$$d = \lambda/2 \sin\theta$$

The presence of a broad peak indicates the given sample is amorphous in nature while sharp peaks indicate a crystalline nature.

The technique has been used to characterize the EPS MSR101 isolated from *Lactobacillus kefiri*. The peaks were observed at 28.67°, 34.75°, and 47.4° with an intensity of 1558.5, 1786.2, and 1000 respectively. The results showed that EPS was partially crystalline in nature (Riaz Rajoka et al., 2019). The conformation and crystallinities of succinoglycan have also been studied using X-ray diffraction technique. This is achieved by comparing the XRD pattern of EPS produced by wild and mutant strains of *Sinorhizobiummeliloti* within the range 2θ = 10–60° using Cu Kα (Hu et al., 2021). Diffraction patterns were integrated, and crystallinity was calculated.

2.8 APPLICATIONS OF EPS

Being natural polymers, EPS have capacities to form pseudoplastic liquids and is biodegradable. They can be used for the production of emulsifiers, stabilizers, microcarriers, moisturizers, viscosifiers, thickeners, and as immobilization matrix. They have high water retention properties and tensile strengths. Their bioactivities have made them versatile for the synthesis of a wide range of commodities in food, pharmaceuticals, cosmetics, adhesives, detergents, oil recovery, brewing, and textile industries as well as for wastewater treatment and remediation (Xiao and Zheng, 2016). Studies have correlated the composition and structure of EPS with their bioproperties, for example, the sugar composition, sulfation, and molecular weight of fucan is responsible for its anticoagulation properties. Anti-complementary properties of fucoidan are directly related to its branching (Clement et al., 2010). EPS from red algae is utilized in the medical industry, while that from green algae and diatoms has been used in wastewater treatment (Xiao and Zheng, 2016). Many properties of medical and pharmaceutical importance of EPS have been explored, and efforts are made to commercially utilize these properties. These properties include their use as prebiotics, antimicrobial agents, anticancer agents, and antioxidative agents (Yildiz and Karatas, 2018). Antioxidant properties are evaluated by in vitro methods like measurement of reducing power, scavenging of superoxide, hydroxyl and 2,2-diphenyl-1-picrylhydrazyl radicals, metal chelating ability, and lipid peroxidation inhibition assay (Liu et al., 2017). Antitumor activity can be accessed by checking the activity of EPS on human carcinoma cell lines. EPS can be modified chemically, and derivatives can be used for inhibition assays. Anti-inflammatory activities can be estimated by checking in vitro membrane stabilizing activity on hypotonic solution induced hemolysis of rat erythrocytes (Mahapatra and Banerjee, 2012). Different applications are described in the next section.

2.8.1 Free radical scavenging activity

The reactive oxygen species (ROS) and free radicals play a vital role in aging, cancer, and heart dysfunction, etc. The EPS has free radical eliminating and antioxidant activity as reported by various research groups. Free radical scavenging activity of EPSs can be measured using 1,1-dipheny-2-picryl-hydrazyl (DPPH). The reaction can be performed using 2 mL of 95% ethanol, 0.1 μM DPPH, and 2 mL of EPS (50–400 mg/L) (Sun et al., 2013). The reaction mixture is incubated at 25°C for 15 min and absorbance is measured at 517 nm. Antioxidant activity can be calculated using the following equation:

$$\text{Scavenging rate (\%)} = [(A_0 - A_1)]/A_0] \times 100$$

Where A_0 is the absorbance of DPPH, and A_1 is the absorbance of the test with EPS.

Exopolysaccharide produced by *Sphingobium yanoikuyae* BBL01 has free radical scavenging activity of 34% at 10 mg/mL (Bhatia et al., 2021). Kefiran is also an example of EPS having antioxidative, immunomodulatory, antimicrobial, and anti-inflammatory properties and thus offers applications in biomedical industries. It is used for the treatment of osteoarthritis. A study has shown that kefiran used in combination with chitosan can show improved antioxidant properties with better tensile strength and puncture strength (Sabaghi et al., 2015). Fractions of homogenous EPS, r-EPS1, and r-EPS2 isolated from *L. plantarum* 70810 have shown antioxidant properties in in vitro experiments on hydroxyl and DPPH radicles (Wang et al., 2014). EPS from *Halolactibacillus miurensis*, named as HMEPS, has also shown antioxidant activity against all free radicals as well as the ability to chelate ferrous ions (Arun et al., 2017).

2.8.2 Cryoprotective activity

Cryoprotectants are commonly used in the food processing industry, to increase the shelf life and quality of food. EPS isolated from microorganisms of extreme environments has also shown cryoprotective behavior.

Protocols of cryoprotective assays have been modified in different studies, but it is based on the harvesting of bacterial cells by centrifugation, followed by freeze-thaw of bacterial cells, and then checking for cell viability either by measuring absorbance in a spectrophotometer or by culture methods like dilution plating (Carrion et al., 2015; Caruso et al., 2019). In one such study, cryoprotection assays against different bacteria were done by growing the test organism on tryptone soya broth to A_{540} nm, and 0.6. 0.1 mL of these bacterial cells aliquots are taken and to it, 0–10% EPS is added, so that the final volume is up to 1 mL. Freeze the vials at −20°C and

−80°C for one week. After one week, take out the vials, thaw them, and measure the number of viable bacteria by a dilution plating method. Fetal bovine serum is used as a positive control. The survival rate is calculated as the percentage of viable cells with reference to unfrozen cells (Carrion et al., 2015). This method has been used to check the cryoprotective activity of EPS isolated from Antarctic *Pseudomonas sp.* ID1 (Carrion et al., 2015).

2.8.3 WATER HOLDING CAPACITY

The EPS can hold water mainly by forming hydrogen bonds. This property makes them more useful in the food and dairy industries. It helps in improving the texture of the food and its rheological properties.

The test is performed by dissolving 0.2 g of EPS in 10 mL of Milli Q water and incubating at 40°C for 10 min for uniform dispersion. Samples are centrifuged at 16400 g for 30 min, and the supernatant is discarded. Pellet is dropped on pre weighed filter paper for drainage of water, then the filter paper is weighed (Ahmed et al., 2013).

Water Holding Capacity (WHC %) = (Total sample weight after absorption/Total dry sample weight)*100

The technique has been used to measure the water holding capacity of EPS produced from bacteria isolated from fermented foods like Idli batter and Kefir (Saravanan and Shetty, 2016). EPS isolated from *Lactobacillus kefiranofaciens* ZW3 showed water holding capacity of 496%, while EPS from *Leuconostoc lactis* KC117496 can hold 117±7.5% (Ahmed et al., 2013; Saravanan and Shetty, 2016).

2.8.4 ANTIMICROBIAL ACTIVITY

With an increase in antimicrobial drug resistance, there is a dire need to find alternative therapeutic options. EPS has been proven to have antibacterial action mainly by causing cell wall disruption, inhibiting mRNA and protein synthesis, and forming a barrier around cells that inhibits the movement of nutrients within bacterial cells (Tokura et al., 1997). It represses the growth of pathogenic microorganisms and also inhibits toxin production from them. Antimicrobial activity is measured using agar well diffusion assay and measuring the diameter of the zone of inhibitions around wells. For measuring the antibacterial activity, indicator microorganisms are procured from microbial banks. The bacteria are grown in LB broth at 37°C and A_{600} is measured. Bacteria are grown till the concentration of 10^7-10^8 CFU/mL is achieved. EPS 5 mg/mL dissolved in deionized water and sterilized using 0.45 μm millipore filter. Spread 100 μL of culture on LB agar plates and make 4 mm diameter wells in it. To each well add 60 μL of EPS sample. Incubate plates at 37°C for 48 h. Measure the zone of inhibition around the wells (Rajoka et al., 2019).

EPS isolated from *L. reuteri* SHA101 and *L. vaginalis* SHA110 have shown strong antibacterial activity against *Salmonella typhimurium, Staphylococcus petrasii subsp., Pragensis,* and *E. coli* (Rajoka et al., 2019). EPS from *L. plantarum* R315 and *Bifidobacterium bifidium* WBIN03 have shown antimicrobial activity against *E. coli, Bacillus cereus, Candida albicans, Staphylococcus aureua, Shigella sonnei,* and *S. typhimurium* (Li et al., 2014).

2.8.5 ANTITUMOR ACTIVITY

The rate of incidence of cancer has increased significantly in the past decade, and chemical-based drugs are used for the treatments. These drugs have high toxicity and side effects. EPS has been explored for its antitumor activities against different cancerous cell lines.

Antitumor activity can be measured by testing the EPS against appropriate cell culture lines and after treatment measuring the inhibitory rates. Rajoka et al. tested the EPS in varying concentrations of 50, 100, 200, 400, and 600 µg/mL against Caco-2 cells. The cells were grown in DMEM medium supplemented with 10% fetal bovine serum and antibiotics at 37°C, under 5% CO_2. After 24 h, the cells were treated with EPS and 5-FU (50 µg/mL) for 12, 24, and 72 h. CCK-8 reagent is added after each treatment and incubated for 4 h. A_{450} was measured, and the inhibitory rate was calculated (Rajoka et al., 2019).

$$\text{Inhibitory rate} = [1- (A_{sample}—A/A_{blank})] *100$$

Where A is the absorbance value at 450 nm.

EPS isolated from *L. reuteri* SHA101 and *L. vaginalis* SHA110 have shown the strong inhibitory ratio of ~91% against Caco-2 cells when used at a concentration of 600 µg/mL (Rajoka et al., 2019). EPS isolated from *Pseudoalteromonas* S-5 and *Trichoderma pseudokoningii* has been shown to inhibit the proliferation of human leukemia K562 cells, by triggering apoptosis (Chen et al., 2015; Huang et al., 2012). Kefiran is a biocompatible EPS proven to have anticancer activity against cervical and hepatocellular cancer (Elsayed et al., 2017). EPS, r-EPS1, and r-EPS2 isolated from *L. plantarum* 70810 showed antitumor activities against Caco-2, BGC-823, and HT-29 cells (Wang et al., 2014).

2.8.6 EMULSIFICATION ACTIVITY

Emulsifiers have been used widely in the cosmetic and food industry to form long-term and stable emulsions. The use of EPS isolated from marine and Antarctic bacteria are a more sustainable solution. They are biodegradable and nontoxic. Emulsification tests can be done by the protocol described by Cooper and Goldenberg or its modification (Cooper and Goldenberg, 1987). The basic principle involves, mixing extracted EPS with commercially available hydrocarbons like n-hexadecane to form emulsions and then checking for EI_{24} index.

$$EI_{24} = (\text{Height of emulsion layer/Total height}) *100$$

To test the emulsifying activity of EPS, to form oil-in-water emulsions can be tested by using a method described by Cirigliano and Carman (Cirigliano and Carman, 1984). In this technique, a known commercial emulsifier (positive control) and test EPS are taken, and their 0.02% (wt/v) solutions are prepared, with 5 mL of these solutions (test and control tubes) mixed with 0.8 mL of several hydrophobic compounds—oils like corn oil, sunflower oil, olive oil, rapeseed, walnut and vegetable oil, cosmetic oil like cetiol V, and hydrocarbons like n-hexa-decane. The same oils and chemicals, without EPS, are also used as controls. Samples were mixed well using a homogenizer. Emulsions were allowed to stand undisturbed at room temperature for 24 h, and turbidity of the lower aqueous layer was measured at A_{540} nm. O.D. of all oils and chemicals without EPS used as controls. Subtract the A_{540} of controls from test absorbance and the final emulsifying activity is calculated.

The emulsifying activity of EPS isolated from Antarctic *Pseudomonas* sp. ID1 and PE12 isolated from *Pseudoalteromonas* sp. strain TG12 have been calculated using this method (Carrion et al., 2015; Gutierrez et al., 2008). The EPS has shown higher emulsifying activities of 1.28 for olive oil, 1.15 for sunflower oil, and 0.66 for corn oil, compared to Arabic gum and xanthan gum (Carrion et al., 2015).

2.8.7 FLOCCULATING ACTIVITY

EPS has shown as a good flocculating agent mainly because of their negative Zeta potentials and polyanionic charges. This property has found wide application in ecological as well as commercial processes like wastewater treatment. Flocculation helps in the development of aggregates and adhesion to the ice surface, thus providing protection to cold environment.

The protocol for measuring the flocculation activity of EPS is as follows:

Prepare 5 g/L solution of EPS and 1% (w/v) solution of calcium chloride. Take 200 μL of EPS, 4 mL of 1% calcium chloride solution, and mix it with 50 mL solution of 5 g/L solution of natural kaolinite (pH-8.0). Kaolinite can be obtained commercially from Sigma-Aldrich. Agitate it for 2 min and let it stand for 5 min at room temperature. Separate the aqueous phase and measure its O.D. at 550 nm using a Spectrophotometer (Sathiyanarayanan et al., 2016).

Calculation of flocculation efficiency (f_e):

$$f_e = \text{b-a/b} * 100$$

Where a = A_{550} for control (distilled water can be used as control)

$$b = A_{550} \text{ for EPS}$$

The EPS concentration present in the system is directly related to its flocculating efficiency. EPS purified from PAMC 28620 has shown maximum flocculating activity 71.17 ± 2.21%, when EPS concentration was 70.0 mg/L (Sathiyanarayanan et al., 2016).

2.8.8 METAL CHELATING ACTIVITY

The metal binding activity of EPS is mainly due to the presence of a wide range of chemical groups on its surface, which can bind with heavy metals. This application has found wide application for their use in heavy metals recovery from industrial wastes as well as for bioremediation.

For measuring metal chelating activity, prepare 50 mg/L solutions of metal salts of Fe^{2+}, Cu^{2+}, Mg^{2+}, Mn^{2+}, Ca^{2+}, Zn^{2+}, etc., in HPLC grade deionized water. Prepare EPS (50 mg/L) samples in distilled water. Mix 4 mL of metal salt solution with 4 mL of EPS solution. Keep the solution at room temperature for 72 h at 150 rev/min. Add 1 volume of ethanol to it, to collect the EPS by centrifuge at 8000 g for 20 min. Remove unbound metal ions from EPS by dialysis against deionized water for overnight. Take 3 mL of aqueous EPS free solution and determine the residual metal ion concentration using inductively coupled plasma atomic/optical emission spectroscopy (Sathiyanarayanan et al., 2016).

$$\text{Metal removal capacity } Q = \text{vol } (C_i\text{-}C_f)/m$$

Where C_i = Initial metal ion concentration in aqueous solution
C_f = Equilibrium level of metal in aqueous solution
m = mass of EPS

Metal chelation activity has been shown by EPS produced by different strains like *Pseudomonas* sp. PAMC28640 (Sathiyanarayanan et al., 2016), *Pseudomonas stutzeri* AS22 (Maalej et al., 2015), and marine bacterium *Sphingobium yanoikuyae* BBL01 (Bhatia et al., 2021). The EPS from *Pseudomonas stutzeri* AS22 showed maximum biosorption for lead at 460 mg/g, followed by cobalt, iron, copper, and cadmium (Maalej et al., 2015).

2.8.9 USE IN NANOPARTICLE SYNTHESIS

Owing to their biodegradability and biocompatibility, microbial EPS are also used as a template for the production of metal nanoparticles. EPS stabilizes the metal nanoparticles and can also be explored as a capping agent. These particles can be used in medicine for drug targeting and drug delivery, to form nanocomposites, in pharmaceutical and cosmetic products, and in agriculture and environment biotechnology.

Silver nanoparticles have been synthesized from bacterial EPS. For this, EPS is purified from bacteria, and 20 mL of it is mixed with 9 mM aqueous solution of silver nitrate solution under stirring conditions. Incubate for 10 h at room temperature. Check visually for the synthesis of colloidal AgNPs when the colourless solution turns to yellow-colored solution. Further, the nanoparticles can be confirmed by Surface Plasmon Resonance by checking peaks at 400–550 nm (Kanmani and Lim, 2013).

The same technique has been used to generate gold and iron nanoparticles.

AgNPs synthesized from EPS isolated from *Lactobacillus rhamnosus* GG, using the previously mentioned technique, have shown antibacterial, antifungal, and antibiofilm activity (Kanmani and Lim, 2013). Kefiran is another example of exopolysaccharide, which can be used along with metal and metal oxide nanoparticles, to be used as nanofibers, nanofillers, and bionanocomposites. They can thus be used in the food packaging industry and further explored for applications in tissue engineering (Jenab et al., 2017).

2.8.10 Biosurfactant activity

EPS produced from microbial sources are seen as potential replacements for chemical counterparts. Biosurfactants are amphipathic surface active compounds, made up of peptides, polysaccharides, cations and anions, and fatty acids which can form micelles. They have found applications in the pharmaceutical, medicine, food, agriculture, and petroleum industry. They are also used for bioremediation processes (Paniagua-Michel et al., 2014).

Biosurfactant activity can be examined using the oil spreading technique.

Take 50 mL of distilled water in a beaker and add 200 µL of crude oil to it. For the test, 20 µL of 0.1% EPS solution is added to the center of this layer. For negative control, add 20 µL of distilled water. For positive control, use 20 µL of 0.1% Trition X 100 or Tween 80 or any standard surfactant. The zone of oil spreading is measured in centimeters macroscopically and indicates a positive surface active test (Suresh Kumar et al., 2007b).

The EPS produced from *Planococcus maitriensis* Anita I has shown the properties of biosurfactants. It gives positive oil spreading test at pH −3, 5.6, and 9, and its efficiency was found equivalent to Tween 80 and better than Triton X 100 (Suresh Kumar et al., 2007b).

2.8.11 Cholesterol removal

EPS has also been used to remove cholesterol and play role in promoting human health, specifically in diseases like coronary heart disease. Probiotics like *L. acidophilus*, *Bifidobacterium bifidium*, and *L. bulgaris* have been shown to reduce cholesterol levels. Cholesterol removal studies can be done by modified method discussed by Gilliland et al. Cholesterol solution (10 mg/mL) is first prepared in ethyl alcohol and sterilized by filtration. This solution is then added to culture media such that the final volume is 100 µg/mL. Inoculate the culture media with test strain (2%) and incubate at 37°C for 16–19 h. Harvest the cells at 10000 g at room temperature (23–25°C) for 20 min. Measure the cholesterol present in the supernatant by spectrophotometrically at 600 nm or by calorimetrically. For calorimetric measurement, a cholesterol standard curve is first prepared, using different concentrations of cholesterol (10–150 µg/mL) (Alp Avci, 2016).

$$\text{Cholesterol removal (A\%)} = 100 - [(B/C) *100]$$

Where
A—Cholesterol removal
B—Amount of cholesterol in inoculated medium (µg/mL)
C—Cholesterol amount in control medium (µg/mL)

The EPS produced by *L. bulgaris* has been proven to reduce cholesterol levels (Tok and Aslim, 2010).

2.9   CONCLUSION

EPS has essential natural roles in mediating host-pathogen interactions, the formation of biofilms by acting as natural adhesives, and protecting microbes against biotic and abiotic stresses, from antibiotics and extreme environmental insults. Additionally, these EPS have found commercial applications in feed, food, pharmaceutical, medical, tissue engineering, and cosmetic industries. EPS can complex metal ions and play role in geochemical cycling. Their function, properties, production, metabolic engineering of production pathways, extraction, characterization, and applications have been studied widely, but challenges still remain. Most of the physiological pathways remain unclear. Structures can be elucidated using a combination of techniques, which may help us to further understand the physiochemical, physiological, and biological properties. High production cost, optimization of synthesis protocols, variations in the quality of end product obtained, and scaling up of the process remain the challenge. Synthetic and systematic biology offers the potential to tailor EPS production with desired functionality.

ACKNOWLEDGMENTS

The authors acknowledge the KU Research Professor Program of Konkuk University, Seoul, South Korea. This research was supported by the C1 Gas Refinery Program through the National Research Foundation of Korea (NRF), funded by the Ministry of Science and ICT (2015M3D3A1A01064882), and by the National Research Foundation of Korea (NRF) [NRF-NRF-2022M3I3A1082545, NRF-2022M3J4A1053702, NRF-2022R1A2C2003138 and NRF-2021R1F1A1050325].

REFERENCES

Ahmed, Z., Wang, Y., Anjum, N., Ahmad, A., Khan, S.T., 2013. Characterization of exopolysaccharide produced by *Lactobacillus kefiranofaciens* ZW3 isolated from Tibet kefir—Part II. Food Hydrocoll. 30, 343–350.

Akdeniz Oktay, B., Bozdemir, M.T., Ozbas, Z.Y., 2022. Optimization of hazelnut husk medium for pullulan production by a domestic *A. pullulans* strain. Prep. Biochem. Biotechnol. 1–14.

Ali, P., Shah, A.A., Hasan, F., Hertkorn, N., Gonsior, M., Sajjad, W., Chen, F., 2020. A glacier bacterium produces high yield of cryoprotective exopolysaccharide. Front. Microbiol. 10, 3096.

Alp Avci, G., 2016. Selection of superior bifidobacteria in the presence of rotavirus. Brazilian J. Med. Biol. Res. 49.

Al-Roomi, F., Al-Sahlany, S., 2022. Identification and characterization of xanthan gum produced from date juice by a local isolate of bacteria *Xanthomonas campestris*. Basrah J. Agric. Sci. 35, 35–49.

Andrew, M., Jayaraman, G., 2022. Molecular characterization and biocompatibility of exopolysaccharide produced by moderately halophilic bacterium *Virgibacillus dokdonensis* from the Saltern of Kumta Coast. Polymers. 14, 3986.

Angelin, J., Kavitha, M., 2020. Exopolysaccharides from probiotic bacteria and their health potential. Int. J. Biol. Macromol. 162, 853–865.

Anguluri, K., La China, S., Brugnoli, M., De Vero, L., Pulvirenti, A., Cassanelli, S., Gullo, M., 2022. Candidate acetic acid bacteria strains for levan production. Polymers. 14.

Aramsangtienchai, P., Kongmon, T., Pechroj, S., Srisook, K., 2020. Enhanced production and immunomodulatory activity of levan from the acetic acid bacterium, *Tanticharoenia sakaeratensis*. Int. J. Biol. Macromol. 163, 574–581.

Arun, J., Selvakumar, S., Sathishkumar, R., Moovendhan, M., Ananthan, G., Maruthiah, T., Palavesam, A., 2017. In vitro antioxidant activities of an exopolysaccharide from a salt pan bacterium *Halolactibacillus miurensis*. Carbohydr. Polym. 155, 400–406.

Bajaj, I.B., Survase, S.A., Saudagar, P.S., Singhal, R.S., 2007. Gellan gum: Fermentative production, downstream processing and applications. Food Technol. Biotechnol. 45, 341–354.

Balasubramanian, B., Ilavenil, S., NA, A.-D., Agastian, P., Choi, K.C., 2019. Isolation and characterization of *Aspergillus* sp. for the production of extracellular polysaccharides by response surface methodology. Saudi J. Biol. Sci. 26, 449–454.

Bhat, I.M., Wani, S.M., Mir, S.A., Masoodi, F.A., 2022. Advances in xanthan gum production, modifications and its applications. Biocatal. Agric. Biotechnol. 42, 102328.

Bhatia, S.K., Gurav, R., Choi, Y.-K., Choi, T.-R., Kim, H.-j., Song, H.-S., Lee, S.M., Park, S.L., Lee, H.S., Kim, Y.-G., 2021. Bioprospecting of exopolysaccharide from marine *Sphingobium yanoikuyae* BBL01: Production, characterization, and metal chelation activity. Bioresour. Technol. 324, 124674.

Bhatia, S.K., Gurav, R., Kim, B., Kim, S., Cho, D.-H., Jung, H., Kim, Y.-G., Kim, J.-S., Yang, Y.-H., 2022. Coproduction of exopolysaccharide and polyhydroxyalkanoates from *Sphingobium yanoikuyae* BBL01 using biochar pretreated plant biomass hydrolysate. Bioresour. Technol. 361, 127753.

Bradford, M.M., 1976. A rapid and sensitive method for the quantitation of microgram quantities of protein utilizing the principle of protein-dye binding. Anal. Biochem. 7, 248–254.

Buksa, K., Kowalczyk, M., Boreczek, J., 2021. Extraction, purification and characterisation of exopolysaccharides produced by newly isolated lactic acid bacteria strains and the examination of their influence on resistant starch formation. Food Chem. 362, 130221.

Cao, F., Bourven, I., Guibaud, G., Rene, E.R., Lens, P.N.L., Pechaud, Y., van Hullebusch, E.D., 2018. Alteration of the characteristics of extracellular polymeric substances (EPS) extracted from the fungus *Phanerochaete chrysosporium* when exposed to subtoxic concentrations of nickel (II). Int. Biodeter. Biodegr. 129, 179–188.

Carrion, O., Delgado, L., Mercade, E., 2015. New emulsifying and cryoprotective exopolysaccharide from *Antarctic Pseudomonas* sp. ID1. Carbohydr. Polym. 117, 1028–1034.

Caruso, C., Rizzo, C., Mangano, S., Poli, A., Di Donato, P., Nicolaus, B., Finore, I., Di Marco, G., Michaud, L., Lo Giudice, A., 2019. Isolation, characterization and optimization of EPSs produced by a cold-adapted Marinobacter isolate from Antarctic seawater. Antarctic Sci. 31, 69–79.

Cerminati, S., Leroux, M., Anselmi, P., Peirú, S., Alonso, J.C., Priem, B., Menzella, H.G., 2021. Low cost and sustainable hyaluronic acid production in a manufacturing platform based on *Bacillus subtilis* 3NA strain. Appl. Microbiol. Biotechnol. 105, 3075–3086.

Chambi, D., Romero-Soto, L., Villca, R., Orozco-Gutiérrez, F., Vega-Baudrit, J., Quillaguamán, J., Hatti-Kaul, R., Martín, C., Carrasco, C., 2021. Exopolysaccharides production by cultivating a bacterial isolate from the hypersaline environment of salar de uyuni (Bolivia) in pretreatment liquids of steam-exploded quinoa stalks and enzymatic hydrolysates of Curupaú sawdust. Fermentation. 7, 33.

Cheaburu-Yilmaz, C.N., Karasulu, H.Y., Yilmaz, O., 2019. Chapter 13 — Nanoscaled dispersed systems used in drug-delivery applications. In: C. Vasile (Ed.), *Polymeric Nanomaterials in Nanotherapeutics*. Elsevier, pp. 437–468.

Chen, G., Qian, W., Li, J., Xu, Y., Chen, K., 2015. Exopolysaccharide of Antarctic bacterium Pseudoaltermonas sp. S-5 induces apoptosis in K562 cells. Carbohydr. Polym. 121, 107–114.

Chen, G., Zhu, Y., Zhang, G., Liu, H., Wei, Y., Wang, P., Wang, F., Xian, M., Xiang, H., Zhang, H., 2019. Optimization and characterization of pullulan production by a newly isolated high-yielding strain *Aureobasidium melanogenum*. Prep. Biochem. Biotechnol. 49, 557–566.

Chen, Y.-P., Li, C., Guo, J.-S., Fang, F., Gao, X., Zhang, P., Li, S., 2013. Extraction and characterization of extracellular polymeric substances in biofilm and sludge via completely autotrophic nitrogen removal over nitrite system. Appl. Biochem. Biotechnol. 169, 526–538.

Chien, C.-Y., Enomoto-Rogers, Y., Takemura, A., Iwata, T., 2017. Synthesis and characterization of regioselectively substituted curdlan hetero esters via an unexpected acyl migration. Carbohydr. Polym. 155, 440–447.

Choi, J.-W., Choi, H.-G., Lee, W.-H., 1996. Effects of ethanol and phosphate on emulsan production by *Acinetobacter calcoaceticus* RAG-1. J. Biotechnol. 45, 217–225.

Chug, R., Mathur, S., Kothari, S.L., Harish, Gour, V.S., 2021. Maximizing EPS production from *Pseudomonas aeruginosa* and its application in Cr and Ni sequestration. Biochem. Biophys. Rep. 100972.

Cirigliano, M.C., Carman, G.M., 1984. Isolation of a bioemulsifier from *Candida lipolytica*. Appl. Environ. Microbiol. 48, 747–750.

Clement, M.-J., Tissot, B., Chevolot, L., Adjadj, E., Du, Y., Curmi, P.A., Daniel, R., 2010. NMR characterization and molecular modeling of fucoidan showing the importance of oligosaccharide branching in its anticomplementary activity. Glycobiology. 20, 883–894.

Cooper, D.G., Goldenberg, B.G., 1987. Surface-active agents from two *Bacillus* species. Appl. Environ. Microbiol. 53, 224–229.

de Sousa Costa, L.A., Inomata Campos, M., Izabel Druzian, J., de Oliveira, A.M., de Oliveira Junior, E.N., 2014. Biosynthesis of xanthan gum from fermenting shrimp shell: Yield and apparent viscosity. Int. J. Polym. Sci. 2014, 273650.

De Vuyst, L., Degeest, B., 1999. Heteropolysaccharides from lactic acid bacteria. FEMS Microbiol. Rev. 23, 153–177.

Donot, F., Fontana, A., Baccou, J.C., Schorr-Galindo, S., 2012. Microbial exopolysaccharides: Main examples of synthesis, excretion, genetics and extraction. Carbohyd. Polym. 87, 951–962.

Drira, M., Elleuch, J., Ben Hlima, H., Hentati, F., Gardarin, C., Rihouey, C., Le Cerf, D., Michaud, P., Abdelkafi, S., Fendri, I., 2021. Optimization of exopolysaccharides production by *Porphyridium sordidum* and their potential to induce defense responses in *Arabidopsis thaliana* against *Fusarium oxysporum*. Biomolecules. 11.

Duan, Z., Tan, X., Zhang, D., Parajuli, K., 2020. Development of thermal treatment for the extraction of extracellular polymeric substances from Microcystis: Evaluating extraction efficiency and cell integrity. Algal Res. 48, 101879.

Elsayed, E.A., Farooq, M., Dailin, D.J., El-Enshasy, H.A., Othman, N.Z., Malek, R.A., Danial, E.N., Wadaan, M.A., 2017. In vitro and in vivo biological screening of kefiran polysaccharide produced by *Lactobacillus kefiranofaciens*. Biomed. Res. (Tokyo). 28, 594–600.

Fang, L., Catchmark, J.M., 2015. Characterization of cellulose and other exopolysaccharides produced from *Gluconacetobacter* strains. Carbohydr. Polym. 115, 663–669.

Fernando, I., Lu, D., Zhou, Y., 2020. Interactive influence of extracellular polymeric substances (EPS) and electrolytes on the colloidal stability of silver nanoparticles. Environ. Sci. Nano. 7, 186–197.

Freitas, F., Alves, V.D., Pais, J., Costa, N., Oliveira, C., Mafra, L., Hilliou, L., Oliveira, R., Reis, M.A.M., 2009. Characterization of an extracellular polysaccharide produced by a *Pseudomonas* strain grown on glycerol. Bioresour. Technol. 100, 859–865.

Freitas, F., Alves, V.D., Reis, M.A.M., 2011. Advances in bacterial exopolysaccharides: From production to biotechnological applications. Trends Biotechnol. 29, 388–398.

Galib, M.A., Abbott, T., Lee, H.-S., 2021. Examination of extracellular polymer (EPS) extraction methods for anaerobic membrane bioreactor (AnMBR) biomass. Sustainability. 13, 12584.

Gansbiller, M., Schmid, J., Sieber, V., 2020. Rheology of sphingans in EPS surfactant systems. Carbohydr. Polym. 248, 116778.

Guangyin, Z., Youcai, Z., 2017. Harvest of bioenergy from sewage sludge by anaerobic digestion. Engineering. https://doi.org/10.1016/B978-0-12-811639-5.00005-X.

Gupta, A., Thakur, I.S., 2016. Study of optimization of wastewater contaminant removal along with extracellular polymeric substances (EPS) production by a thermotolerant *Bacillus* sp. ISTVK1 isolated from heat shocked sewage sludge. Bioresour. Technol. 213, 21–30.

Gurav, R., Bhatia, S.K., Choi, T.-R., Hyun Cho, D., Chan Kim, B., Hyun Kim, S., Ju Jung, H., Joong Kim, H., Jeon, J.-M., Yoon, J.-J., Yun, J., Yang, Y.-H., 2022. Lignocellulosic hydrolysate based biorefinery for marine exopolysaccharide production and application of the produced biopolymer in environmental clean-up. Bioresour. Technol. 359, 127499.

Gutierrez, T., Shimmield, T., Haidon, C., Black, K., Green David, H., 2008. Emulsifying and metal ion binding activity of a glycoprotein exopolymer produced by *Pseudoalteromonas* sp. strain TG12. Appl. Environ. Microbiol. 74, 4867–4876.

Hansma, H.G., Pietrasanta, L.I., Auerbach, I.D., Sorenson, C., Golan, R., Holden, P.A., 2000. Probing biopolymers with the atomic force microscope: A review. J. Biomat. Sci. Polym. 11, 675–683.

Hay, I.D., Ur Rehman, Z., Ghafoor, A., Rehm, B.H.A., 2010. Bacterial biosynthesis of alginates. J. Chem. Technol. Biotechnol. 85, 752–759.

Hu, Y., Kim, Y., Hong, I., Kim, M., Jung, S., 2021. Fabrication of flexible pH-responsive agarose/succinoglycan hydrogels for controlled drug release. Polymers. 13, 2049.

Huang, T., Lin, J., Cao, J., Zhang, P., Bai, Y., Chen, G., Chen, K., 2012. An exopolysaccharide from *Trichoderma pseudokoningii* and its apoptotic activity on human leukemia K562 cells. Carbohydr. Polym. 89, 701–708.

Ibrahim, H.A.H., Abou Elhassayeb, H.E., El-Sayed, W.M.M., 2022. Potential functions and applications of diverse microbial exopolysaccharides in marine environments. J. Gene. Eng. Biotechnol. 20, 151.

Ispirli, H., Sagdic, O., Yilmaz, M.T., Dertli, E., 2019. Physicochemical characterisation of an glucan from *Lactobacillus reuteri* E81 as a potential exopolysaccharide suitable for food applications. Process Biochem. 79, 91–96.

Izawa, N., Hanamizu, T., Iizuka, R., Sone, T., Mizukoshi, H., Kimura, K., Chiba, K., 2009. *Streptococcus thermophilus* produces exopolysaccharides including hyaluronic acid. J. Biosci. Bioeng. 107, 119–123.

Jazini, M.H., Fereydouni, E., Karimi, K., 2017. Microbial xanthan gum production from alkali-pretreated rice straw. RSC Adv. 7, 3507–3514.

Jenab, A., Roghanian, R., Emtiazi, G., Ghaedi, K., 2017. Manufacturing and structural analysis of antimicrobial kefiran/polyethylene oxide nanofibers for food packaging. Iranian Polym. J. 26, 31–39.

Jeong, J.-P., Kim, Y., Hu, Y., Jung, S., 2022. Bacterial succinoglycans: Structure, physical properties, and applications. Polymers. 14, 276.

Jin, H., Jeong, Y., Yoo, S.-H., Johnston, T.V., Ku, S., Ji, G.E., 2019. Isolation and characterization of high exopolysaccharide-producing *Weissella confusa* VP30 from young children's feces. Microb. Cell. Fact. 18, 110.

Jumma Kareem, A., Abdul Sattar Salman, J., 2019. Production of dextran from locally *Lactobacillus* Spp. isolates. Rep. Biochem. Mol. Biol. 8, 287–300.

Kanamarlapudi, S.L.R.K., Muddada, S., 2017. Characterization of exopolysaccharide produced by *Streptococcus thermophilus* CC30. BioMed Res. Int. 2017, 4201809.

Kanmani, P., Lim, S.T., 2013. Synthesis and structural characterization of silver nanoparticles using bacterial exopolysaccharide and its antimicrobial activity against food and multidrug resistant pathogens. Process Biochem. 48, 1099–1106.

Kekez, B., Gojgic-Cvijovic, G., Jakovljevic, D., Pavlovic, V., Beskoski, V., Popovic, A., Vrvic, M.M., Nikolic, V., 2017. Synthesis and characterization of a new type of levan-graft-polystyrene copolymer. Carbohydr. Polym. 154, 20–29.

Kim, J., Hwang, J., Seo, Y., Jo, Y., Son, J., Choi, J., 2017. Engineered chitosan-xanthan gum biopolymers effectively adhere to cells and readily release incorporated antiseptic molecules in a sustained manner. J. Ind. Eng. Chem. 46, 68–79.

Kruger, N.J., 2009. The bradford method for protein quantitation. In: J.M. Walker (Ed.), *The Protein Protocols Handbook*. Humana Press, pp. 17–24.

Kumar, S.A., Mody, K., Jha, B., 2007a. Bacterial exopolysaccharides—a perception. J. Basic. Microbiol. 47, 103–117.

Kumar, S.A., Mody, K., Jha, B., 2007b. Evaluation of biosurfactant/bioemulsifier production by a marine bacterium. Bull. Environ. Contam. Toxicol. 79, 617–621.

Lakra, A.K., Domdi, L., Tilwani, Y.M., Arul, V., 2020. Physicochemical and functional characterization of mannan exopolysaccharide from *Weissella confusa* MD1 with bioactivities. Int. J. Biol. Macromol. 143, 797–805.

Laws, A., Gu, Y., Marshall, V., 2001. Biosynthesis, characterisation, and design of bacterial exopolysaccharides from lactic acid bacteria. Biotechnol. Adv. 19, 597–625.

Li, H., Jiao, X., Sun, Y., Sun, S., Feng, Z., Zhou, W., Zhu, H., 2016a. The preparation and characterization of a novel sphingan WL from marine *Sphingomonas* sp. WG. Sci. Rep. 6, 37899.

Li, S., Huang, R., Shah, N.P., Tao, X., Xiong, Y., Wei, H., 2014. Antioxidant and antibacterial activities of exopolysaccharides from *Bifidobacterium bifidum* WBIN03 and *Lactobacillus plantarum* R315. J. Dairy. Sci. 97, 7334–7343.

Li, Y., Guo, S., Zhu, H., 2016b. Statistical optimization of culture medium for production of exopolysaccharide from endophytic fungus *Bionectria ochroleuca* and its antitumor effect in vitro. Excli. J. 15, 211–220.

Liu, J., Wang, X., Pu, H., Liu, S., Kan, J., Jin, C., 2017. Recent advances in endophytic exopolysaccharides: Production, structural characterization, physiological role and biological activity. Carbohydr. Polym. 157, 1113–1124.

Liu, T., Zhou, K., Yin, S., Liu, S., Zhu, Y., Yang, Y., Wang, C., 2019. Purification and characterization of an exopolysaccharide produced by *Lactobacillus plantarum* HY isolated from home-made Sichuan Pickle. Int. J. Biol. Macromol. 134, 516–526.

Liu, Y., Gu, Q., Ofosu, F.K., Yu, X., 2015. Isolation and characterization of curdlan produced by *Agrobacterium* HX1126 using Î±-lactose as substrate. Int. J. Biol. Macromol. 81, 498–503.

Maalej, H., Hmidet, N., Boisset, C., Buon, L., Heyraud, A., Nasri, M., 2015. Optimization of exopolysaccharide production from *Pseudomonas stutzeri* AS22 and examination of its metal-binding abilities. J. Appl. Microbiol. 118, 356–367.

Maeda, K., Okuda, Y., Enomoto, G., Watanabe, S., Ikeuchi, M., 2021. Biosynthesis of a sulfated exopolysaccharide, synechan, and bloom formation in the model cyanobacterium *Synechocystis* sp. strain PCC 6803. eLife. 10, e66538.

Mahapatra, S., Banerjee, D., 2012. Structural elucidation and bioactivity of a novel exopolysaccharide from endophytic *Fusarium solani* SD5. Carbohydr. Polym. 90, 683–689.

Mahapatra, S., Banerjee, D., 2016. Production and structural elucidation of exopolysaccharide from endophytic *Pestalotiopsis* sp. BC55. Int. J. Biol. Macromol. 82, 182–191.

Mancuso Nichols, C.A., Garon, S., Bowman, J.P., Raguénès, G., Guezennec, J., 2004. Production of exopolysaccharides by Antarctic marine bacterial isolates. J. Appl. Microbiol. 96, 1057–1066.

Masuko, T., Minami, A., Iwasaki, N., Majima, T., Nishimura, S.-I., Lee, Y.C., 2005. Carbohydrate analysis by a phenol—sulfuric acid method in microplate format. Anal. Biochem. 339, 69–72.

Miller, G.L., 1959. Use of dinitrosalicylic acid reagent for determination of reducing sugar. Anal. Chem. 31, 426–428.

Mohd Nadzir, M., Nurhayati, R.W., Idris, F.N., Nguyen, M.H., 2021. Biomedical applications of bacterial exopolysaccharides: A review. Polymers. 13, 530.

Moradi, M., Guimaraes, J.T., Sahin, S., 2021. Current applications of exopolysaccharides from lactic acid bacteria in the development of food active edible packaging. Curr. Opin. Food. Sci. 40, 33–39.

Mueller, L.N., de Brouwer, J.F.C., Almeida, J.S., Stal, L.J., Xavier, J.O.B., 2006. Analysis of a marine phototrophic biofilm by confocal laser scanning microscopy using the new image quantification software PHLIP. BMC Ecol. 6, 1.

Nacher-Vazquez, M., Ballesteros, N., Canales, A., Rodriguez Saint-Jean, S., Perez-Prieto, S.I., Prieto, A., Aznar, R., Lopez, P., 2015. Dextrans produced by lactic acid bacteria exhibit antiviral and immunomodulatory activity against salmonid viruses. Carbohydr. Polym. 124, 292–301.

Nguyen, P.-T., Nguyen, T.-T., Bui, D.-C., Hong, P.-T., Hoang, Q.-K., Nguyen, H.-T., 2020. Exopolysaccharide production by lactic acid bacteria: The manipulation of environmental stresses for industrial applications. AIMS Microbiol. 6, 451–469.

Okoro, O.V., Gholipour, A.R., Sedighi, F., Shavandi, A., Hamidi, M., 2021. Optimization of exopolysaccharide (EPS) production by *Rhodotorula mucilaginosa* sp. GUMS16. Chem. Eng. 5, 39.

Ozturk Urek, R., Ilgin, S., 2019. Production and partial characterization of the exopolysaccharide from Pleurotus sajor caju. Annals Microbiol. 69, 1201–1210.

Padmanabhan, A., Shah, N.P., 2020. Structural characterization of exopolysaccharide from *Streptococcus thermophilus* ASCC 1275. J. Dairy. Sci. 103, 6830–6842.

Paniagua-Michel, J.D.J.S., Olmos-Soto, J., Morales-Guerrero, E.R., Kim, S.-K., 2014. Chapter 11 — Algal and microbial exopolysaccharides: New insights as biosurfactants and bioemulsifiers. In: *Advances in Food and Nutrition Research*, Vol. 73. Academic Press, pp. 221–257.

Pham, D., Ivanova, E., Wright, J., Nicolau, D. 2003. AFM analysis of the extracellular polymeric substances (EPS) released during bacterial attachment on polymeric surfaces. SPIE. Proc. SPIE. 4962, 151–159.

Prajapati, D., Bhatt, A., Gupte, A., 2022. Production, optimization, partial-purification and pyrolysis kinetic studies of exopolysaccharide from a native brown-rot fungi *Fomitopsis meliae* AGDP-2. Bioresour. Technol. Rep. 17, 100948.

Rajoka, M.S.R., Mehwish, H.M., Hayat, H.F., Hussain, N., Sarwar, S., Aslam, H., Nadeem, A., Shi, J., 2019. Characterization, the antioxidant and antimicrobial activity of exopolysaccharide isolated from poultry *Origin Lactobacilli*. Probiotics Antimicrob. Proteins. 11, 1132–1142.

Rath, T., Rühmann, B., Schmid, J., Sieber, V., 2022. Systematic optimization of exopolysaccharide production by *Gluconacetobacter* sp. and use of (crude) glycerol as carbon source. Carbohydr. Polym. 276, 118769.

Riaz Rajoka, M.S., Mehwish, H.M., Fang, H., Padhiar, A.A., Zeng, X., Khurshid, M., He, Z., Zhao, L., 2019. Characterization and anti-tumor activity of exopolysaccharide produced by *Lactobacillus kefiri* isolated from Chinese kefir grains. J. Functional. Foods. 63, 103588.

Robyt, J.F., Yoon, S.-H., Mukerjea, R., 2008. Dextransucrase and the mechanism for dextran biosynthesis. Carbohydr. Res. 343, 3039–3048.

Sabaghi, M., Maghsoudlou, Y., Habibi, P., 2015. Enhancing structural properties and antioxidant activity of kefiran films by chitosan addition. Food Struct. 5, 66–71.

Saleh Amer, M., Zaghloul, E.H., Ibrahim, M.I.A., 2020. Characterization of exopolysaccharide produced from marine-derived *Aspergillus terreus* SEI with prominent biological activities. Egyp. J. Aqua. Res. 46, 363–369.

Santander, J., Martin, T., Loh, A., Pohlenz, C., Gatlin Iii, D.M., Curtiss Iii, R., 2013. Mechanisms of intrinsic resistance to antimicrobial peptides of *Edwardsiella ictaluri* and its influence on fish gut inflammation and virulence. Microbiology. 159, 1471.

Saravanan, C., Shetty, P.K.H., 2016. Isolation and characterization of exopolysaccharide from *Leuconostoc lactis* KC117496 isolated from idli batter. Int. J. Biol. Macromol. 90, 100–106.

Sarwat, F., Ul Qader, S.A., Aman, A., Ahmed, N., 2008. Production & characterization of a unique dextran from an indigenous *Leuconostoc mesenteroides* CMG713. Int. J. Biol. Sci. 4, 379–386.

Sathiyanarayanan, G., Bhatia, S.K., Kim, H.J., Kim, J.-H., Jeon, J.-M., Kim, Y.-G., Park, S.-H., Lee, S.H., Lee, Y.K., Yang, Y.-H., 2016. Metal removal and reduction potential of an exopolysaccharide produced by Arctic psychrotrophic bacterium *Pseudomonas* sp. PAMC 28620. RSC Adv. 6, 96870–96881.

Schmid, J., 2018. Recent insights in microbial exopolysaccharide biosynthesis and engineering strategies. Curr. Opin. Biotechnol. 53, 130–136.

Senthilkumar, M., Amaresan, N., Sankaranarayanan, A., 2021. Extraction of extracellular polymeric substances (EPS) from bacteria. In: M. Senthilkumar, N. Amaresan, A. Sankaranarayanan (Eds.), *Plant-Microbe Interactions: Laboratory Techniques*. Springer, pp. 135–137.

Shen, J.-W., Shi, C.-W., Xu, C.-P., 2013. Exopolysaccharides from *Pleurotus pulmonarius*: Fermentation optimization, characterization and antioxidant activity. Food Technol. Biotechnol. 51, 520–527.

Singh, R.S., Saini, G.K., Kennedy, J.F., 2009. Downstream processing and characterization of pullulan from a novel colour variant strain of *Aureobasidium pullulans* FB-1. Carbohydr. Polym. 78, 89–94.

Solmaz, K.B., Ozcan, Y., Mercan Dogan, N., Bozkaya, O., Ide, S., 2018. Characterization and production of extracellular polysaccharides (EPS) by *Bacillus Pseudomycoides* U10. Environments. 5, 63.

Sran, K.S., Sundharam, S.S., Krishnamurthi, S., Roy Choudhury, A., 2019. Production, characterization and bio-emulsifying activity of a novel thermostable exopolysaccharide produced by a marine strain of *Rhodobacter johrii* CDR-SL 7Cii. Int. J. Biol. Macromol. 127, 240–249.

Strathmann, M., Wingender, J., Flemming, H.-C., 2002. Application of fluorescently labelled lectins for the visualization and biochemical characterization of polysaccharides in biofilms of *Pseudomonas aeruginosa*. J. Microbiol. Methods. 50, 237–248.

Sugumaran, K.R., Gowthami, E., Swathi, B., Elakkiya, S., Srivastava, S.N., Ravikumar, R., Gowdhaman, D., Ponnusami, V., 2013. Production of pullulan by Aureobasidium pullulans from Asian palm kernel: A novel substrate. Carbohydr. Polym. 92, 697–703.

Sugumaran, K.R., Ponnusami, V., 2017. Review on production, downstream processing and characterization of microbial pullulan. Carbohydr. Polym. 173, 573–591.

Sun, X., Hao, L., Ma, H., Li, T., Zheng, L., Ma, Z., Zhai, G., Wang, L., Gao, S., Liu, X., Jia, M., Jia, L., 2013. Extraction and in vitro antioxidant activity of exopolysaccharide by *Pleurotus eryngii* SI-02. Brazilian J. Microbiol. 44, 1081–1088.

Tok, E., Aslim, B., 2010. Cholesterol removal by some lactic acid bacteria that can be used as probiotic. Microbiol. Immunol. 54, 257–264.

Tokura, S., Ueno, K., Miyazaki, S., Nishi, N., 1997. Molecular weight dependent antimicrobial activity by Chitosan. Macromol. Symp. 120, 1–9.

Wang, K., Li, W., Rui, X., Chen, X., Jiang, M., Dong, M., 2014. Structural characterization and bioactivity of released exopolysaccharides from *Lactobacillus plantarum* 70810. Int. J. Biol. Macromol. 67, 71–78.

Wang, Y., Zhang, R., Duan, J., Shi, X., Zhang, Y., Guan, F., Sand, W., Hou, B., 2022. Extracellular polymeric substances and biocorrosion/biofouling: Recent advances and future perspectives. Int. J. Mol. Sci. 23.

Wang, Z., Wu, J., Zhu, L., Zhan, X., 2017. Characterization of xanthan gum produced from glycerol by a mutant strain *Xanthomonas campestris* CCTCC M2015714. Carbohydr. Polym. 157, 521–526.

Woo, J.E., Seong, H.J., Lee, S.Y., Jang, Y.-S., 2019. Metabolic engineering of *Escherichia coli* for the production of hyaluronic acid from glucose and galactose. Front. Bioeng. Biotechnol. 7.

Xiao, R., Zheng, Y., 2016. Overview of microalgal extracellular polymeric substances (EPS) and their applications. Biotechnol. Adv. 34, 1225–1244.

Yildiz, H., Karatas, N., 2018. Microbial exopolysaccharides: Resources and bioactive properties. Process Biochem. 72, 41–46.

Yun, E.J., Cho, Y., Han, N.R., Kim, I.J., Jin, Y.-S., Kim, K.H., 2021. Increased production of colanic acid by an engineered *Escherichia coli* strain, mediated by genetic and environmental perturbations. Appl. Biochem. Biotechnol. 193, 4083–4096.

Zakeri, A., Rasaee, M.J., Pourzardosht, N., 2017. Enhanced hyluronic acid production in *Streptococcus zooepidemicus* by over expressing HasA and molecular weight control with Niscin and glucose. Biotechnol. Rep. 16, 65–70.

Zeidan, A.A., Poulsen, V.K., Janzen, T., Buldo, P., Derkx, P.M.F., Øregaard, G., Neves, A.R., 2017. Polysaccharide production by lactic acid bacteria: From genes to industrial applications. FEMS Microbiol. Rev. 41, S168–S200.

Zhang, P., Fang, F., Chen, Y.-P., Shen, Y., Zhang, W., Yang, J.-X., Li, C., Guo, J.-S., Liu, S.-Y., Huang, Y., 2014. Composition of EPS fractions from suspended sludge and biofilm and their roles in microbial cell aggregation. Chemosphere. 117, 59–65.

Zhang, S., Chen, K.Y., Zou, X., 2021. Carbohydrate-protein interactions: Advances and challenges. Commun. Inf. Syst. 21, 147–163.

Zhang, T., Fang, H.H.P., 2001. Quantification of extracellular polymeric substances in biofilms by confocal laser scanning microscopy. Biotechnol. Lett. 23, 405–409.

Zhang, Z., Khan, N.M., Nunez, K.M., Chess, E.K., Szabo, C.M., 2012. Complete monosaccharide analysis by high-performance anion-exchange chromatography with pulsed amperometric detection. Anal. Chem. 84, 4104–4110.

Zheng, L.P., Zou, T., Ma, Y.J., Wang, J.W., Zhang, Y.Q., 2016. Antioxidant and DNA damage protecting activity of exopolysaccharides from the endophytic bacterium *Bacillus cereus* SZ1. Molecules. 21.

3 Advances in fermentation technology for EPS production

Mamta Devi, Kalpana Thakur,
Deeksha Kumari, Manya Behl,
Ranju Kumari Rathour,
Arvind Kumar Bhatt, and Ravi Kant Bhatia

3.1 INTRODUCTION

Microorganisms, the most abundant and largest community in the biological system of planet "Earth." The microbes are reported to be present in almost every ecosystem of nature including air, water, soil, water, cold desert, hot springs, etc. These microscopic cell factories have been reported to synthesize wide varieties of industrially imperative goods and multifaceted polysaccharides intracellular as well as extracellular. Extracellular exopolysaccharides (EPS) are one such organic polymer synthesized by microbial cell to survive under harsh environment conditions (Kambourova et al., 2015).

EPS, a high molecular weight polymeric saccharide, primarily consists of repeated subunits of monomeric sugars. Besides carbohydrates, EPS also have phosphate, acetate, succinate, and pyruvate as non-carbohydrate moieties. EPS is generally secreted as an integral component of biofilm formed by bacteria. The EPS production quantity is directly proportional to exposure of microorganism to harsh conditions (Dave et al., 2016). Based on the presence of units of monosaccharides, EPS are divided into two types, that is, homo-EPS (single monosaccharide unit) and hetero-EPS (different monosaccharide units) (Abarquero et al., 2022). In bacterial system, EPS act as an important factor of immune response as it helps formation of biofilm as defense mechanism to survive under harsh environmental conditions (pH, temperature, and osmotic pressure) (Castro-Bravo et al., 2017), helps in cell-to-cell interaction (Lynch et al., 2018), protect microbes from toxic substances (lysozymes, enzymes, antibiotics, metal ions, and bile salts) and prevent bacteria from the attack of phages (Abdalla et al., 2021). The polysaccharides synthesized from microbial sources are advantageous over synthetic ones as these are biodegradable/compatible and nontoxic. Being industrial important polysaccharide, EPS production has been reported from a variety of organisms (prokaryotes as well as eukaryotes) (Hamidi et al., 2022). The prime sources of EPS are plants. The cell wall of plants is the store house of EPS where these are present in the form of polymeric cellulose and lignin. But due to limitation of plants as substrate, new opportunities for EPS production has been explored. In the past few decades, numerous bacteria and fungi are reported to produce polysaccharides, that is, storage (glycogen), capsular (o-antigen), and extracellular (Schmid et al., 2015). Among all, EPS were mainly reported from marine bacteria *Halomonas* sp., *Pseudomonas* sp. and lactic acid bacteria (LAB) (*Lactobacillus, Lactococcus*), probiotic bacteria (*Bifidobacterium, Pediococcus* and *Leuconostoc* sp.), *Enterobacter cloacae, B. coagulans*, etc (Mbye et al., 2020). The probiotic bacteria need to survive in the human body at very low pH, higher bile salts concentration and gastric juices. Thus, they synthesize EPS to survive these variable conditions (pH, temperature, antibiotics resistance) (Angelin and Kavitha, 2020; Nguyen et al., 2020; Shyam et al., 2021). Besides this, EPS were also reported to form fungi, mainly basidiomycetes, filamentous fungi, and yeasts, that is, *Absidia* sp., *Acremonium* sp., *Agaricus* sp., *Alternaria* sp., *Aspergillus* sp., *Cladosporium* sp., and many more (Mahapatra and Banerjee, 2013).

In microbial metabolism, polysaccharides synthesis takes place by four different mechanisms, that is, ABC dependent, synthase-dependent, Wzx/Wzy-dependent, and glycosyltransferase (GT's) polymerizing genes. The precursors of EPS biosynthetic exists in cell while its polymerization takes place in cell envelope (Rana and Upadhyay, 2020; Schmid et al., 2015). EPS coding genes were identified in *S. thermophiles Sfi*6 and *L. lactis* NIZO B40 for the first time. The size of these gene clusters varies from 12–15 kb depending upon type of source. The yield of EPS production depends upon environmental/ecological condition and interaction of microbe with surrounding (Madhuri and Prabhakar, 2014). Besides this, extraction methods also greatly influence the yield. Various physical, chemical, as well as biological methods have been employed to extract EPS (Galib et al., 2021).

The major limitation of microbial EPS is its production/extraction cost and substrate specificity. But the increasing research interest in sustainable utilization of natural resources as substrate especially for biofuel production has also enhanced the production of EPS and its utilization in industrial applications. Thus, in this chapter authors try to provide a holistic view on the sources, microbial biosynthetic pathways of EPS, and industrial production of EPS. Further, diverse perceptions affecting the extraction process, industrial applicability, bottlenecks, and future possibilities to enhance EPS production to exploit it for multifaceted usage in industries have also been emphasized.

3.2 MICROBIAL SOURCES OF EXOPOLYSACCHARIDES

EPS are metabolic products secreted by producers from various taxonomic groups, having a new or different molecular arrangement. The first evidence of EPS synthesis by microbes dated back to the 1980s (Whitfield, 1988). As stated previously that EPS are defense polysaccharide on the cell membrane of microbial cell produced to survive harsh environment conditions (McSwain et al., 2005). Majority of microbes from each category, that is, bacteria, fungi, are reported by researchers for production of these polysaccharides as long chain polymers of sugars like glucose, galactose, and rhamnose (Ghosh and Maiti, 2016; Xiao et al., 2021). However, based on their site of synthesis, these are divided into three main groups: extracellular, intracellular, and structural polysaccharides. Among these, structural polysaccharide is found as the key component of primary part of cell walls (Schmid et al., 2015; Freitas et al., 2011). Production of EPS from microbial sources is of greater significance as microbial EPS often has shorter production time than other natural sources and can be easily extracted from fermentation broth. Further, to make the process more economic, the waste biomass (agro-industrial, forest waste) can be used as feedstock for microbial fermentation. Additionally, polysaccharides exhibit a wider range of structural and physicochemical characteristics. They can be used for a variety of purposes, such as modifying the rheology of aqueous systems or as medicinal agents (Donot et al., 2012; Rehm, 2010; Moscovici, 2015).

In the past, EPS are reported from eubacteria, archaebacteria, fungi, algae, as well as from lower order phytoplankton that belongs to any kind of temporal zone (thermophilic, mesophilic, or halophilic) (Delattre et al., 2016; Moscovici, 2015). Bacteria also produce EPSs, which can either be homopolymeric or heteropolymeric or have a wide variety of chemical configurations (Nwodo et al., 2012). EPS producing microbes were reported in a variety of ecological inches by researchers. In a study of deep-sea hydrothermal vents, geothermal springs, saline lakes, and Antarctic habitats have been used as a potential source for isolation of EPS producing microbes (Freitas et al., 2011). According to Fazenda et al. (2008), EPS yield is directly proportional to C/N ration of medium; that means the higher the C/N ratio, the more will be the EPS accumulation.

3.2.1 BACTERIA

Bacterial species, especially of *Streptococcus*, *Xanthomonas*, and *Acetobacter* genera have reported to synthesize polysaccharides. Dextran of bacterial origin, reported to be produced using lactic acid bacteria (*Pediococcus, and Lactobacillus*), *Leuconostoc, Streptococcus*, and *Weisella*, was the first commercially produced water soluble EPS. LABs are the model organism for commercial manufacture of EPS as they are safe and also significantly influence the fermentation process. The LAB along with EPS produces other metabolites (hydrogen peroxide, lactic acid, acetic acid) and low molecular weight compound (diacetyl, fatty acids, and reuterin) having antibacterial and antifungal potential (Ahmad et al., 2015). *Alcaligenes* and *Agrobacterium* species also produce curdlan, another glucan. It is polymer, that is, insoluble in water and form gel when dissolved in it. The Food and Drug Administration (FDA) has approved both dextran and curdlan to be utilized in foodstuff, cosmetics, and clinical procedures (Yang et al., 2016). *Glucanoacetobactera, Rhizobium*, and *Acetobacter* produce water insoluble cellulose through the polymerization of UDP-glucose nucleotide sugar molecules. Physiochemically, cellulose has low density, high mechanical, as well as water adsorption ability (Chawla et al., 2009; Gomes et al., 2013). The EPS produced by bacteria like *K. pneumoniae, P. oleovorans, Enterobacter* A47, *E. coli*, and others have a peculiar structure and properties due to presence of multiple subunits of monomeric sugars like glucose, galactose, fucose, mannose, and others (Freitas et al., 2009; Patel and Prajapati, 2013). Extremophillic bacteria like *Pseudoalteromonas* sp., *Halomonas* sp., *Bacillus* sp., and *Vibrio* sp., etc., isolated from severe habitats, especially from Antarctic, salty, and geothermal regions, were also found to produce EPS having flocculate and metal binding ability (Li et al., 2008). As per studies, the thermophilic EPS fructo-fucan produced by *B. licheniformis* strain T4 has antiviral and anti-cytotoxic properties. A complex polysaccharide with emulsifying properties is produced by *Aeribacillus pallidus* 418 (Gugliandolo et al., 2013; Spano et al., 2013). HA-like EPSs synthesized from the sea bacterium *Vibrio diabolicus* find its usage in tissue regeneration and speed up the collagen fibrillation process (in vitro) (Senni et al., 2011).

Additionally, Lin and co-workers (2010) reported that using microbial consortia can enhance EPS yield and excreted EPS up to 10% of the mass of the granules. According to Yang and colleagues (2014), an increased interest has been found in the scientific community to determine the EPS recovery from granular sludge as a large volume of this aerobic sludge is produced in various operations across the world. And genera like *Pseudomonas, Clostridium, Thauera,* and *Arthrobacter* were prominently observed in this sludge. Yang et al. (2014) also reported alginate-like EPS (ALE) that resemble alginate from this sludge under both aerobic condition, while Li et al. (2016a) reported EPS from the same under anaerobic conditions. Lin et al. (2010) also reported high GG content ALE from the same kind of sludge having highest hydrophobic ability.

The greatest beneficial use of EPSs from LAB is enhancement of the rheology, influence texture, and feel of fermented milk goods like cheese, yoghurt, cream, desert. *Streptococcus, Lactococcus, Lactobacillus, Leuconosto, Pediococcus,* and *Weissellale* are some of the LAB strains that are frequently utilized to produce EPSs (Patel and Prajapati, 2013). LAB can produce either hetero polysaccharides or homo polysaccharides depending on their chemical makeup. Regarding the monomer content molecular mass, and structural characteristics, LAB produces a large variety of EPS. LAB are capable of producing EPSs with a variety of various shapes without posing any health risks as considered as safe microorganisms (GRAS—Generally Recognized As Safe) (Polak-Berecka et al., 2014). According to Surayot et al. (2014), EPSs have a significant impact on LAB cell surface physicochemical features, helping to protect bacterial cells from phage attacks, phagocytosis, dehydration, adverse environmental effects, and antibiotics. EPSs participate in the structural elements of the extracellular matrix, which serves as a container for cells as they create a cell membrane (Marvasi et al., 2010). According to study by Patel et al. (2012), several bacterial strains, *L. fermentum, L. sakei,* and *L. hilgardii,* produced (1,6) glucans that were ramified at position 3 by dextran residues, and least at locations 2 and 4. Different sources of EPS and their application has been summarized in Table 3.1.

3.2.2 FUNGI AND YEASTS

Fungi of genera *Aureobasidium, Candida,* and *Cryptococcus,* and many others, produces EPS depending on the strain employed, the components of the media used, and the physical conditions provided during the fermentation process. Uonic acids, non-sugar substituents, and D-mannose, both alone and in conjunction with galactose and xylose are the major polymers reported from fungi (Osinska-Jaroszuk et al., 2015). Pullulan, a starchy EPS produced from *Aureobasidium pullulans,* used as polymeric systems for producing hydrogels and nanogels. This linear polymer of α-glucan has greater adhesiveness and mechanical strength (Prajapati et al., 2013). Hayashibara

TABLE 3.1

Different sources of EPSs and their applications

Microorganism	Polymer	Sugar monomers	Application	References
Bacteria				
Agrobacterium sp. *Alcaligenes* sp.	Curdlan	Glucose	Curdle gels mainly used as thickener in food industry, texturizer, and stabilizer in cosmetics and medicine	Yang et al., 2016; Cimini et al., 2012; Saha et al., 2020
A. Vinelandii *P. aeruginosa*	Alginate	Mannuronic acid and guluronic acid	Alginate based hydrogel can be used as wound healing additive, drug delivery agents in biomedicine, and also used in food and feed	Oner et al., 2016
B. subtilis *H. smyrensi* *Z. mobilis*	Levan	D-fructo-furanoside	Prebiotic agent that lowers cholesterol, bio-thickener used in the food industry, and antitumor properties	
Enterobacter A47	FucoPol	Fucose, galactose, glucose, and glucuronic acid	Can be used as alternative cryopreservative in biotechnological, supplement in food, antioxidant photo-protective in sunscreens	Guerreiro et al., 2021; Vázquez-González et al., 2022
K. pneumoniae	Fucogel	4-O-acetyl-galacturonic acid galactose and fucose	This polymer is used as antiaging and moisturizing agent in medicine and cosmetics	Guerreiro et al., 2021
Acetobacter sp. *Rhizobium* sp. *Sarcina* sp. *Lactobacillus* sp.	Bacterial cellulose	Glucose	Synthase dependent product used as supplement, dietary fiber and preservative in food, medicine, acoustics	Chawla et al., 2009; Liu et al., 2017; Guo et al., 2021
Leuconostoc sp. *Streptococcus* sp. *P. oleovorans*	Dextran	Glucose	As an adjuvant, emulsifier, carrier, stabilizer, plasma substitute, column matrix in biomaterial production, and in oil recovery and healthcare	Patel and Prajapati, 2013; Díaz-Montes, 2021

(Continued)

TABLE 3.1 (Continued)
Different sources of EPSs and their applications

Microorganism	Polymer	Sugar monomers	Application	References
Sphingomonas paucimobilis	Galacto-Pol	Galactose, Glucose, Mannose, Rhamnose	Used in film-forming, emulsifying agent, and flocculating solutions in aqueous environments	Freitas et al., 2009
Streptococcus zooepidemicus	Gellan	Glucose, Rhamnose, Glucornic acid	Used as a thickener and binder in food industry. Stabilizer in water-based gels for jelly making. Replaces gelatin in dairy products and vegan items.	Prajapati et al., 2013
Xanthomonas sp.	Hyaluronic acid	Glucornic acid, N-Acetylglucosamine	Used as wound healing and scar reducing additive in medicine, moisturizer in lotions, lubricant in ointments and serums	Zhang et al., 2006; West et al., 2021
Fungi				
Aureobasidium pullulans	Xanthan	Glucose, Mannose, Glucornic acid	Used in oil recovery and as viscosifying agent	Prajapati et al., 2013
Aspergillus sp. *Y16 Aspergillus fumigatus*	Pullulan	Glucose, Mannose, Galactose	In pharmaceuticals used in tissue engineering, drug and gene delivery agent, film-forming agent, backbone of hydro and nanogels	Muthusamy et al., 2022
Botryosphaeria sp. *Candida albican*	Galactosaminoglucan Botryosphaeran	Galactose Glucose	Utilized in petroleum, bioremediation, food, biomedical, cosmetic, textile, and pharmaceutical	Prajapati et al., 2013
P. ostreatus	Chitin and Chitosan	Glucose, Mannose	Biomedical industry (drugs delivery, wound healing, and tissue regeneration)	Cheng et al., 2011
Sclerotium sp.	Pleuran	Glucose, Mannose, Galactose	Acts as antitumor and immunomodulator	Mahapatra and Banerjee, 2012
S. commune	Scleroglucan	Glucose	Improve oil recovery, food additives, medicine delivery, and cosmetic as a thickening, gelling, and stabilizing agent	Mahapatra and Banerjee, 2012

Microalgae				
A. platensis	Schizophyllan	Glucose	It is used in medicines as antitumor agent to cure cancer, especially in cervical cancer.	Rehm, 2010
C. stigmatophora	Spirulan	Glucose, fucose, xylose, Rhamnos, fructose	Major substitute for protein in aquafeeds or as a partial supplement. Also used as pharmaceutical additives and for nutritive purposes (artificial sweetener, emulsifying agent, drug, vaccines, etc.).	Di Pippo et al., 2013
P. cruentum Tetraselmis sp. C. vulgaris	Galactan	Glucose, fucose, xylose	Anti-inflammatory, immunomodulator	Soanen et al., 2016;
Euglena gracilis	B-(1,3)-glucan	Glucose	In the treatment of cardiovascular disease, allergies, and obesity	Zhang et al., 2019
	β-glucan		Supplement in food, paper, textile, pharmaceutical, and cosmetics industries. Also helps in remediation of heavy metals from wastewater and soil.	Yoshida et al., 2016

Co., Ltd., a private firm in Japan, has commercialized this starchy polymer for direct consumption as an edible coating or as a food thickening agent. Scleroglucanis another commercially available EPS extracted from a plant pathogen, *Sclerotium* sp. It was firstly commercialized in early 1970s under a variety of trade names. Actigum, Clearogel, Polytetran, and Polytran FS are commonly used scleroglucans additives in food, drugs, cosmetics, and refineries (Schmid et al., 2015; Taskin et al., 2010).

EPS with bioactive qualities such as antioxidant, anticancer, immune stimulating, and antibacterial capabilities are secreted by a number of higher fungi from the genera *Ganoderma*, *Poria*, and *Tremella* (Osinska-Jaroszuk et al., 2015). Studies indicate that endophytic fungi produce EPS to survive under harsh environment caused by interactions of plants with endophytes. These biopolymers are distinguished by new topologies that demonstrate biological activity (Liu et al., 2017). Rhamnogalactan is one such pectin polysaccharide produced by *Fusarium solani*. SD5 can be used as curing agent during inflammatory and allergic reactions (Mahapatra and Banerjee, 2012). Similarly, Mahapatra and Banerjee (2012) also reported the glucans and galactomannan from *Diaporthe* sp. strains JF766998 and *Aspergillus* sp. Y16 with antitumor activity.

3.2.3 MICROALGAE

Production of microalgae and its utilization for production of industrial product is an emerging sector, and its production estimated to grow by 10% every economic year globally. Production of high-value pigments, antioxidant polymers, nutraceuticals, fish meal, and organic fertilizers are main area of applications of microalgae. *A. platensis, C. vulgaris, Aphanizomenon, D. salina*, and *Porphyridium cruentum* are the principal species that produce EPS (Xiao et al., 2018). Homo as well as hetero-polymers of glucose or galactose and other sugar monomers, represent two examples of the complex chemical structures that define the characteristics of microalgae EPS. Algal EPS frequently contain uronic acids and sulfate, which further adds to their unique characteristics. Although, a number of microalgae secrete significant EPS. The initial perception of them, however, is that they are by-products of other processes (Delattre et al., 2016). Unlike others, the use of EPS from algal sources is still insufficient, and researchers in recent past have begun to take into account their potential as sources of profitable polysaccharides. In spite of this, EPS from some microalgae *Porphyridium* sp. have been widely used in cosmetics.

3.3 BIOSYNTHETIC PATHWAYS FOR EXOPOLYSACCHARIDES PRODUCTION

Microbial EPS biosynthesis is a multifaceted process, depends on the environment conditions and microbes being used as source. Many gene clusters that code for certain enzymes and carrier proteins play a role in the various steps of EPS production. Microorganisms that produce EPSs use sugars (dextrose, fructose, glucose) as energy, ammonium salts and amino acids as nitrogen source (Zhang et al., 2016). EPSs production by microorganisms primarily depends on how much carbon and nitrogen are present in the environment. Although increased exopolysaccharides synthesis is induced by low nitrogen content. For maximal EPSs synthesis, the ideal C/N ratio is

10:1. According to Rehm's (2010) C/N ratio in growth culture media accounted up to 33.84% of carbon and just 0.88% nitrogen form total EPS produced by *Alteromonas* sp. JL281. Wzx/Wzy-dependent, ATP-binding cassette (ABC) transporter dependent, synthase-dependent and extracellular synthesis are four different pathways used by source microorganism for production of EPS by single sucrase protein (Donlan and Costerton, 2002). In majority of biosynthetic pathways, the precursor molecules are formed inside the cell by a variety of enzymatic reactions required for elongation of the polymer strands and for production of activated sugars, sugar acids, etc. The polymeric strand is directly extended for extracellular fabrication by the adding up monosaccharide formed by the cleavage of di- or trisaccharides (Schmid et al., 2015).

3.3.1   Wzx/Wzy-dependent pathway

Wzx/Wzy-dependent pathway commonly used by members of *Streptococcus, Lactobacillus,* and *Leuconostoc* generas for the production of polysaccharides, in a five-step reaction process: phosphorylation, activation of monosaccharides, synthesis of repeating units, polymerization, and release of EPS (Islam and Lam, 2014). Activated sugar molecule (UDP-Glc) joined to the undecaprenyl phosphate act as a lipid carrier and form phosphate linkage, which is initially involved in the production of lipopolysaccharide (LPS) and group 1 capsular polysaccharide (CPS) from *E. coli.* More sugar units gradually linked together by glucosyltransferases (GTs) to create repeating units. A Wzx flippase protein traverse the cytoplasmic membrane before being exported to the cell surface and polymerizes the translocated oligosaccharide (Cuthbertson et al., 2009; Morona et al., 2009; Islam and Lam, 2014). Additionally, to transfer the polymerized units to the surface of cell, another protein molecules, polysaccharide co-polymerase (PCP), and outer membrane polysaccharide export (OPX) families are also required (Cuthbertson et al., 2009). All polysaccharides produced via this way have a highly diversified sugar pattern (4–5) present in their chemical structure and thus categorized as heteropolymers (e.g., xanthan). The genes code for the Wzx and Wzy proteins present in EPS operons system of utilizing bacteria (Morona et al., 2009). Figure 3.1 showed schematic diagram of Wzx/Wzy-dependent pathway.

3.3.2   ABC transporter-dependent pathway

Capsular polysaccharides (CPS) are synthesized by formation of glycosydic linkage inside the cytoplasm via the action of glycosyltransferase (Whitney and Howell, 2013). This complex of ABC transporters, which resembles a tripartite efflux pump, is found on the cytoplasmic head of the inner and periplasmatic membrane. Wzx/Wzy, PCP, and OPX proteins function in ABC pathways similarly to OPX and PCP proteins. Glycosyltransferase (GT) is employed to produce polysaccharides, and their translocation to the cell surface is aided by a tripartite efflux pump complex. One GT-containing operator leads to formation of homopolymers, while multiple GTs used during assembly process produces heteropolymers (Morona et al., 2009). At the reducing terminus, conserved poly-2-keto-3-deoxyoctulosonic acid linked glycolipid and phosphatidyl glycerol is essentially required for production of CPS of this mechanism (Willis and Whitfield, 2013). Figure 3.2 gives a pictographic demonstration of this pathway.

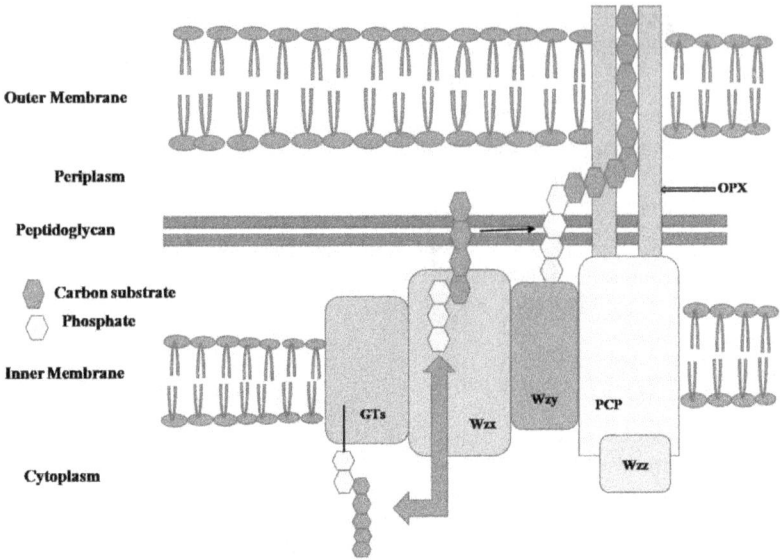

FIGURE 3.1 Wzx/Wzy-dependent pathway for biosynthesis of EPS

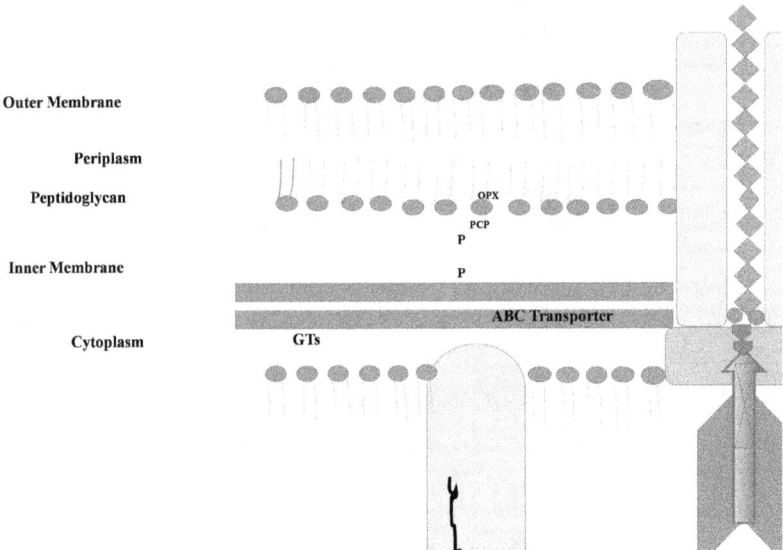

FIGURE 3.2 ABC transporter-dependent pathway

3.3.3 SYNTHASE-DEPENDENT PATHWAY

The entire polymer strands either via membranes or by cell wall are independent of a flip-pase in this biosynthetic process. Both vertebrate and microbial systems use this pathway for the synthesis of polysaccharides. The synthase-dependent EPS production was mostly studied in gram-negative bacteria for polymerization and translocation of alginates, cellulose, and poly-D-N-acetylglucosamine (PNAG) (Rehm, 2010). For example, in curdlan biosynthesis, the polymer exclusively contains -(1–3)-linked glucose. Bacterial cellulose is another example of a strong homopolymer because it exclusively contains -(1–4)-linked glucose units. It is common practice to assemble homopolymers using synthase-dependent routes since they only need one kind of sugar precursor (Hubbard, 2012). *Hyaluronan (HA) is unbranched, multifunctional* extracellular linear *polysaccharide repeating units of D-N-acetylglucosamine (GlcNAc) and D-glucuronic acid. Vertebrates predominantly produce hyaluronan/Hyaluronic acid (HA) to facilitate an array of physical functions, such as cell adhesion, relocation, and segregation* (Chung et al., 2009). Hyaluronan synthase (HAS), a cell membrane protein, produces HA polymer on the cytosolic side. Andhare et al. (2009) reported that HAS from *Streptococcus equisimilis* is comparable with human HAS used for synthesis of HA. In order to manufacture and transport HA, pure *S. equisimilis* HAS was reconstituted into proteo-liposomes. Homopolymer secretion is independent to lipid acceptor molecule (Rehm, 2010).

The entire polymer strands are secreted in this pathway transversely across the cell membranes and wall. The transfer of monomeric subunits is independent to flippase activity, and a single synthase protein subunit extents the envelope and responsible for polymerization as well as translocation. This whole process is directed by cell surface enzyme glycosyl transferase, and homopolymer EPS are secreted along the cell wall and membranes (Czaczyk, 2007). Figure 3.3 schematically represent the synthase-dependent pathway.

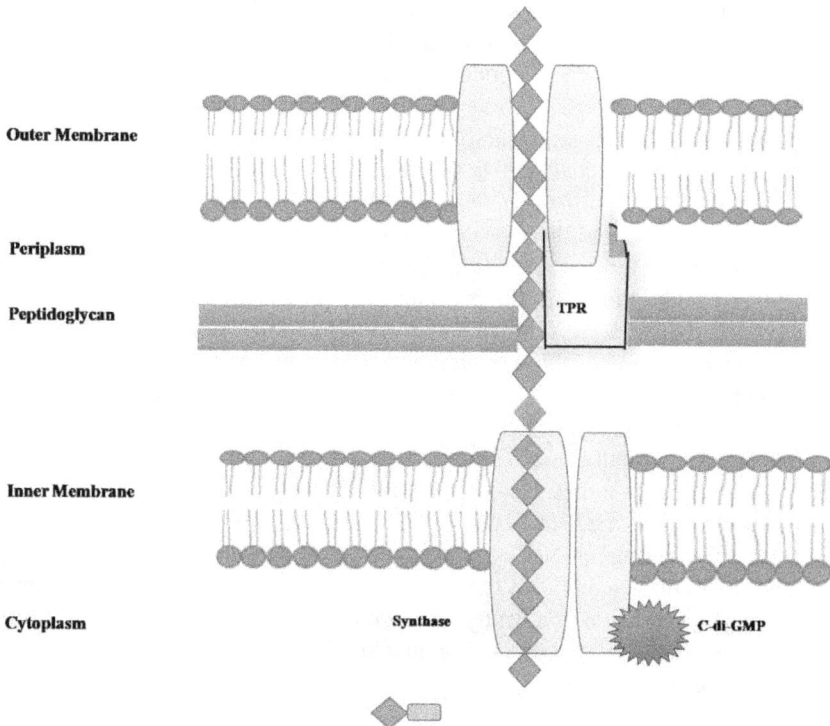

FIGURE 3.3 Synthase-dependent pathway

3.3.4 EXTRACELLULAR SYNTHESIS/THE SUCRASE-MEDIATED PATHWAY

The last process for extracellular biosynthesis is the sucrase-mediated pathway. It involves first turning a disaccharide molecule into a monosaccharide and adding it to a polysaccharide chain in an extracellular setting to start a polymerization event. Central cellular carbon metabolism of cell does not involve secretion of EPS by this method, and EPS thus formed having structural variation. Glycansucrase catalyzes the transfer of monosaccharides to an acceptor molecule. Energy released during sugars hydrolyses is used to transfer the glycosyl residue to developing homo polysaccharide (Kumar et al., 2011). Because of their distinct and different qualities from those of polysaccharides obtained from natural sources, many microbial EPS have significant uses in R&D. Recent years have seen a rise in the demand for natural polymers for a variety of industrial uses, which has sparked interest in EPS synthesis from new sources.

3.4 COMMERCIALLY PRODUCED EXOPOLYSACCHARIDES

Industrial variety of fungi are used for synthesis of number of (1,3), (1,6)-glucans, and polymeric carbohydrates in submerged as well as soil state culture conditions. These glucans, comprising scleroglucan and schizophyllan, having (1,6)-linkaged glucose units and a (1,3)-linked glucose backbone (Seviour et al., 2012). Numerous have proven to be potent non-self-immuno-modulators that can boost human immune systems and help the body fight off cancer cells, microbial infections, and other serious illnesses (Chen and Seviour, 2007). The linear glucan, pullulan, has been reported to commercially produce from a fungus, *Aureobasidium pullulans* (Singh et al., 2008). Because of its several useful physicochemical properties, pullulan find its uses in a variety of industry with maximum utility in food pharmaceutical sectors (Shingel et al., 2004). Another commercially available EPS, that is, curdlan is produced from *Agrobacterium* sp. having rheological and thermal gelling properties. And due to these properties, this is being used as a thickening and gelling agent in the food processing and immuno-modulatory activity (McIntosh et al., 2005). Besides this, it also faces a competition in front of other conventionally utilized polycarbohydrates. Xanthan makes up to 6% of the worldwide EPS market and is the only substantial EPS polysaccharide generated from bacteria. It is highly viscous and pseudoplastic in solutions even at low concentrations. These characteristics make it more superior to other existing gums (Zannini et al., 2016). Currently, the bulk of commercially used carbohydrates are obtained from natural sources such as plant or algal sources. However, EPS made by bacteria have certain distinct advantages over those made by plants or algae. Bacterial EPS can be manufactured using renewable resources, and regulated fermentation can yield a product with high reproducibility, quality, and titer (Moscovici, 2015). In addition to its numerous other uses, xanthan is frequently employed in food, ceramic, and agricultural production as a thickening, suspending, and stabilizing ingredient (Garcia-Ochoa et al., 2000). Different commercially available EPS and their industrial applications have been summarized in Table 3.2.

TABLE 3.2
Different applications of exopolysaccharides in various industries

Type of EPS	Applications	Role	References
Dextran	• Food industry • Pharmaceutical industry	• Increases viscosity and moisture retention duration. • Gelling agent in gums, jellies, and candies. • A potential prebiotic agent in frozen items. • Anti-biofilm activity, dextran-chitosin gel is used as substitute to plasma due to its adhesive characteristics.	Zhang et al., 2012; Mende et al., 2020 Kulicke and Heinze, 2005
Xanthan	• Food industry • Cosmetic industry • Pharmaceutical industry	• Thickener and stabilizer in *frozen foods*. • Enhances freeze thaw reliability. • Protect the product from degradation. • Used in cosmetics baby care products, hair and nail colors, dental hygiene and skin conditioners. • Carriers for drugs, intra-articular injections. • Immune system adjuvants as xanthan-based hydrogels having anticancer and wound-healing property.	Palaniraj and Jayaraman, 2011 Paquet et al., 2014 Imeson, 2010; Kumar et al., 2007
Pullulan	• Food industry	• Used as an edible coating as it is safe to consume. • Protect the food from degeneration, attack of molds, and maintain softness of food.	Eroglu et al., 2014
Gellan	• Food industry • Cosmetic industry • Pharmaceutical industry	• Enhance the binding efficacy, stability, and texture of processed foods. • In cosmetics, increases viscosity and stability of powders, eye and hair products are among other uses. • Implant for insulin delivery, nasal formulations, bone regrowth through tissue engineering. • Gellan-hydrogels can be injected to cure a bone deformity.	Zhang et al., 2021; Srisuk et al., 2018 Zia et al., 2018

(Continued)

TABLE 3.2 (Continued)
Different sources of EPSs and their applications

Type of EPS	Applications	Role	References
Curdlan	• Food industry • Pharmaceutical industry	• *Additive to prebiotics to prevent it from hydrolyze.* • *Improves the softness, adaptability, and durability of packed, canned, and frozen foods.* • Hydrogels help in protein delivery and wound healing. • Curdlan sulfate has antioxidant and antitumor potential and is used to treat severe malaria, hepatitis B infection.	Cai and Zhang, 2017; Nishinari, 2007
Scleroglucan	• Food industry	• Stabilizer in frozen foods, aerated desserts, creams, and sauces. • Prevent the foods from discoloration and hardness without affecting properties.	Survase et al., 2007; Schmid et al., 2015
Cellulose	• Cosmetic industry • Pharmaceutical industry	• Wound healing, skin repair, skin-moisturizing materials, and base for artificial nails. • Cellulose of microbial origan can also be utilized to make face masks. • Antimicrobial and wound healing potential.	Ioelovich, 2008; Pértile et al., 2012 Andhare et al., 2009
Hyaluronic acid	• Cosmetic industry • Pharmaceutical industry	• Essential ingredient in cosmetics as used in skin hydrating moisturizer, antiaging lotions and lipsticks. • Induces the expression of inflammatory mediators, inhibits tumor growth, and encourages angiogenesis. • Water-soluble EPS that exhibits a highly non-Newtonian behavior and behaves like a gel.	Liu et al., 2012 Zhang et al., 2006 Kim and Kumar, 2014
Levan	• Pharmaceutical industry	• Aldehyde groups of levan has anticancer qualities. • Improve the adhesion properties of live cells. • Hypocholesterolemic agent also having antitumor properties.	Taylan et al., 2019
Alginate	• Pharmaceutical industry	• An anionic polysaccharide works effectively as a tablet dissolving agent. • Enhance thickness and stability of pharmaceutical syrups and lotions. • Act as anti-acid and cure acidity.	Xiao et al., 2018

3.4.1   SCALE-UP STRATEGIES AND PROCESS PARAMETERS

Commercial production and synthesis of microbial EPS are quite difficult since so many variables must be taken into account. The ultimate goal is to acquire the maximum yield and to optimize the process. It is advised that yields should fall between 10–15 g/L in order to produce economically viable EPS to use it as an ingredient in food (De Vuyst and Degeest, 1999). The selection of a suitable strain or strain co-culture holds a great importance. Other factors include the ideal growing circumstances and parameters, including media composition, micronutrients, and fermentation setup that is optimized to provide the optimum atmosphere for selected strain. The hydrodynamics of the fermenter needs to be studied well because the broth's rheological qualities change during cultivation. The heterogeneities rise with reactor volume on the leveling up of process from flask level to the commercial level due to the greater physical separation between the impellers and vessel walls. On the standardization of the large-scale production operations, selecting the generating strain is essential to obtain high quantity of polymers with reliable yield. To ensure consistency in product output and quality as well as to adhere to legal regulations, it is crucial to monitor and regulate the operations carried out for the EPS production.

The typical approach for any commercial level production is fermentation. It is carried out at flask level using batch shaking flask systems to observe and study factors such as ingredients of medium. However, in case of highly viscous culture media with sampling and its analysis need consideration. A number of variables play a significant role to maximize EPS production, especially low nitrogen and high carbon substrate concentration in production medium (Fazenda et al., 2008; Seviour et al., 2010). In order to decrease the expenses of production, handle competitive market, enhance yields and expensive substrates are compared using statistically based experimental design methods (Singh et al., 2008). This allows for the possible accommodation of interactions such as those between the pH of medium, temperature, dissolved oxygen level, agitation speed, and nitrogen source used in the medium. For the same objective, some techniques that do not involve statistical methods like the rotating simplex and uniform design methods are being used (Xu et al., 2018). However, it is crucial to select the parameters to optimize and later assess their impact on the process (Fazenda et al., 2008). It is also important to note that in shaking flasks, pH and O_2 levels cannot be monitored or controlled. In many organisms, both have a crucial role in the synthesis of exopolysaccharides (Gibbs et al., 2000; Seviour et al., 2010), and both effects fluctuate throughout every shake flask batch culture. In order to get the maximum output, any screening procedure should ideally use tiny fermenters with characteristics like aspect ratios, mixing times, etc., similar to the final reactor selected for large-scale production.

3.4.2   EFFECT OF VARIOUS PHYSICOCHEMICAL FACTORS
ON EXOPOLYSACCHARIDES PRODUCTION

The process parameters greatly influence the end product range. These must be maintained at a specific range to ensure the consistency, replicability, robustness, and performance of any process. These parameters either affect the microbial growth or alter the metabolic pathways of synthesis of any product. Similarly, in the past researchers has also analyzed the effect of physicochemical factors on the production and yield of EPS. The effect of some of these physicochemical parameters on EPS production has been discussed in this section.

a.) *Effect of carbon sources*: Carbon source greatly affects the manufacture of EPS as carbon source directly influences the growth pattern of fermenting microbe. EPS-producing bacteria can thrive across a vast range of carbon sources, including dairy and starch-based media, sucrose, glucose, galactose, fructose, lactose, maltose, and mannose (Macedo et al., 2002). Yuksekdag and Aslim (2008) in a study reported that excess of sugars induce the production of EPS. The most typical carbon source for bacterial EPS synthesis is sucrose. Seesuriyachan et al. (2012) used sucrose in an MRS medium to maximize the yield of EPS from *Lactobacillus confusus* TISTR 1498. Additionally, maximum EPS was produced by *Lactobacillus* strains *L. delbrueckiibulgaricus, L. helveticus*, and *L. casei* in a fermentation medium containing 20% sucrose as source of energy (Hussein et al., 2015). Similarly, the production of EPSs by *Fructilactobacillus sanfranciscensis* LTH2590 has been reported to increase as sucrose concentration increased, roughly from 40 g/L to 160 g/L (Korakli et al., 2003). *Leuconostoc mesenteroides* NRRL B-1299 produced more dextran when the culture medium had a sucrose concentration of over 5 g/L (Dols et al., 1997). The high value of sugars present, like sucrose, is beneficial for the formation of EPSs. According to another study, increasing glucose concentration also increased the formation of EPS in *Streptococcus thermophilus* (W22) and *L. delbrueckii* (Yuksekdag and Aslim, 2008). It has also been demonstrated that the presence of excess sugar in the medium improved EPS formation from *L. rhamnosus* and *L. casei* (Gamar et al., 2003). To make the process more effective, commercial sugars can be replaced with inexpensive substrates, including agro-industrial wastes (molasses, cheese whey, or glycerol), which are reported to produce a sufficient amount of a number of bacterial EPSs (Huu et al., 2014). Bacteria cultured on glycerol waste from the biodiesel sector were used to manufacture GalactoPol and FucoPol. This inexpensive substrate serves as an appropriate carbon source for the synthesis of both polymers (Freitas et al., 2010). Carbon source chosen for optimal EPS production is species and strain-dependent and also the amount of source of carbon affects both growth and EPS production. Li et al. (2016b) reported that *S. thermophilus* has shown a predilection for sucrose as a carbon source, producing 108 mg/L of EPS, followed by glucose and lactose. In a similar study, Imran et al. (2016) reported 956 mg/L EPS when the fermentation medium was supplemented with glucose. Similarly, Tallon et al. (2003) in another study reported 140 mg/L EPS in lactose-based medium.

b.) *Effect of nitrogen sources*: The microbial production of EPS is highly stimulated by nitrogen limited medium (Marshall et al., 2009). Increased EPS yield was recorded from *Lactobacillus delbrueckii*s sp. *Bulgaricus* when medium was nitrogen limited (Gracia-Garibay and Marshall, 2008). For LAB to grow at their best, a variety of amino acids are necessary, and lacking asparagine, glutamine, or threonine can hinder growth (Farnworth et al., 2006). Although, they are not utilized directly in the formation of EPS, but amino acids can be used as sources of carbon and nitrogen in the growth medium. Addition of amino acids to *S. thermophilus* and *L. rhamnosus* medium does not affect the EPS yield, while in case of *L. delbrueckii*, addition of amino acids positively affects the EPS production (Degeest et al., 2002). According to Macedo et al. (2002), whey and yeast extract-based medium achieved significant EPS yields of 2775 mg/L when additional amino acids and salts were added.

c.) *Temperature*: The formation, yield, and type of microbial EPS are genetically determined by the properties, culture circumstances, and media components (Nguyen et al., 2016). High temperature has been recorded to affect EPS production in LAB. Nguyen et al. (2020) has claimed that nonlethal thermal stress increases EPS production, and the viability of *B. bifidum* has been shown to be improved. The ideal temperature range for EPS synthesis varies with fermenting microbes. *L. lactis*, *L. rhamnosus*, and *L. plantarum* produces maximum EPS in temperature ranging between 18 and 25°C (Sørensen et al., 2022; Zhang et al., 2021). For *S. thermophilus*, the temperature range for optical production was reported to be between 32 and 42°C (Zisu and Shah, 2003).

d.) pH: There is a significant strain-dependent component to the specific pH and temperature dependency. pH alters the ionic structure and configuration of the cell wall. In a study Nguyen et al. (2020) performed EPS production on varied pH to study its effect and observed maximum production at pH 4.5 from *Lactobacillus helveticus* ATCC 15807 (Nguyen et al., 2020). Similarly, EPS production by native *L. salivarius* UCO 979C-1 strain was 450 mg/L at pH 6.4 (Sanhueza et al., 2015). Increased EPS yield has been found to be in the presence of pH control as compared to acidified batches (Macedo et al., 2002). In a study, 1029 mg/L of EPS was obtained from *S. Thermophilus* at pH 5.5 after 24 h, but the yield was only 491 mg/L in a similar fermentation without pH control. Additionally, it has been observed that batches with pH control have a higher EPS output than batches with acidification (Zisu and Shah, 2003).

e.) *Effect of osmotic pressure*: EPS synthesis, especially on the cell wall, can be stimulated in the presence of chemicals inducing high osmotic pressures, such as NaCl. Seesuriyachan et al. (2012) stated that *L. confusus* TISTR 1498's EPS synthesis was independent of biomass and that the stress of a high NaCl concentration might influence EPS production in solid state. Similar to this, *Leuconostoc mesenteroides/pseudomesenteroides* 406 produced the most EPS when 5% NaCl was present (Grosu-Tudor et al., 2014). In contrast, NaCl was found to suppress the formation of EPS in *L. helveticus* ATCC 15807 (Zhang et al., 2021).

f.) *Effect of aeration and agitation:* The majority of bacteria produces EPS under aerobic environments. Maximal aeration is necessary for the best production of some EPSs, whereas microaerophilic conditions are best for the maximum production of EPSs. GalactoPol and FucoPol are two examples of bacterial EPS synthesized in the presence of limited nitrogen and oxygen levels (Freitas et al., 2011). Muller and Alegre (2006) in a study examined the impact of aeration on EPS production by *Paenibacillus spessential*. And observed that under anaerobic conditions, the cell population does not multiply or create EPS. A considerable rise in the EPS content results from vigorous aeration during fermentation. Previously Gracia-Ochoa et al. (2000) also suggested that agitation might enhance the features of mass transfer with respect to substrates, products, and oxygen, which could help microbial cells grow and function more effectively. The results of Rafigh et al. (2014) also showed that the greatest output of curdlan gum and biomass from *P. polymyxa* ATCC 21830 was greatly influenced by agitation speed ranging from 120 to 150 rpm. However, curdlan gum and biomass levels were found to be lower at 180 rpm, which can be attributed to bacterial fragmentation caused by various shearing mechanisms (Liang and Wang, 2015). Higher agitation speeds (600 rpm) have been used for the generation of curdlan from some other kinds of bacteria, such as *Agrobacterium* sp. (Riaz et al., 2020).

3.4.3 REACTORS USED FOR THE PRODUCTION AT INDUSTRIAL LEVEL

Reactors are important for the production of EPS. Among all, stirred tank reactors are the most preferred reactor system for the synthesis of EPS at both the scientific and industrial scales. Microbes produce EPS majorly under aerobic conditions, so the structure of reactor, position, design, shape, and length of impellers play a significant role. The design of reactor may vary with microorganism as physiological conditions and cultural morphology required for growth of fungi are different from that of bacteria (Garcia-Ochoa and Gomez, 2009).

Continuous Stirred Tank Reactor (CSTR) is facilitated with continuous input of nutrient media and output of products simultaneously. In CSTR, radial flow impellers provide turbulence that ensures uniform mixing of the broth, facilitating the mass transfer of nutrients such to the cells (Figure 3.4). The shear stress, which may affect the quality of end product, rises as the agitation does as well. The disadvantages of CSTRs include compromising structure, smaller dispersal systems, multiplied impeller systems with maximum agitation, high-shear rushton turbines, and lower mixing efficiencies. In order to overcome these disadvantages, a large-diameter single-bladed vessel with low-shear axial flow impellers, slow agitation speeds operating system is required (Garcia-Ochoa and Gomez, 2009). Orr et al. (2009) have reported the production of pullulan using CSTR, and a high yield in EPS production was observed depending upon the agitation rate.

In *Airlift reactor*, agitation and mixing of the fluid is carried out with the aid of airlift pump. It is a gas liquid bioreactor that works on the principle of a draught tube. In this reactor, condensed oxygen is injected from the bottom of a discharge pipe for facilitating aeration and agitation. Turbulence due to fluid flow results in the proper mixing of the liquid. The vessel of the reactor has a central draft tube. The upward passage of air having low velocity causes circulatory flow in the reactor with low

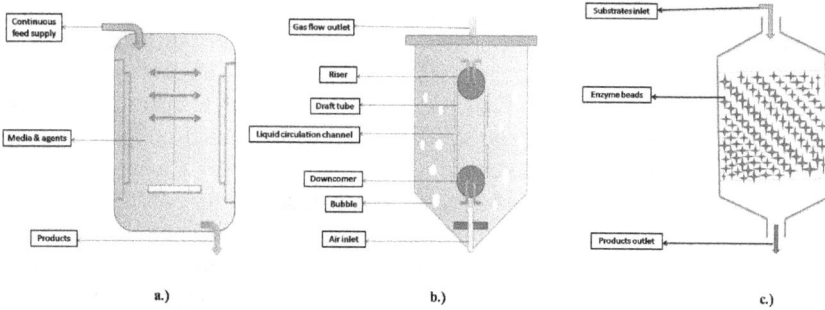

FIGURE 3.4 Different types of bioreactors: (a) continuous stirred tank reactor (CSTR), (b) airlift reactor, (c) continuous flow packed bed reactor

energy consumption. The volumetric mass transfer coefficient makes the rate of airflow. It consists of a cylindrical cross type having an internal or external loop riser in the column of the gas draft tube. In the case of Internal loop risers, a sparger ring is used to feed air at the bottom of a central draft tube, which circulates air bubbles and liquid. External loop risers sparge bubbles outside the draft tube and have an adjustable configuration that aids in providing the required oxygen concentration. Since its flow is outside, it causes turbulence near the heat jacket, which leads to better heat transfer efficiency. The agitation with the gas flow leads to low shear and tear of the cells. It is a suitable choice for aerobic cultures because of the high coefficient of oxygen mass transfer (Castillo et al., 2015). In a study Choi et al. (2007) used airlift bioreactor for production EPS from *Fomitopsispinicola* sp. and observed 4.4 g/L EPS in the fermentation broth.

Packed bed bioreactors have also been used for the microbial production of EPS. A packed bed bioreactor constitutes a bed of porous or homogenous nonporous solid particles supporting biocatalyst. Immobilized proteins move on the packed bed which is facilitated with fluidic nutrients. Metabolites/products formed by the reaction were released into this mobile fluid and later removed in outflow. The flow can be both upward and downward, but due to limited velocity, downward flow preferred more in comparison to upward flow. These columns hold the potential of whole cell immobilization. Besides this, change occurs in the porosity of operation and is the major limitation of this system that alters the flow rate. This can cause turbulence in the bed leading to back mixing, which can bring changes in the properties of fermentation. Concentration and compressibility of solids are highly influenced by the depth of the bed, which further influences oxygen passage and fluid flow rate throughout the column (Dishisha et al., 2012). An increase in the bed depth leads to a slowdown flow of nutrients. It comes across an issue of pH gradients as its control is quite difficult. Packed beds are used for the product inhibited reactions. Belgrano et al. (2018) in a study used packed bed reactor for EPS production using *Propionibacterium freudenreichii*. In another study, *Z. mobilis* B-14023 immobilized Ca-alginate beads were used as substrate for cost-effective production of levan (EPS) in packed bed.

However, it was also observed that prolonged fermentations reduces the pressure in reactor, which leads to the bursting of Ca-alginate gel beads (Silbir et al., 2014).

The microbes of the thermophilic range have also been used for the industrial production of EPS. Thermophilic microbes harbor a number of benefits, including speedy cultivation, reduced contamination, good mass transfer, and broth with decreased viscosity due to their fast cell growth rates at high temperatures. Rastogi et al. (2010) reported highly stable lignocellulolytic enzymes from thermali stable *Geobacillus* sp. WSUCF1. These enzymes convert complex saccharide (di, tri, poly) to monomeric sugars, which serves as carbon source for production of EPS. Additionally, significant investments are needed for heat exchangers, air compressors, and fermenters, which must be particularly constructed to resist high temperatures. EPS production began with a logarithmic phase and increased until the late stationary phase, with a maximum production level being reached in the stationary phase. Wang et al. (2021) reported 526 mg/L of EPS with 803 mg/L DCW (dry cell weight) after 10 h of incubation, that is, and 0.66 mg EPS/mg DCW. During the exponential phase, the maximum specific growth rate was 0.69 h/L.

3.5   EXTRACTION AND PURIFICATION

Exopolysaccharides, as mentioned previously, are member of defense mechanism and prevent host cell from infections and pathogenic organisms. These molecules mainly present on cell surface and exterior of cell, and because of this, these can be easily extracted and purified. EPS binds to the matrix with different interaction (hydrophobic, electrostatic), H-bonds, and vander Waal forces (Christensen and Characklis, 1990). Some common methods of extraction and purification have been summarized in Table 3.3.

3.5.1   PHYSICAL METHODS

EPS can be extracted from fermentation broth by different physical methods like ultrasonic treatment, centrifugation, membrane-based method, and microwave treatment.

3.5.1.1   Ultrasonic treatment

Ultrasonication is a technique that depolymerize the high molecular weight macromolecules to low weight monomers by breaking the chemical bonds between certain polymers. It was most preferred and a very beneficial approach for depolymerization because it depolymerize the polymer without altering the polymer's chemical composition. But the major limitation of this treatment is requirement of time and power (Zheng et al., 2008; Sheng et al., 2006). In the past this technique was effectively used for extraction of EPS from fermentation broth. Jia et al. (2012) employed ultrasonication to examine the extraction of novel exopolysaccharide WL-26 produced by *Sphingomonas* sp. ATCC 31555. Recently, Yatipanthalawa et al. (2022) also used this method for extraction of extracellular EPS from *Navicula* sp.

3.5.1.2   Centrifugation

Water soluble or somewhat slime polysaccharides are commonly separated from cell biomass by centrifugation (Hebbar et al., 1992; Troch et al., 1992). However, this technique is unable to extract the EPS molecules tightly bounded to the microbial

TABLE 3.3
Common methods of extraction and purification of exopolysaccharides

Processing steps	Common methods	Mechanism	References
Detachment of EPS	Ultrasound, heating, NaOH treatment	The loosen out the EPS from the matrix and get removed under sudden pressure created by sonication.	Sheng et al., 2006
Cell separation	Centrifugation, microfiltration	Centrifugal forces break the bond and separate off membrane bounded EPS from the cell surface	Troch et al., 1992
Protein removal	Enzymatic hydrolysis, TCA precipitation	Hydrolyzing enzymes ruptured and digested target metabolite	Sesay et al., 2006
Concentration	Microfiltration, Ultrafiltration	Semipermeable membrane in microfiltration induces a certain amount of pressure to segregate solutes from solution	Schachman, 2013
Purification	Diafiltration, dialysis, precipitation, chromatography	Separation and purification of the target product	Balti et al., 2018
Drying	Freeze drying, spray drying	A vacuum chamber is used to sublime polysaccharides from their solid state into the vapor phase after they have been frozen to form ice crystals	Ma et al., 2018

mass (Novak and Haugan, 1981; Frolund et al., 1996). After centrifugation, in order to separate the polysaccharide from the cell biomass and other metabolites, other extraction techniques were employed. Generally, the fermentation broth rich in EPS and cell biomass spin at lower rpm to remove cell biomass and other heavy debris and thereafter again centrifuge at higher rpm for a longer time to get these constituents (Frolund et al., 1996). Subramanian et al. (2010) isolated six EPS producing bacterial isolates (BS2, BS8, BS9, BS11, BS15, and BS25) from municipal waste, employed them for EPS production under standard conditions, and used then centrifuged the production medium to extract the EPS.

3.5.1.3 Membrane based methods

This technique is effectively used to separate EPS from low molecular weight metabolites synthesized by source organism using fermentation as by-product. The synthetic membranes of varied pore size (depending upon the size of metabolite) were used to filter these molecules present in fermentation broth. Polyether sulfone membranes 300 kDa with180± 30 L/h/^{m2}/bar permeability, and an average surface area of 0.1 m^2 was most commonly used synthetic membrane because the molecular weight of EPS is in range 2–7 106 Da. Like dialysis this technique is also effective to enhance purity of extracted EPS as also removes the unwanted salts (Gresh et al., 2002). Patel et al. (2013) employed this method for extraction of EPS. For this the authors firstly diluted the native solution from 1.4 g/L GlcEq concentration to 0.6 g/L, filterated it from a membrane and observed pure EPS.

3.5.1.4 Microwave treatment

In this method, microwaves were applied to the solution that causes variations in the electromagnetic field at high frequency, which ultimately increases the temperature and rupture the membrane to penetrate in material. These radiations speed up the mass transfer and dipolar rotation by generating friction in the molecules. This treatment can enhance the extraction rate by encouraging the release of intracellular compounds and extracting the naturally occurring biologically active components, such as polysaccharides found in cell membrane. Due to the lower energy and solvent consumption, it is also more environmentally friendly (Tsubaki et al., 2016). In order to extract exopolysaccharides from *Arthrospiraplatensis* using the idea of green chemistry, Silva et al. (2018) employed the microwave treatment to the cell suspension.

3.5.2 Chemical methods

This method involves the use of several chemicals as an extracting agent. These agents break the tight bounds formed between the bacterial cell and EPS matrix, enabling the release of EPS to the media.

3.5.2.1 Formaldehyde/NaOH treatment

NaOH and other alkaline chemicals are used as extracting agents. These alkaline molecules alter the ionization state of bacterial cell wall by changing the pH. The isoelectric point of most of bacterial cell is 4.0–6.0, addition of alkali to the medium increases the pH above 6.0 and the charge group (carboxyl) present on the surface of cell get ionized. The configuration of cell wall constituents changes, bond get break that facilitate the extraction EPS to medium (Felz et al., 2016). It was observed that at pH levels greater than 9.0 the covalent disulfide bonds in glycoproteins may be disrupted, facilitating the extraction of these substances (Zayas, 1997). Emerson and Ghiorse (1993) also reported that mercaptoethanol or other alkali substances also work in the same manner and break or loosen up the disulfide bonds. 1–9 N NaOH was most commonly employed to change the pH of medium for extraction of desired product. Dai et al. (2016) extracted the EPS from *Shewanella oneidensis* MR-by using NaOH treatment.

3.5.2.2 EDTA

Ethylenediaminetetraacetic acid is a chelating agent used to extract the exopolysaccharides from a variety of bacterium. Addition of EDTA causes exchange of ions, and divalent ions became monovalent. This exchange of ions increase the solubility of EPS towards water by enhancing the repulsive forces between EPS constituents and cell membrane. The removal of divalent calcium and magnesium (Ca^{2+} and Mg^{2+}) from EPS disintegrate the cross linkage formed between charged groups (Sun et al., 2015). In a study, hot water and α-amylase was added to ginseng roots for extraction of EGP (6.0% yield), which was then extracted using EDTA (Sun et al., 2015). In another study Felz et al. (2016) *also treated fermentation broth separately with EDTA, NaOH, and Na_2CO_3 to extract EPS bounded to the cell membrane.* In several studies, the combination of high pH and ion exchange has been applied by using NH_4OH and EDTA (Sato and Ose, 1984).

3.5.2.3   Cation exchange resin (CER)

EDTA and other chelating agents acts specifically on divalent ions, but in order to remove divalent as well as multivalent cations, cation exchange resin (CER) were commonly used. These resin are a type/category of recyclable cation-binding agent. CER are generally employed to extract EPS, and Redmile-Gordon et al. (2014) in a study demonstrated the applicability of this system for extraction of EPS from sandy soil. However, the applicability of this method for extraction of EPS from acidic soil (Ultisols) it was still unknown. This is because EPS bound to Fe^{3+} might be more difficult to extract by CER due to presence of strong trivalents than EPS bound to Mg^{2+} and/or Ca^{2+} (Wilen et al., 2003). CER was more selective for removing Mg^{2+} and Ca^{2+} rich EPS from activated sludge, according to research by Park and Novak (2007), while sodium sulfide (SS) was more selective for extracting Fe-containing EPS. CER and SS have previously been studied by researchers for extraction of divalent and multivalent cation associated EPS (Wilen et al., 2003; Park and Novak et al., 2007). A higher concentration of NaCl was also used as a cation exchanger. Adhesive exopolymers from *P. putida* and *P. fluorescens* have been extracted by using NaCl (Read and Costerton, 1987; Christensen and Characklis, 1990).

3.5.2.4   Ethanol precipitation

Ethanol is also used as chemical agent for extraction of EPS. Ethanol concentrates the EPS present in the medium by precipitating the polymer prior to dialyses. This method of extracting polysaccharides is widely used, despite the disadvantages of using a number of solvents that may cause loss of some polysaccharide, resulting in low production and being expensive (Zhou et al., 2018; Khoo et al., 2020). Various advanced technologies to obtain polysaccharides are recently used to make the extraction and downstreaming of EPS an economic and cost-effective process, including the following:

Aqueous two-phase extraction (ATPE)

ATPE is a liquid-liquid extraction technology that has gained attention due to its considerable potential/applicability to separate, extract, purify, enrich, and concentrate proteins. In this method two immiscible liquids are separated from each other. At a critical point where two liquids became immiscible, a layer of interface will form between them (Figure 3.5). The characteristics phase results in physicochemical interactions between the elements (phase-forming elements), and the targeted product can more efficiently be transported to the top or bottom phase, depending on the components' selectivity (Khoo et al., 2020). To achieve the maximum extraction

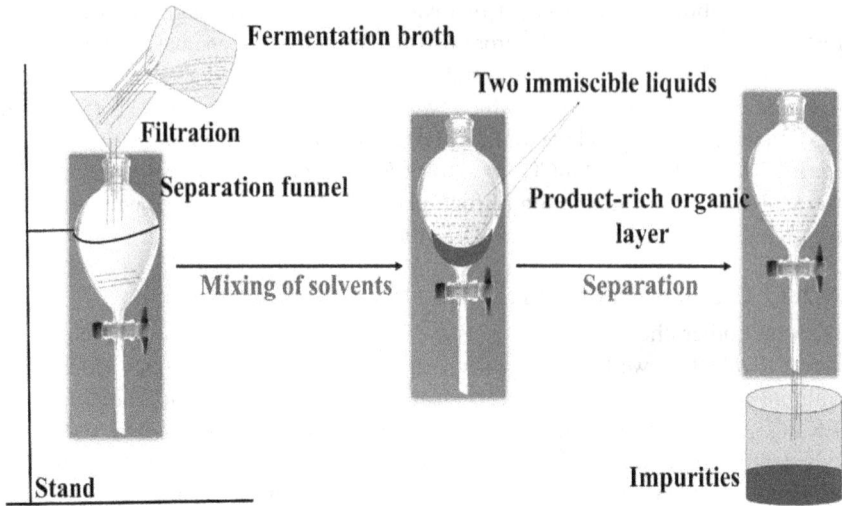

FIGURE 3.5   Aqueous two-phase methods of extraction

using ATPE, variables such as the selection and concentration of phase-forming elements, pH, temperature, and recycling possibilities must be considered (Khoo et al., 2020). This approach is widely used for isolation and extraction of biomolecules because it is easy to use, works under mild conditions, ecofriendly, user/product friendly, as well as economic. However, the type and concentration of liquid and pH of the system may affect the distribution, recovery, yield, and partition coefficients (Zhou et al., 2018). It can be easily employed to scale up practices with possibilities of a high recovery yield. Gao et al. (2020) extract polysaccharides from *Camellia oleifera* Abel. seed cake with 86.91%, recovery yield using ATPE. In a study Hui et al. (2015) also employed ethanol-based ATPE method for extracting EPS produced by *Potentillaanserine* and proves its authenticity.

3.5.2.5   Ultrasonic microwave synergistic extraction technology (UMSE)

UMSE is innovative, most preferred, effective, rapid, novel, and cost effective in comparison to ATPE enzyme-assisted extraction and PEFAE. This is a combination of ultrasonic and ultrasonic and microwave treatment in which both behave synergistically. Here microwaves allow easy penetration of cell wall and ultrasonic maintain cavitation (Chemat et al., 2011; Mustapa et al., 2015; Zhang et al., 2018). Figure 3.6 shows the schematic representation of UMSE method of extraction. Zeng et al. (2015) used this combination method to study the effect of UMSE on extraction yield and observed increase in extraction of polysaccharides of *Fortunellamargarita* from fermentation broth. It was observed that a high extraction yield of polysaccharide in less processing time was achieved using UMSE from *G. lucidum*, *M. conica*, and *T. orientalis* (Huang and Ning, 2010; Xu et al., 2018). Zhang et al. (2014) further add an element to properties of UMSE and observed that DPPH free radical scavenging efficiency, antioxidant activity, improves.

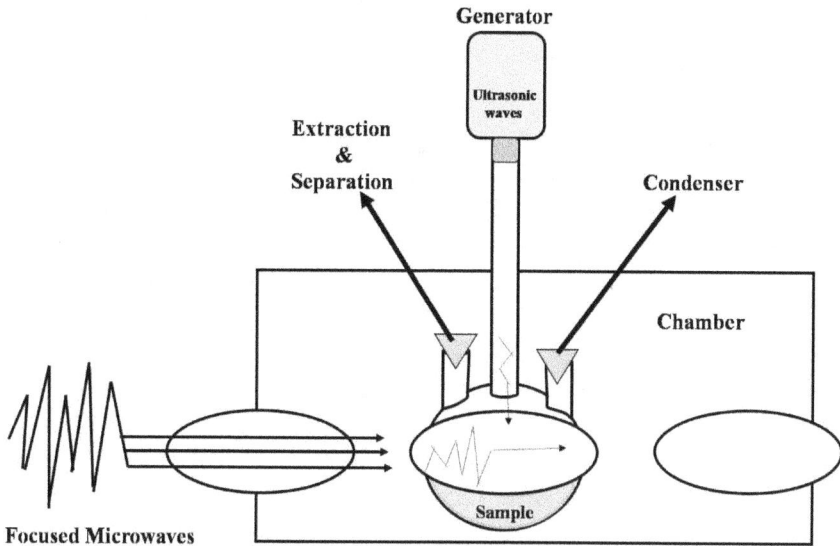

FIGURE 3.6 Ultrasonic microwave synergistic extraction

3.5.2.6 Pulsed electric field assisted extraction technology (PEFAE)

This is an electroporation technique in which a high voltage pulse (10–100 kV/cm) is generated and employed to target system for a very short duration of time (micro to milliseconds) (Figure 3.7). The high voltage pulse when applied enhances the permeability and porosity by breaking and changing configuration of local membrane. As a result, plant tissues become more permeable, and the intercellular components can be extracted easily (Xue and Farid, 2015). Very little energy consumption (1–20 kJ/kg) and processing time (micro-milli seconds) are two major properties of this method along with low working temperature (Liu et al., 2016; Parniakov et al., 2014). Besides this, production of a high-intensity electric field, uniform treatment throughout the treatment chamber *mongolicum* were purified using this method, and use of target friendly electrodes are three important features of PEFAE. Although, optimization of time, electric field intensity, and temperature are still needed to enhance the effectiveness and productivity of system (Xue and Farid, 2015). The use of moderate to low intensity (≤25 kV/cm) is preferred to prevent the target product from degrading (Liu et al., 2016).

The vertical treatment chamber are preferred more to control the internal temperature, liquid residence duration, and local conductivity of the product. Use of vertical chamber also prevents liquid stratification inside the chamber. Employing electrodes with uniform surfaces and round edges reduces the electron emission and protects from the electric field formed near sharp edges, and ultimately prevents the product from electrolysis (Xue and Farid, 2015). Anwar et al. (2021) extracted water-soluble non-starch polysaccharide from taro (*Colocasia esculenta*) peel using a pulsed electric field. Najim and Aryana (2013) observed rise in growth rate when *Lactobacillus* strains *L. acidophilus* LA-K and *L. delbrueckii* LB-12 were exposed to PEF. Similarly,

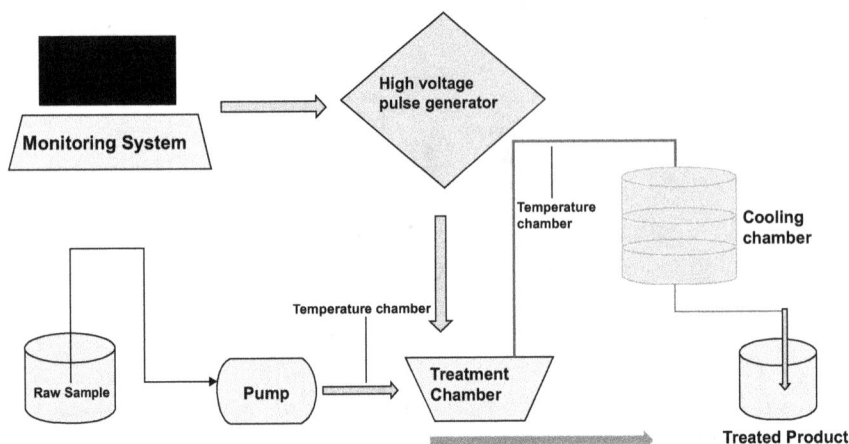

FIGURE 3.7   Pulsed electric field assisted extraction

L. acidophilus BT 1088 grew 4.49–21.25% faster when exposed to PEF at 7.5 kV/cm for 3.5 ms. The ability of the lactic acid bacteria to flourish and produce a sweet cherry fermented juice was also examined by Sotelo et al. (2018) in relation to a PEF treatment. In comparison to the conventional methodologies, the modern extraction methods were more effective, fast, and provide higher yield polysaccharides.

3.5.3   ENZYMATIC METHODS

Like chemical and physical methods of EPS extraction, enzymatic methods also employed to extract EPS by destabilizing bioaggregates. The enzymes do not alter the 3-D structure of polysaccharides and product thus formed is more bioactive. Use of extractive protein makes the process more specialized as they are substrate specific, easy to use/handle, eco-friendly, effective, and energy-efficient (Lima Santos and Modesto, 2018) (Figure 3.8). The glucanase dsrWC enzyme was isolated from *W. cibaria* CMU using human saliva. This enzyme is used for extraction of glucans as it acts on the 1–6 linkage in its glucans. Plasmids of the *Pediococcus* genus with the dps glycosyltransferase gene, which encodes EPS, are associated with EPS production. This gene causes the creation of Gtf, which polymerizes UDP-glucose to produce glucans (Kang et al., 2009). Pham et al. (2000) extracted the exopolysaccharides from *L. rhamnosus* R by enzymatic degradation.

Lin et al. (2018) extracted the polysaccharide from *Penthorum* Chinese Pursh by using the enzymatic method. Some commonly used EPS degrading/extracting enzymes are as follows:

- Amylase: Starch is broken down into the monomeric sugar by using this enzyme. Amylose and amylopectin are the main polymers of starch. Molobela et al. (2010) used amylases for the removal of biofilm and degradation of EPS produced by *Pseudomonas fluorescens* bacteria.

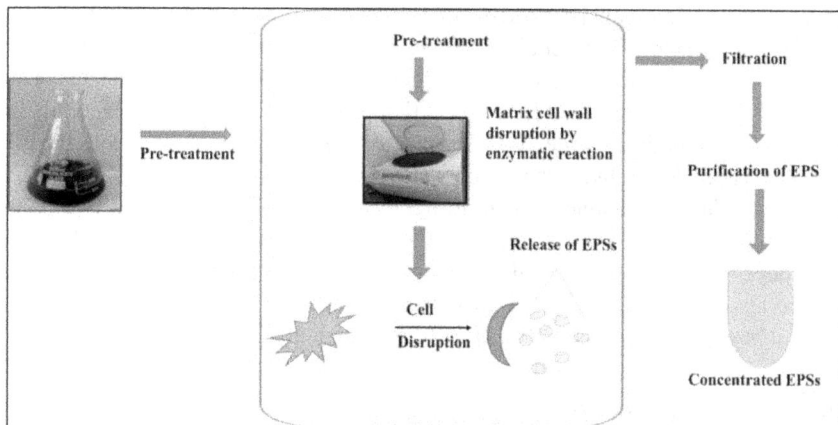

FIGURE 3.8   Enzymatic extraction of exopolysaccharides

- Nucleases: The main function of this enzyme is to cleave single and double strand in target molecules by cleaving the phosphodiester linkages between nucleotides of nucleic acids (Kaplan et al., 2018).
- Protease: This enzyme breaks the peptide bonds formed between two adjacent amino acid residues in a protein molecule and forms shorter peptides and amino acids.

3.6   BOTTLENECKS IN THE PRODUCTION OF EPS

To synthesize microbial EPS at pilot as well as commercial scale, the most important thing is complete knowledge about its structural configurations, physicochemical properties and downstreaming strategies. Downstream processing of EPS fermentation is most challenging and may hinder in its effective extraction that further affects the yield of overall process. Physicochemical properties mainly affect the applicability of EPS in various industries especially in pharmaceutical or cosmetic, where purity is more important than production rates and cost (Freitas et al., 2017). With advancements in sciences and modernizations in technology, certain improvements have been down to enhance purity and yield, but still R&D efforts are required to determine their safety and bioavailability in humans for usage in biomedical applications. In order to further expand the opportunities for production of microbial EPS at commercial level, more studies on major bottlenecks, biosynthesis, manufacturing, and recovery are required. The major drawbacks associated with EPS production are as follows:

3.6.1   PROBLEMS ASSOCIATED WITH GROWTH OF EPS
PRODUCING MICROBES IN BIOREACTORS

The microbes are very versatile in nature and behave differently in varied environmental conditions. The rheological characters of the source organism vary as a result of EPS formation from non-viscous Newtonian behavior to extremely viscous non-Newtonian behavior. This is dominant process parameter and most important for growth of organism in reactor system (Gibbs et al., 2000; Seviour et al., 2010; Kang et al., 2001). The fermentation broth is less viscous, that is, comparable to water and homogeneous in the beginning of culture based EPS production process. It is easy to manage this low viscous liquid in both lab scale as well as reactor scale fermentation but with the time the cell starts dividing and viscosity of system starts increasing. With the increasing thickness of fluid, it became difficult to track the changes that occurred in medium during process in large scale reactors (McNeil and Harvey, 1993). Besides this other strain, dependent physicochemical conditions (temperature and pH) and external environment also affect the overall production and is difficult to maintain in bioreactors though it has specific proofs for the same (Sørensen et al., 2022).

3.6.2 CHALLENGES IN EPS FERMENTATIONS

The consistency, controlled quality throughout batches, and yield of product solely depend on fermentation process, and thus its regulation and maintenance is crucial. Major challenge in fermentative production of EPS is to check and control the whole process. It is a tedious procedure with multiple steps. The hurdles start from the beginning in the form of improper mixing of ingredients as majority of the culture medium is found in the periphery zone, and only very little volume of culture is present near impeller which slows down the rate. Similarly, with increasing microbial growth, heterogenicity also increases and acts as barrier in uniform transfer of nutrient (Shyam et al., 2021). Besides improper mixing, another major issue is to maintain proper exchange of gases. The lower the gaseous exchange, the lower will be the oxygen transfer, and ultimately yield of product will also be less. It is difficult to maintain or enhance gas flow because the gas would channel through the fermenter's well-mixed, low viscosity zone and would just feed the impeller (Matthews, 2008). Therefore, even though the additional power inputs can be added for proper mixing at the stack of overall cost of the product, the net positive benefit will be little.

Secondly, viscosity of culture medium or fermentation broth also limits the heat transmission and momentum inside the reactor and lowers down the overall yield (Rosalam and England, 2006). The limitation associated with fermentative production of EPS with reference to versatile nature of source organism, alteration in physiochemical parameters, higher production cost make the process compromised. Although these limitations can be reduced using continuous and fed batch fermentation (Søresen et al., 2022). Thus it was suggested to use optimal fermentation in reactor scale. Li et al. (2016) in a study reported 4.2 fold higher production of EPS from *S. thermophilus* under optimized fermentation. Nguyen et al. (2020) in their study also reported that physical stress and improper nutrition during fermentation negatively affects the EPS production rate.

3.6.3 SCALE EFFECTS

With the increase in production level, issues with the rheological characteristics of the culture media become more difficult. With increase in volume of reactors, heterogeneities and complexities also increase, which further decreased the power inputs by increasing the distance between the impellers and walls of vessel (McNeil and Harvey, 1993; Suh et al., 1990). Physical nature of culture media, properties of vessels, and requirements of process make the scale-up process more difficult. Besides this, adhesion is another limiting factor to scale up EPS production from 10 L to a thousand liters reactor (Gill et al., 2008). However, advanced scale-up techniques of predictive modeling from lab to micro scale reactors promise effective and relevant scale transfer to EPS process in some microbial cell culture.

3.6.4 CULTURE MORPHOLOGY AND EPS

The morphology of source microbes influences the whole process and acts as rate limiting factor. As the previous research reported, majority of useful EPS are derived from filamentous fungi and observed that filamentous microbial cultures add more complication and complexities to overall process (Giavasis et al., 2000; Fazenda et al., 2008; Seviour et al., 2010). The hyphae of filamentous fungi aggregate together during processing and form either hyphal net or pellets in culture medium, which further obstruct the downstreaming (Papagianni, 2004). Moreover, the fungi based EPS production media became more viscous due to EPS production and fungal growth and abrupt in uniform distribution of nutrients and oxygen throughout the chamber. The abrupt distribution of nutrient further slows down the growth, rate, and yield of EPS production in continuous as well as in batch fermentation (Papagianni, 2004). The physiological conditions of reactors also affect the cellular morphology (shape and size) of source organism, especially in case of fungi. The agitation rate leads to altering the shape of fungal cells. The lower the agitation rate, the lower the yield and vice-versa. McNeil and Kristiansen (1987) in a study reported that *A. pullulans* produced more pullulan of low molecular weight when medium was agitated vigorously. The cell became unicellular that enhances the yield, but vigorous mixing causes conformational changes that leads to reduced molecular weight. In a similar study it was observed that the production and yield was higher in cases of highly filamentous fungi *Ganoderma lucidum*, but extra manual monitoring with reference to blending of inoculums, fed-batch cultures, specific feed control, and higher downstreaming strategies are needed (Fazenda, 2008; Fazenda et al., 2008).

3.7 CONCLUSION AND FUTURE PROSPECTIVE

Exopolysaccharides are a diverse group of polymeric sugar subunits. Production of EPS by microorganisms is a strategy to survive under varied environmental conditions. It protects the microbial source from predators, harsh environment, and antibiotic attack, facilitates microbial metabolism, and entraps the nutrients. Besides this, EPS enhances the aggregation, fertility, and quality of the soil. These are biocompatible and complex substances used in a variety of industries. EPS are the major constituent of microbial biofilm and are well localized and well stable. The biosynthesis mainly involves four major systems and can be extracted by different physicochemical as well as biological methods. Lactic acid producing bacteria are the main source of EPS as they improved the functionality and quality of these polysaccharides. Although there are still some limitations in EPS production with reference to their yield and quantity.

For this, there is a need to understand the EPS producing genes and their regulatory mechanisms. Furthermore, the blending of advanced technological interventions and high-throughput techniques with classical microbiology will help in the synthesis of new EPS with improved functions, structures, quality, quantity, and yield.

ACKNOWLEDGMENTS

Authors would like to express their sincere and heartfelt gratitude to the Department of Biotechnology, Himachal Pradesh University, Shimla, India.

REFERENCES

Abarquero, D., Renes, E., Fresno, J.M., Tornadijo, M.E., 2022. Study of exopolysaccharides from lactic acid bacteria and their industrial applications: A review. Int. J. Food Sci. Technol. 57, 16–26.

Abdalla, A.K., Ayyash, M.M., Olaimat, A.N., Osaili, T.M., Al-Nabulsi, A.A., Shah, N.P., Holley, R., 2021. Exopolysaccharides as antimicrobial agents: Mechanism and spectrum of activity. Front. Microbiol. 12, 1182.

Ahmad, N.H., Mustafa, S., Che-Man, Y.B., 2015. Microbial polysaccharides and their modification approaches: A review. Int. J. Food. Prop. 18(2), 332–347.

Andhare, R.A., Takahashi, N., Knudson, W., Knudson, C.B., 2009. Hyaluronan promotes the chondrocyte response to BMP-7. Osteoarthr. Cartil. 17(7), 906–916.

Angelin, J., Kavitha, M., 2020. Exopolysaccharides from probiotic bacteria and their health potential. Int. J. Biol. Macromol. 162, 853.

Anwar, M., Babu, G., Bekhi, A., 2021. Utilization of ultrasound and pulse electric field for the extraction of water-soluble non-starch polysaccharide from taro (*Colocasia esculenta*) peel. Innov. Food Sci. Emerg. Technol. 70, 102691.

Balti, R., Brodu, N., Gilbert, M., Le Gouic, B., Le Gall, S., Sinquin, C., Masse, A., 2018. Concentration and purification of *Porphyridiumcruentum* exopolysaccharides by membrane filtration at various cross-flow velocities. Process Biochem. 74, 175–184.

Belgrano, F.D.S., Verçoza, B.R.F., Rodrigues, J.C.F., Hatti-Kaul, R., Pereira Jr, N., 2018. EPS production by *Propionibacterium freudenreichii* facilitates its immobilization for propionic acid production. J. Appl. Microbiol. 125(2), 480–489.

Cai, Z., Zhang, H., 2017. Recent progress on curdlan provided by functionalization strategies. Food Hydrocoll. 68, 128–135.

Castillo, N.A., Valdez, A.L., Farina, J.I., 2015. Microbial production of scleroglucan and downstream processing. Front. Microbiol. 6, 1106.

Castro-Bravo, N., Hidalgo-Cantabrana, C., Rodriguez-Carvajal, M.A., Ruas-Madiedo, P., Margolles, A., 2017. Gene replacement and fluorescent labeling to study the functional role of exopolysaccharides in *Bifidobacteriumanimalis* subsp. *lactis*. Front. Microbiol. 8, 1405.

Chawla, P.R., Bajaj, I.B., Survase, S.A., Singhal, R.S., 2009. Microbial cellulose: Fermentative production and applications. Food Technol. Biotechnol. 47(2), 107–124.

Chemat, F., Zill-e-Huma, Khan, M.K., 2011. Applications of ultrasound in food technology: Processing, preservation and extraction. *UltrasonSonochem*. 8, 813–835.

Chen, J., Seviour, R., 2007. Medicinal importance of fungal β-(1→3), (1→6)-glucans. Mycol. Res. 111(6), 635–652.

Cheng, K.C., Demirci, A., Catchmark, J.M., 2011. Pullulan: Biosynthesis, production, and applications. Appl. Microbiol. Biotechnol. 92(1), 29–44.

Choi, D.B., Maeng, J.M., Ding, J.L., Cha, W.S., 2007. Exopolysaccharide production and mycelial growth in an air-lift bioreactor using Fomitopsis pinicola. J. Microbiol. Biotechnol. 17(8), 1369–1378.

Christensen, B.E., Characklis, W.G., 1990. Physical and chemical properties of biofilms. In: W.G. Characklis, K.C. Marshall (Eds.), *Biofilms*. Wiley, pp. 93–130.

Chung, C., Beecham, M., Mauck, R.L., Burdick, J.A., 2009. The influence of degradation characteristics of hyaluronic acid hydrogels on in vitro neocartilage formation by mesenchymal stem cells. Biomaterials. 30(26), 4287–4296.

Cimini, D., DeRosa, M., Schiraldi, C., 2012. Production of glucuronic acid-based polysaccharides by microbial fermentation for biomedical applications. Biotechnol. J. 7, 237–250.

Cuthbertson, L., Mainprize, I.L., Naismith, J.H., Whitfield, C., 2009. Pivotal roles of the outer membrane polysaccharide export and polysaccharide copolymerase protein families in export of extracellular polysaccharides in gram-negative bacteria. Microbiol. Mol. Biol. Rev. 73, 155–177.

Czaczyk, K., Myszka, K., 2007. Biosynthesis of extracellular polymeric substances (EPS) and its role in microbial biofilm formation. Pol. J. Environ. Stud. 16(6), 799–806.

Dai, Y.F., Xiao, Y., Zhang, E.H, Liu, L.D., Qiu, L., Gurumurthy, D.M., Chen, B.L., Zhao, F., 2016. Effective methods for extracting extracellular polymeric substances from *Shewanella oneidensis* MR-1. Water Sci. Technol. 23, 1–10.

Dave, S.R., Vaishnav, A.M., Upadhyay, K.H., Tipre, D.R., 2016. Microbial exopolysaccharide—An inevitable product for living beings and environment. J. Bacteriol. Mycol. 2.

Degeest, B., Mozzi, F., De Vuyst, L., 2002. Effect of medium composition and temperature and pH changes on exopolysaccharide yields and stability during *Streptococcus thermophilus* LY03 fermentations. Int. J. Food Microbiol. 79, 161–174.

Delattre, C., Pierre, G., Laroche, C., Michaud, P., 2016. Production, extraction and characterization of microalgal and cyanobacterial exopolysaccharides. Biotechnol. Adv. 34, 1159–1179.'

De Vuyst, L., Degeest, B., 1999. Heteropolysaccharides from lactic acid bacteria. FEMS Microbiol. Rev. 23(2), 153–177.

Díaz-Montes, E., 2021. Dextran: Sources, structures, and properties. Polysaccharides. 2(3), 554–565.

Di Pippo, F., Ellwood, N.T.W., Gismondi, A., Bruno, L., Rossi, F., Magni, P., De Philippis, R., 2013. Characterization of exopolysaccharides produced by seven biofilm forming cyanobacterial strains for biotechnological applications. J. Appl. Phycol. 25, 1697–1708.

Dishisha, T., Alvarez, M.T., Hatti-Kaul, R., 2012. Batch- and continuous propionic acid production from glycerol using free and immobilized cells of *Propionibacterium acidipropionici*. Bioresour. Technol. 118, 553–562.

Dols, M., Remaud-Simeon, M., Monsan, P.F., 1997. Dextransucrase production by *Leuconostoc mesenteroides* NRRL B-1299, comparison with *L. mesenteroides* NRRL B-512F. Enzyme Microb. Technol. 20, 523–530.

Donlan, R.M., Costerton, J.W., 2002. Biofilms: Survival mechanisms of clinically relevant microorganisms. Clin. Microbiol. Rev. 15(2), 167–193.

Donot, F., Fontana, A., Baccou, J.C., Schorr-Galindo, S., 2012. Microbial exopolysaccharides: Main examples of synthesis, excretion, genetics and extraction. Carbohydr. Polym. 87, 951–962.

Emerson, D., Ghiorse, W.C., 1993. Role of disulphide bonds in maintaining the structural integrity of the sheath of *Leptothrix discophora* SP-6. J. Bacteriol. 175, 7819–7827.

Eroglu, E., Torun, M., Dincer, C., Topuz, A., 2014. Influence of pullulan-based edible coating on some quality properties of strawberry during cold storage. Packag. Technol. Sci. 27(10), 831–838.

Farnworth, E.R., Champagne, C.P., Van Calsteren, M.R., 2006. Exopolysaccharides from lactic acid bacteria: Food uses, production, chemical structures, and health effects. In: *Probiotics in Food Safety and Human Health*. Robert E.C Wildman (Ed.). CRC Taylor & Francis Group.

Fazenda, M.L., 2008. Submerged culture fermentation of the basidiomycete fungus ganoderma lucidum for biomass formation. PhD thesis. University of Strathclyde.

Fazenda, M.L., Seviour, R., McNeil, B., Harvey, L.M., 2008. Submerged culture fermentation of higher fungi: The macro fungi. Adv. Appl. Microbiol. 63, 33–103.

Felz, S., Al-Zuhairy, S., Aarstad, O.A., van Loosdrecht, M.C., Lin, Y.M., 2016. Extraction of structural extracellular polymeric substances from aerobic granular sludge. JoVE. 115, e54534.

Freitas, F., Alves, V.D., Pais, J., Carvalheira, M., Costa, N., Oliveira, R., Reis, M.A.M., 2010. Production of a new exopolysaccharide (EPS) by *Pseudomonas oleovorans* NRRL B-14682 grown on glycerol. Process Biochem. 45(3), 297–305.

Freitas, F., Alves, V.D., Pais, J., et al., 2009. Characterization of an extracellular polysaccharide produced by a *Pseudomonas* strain grown on glycerol. Biores. Technol. 100, 859–865.

Freitas, F., Alves, V.D., Reis, M.A., 2011. Advances in bacterial exopolysaccharides: From production to biotechnological applications. Trends Biotechnol. 29(8), 388–398.

Freitas, F., Torres, C.A., Reis, M.A., 2017. Engineering aspects of microbial exopolysaccharide production. Biores. Technol. 245, 1674–1683.

Frolund, B., Palmgren, R., Keiding, K., Nielsen, P.H., 1996. Extraction of extracellular polymers from activated sludge using a cation exchange resin. Wat. Res. 30, 1749–1758.

Galib, M.A., Abbott, T., Lee, H.S., 2021. Examination of extracellular polymer (EPS) extraction methods for anaerobic membrane bioreactor (anmbr) biomass. Sustain. 13, 12584.

Gamar, L., Blondeau, K., Simonet, J.M. 2003. Physiological approach to extracellular polysaccharide production by *Lactobacillus rhamnosus* strain C83. J. Appl. Microbiol. 83, 281–287.

Gao, C., Cai, C., Liu, J., Wang, Y., Chen, Y., Wang, L., Tan, Z., 2020. Extraction and preliminary purification of polysaccharides from *Camellia oleifera* Abel. seed cake using a thermo separating aqueous two-phase system based on EOPO copolymer and deep eutectic solvents. Food Chem. 313, 126164.

Garcia-Garibay, M., Marshall, V., 2008. Polymer production by *Lactobacillus delbrueckii* ssp. J. Appl. Microbiol. 70, 325–328.

Garcia-Ochoa, F., Gomez, E., 2009. Bioreactor scale-up and oxygen transfer rate in microbial processes: An overview. Biotechnol. Adv. 27, 153–176.

Garcia-Ochoa, F., Santos, V.E., Casas, J.A., Gomez, E., 2000. Xanthan gum: Production, recovery, and properties. Biotechnol. Adv. 18, 549–579.

Ghosh, P.K., Maiti, T.K., 2016. Structure of extracellular polysaccharides (EPS) produced by Rhizobia and their functions in legume–bacteria symbiosis:—A review. Achiev. Life Sci. 10(2), 136–143.

Giavasis, I., Harvey, L.M., McNeil, B., 2000. Gellan gum. Crit. Rev. Biotechnol. 20(3), 177–211.

Gibbs, P.A., Seviour, R.J., Schmid, F., 2000. Growth of filamentous fungi in submerged culture: Problems and possible solutions. Crit. Rev. Biotechnol. 20, 17–48.

Gill, N.K., Appleton, M., Baganz, F., Lye, G.J., 2008. Design and characterisation of a miniature stirred bioreactor system for parallel microbial fermentations. Biochem. Eng. J. 39, 164–176.

Gomes, F.P., Silva, N.H.C.S., Trovatti, E., et al., 2013. Production of bacterial cellulose by *Gluconacetobacter sacchari* using dry olive mil residue. Biomass Bioenerg. 55, 205–211.

Gresh, N., Policar, C., Giessner-Prettre, C., 2002. Modeling copper (I) complexes: SIBFA molecular mechanics versus ab initio energetics and geometrical arrangements. J. Phys. Chem. 106(23), 5660–5670.

Grosu-Tudor, S.S., Stancu, M.M., Pelinescu, D., Zamfir, M., 2014. Characterization of some bacteriocins produced by lactic acid bacteria isolated from fermented foods. World J. Microbiol. Biotechnol. 30, 2459–2469.

Guerreiro, B.M., Silva, J.C., Lima, J.C., Reis, M.A., Freitas, F., 2021. Antioxidant potential of the bio-based fucose-rich polysaccharide fucopol supports its use in oxidative stress-inducing systems. Polymers. 13(18), 3020.

Gugliandolo, C., Spano, A., Lentini, V., Arena, A., Maugeri, T.L., 2013. Antiviral and immune modulatory effects of a novel bacterial exopolysaccharides of shallow marine vent origin. J. Appl. Microbiol. 116, 1028–1034.

Guo, L., Liu, H., Peng, F., Qi, H., 2021. Efficient and portable cellulose-based colorimetric test paper for metal ion detection. Carb. Pol. 274, 118635.

Hamidi, M., Okoro, O.V., Milan, P.B., Khalili, M.R., Samadian, H., Nie, L., Shavandi, A., 2022. Fungal exopolysaccharides: Properties, sources, modifications, and biomedical applications. Carbohyr. Polym. 284, 119152.

Hebbar, K.P., Gueniot, B., Heyraud, A., Colin-Morel, P., Heulin, T., Blandreau, J., Rinaudo, M., 1992. Characterization of exopolysaccharides produced by *Rhizobacteria*. Appl. Microbiol. Biotechnol. 38, 248–253.

Huang, S.Q., Ning, Z.X., 2010. Extraction of polysaccharide from *Ganoderma lucidum* and its immune enhancement activity. Int. J. Biol. Macromol. 47(3), 336–341.

Hubbard, C., McNamara, J.T., Azumaya, C., Patel, M.S., Zimmer, J., 2012. The hyaluronan synthase catalyzes the synthesis and membrane translocation of hyaluronan. J. Mol. Biol. 418(1–2), 21–31.

Hussein, M.D., Ghaly, M., Osman, M., et al., 2015. Production and prebiotic activity of exo-polysaccharides derived from some probiotics. Egypt. Pharm. J. 14, 1–9.

Huu, N., Razafindralambo, H., Blecker, C., et al., 2014. Stochastic exposure to sub-lethal high temperature enhances exopolysaccharides (EPS) excretion and improves *Bifidobacterium bifidum* cell survival to freeze-drying. Biochem. Eng. J. 88, 85–94.

Imeson, A., 2010. *Food Stabilisers, Thickeners and Gelling Agents*. Wiley.

Imran, M.Y.M., Reehana, N., Jayaraj, K.A., Ahamed, A.A.P., Dhanasekaran, D., Thajuddin, N., Alharbi, N.S., 2016. Statistical optimization of exopolysaccharide production by *Lactobacillus plantarum* NTMI05 and NTMI20. Int. J. Biol. Macromol. 93, 731–745.

Ioelovich, M., 2008. Cellulose as a nanostructured polymer: A short review. BioRes. 3(4), 1403–1418.

Islam, S.T., Lam, J.S., 2014. Synthesis of bacterial polysaccharides via the Wzx/Wzy-dependent pathway. Can. J. Microbiol. 60(11), 697–716.

Jia, W., Zhu, L., Zhang, J.S., Zhan, X.B., Bai, Y.Y., Zhou, S., Liu, Y.F., Zheng, Z.Y., Lin, C.C. 2012. Ultrasound-assisted degradation of a new bacterial exopolysaccharide WL-26 from *Sphingomonas* sp. Afr. J. Biotechnol. 11(79), 14474–14481.

Kambourova, M., Oner, E.T., Poli, A., 2015. Exopolysaccharides from prokaryotic microor-ganisms-promising sources for white biotechnology processes. Indus. Biorefin. White Biotechnol. 523–554.

Kang, H., Oh, J., Kim, D., 2009. Molecular characterization and expression analysis of the glucansucrase DSRWC from *Weissella cibaria* synthesizing a α-1$\rightarrow$6) glucan. FEMS Microbiol. Lett. 292(1), 33–41.

Kang, X., Wang, H., Wang, Y., Harvey, L.M., McNeil, B., 2001. Hydrodynamic characteristics and mixing behaviour of *Sclerotium glucanicum* culture fluids in an airlift reactor with an internal loop used for scleroglucan production. J. Ind. Microbiol. Biotechnol. 27, 208–214.

Kaplan, J.B., Mlynek, K.D., Hettiarachchi, H., Alamneh, Y.A., Biggemann, L., Zurawski, D. V., et al., 2018. Extracellular polymeric substance (EPS)-degrading enzymes reduce staphylococcal surface attachment and biocide resistance on pig skin in vivo. PLoS One. 13(10), e0205526.

Khoo, K.S., Leong, H.Y., Chew, K.W., Lim, J.W., Ling, T.C., Show, P.L., et al., 2020. Liquid biphasic system: A recent bio separation technology. Processes, 8(149), 2–22.

Kim, Y., Kumar, S., 2014. CD44-mediated adhesion to hyaluronic acid contributes to mechanosensing and invasive motility. Mol. Cancer Res. 12(10), 1416–1429.

Korakli, M., Pavlovic, M., Gänzle, M.G., Vogel, R.F., 2003. Exopolysaccharide and kestose production by *Lactobacillus sanfranciscensis* LTH2590. Appl. Environ. Microbiol. 69(4), 2073–2079.

Kulicke, W.M., Heinze, T., 2005. Improvements in polysaccharides for use as blood plasma expanders. Macromo. Sym. 231(1), 47–59.

Kumar, M.A., Anandapandian, K.T.K., Parthiban, K., 2011. Production and characterization of exopolysaccharides (EPS) from biofilm forming marine bacterium. Braz. Arch. Biol. Technol. 54, 259–265.

Kumar, R., Srivastava, A., Behari, K., 2007. Graft copolymerization of methacrylic acid onto xanthan gum by Fe^{2+}/H_2O_2 redox initiator. J. Appl. Polym. Sci. 105(4), 1922–1929.

Li, D., Li, J., Zhao, F., Wang, G., Qin, Q., Hao, Y., 2016a. The influence of fermentation condition on production and molecular mass of eps produced by *Streptococcus thermophilus* 05-34 in milk-based medium. Food Chem. 197, 367–372.

Li, P., Li, T., Zeng, Y., Li, X., Jiang, X., Wang, Y., Xie, T., Zhang, Y., 2016b. Biosynthesis of xanthan gum by *Xanthomonas campestris* LRELP-1 using kitchen waste as the sole substrate. Carbohydr. Pol. 151, 684–691.

Li, W.W., Zheng, P., Zhang, M., Zeng, Z., Wang, Z., Ding, A., Ding, K., 2016. Hydrophilicity/ hydrophobicity of anaerobic granular sludge surface and their causes: An in situ research. Bioresour. Technol. 220, 117–123.

Li, W.W., Zhou, W.Z., Zhang, Y.Z., Wang, J., Zhu, X.B., 2008. Flocculation behavior and mechanism of an exopolysaccharides from the deep-sea psychrophilic bacterium *Pseudoalteromonas* sp. SM9913. Biores. Technol. 99, 6893–6899.

Liang, T.W., Wang, S.L., 2015. Recent advances in exopolysaccharides from *Paenibacillus* spp.: Production, isolation, structure, and bioactivities. Mar. Drugs. 13(4), 1847–1863.

Lima Santos, F., Modesto, A.G., 2018. Biotechnological challenges and perspectives of using exopolysaccharides. J. Anal. Pharma. Res. 7.

Lin, L.M., de Kreuk, M., van Loosdrecht, M.C.M., Adin, A., 2010. Characterization of alginate-like exopolysaccharides isolated from aerobic granular sludge in pilot-plant. Wat. Res. 44, 3355–3364.

Lin, L.M., Zhao, L.J., Deng, J., Xiong, S.H., Tang, J., Li, Y.M., Liao, D.F., 2018. Enzymatic extraction, purification, and characterization of polysaccharides from Penthorum chinense Pursh: Natural antioxidant and anti-inflammatory. BioMed Res. Int. 2018, 13.

Liu, C., Sun, Y., Mao, Q., Guo, X., Li, P., Liu, Y., et al., 2016. Characteristics and antitumor activity of *Morchella esculenta* polysaccharide extracted by pulsed electric field. Int. J. Mol. Sci. 17(6), 986.

Liu, J., Luo, J., Ye, H., Zeng, X., 2012. Preparation, antioxidant and antitumor activities *in vitro* of different derivatives of levan from endophytic bacterium *Paenibacilluspolymyxa* EJS-3. Food Chem. Toxicol. 50, 767–772.

Liu, J., Wang, X., Pu, H., Liu, S., Kan, J., Jin, C., 2017. Recent advances in endophytic exopolysaccharides: Production, structural characterization, physiological role and biological activity. Carbohydr. Polym. 157, 1113–1124.

Lynch, K.M., Zannini, E., Coffey, A., Arendt, E.K., 2018. Lactic acid bacteria exopolysaccharides in foods and beverages: Isolation, properties, characterization, and health benefits. Annu. Rev. Food Sci. Technol. 9, 155–176.

Ma, H., Wei, J.-Q., Su, D., Guo, T., Feng, Z.-P., Zhang, J.-B., Tan, S.-B., Wei, L.-Y., 2015. Aqueous two phase extraction of exopolysaccharides. Adv. J. Food. Sci. Technol. 9(10), 807–811.

Ma, Q., Santhanam, R.K., Xue, Z., Guo, Q., Gao, X., Chen, H., 2018. Effect of different drying methods on the physicochemical properties and antioxidant activities of mulberry leaves polysaccharides. Int. J. Biol. Macromol. 119, 1137–1143.

Macedo, M.G., Lacroix, C., Gardner, N.J., Champagne, C.P., 2002. Effect of medium supplementation on exopolysaccharide production by *Lactobacillus rhamnosus* RW-9595M in whey permeate. Int. Dairy J. 12, 419–426.

Madhuri, K.V., Vidya Prabhakar, K., 2014. Microbial exopolysaccharides: Biosynthesis and potential applications. Orien. J. Chem. 30, 1401–1410.

Mahapatra, S., Banerjee, D., 2012. Structural elucidation and bioactivity of a novel exopolysaccharide from endophytic Fusarium solani SD5. Carbohydr. Polym. 90, 683–689.

Mahapatra, S., Banerjee, D., 2013. Fungal exopolysaccharide: Production, composition and applications. Microbiol. Insig. 6, 1.

Marshall, V.M., Cowie, E.N., Moreton, R.S., 2009. Analysis and production of two exopolysaccharides from *Lactococcus lactis* subsp. *cremoris* LC330. J Dairy Res. 62, 621–628.

Marvasi, M., Visscher, P., Casillas-Martinez, L., 2010. Exopolymeric substances (EPS) from *Bacillus subtilis*: Polymers and genes encoding their synthesis. FEMS Microbiol. Lett. 313, 1–9.

Matthews, G., 2008. Fermentation equipment selection: Laboratory scale bioreactor design considerations. In: Brain McNeil and Linda M. Harvey (Eds.), *Practical Fermentation Technology*, pp. 3–36. Wiley.

Mbye, M., Baig, M.A., AbuQamar, S.F., et al., 2020. Updates on understanding of probiotic lactic acid bacteria responses to environmental stresses and highlights on proteomic analyses. Compr. Rev. Food Sci. Food Saf. 19, 1110–1124.

McIntosh, M., Stone, B.A., Stanisich, V.A., 2005. Curdlan and other bacterial (1 -> 3)-beta-D-glucans. Appl. Microbiol. Biotechnol. 68, 163–173.

McNeil, B., Harvey, L.M., 1993. Viscous fermentation products. Crit. Rev. Biotechnol. 13, 275–304.

McNeil, B., Kristiansen, B., 1987. Influence of impeller speed upon the pullulan fermentation. Biotechnol. Lett. 9, 101–104.

McSwain, B.S., Irvine, R.L., Hausner, M., Wilderer, P., 2005. Composition and distribution of extracellular polymeric substances in aerobic flocs and granular sludge. Appl. Environ. Microbiol. 71(2), 1051–1057.

Mende, S., Jaros, D., Rohm, H., 2020. Dextran modulates physical properties of rennet-induced milk gels. Int. J. Food Sci. Technol. 55(4), 1407–1415.

Molobela, I.P., Cloete, T.E., Beukes, M., 2010. Protease and amylase enzymes for biofilm removal and degradation of extracellular polymeric substances (EPS) produced by *Pseudomonas fluorescens* bacteria. Afr. J. Microbiol. Res. 4(14).

Morona, R., Purins, L., Tocilj, A., Matte, A., Cygler, M., 2009. Sequence-structure relationships in polysaccharide co-polymerase (PCP) proteins. Trends Biochem. Sci. 34(2), 78–84.

Moscovici, M., 2015. Present and future medical applications of microbial exopolysaccharides. Front. Microbiol. 6, 1012.

Muller, J.M., Alegre, R.M., 2006. Alginate production by Pseudomonas mendocina in a stirred draft fermenter. World J. Microbiol. Biotechnol. 23, 691–695.

Mustapa, A.N., Martin, A., Gallego, J.R., Mato, R.B., Cocero, M.J., 2015. Microwave-assisted extraction of polyphenols from *Clinacanthus nutans* Lindau medicinal plant: Energy perspective and kinetics modeling. Chem Eng. Process. 97, 66–74.

Muthusamy, S., Anandharaj, S.J., Kumar, P.S., Meganathan, Y., Vo, D.V.N., Vaidyanathan, V.K., Muthusamy, S., 2022. Microbial pullulan for food, biomedicine, cosmetic, and water treatment: A review. Environ. Chem. Let. 1–36.

Najim, N., Aryana, K.J., 2013. A mild pulsed electric field condition that improves acid tolerance, growth, and protease activity of *Lactobacillus acidophilus* LA-K and *Lactobacillus delbrueckii* subspecies *bulgaricus* LB-12. J. Dairy Sci. 96(6), 3424–3434.

Nguyen, H.T., Truong, D.H., Kouhounde, S., et al., 2016. Biochemical engineering approaches for increasing viability and functionality of probiotic bacteria. Int. J. Mol. Sci. 17, 867.

Nguyen, P.T., Nguyen, T.T., Bui, D.C., Hong, P.T., Hoang, Q.K., Nguyen, H.T., 2020. Exopolysaccharide production by lactic acid bacteria: The manipulation of environmental stresses for industrial applications. AIMS. Microbiol. 6(4), 451–469.

Nishinari, K., 2007. Rheological and related studies on industrially important polysaccharides and proteins. J. Central South Univ. Technol. 14, 498–504.

Novak, J.T., Haugan, B.E., 1981. Polymer extraction from activated sludge. J. Wat. Pollut. Control. Fed. 53, 1420–1424.

Nwodo, U.U., Green, E., Okoh, A.I., 2012. Bacterial exopolysaccharides: Functionality and prospects. Int. J. Mol. Sci. 13(11), 14002–14015.

Oner, E.T., Hernandez, L., Combie, J., 2016. Review of Levan polysaccharide: From a century of past experiences to future prospects. Biotechnol. Adv. 34(5), 827–844.

Orr, D., Zheng, W., Campbell, B.S., McDougall, B.M., Seviour, R.J., 2009. Culture conditions affect the chemical composition of the exopolysaccharide synthesized by the fungus *Aureobasidium pullulans*. J. Appl. Microbiol. 107, 691–698.

Osinska-Jaroszuk, M., Jarosz-Wilkołazka, A., Jaroszuk-Sciseł, J., et al., 2015. Extracellular polysaccharides from *Ascomycota* and *Basidiomycota*: Production conditions, biochemical characteristics, and biological properties. World J. Microbiol. Biotechnol. 31, 1823–1844.

Palaniraj, A., Jayaraman, V., 2011. Production, recovery and applications of xanthan gum by *Xanthomonas campestris*. Food Eng. 106(1), 1–12.

Papagianni, M., 2004. Fungal morphology and metabolite production in submerged mycelial processes. Biotechnol. Adv. 22, 189–259.

Paquet, É., Bédard, A., Lemieux, S., Turgeon, S.L., 2014. Effects of apple juice-based beverages enriched with dietary fibres and xanthan gum on the glycemic response and appetite sensations in healthy men. Bioact. Carbohydr. Diet. Fibre. 4(1), 39–47.

Park, C., Novak, J.T., 2007. Characterization of activated sludge extracellular polymers using several cation-associated extraction methods. Water Res. 41(8), 1679–1688.

Parniakov, O., Lebovka, N.I., Van Hecke, E., Vorobiev, E., 2014. Pulsed electric field assisted pressure extraction and solvent extraction from mushroom (*Agaricus bisporus*). Food Bioproc Tech. 7(1), 174–183.

Patel, A., Laroche, C., Marcati, A., Ursu, A.V., Jubeau, S., Marchal, L., Petit, E., Djelveh, G., Michaud, P., 2013. Separation and fractionation of exopolysaccharides from *Porphyridium cruentum*. Biores. Tech. 145, 345–350.

Patel, A., Lindström, C., Patel, A., Prajapati, J.B., Holst, O., 2012. Probiotic properties of exopolysaccharides producing lactic acid bacteria isolated from vegetables and traditional Indian fermented foods. Int. J. Fermented Foods. 1(1), 87–102.

Patel, A., Prajapati, J., 2013. Food and health applications of exopolysaccharides produced by Lactic acid Bacteria. Adv. Dairy Res. 1, 1–7.

Pértile, R., Moreira, S., Andrade, F., Domingues, L., Gama, M., 2012. Bacterial cellulose modified using recombinant proteins to improve neuronal and mesenchymal cell adhesion. Biotechnol. Prog. 28(2), 526–532.

Pham, P.L., Dupont, I., Roy, D., Lapointe, G., Cerning, J., 2000. Production of exopolysaccharide by *Lactobacillus rhamnosus* R and analysis of its enzymatic degradation during prolonged fermentation. Appl. Environ. Microbiol. 66(6), 2302–2310.

Polak-Berecka, M., Wasko, A., Paduch, R., Skrzypek, T., Sroka-Bartnicka, A., 2014. The effect of cell surface components on adhesion ability of *Lactobacillus rhamnosus*. Antonie Van Leeuwenhoek. 106(4), 751–762.

Prajapati, V.D., Jani, G.K., Zala, B.S., Khutliwala, T.A., 2013. An insight into the emerging exopolysaccharide gellan gum as a novel polymer. Carbohydr. Polym. 93(2), 670–678.

Rafigh, S.M., Yazdi, A.V., Vossoughi, M., Safekordi, A.A., Ardjmand, M., 2014. Optimization of culture medium and modeling of curdlan production from *Paenibacillus polymyxa* by RSM and ANN. Int. J. Biol. Macromol. 70, 463–473.

Rana, S., Upadhyay, L.S.B., 2020. Microbial exopolysaccharides: Synthesis pathways, types and their commercial applications. Int. J. Biol. Macromol. 157, 577–583.

Rastogi, G., Bhalla, A., Adhikari, A., Bischoff, K.M., Hughes, S.R., Christopher, L.P., et al., 2010. Characterization of thermostable cellulases produced by *Bacillus* and *Geobacillus* strains. Bioresour. Technol. 101, 8798–8806.

Read, R.R., Costerton, J.W., 1987. Purification and characterization of adhesive exopolysaccharides from *Pseudomonas putida* and *Pseudomonas fluorescens*. Can. J. Microbiol. 33, 1080–1090.

Redmile-Gordon, M.A., Brookes, P.C., Evershed, R.P., Goulding, K.W.T., Hirsch, P.R., 2014. Measuring the soil-microbial interface: Extraction of extracellular polymeric substances (EPS) from soil biofilms. Soil Biol. Biochem. 72, 163–171.

Rehm, B.H.A., 2010. Bacterial polymers: Biosynthesis, modifications and applications. Nat. Rev. Microbiol. 8, 578–592.

Riaz, M.S., Wu, Y., Mehwish, H.M., et al., 2020. *Lactobacillus* exopolysaccharides: New perspectives on engineering strategies, physiochemical functions, and immunomodulatory effects on host health. Trends Food Sci. Technol. 103, 36–48.

Rosalam, S., England, R., 2006. Review of xanthan gum production from unmodified starches by *Xanthomonas camprestris sp*. Enzyme Microb. Technol. 39, 197–207.

Saha, T., Hoque, M.E., Mahbub, T., 2020. Biopolymers for sustainable packaging in food, cosmetics, and pharmaceuticals. In: *Advanced Processing, Properties, and Applications of Starch and Other Bio-Based Polymers*. Elsevier, pp. 197–214.

Sanhueza, E., Paredes-Osses, E., Gonzalez, C., et al., 2015. Effect of pH in the survival of *Lactobacillus salivarius* strain UCO_979C wild type and the pH acid acclimated variant. Electron. J. Biotechnol. 18, 343–346.

Sato, T., Ose, Y., 1984. Floc-forming substances extracted from activated sludge with ammonium hydroxide and EDTA solutions. Wat. Sci. Tech. 17(5), 17–528.

Schachman, H.K., 2013. *Ultracentrifugation in Biochemistry*. Elsevier.

Schmid, J., Sieber, V., Rehm, B., 2015. Bacterial exopolysaccharides: Biosynthesis pathways and engineering strategies. Front. Microbiol. 6, 496.

Seesuriyachan, P., Kuntiya, A., Hanmoungjai, P., et al., 2012. Optimization of exopolysaccharide overproduction by *Lactobacillus confusus* in solid state fermentation under high salinity stress. Biosci. Biotechnol. Biochem. 76, 912–917.

Senni, K., Pereira, J., Gueniche, F., et al., 2011. Marine polysaccharides: A source of bioactive molecules for cell therapy and tissue engineering. Mar. Drugs. 9(9), 1664–1681.

Sesay, M.L., Ozcengiz, G., Sanin, F.D., 2006. Enzymatic extraction of activated sludge extracellular and implications on biofloccuiation. Water Res. 40, 1359–1366.

Seviour, R.J., Schmid, F., Campbell, B.C., 2010. *Polysaccharides in Medicinal and Pharmaceutical Application.* V. Popa (Ed.). Smithers Rapra.

Seviour, T.W., Yuan, Z., van Loosdrecht, M.C.M., Lin, Y., 2012. Aerobic sludge granulation: A tale of two polysaccharides? Wat. Res. 46, 4803–4813.

Sheng, G.P., Yu, H.Q., Yue, Z., 2006. Factors influencing the production extracellular polymeric substances by *Rhodopseudomonasacidophila.* Int. Biodeter. Biodegr. 58, 89–93.

Shingel, K.I., 2004. Current knowledge on biosynthesis, biological activity, and chemical modification of the exopolysaccharide, pullulan, Carbohyd. Res. 339, 447–460.

Shyam, P.K., Rajkumar, P., Ramya, V., Sivabalan, S., Kings, A.J., Miriam, L.R.M., 2021. Exopolysaccharide production by optimized medium using novel marine *Enterobactercloacae* MBB8 isolate and its antioxidant potential. Carbohy. Polym. Technol. Appl. 2, 100070.

Silbir, S., Dagbagli, S., Yegin, S., Baysal, T., Goksungur, Y., 2014. Levan production by *Zymomonasmobilis* in batch and continuous fermentation. Carbohydr. Polym. 99, 454–461.

Silva, A.D.S., Magalhaes, W.T., Moreira, L.M., Rocha, M.V.P., Bastos, A.K.P., 2018. Microwave-assisted extraction of polysaccharides from Arthrospira (Spirulina) platensis using the concept of green chemistry. Algal Res. 35, 178–184.

Singh, R.S., Saini, G.K., Kennedy, J.F., 2008. Pullulan: Microbial sources, production and applications. Carbohy. Polym. 73, 515–531.

Soanen, N., Da Silva, E., Gardarin, C., Michaud, P., Laroche, C., 2016. Improvement of exopolysaccharide production by Porphyridiummarinum. Biores. Technol. 213, 231–238.

Sorensen, H.M., Rochfort, K.D., Maye, S., MacLeod, G., Brabazon, D., Loscher, C., Freeland, B., 2022. Exopolysaccharides of lactic acid bacteria: Production, purification and health benefits towards functional food. Nutrients. 14(14), 2938.

Sotelo, K.A., Hamid, N., Oey, I., Pook, C., Gutierrez-Maddox, N., Ma, Q., Lu, J., 2018. Red cherries (Prunus avium var. Stella) processed by pulsed electric field–Physical, chemical and microbiological analyses. Food Chem. 240, 926–934.

Spano, A., Gugliandolo, C., Lentini, V., Maugeri, T.L., Anzelmo, G., Poli, A., Nicolaus, B., 2013. A novel EPS-producing strain of *Bacillus licheniformis* isolated from a shallow vent off Panarea Island (Italy). Curr. Microbiol. 67, 21–29.

Srisuk, P., Berti, F.V., da Silva, L.P., Marques, A.P., Reis, R.L., Correlo, V.M., 2018. Electroactive gellan gum/polyaniline spongy-like hydrogels. ACS Biomat. Sci. Eng. 4(5), 1779–1787.

Subramanian, B., Yan, S., Tyagi, S., Surampalli, R.Y., 2010. Extracellular polymeric substances (EPS) producing bacterial strains of municipal wastewater sludge: Isolation, EPS characterization and performance for sludge settling and dewatering. Water Res. 44(7), 2253–2266.

Suh, I.S., Herbst, H., Schumpe, A., Deckwer, W.D., 1990. The molecular weight of xanthan polysaccharide produced under oxygen limitation. Biotechnol. Lett. 12, 201–206.

Sun, L., Wu, D., Ning, X., Yang, G., Lin, Z.H., Tian, M.H., Zhou, Y.F., 2015. α-Amylase-assisted extraction of polysaccharides from Panax ginseng. Int. J. Biol. Macromol. 75, 152–157.

Surayot, U., Wang, J., Seesuriyachan, P., Kuntiya, A., Tabarsa, M., Lee, Y., You, S., 2014. Exopolysaccharides from lactic acid bacteria: Structural analysis, molecular weight effect on immunomodulation. Int. J. Biol. Macromol. 68, 233–240.

Survase, S.A., Saudagar, P.S., Bajaj, I.B., Singhal, R.S., 2007. Scleroglucan: Fermentative production, downstream processing and applications. Food Technol. Biotechnol. 45, 107–118.

Tallon, R., Bressollier, P., Urdaci, M.C., 2003. Isolation and characterization of two exopolysaccharides produced by *Lactobacillus plantarum* EP56. Res. Microbiol. 154, 705–712.

Taskin, M., Erdal, S., Canli, O., 2010. Utilization of waste loquat (Eriobotrya Japonica Lindley) kernels as substrate for scleroglucan production by locally isolated Sclerotium rolfsii. Food Sci. Biotechnol. 19, 1069–1075.

Taylan, O., Yilmaz, M.T., Dertli, E., 2019. Partial characterization of a levan type exopolysaccharide (EPS) produced by Leuconostoc mesenteroides showing immunostimulatory and antioxidant activities. Int. J. Biol. Macromol. 136, 436–444.

Troch, P., Philip-Hollingworth, S., Orgambide, G., Dazzo, F.B., Vanderleyden, J., 1992. Analysis of extracellular polysaccharides isolated from *Azospirillum brasilense* wild type and mutant strains. Symbiosis. 13, 229–241.

Tsubaki, S., Oono, K., Hiraoka, M., Onda, A., Mitani, T., 2016. Microwave-assisted hydrothermal extraction of sulfated polysaccharides from Ulva spp. and *Monostroma latissimum*. Food Chem. 210, 311–316.

Vázquez-González, Y., Prieto, C., Stojanovic, M., Torres, C.A., Freitas, F., Ragazzo-Sánchez, J.A., Lagaron, J.M., 2022. Preparation and characterization of electrospun polysaccharide fucopol-based nanofiber systems. Nanomaterials. 12(3), 498.

Wang, Y., Wang, J., Liu, Z., Huang, X., Fang, F., Guo, J., Yan, P., 2021. Effect of EPS and its forms of aerobic granular sludge on sludge aggregation performance during granulation process based on XDLVO theory. Sci. Total Environ. 795, 148682.

West, B.J., Alabi, I., Deng, S., 2021. A face serum containing bakuchiol, palmitoyl tripeptide-38, hydrolyzed hyaluronic acid and a polyherbal and vitamin blend improves skin quality in human volunteers and protects skin structure in vitro. Preprints. 2021060580.

Whitfield, C., 1988. Bacterial extracellular polysaccharides. Can. J. Microbiol. 34, 415–420.

Whitney, J.C., Howell, P.L., 2013. Synthase-dependent exopolysaccharide secretion in Gram-negative bacteria. Trends Microbiol. 21(2), 63–72.

Wilen, B.M., Jin, B., Lant, P., 2003. The influence of key chemical constituents in activated sludge on surface and flocculating properties. Water Res. 37(9), 2127–2139.

Willis, L.M., Whitfield, C., 2013. Structure, biosynthesis, and function of bacterial capsular polysaccharides synthesized by ABC transporter-dependent pathways. Carbohydr. Res. 378, 35–44.

Xiao, M., Fu, X., Wei, X., Chi, Y., Gao, W., Yu, Y., Liu, Z., Zhu, C., Mou, H., 2021. Structural characterization of fucose-containing disaccharides prepared from exopolysaccharides of *Enterobacter sakazakii*. Carbohydr. Polym. 252, 117139.

Xiao, R., Yang, X., Li, M., Li, X., Wei, Y., Cao, M., Zheng, Y., 2018. Investigation of composition, structure and bioactivity of extracellular polymeric substances from original and stress-induced strains of *Thraustochytrium striatum*. Carbohydr. Polym. 195, 515–524.

Xu, N., Sun, Y.H., Guo, X.L., Liu, C., Mao, Q., Hou, J.M., 2018. Optimization of ultrasonic-microwave synergistic extraction of polysaccharides from *Morchellaconica*. J. Food Process. Preserv. 42(2), 1–7.

Xue, D., Farid, M.M., 2015. Pulsed electric field extraction of valuable compounds from white button mushroom (*Agaricus bisporus*). Innov. Food. Sci. Emer. 29, 178–186.

Yang, M., Zhu, Y., Li, Y., Bao, J., Fan, X., Qu, Y., Wang, Y., Hu, Z., Li, Q., 2016. Production and optimization of curdlan produced by *Pseudomonas* sp. QL212. Int. J. Biol. Macromol. 89, 25–34.

Yang, Y.C., Liu, X., Wan, C., Sun, S., Lee, D.J., 2014. Accelerated aerobic granulation using alternating feed loadings: Alginate-like exopolysaccharides. Biores. Technol. 171, 360–366.

Yatipanthalawa, B.S., Ashokkumar, M., Scales, P.J., Martin, G.J., 2022. Ultrasound-assisted extracellular polymeric substance removal from the Diatom *Navicula* sp.: A route to functional polysaccharides and more efficient algal biorefineries. ACS Sust. Chem. Eng. 10(5), 1795–1804.

Yoshida, Y., Tomiyama, T., Maruta, T., Tomita, M., Ishikawa, T., Arakawa, K., 2016. De novo assembly and comparative transcriptome analysis of *Euglena gracilis* in response to anaerobic conditions. BMC Genom. 17(1), 1–10.

Yuksekdag, Z.N., Aslim, B., 2008. Influence of different carbon sources on exopolysaccharide production by *Lactobacillus delbrueckii* subsp. *bulgaricus* (B3, G12) and *Streptococcus thermophilus* (W22). Braz. Arch. Biol. Technol. 51, 581–585.

Zannini, E., Waters, D.M., Coffey, A., Arendt, E.K., 20165. Production, properties, and industrial food application of lactic acid bacteria-derived exopolysaccharides. Appl. Microbiol. Biotechnol. 100, 1121–1135.

Zayas, J.F., 1997. *Functionality of Proteins in Food*. Springer.

Zeng, H., Zhang, Y., Lin, S., Jian, Y., Miao, S., Zheng, B., 2015. Ultrasonic-microwave synergistic extraction (UMSE) and molecular weight distribution of polysaccharides from *Fortunella margarita* (Lour.) Swingle. Sep. Puri. Technol. 144, 97–106.

Zhang, J., Cao, Y., Wang, J., Guo, X., Zheng, Y., Zhao, W., Yang, Z., 2016. Physicochemical characteristics and bioactivities of the exopolysaccharide and its sulfated polymer from *Streptococcus thermophilus* GST-6. Carbohydr. Polym. 146, 368–375.

Zhang, J., Ding, X., Yang, L., Kong, Z., 2006. A serum-free medium for colony growth and hyaluronic acid production by *Streptococcus zooepidemicus* NJUST01. Appl. Microbiol. Biotechnol. 72(1), 168–172.

Zhang, J., Liu, L., Chen, F., 2019. Production and characterization of exopolysaccharides from *Chlorella zofingiensis* and *Chlorella vulgaris* with anti-colorectal cancer activity. Int. J. Biol. Macromol. 134, 976–983.

Zhang, J., Wen, C., Qin, W., Qin, P., Zhang, H., Duan, Y., 2018. Ultrasonic-enhanced subcritical water extraction of polysaccharides by two steps and its characterization from *Lentinus edodes*. Int. J. Biol. Macromol. 118, 2269–2277.

Zhang, J.B., Wu, N.N., Yang, X.Q., He, X.T., Wang, L.J., 2012. Improvement of emulsifying properties of Maillard reaction products from β-conglycinin and dextran using controlled enzymatic hydrolysis. Food Hydrocoll. 28(2), 301–312.

Zhang, X., Li, Z., Pang, S., Jiang, B., Yang, Y., Duan, Q., Zhu, G., 2021. The impact of cell structure, metabolism and group behavior for the survival of bacteria under stress conditions. Arch. Microbiol. 2, 431–441.

Zhang, Z., Lv, G., He, W., Shi, L., Pan, H., Fan, L., 2014. Effects of extraction methods on the antioxidant activities of polysaccharides obtained from *Flammulinavelutipes*. Carbohydr. Polym. 98(2), 1524–1531.

Zheng, X.T., Wu, D.M., Meng, Q.Y., Wang, K.J., Liu, Y., Wan, L., Ren, D.Y., 2008. Mechanical properties of low density polyethylene/nanomagnesium hydroxide composites prepared by an in situ bubble stretching method. J. Polym. Res. 15, 59–65.

Zhou, S., Wu, X., Huang, Y., Xie, X., Lin, Y., Fan, H., et al., 2018. Microwave-assisted aqueous two-phase extraction of alkaloids from Radix *Sophorae Tonkinensis* with an ethanol/Na2HPO4 system: Process optimization, composition identification and quantification analysis. Ind. Crops. Prod. 122, 316–328.

Zia, K.M., Tabasum, S., Khan, M.F., Akram, N., Akhter, N., Noreen, A., Zuber, M., 2018. Recent trends on gellan gum blends with natural and synthetic polymers: A review. Int. J. Biol. Macromol. 109, 1068–1087.

Zisu, B., Shah, N.P., 2003. Effects of PH, temperature, supplementation with whey protein concentrate, and adjunct cultures on the production of exopolysaccharides by *Streptococcus thermophilus* 1275. J. Dairy Sci. 86, 3405–3415.

4 Molecular biology of exopolysaccharides production

Monika Thakur, Amit Kumar Rai, and Sudhir P. Singh

4.1 INTRODUCTION

Exopolysaccharides (EPS) are large molecular weight carbohydrate macromolecules that are synthesized naturally by a wide range of microorganisms, mostly to thrive in harsh conditions. EPS are essential for defending against abiotic stressors, antimicrobials, phagocytosis, cell recognition, viral invasion, and desiccation (Kavitake et al., 2022). The microbial EPSs serve as alternative biomaterials and have a wide range of commercial uses, spanning cosmetology, pharmaceuticals, secondary processing, fabrics, cleaners, lubricants, mining, sewage treatment, and dietary supplements. Levan, dextran, inulin, pullulan, glucan, galactan, and mannan are a few of the polysaccharides that some notable bacterial species are renowned for producing. The complex and diverse chemical structure confers specialized physicochemical and rheological properties to the EPS produced by bacteria (Ates, 2015; Kirtel et al., 2017). This variation is represented in the glycosyl transferases, branching enzymes, polymerases, and other enzymes involved in the insertion of substituents or alteration of the sugar moieties in the corresponding polymers that are produced during biosynthesis. These include *Pediococcus, Lactobacillus, Lactococcus, Weissella, Enterococcus, Streptococcus, Leuconostoc,* etc. (Wu et al., 2021; Sharma et al., 2019; Jadaun et al., 2019; Soumya and Nampoothiri, 2021).

Microbial biopolymers are preferable over chemical polymers, mainly in biomedical usage, and processability, and are environment friendly (Keshavarz and Roy, 2010; Sharma et al., 2021). Bacterial EPS are cost-competitive to that of plant and algal-origin, which are influenced by different climatic and geographical external factors. The industrial processing of microbial EPS production has bright scope for a quick and high yield under controlled conditions (Kaur et al., 2014). Although there is an increasing demand for microbial EPSs in commercial and medical settings, only a few bacterial EPSs have been successful in attaining commercial success, for example, xanthan gum and levan. The establishment of EPSs in the polymer market requires economical production costs at a pilot scale. The growing importance of EPSs in industrial and medical biotechnology necessitates in-depth studies on the biosynthetic mechanism and molecular strategy for metabolic regulation and subsequently enhancing the microbial production of polymers (Barrett et al., 2006).

Molecular biology seeks to provide thorough biochemical accounts of organisms. It has been possible to find new EPS assembly routes and comprehend the fundamentals of EPS creation via strategies involving functional genomics, genome sequencing, protein sequence analysis, and artificial intelligence (Schmid and Sieber, 2015). The macromolecular framework of a cell can be understood using a modelling strategy that integrates omics datasets with simulations of variable expression and enzyme activity (Lerman et al., 2012). To accomplish this, it is possible to comprehend and forecast the metabolic potential of microbial communities by genome-based and genome-scale metabolic reconstructions. Thus, molecular biology advances the relationship between metabolism and microbial EPS production and enhances the metabolic performance of microbial hosts for industrial applications (Ates, 2015). Figure 4.1 represents the molecular biology applications in exopolysaccharides' biosynthesis.

EPS-producing bacteria can be identified based on gene cluster encoding enzymes in the prototype genomes. In contrast to polymers like dextran and levan, biosynthesized by a single enzymatic step, EPS operons may contain 20 or more genes (e.g., welan), which correspond to the variance in polysaccharide structure (Wang et al.,

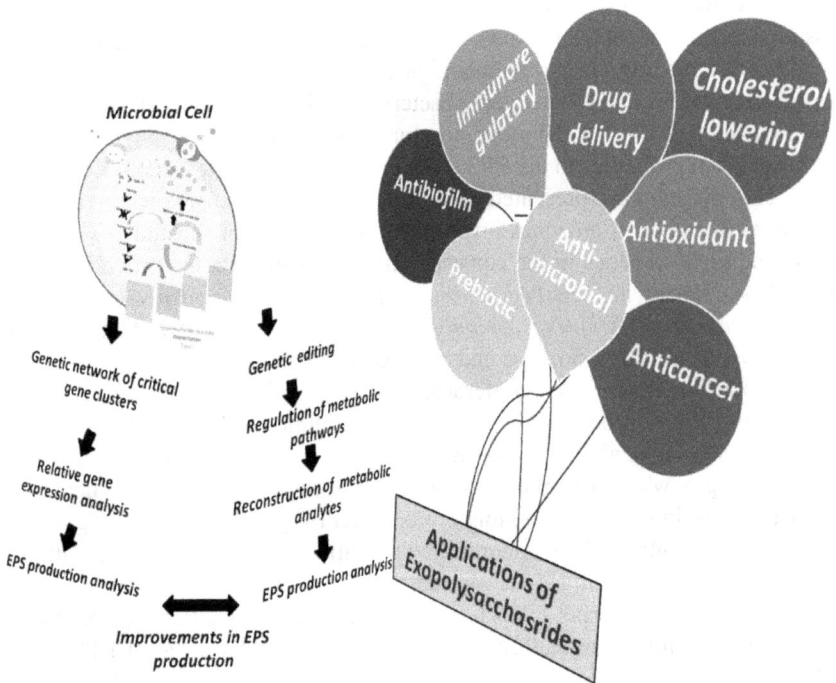

FIGURE 4.1 Molecular biology approaches for genetic improvement in the microbial production of EPS

2012; Tao et al., 2012). Despite the fact that many of the gene clusters involved in EPS production have been decoded, the mode of action and the catalytic traits associated with these genes have not yet been thoroughly characterized (Schmid and Sieber, 2015). Next-generation sequencing and artificial intelligence offer a molecular understanding of EPS biosynthetic pathways and execute the tailoring of enzymes and genome engineering for EPS production. RAST (rapid annotations using subsystems technology) and InterProScan are automated genome identification technologies that enable the simultaneous manipulation of several genomes (Aziz et al., 2008; Quevillon et al., 2005).

Currently, several bacteria have been recognized as microbial biopolymer makers that can be recovered from the fermentation media or found adhering to the cell surface. In adaptation to specific climate impacts, bacteria utilize these microbial biopolymers as storage materials. According to their biological roles, microbial polysaccharides can be broadly categorized as extracellular bacterial polysaccharides (e.g., sphingan, levan, alginate, xanthan, pullulan, cellulose, etc.), and intracellular storage polysaccharides (e.g., glycogen) (Schmid and Sieber, 2015). The capsular polysaccharides are intermediate between these two types of polysaccharides; they are extracellular yet remain attached to the cell wall, making them difficult to remove.

The majority of the discussion in this article will center on the extracellular bacterial polysaccharides that are secreted following a number of enzymatic changes inside the cell. An overview of EPS' applications is also presented in the chapter. Following that, a thorough analysis of the synthetic mechanism is given, including the most important genetic, biochemical, and structural facets of the catalysts involved. The studies of parameter optimization for different EPS biosynthesis by the intact bacteria or the enzyme, in particular, are covered in this chapter. The biosynthetic processes for the high-value EPSs like curdlan, alternan, levan, pullulan, and dextran have been explained.

4.2 ENZYMATIC SYNTHESIS OF EXOPOLYSACCHARIDES

The majority of genes involved in producing polysaccharide precursors share substantial similarity at the nucleotide and amino acid levels (Thorne et al., 2000). The EPS synthesis-related enzymes have been largely classified into four groups. Hexokinase, which phosphorylates glucose (Glc) to glucose-6-phosphate (Glc-6-P), belongs to the first class of enzymes. The second group is necessary to catalyze the transformation of sugar nucleotides. The enzyme uridine-5′diphosphate (UDP)-glucose pyrophosphorylase transforms Glc-1-P into UDP-Glc and is one of the key biocatalysts in the synthesis of EPS. Glycosyltransferases (GTFs), which are generally present in the periplasmic membrane of cells, transfer the sugar residue from a donor molecule to a saccharide or non-saccharide acceptor molecule. More than 94 GTF families were identified in the Carbohydrate-Active EnZymes (CAZy) database (www.cazy.org). The final class, which exists outside of the cell wall, is involved in the polymerization of the biomolecules (Kumar et al., 2007).

Inside the cell, the precursor molecules are converted into activated sugars or sugar acids, which move essentially via three different ways, for example, ATP hydrolysis coupled to ATPase sugar translocation, coupled to the movement of ions and other solutes, and another system is phosphoenolpyruvate (PEP) transport system (Schmid and Sieber, 2015; Cescutti et al., 2010). GTFs connect activated sugars to a lipid carrier in a particular order since a repeating unit incorporates via Wzy protein (Freitas et al., 2011). A different method involves the extension of the polymer strand in the extracellular synthesis route by adding monosaccharides generated by the cleavage of di- or trisaccharides (Figure 4.2).

Most of the bacterial EPSs are formed intracellularly and then exported to the extracellular space; however, some homopolysaccharides, such as dextran, levan, and mutan, are produced outside of the cells by the activity of secreted enzymes that further convert these substrates into the polymer (Ates, 2015; Sharma et al., 2016). The biosynthesis of EPSs such as heteropolysaccharides (glycosaaminoglycans, agar,

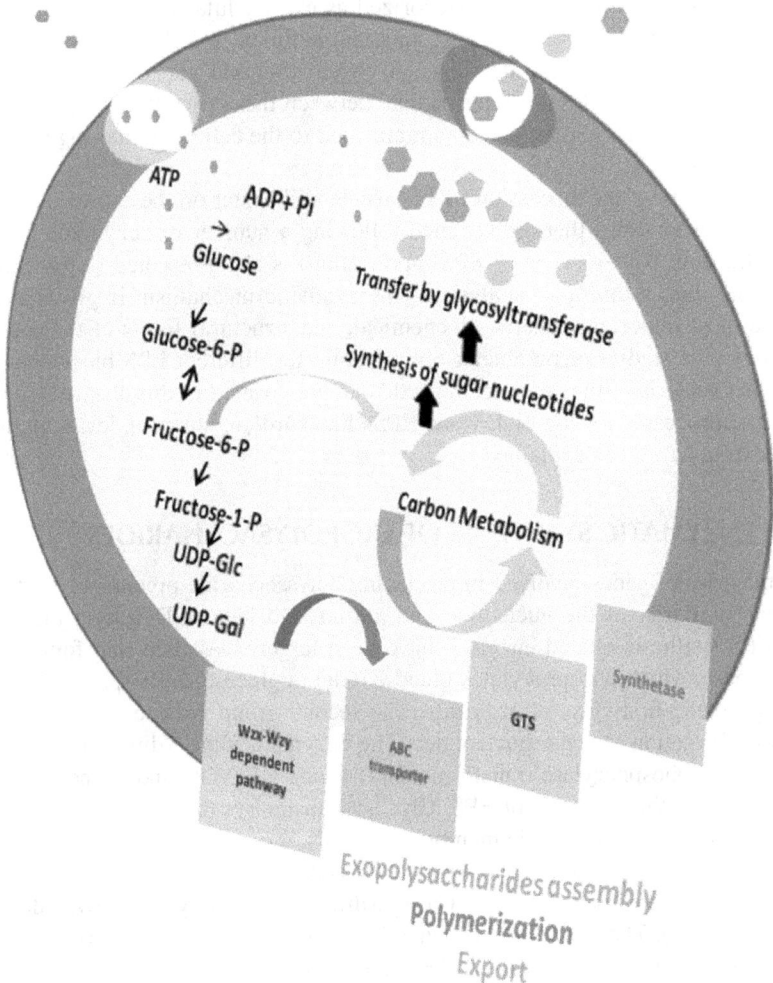

FIGURE 4.2 Metabolic pathway for the production of microbial EPS

mucopolysaccharides, pectins, and peptidoglycan) and homopolysaccharides occur in two different ways. GTFs catalyze heteropolysaccharides biosynthesis in intercellular space, and the only step of polymerization occurs in the extracellular environment. The first step of sugar uptake is driven via an active transport pathway, which is further catabolized in the cytoplasm through glycolysis, resulting in the formation of sugar nucleotides. The phosphorylation of sugars results in the biosynthesis of activated precursors that connects to lipid carriers and finally secretes into the extracellular environment (Boels et al., 2001; Freitas et al., 2011). Homopolysaccharides (dextran, levan, and inulin) are synthesized in the extracellular environment via different approaches. This includes the glycansucrase class that utilizes sucrose as a donor substrate (Yu et al., 2022). The glycansucrase's catalytic system executes the breakdown of sucrose to produce a glucosyl-enzyme precursor. In the case of hydrolytic action, water serves as a receptor and hydrolyzes to glucose; whereas in transglycosylation the glucosyl moiety acts as accepting sugar, and fructose is released. Sucrose treated with glycansucrase generates the polymer dextran (Robyt et al., 2008). The degree of homopolysaccharide branching is impacted by the conformation of the enzyme. Different types of glucansucrases have been characterized, forming different glycosidic linkage types. Dextransucrase catalyzes polymer-containing sugar units linked via α-1, 6 linkage, and a few α-1, 2, α-1, 3, and α-1, 6 branches. Mutansucrases catalyzes the synthesis of mutan containing α-1, 3 linked residues. The polymer alternan, synthesized by alternansucrase, contains alternate α-1, 6 and α-1, 3 glucosidic linkages. Reuteransucrase produces reuteran containing α 1, 4 and α 1, 6 linkages (van Hijum et al., 2006). Transfructosidases like levansucrase and inulosucrase produce fructans, levan, and inulin, respectively. Fructosyltransferase catalyzes the hydrolysis of sucrose and transfers fructosyl onto fructans polymerized chains (Jadaun et al., 2019).

4.3   DEXTRAN

Dextrans are homopolysaccharides of glucose with a significant proportion of consecutive α-1, 6-linkages in their main chains. D-glucans have side chains with branching connections of type (α-I, 3), and occasionally linkages of type (α-I, 4) or (α-I, 2), or both. Dextran produced from dextransucrase(s) from different microbial sources may have a slightly different structure based on the proportion of glycosyl linkage types (Sharma et al., 2019). Clinical dextran, also known as blood-plasma dextran, is 40,000 to 100,000 Da (Robyt et al., 2008). Dextran with a low molecular weight is useful in the treatment of catastrophic damage. Dextran with molecular weights of 70, 60, and 40 KDa is offered for replenishing mild blood losses (Nessens et al., 2005).

High-quality dextran for clinical usage is made by *L. mesenteroides* B-512F dextransucrase because it contains a high percentage (~95%) of (α-1, 6)-glycosidic linkages (Robyt et al., 2008; de Belder, 1996). Dextran with a high percentage of α-1, 6 linkages possess low antigenicity, high water solubility, and high biological stability in the human bloodstream that makes it acceptable for use as a blood-plasma substitute. Moreover, clinical dextran could be used in a disaster to replace blood plasma and save lives. High molecular weight dextrans with low percentages of

(α 1, 6)-linkages in dextrans are associated with a higher prevalence of anaphylactic attack (de Belder, 1996).

The multi-domain dextransucrase has been characterized from *Leuconostoc, Streptococcus,* and *Lactobacillus* species. Dextransucrase from these sources is responsible for the vast majority of dextrans production, utilizing sucrose as a substrate (Bounaix et al., 2010; Naessens et al., 2005; Takagi et al., 1994). Dextran dextrinase is a transglucosidase found in *Gluconobacter* strains, useful in metabolizing maltodextrin as a substrate for dextrans production. The dextran generated by *L. mesenteroides* NRRL B512F dextransucrase has drawn the greatest attention and is sufficient to produce it at the commercial level. This dextransucrase generates a high molecular weight, soluble dextran that is released in relatively significant amounts into the culture supernatant with the least amount of associated interfering enzymes, such as levansucrase and invertase.

Dextransucrase, in the lack of an acceptor molecule, translocates d-glucopyranosyl units from sucrose to an expanding glucan chain, resulting in the synthesis of the polymer dextran (Figure 4.3). However, with the advent of an acceptor molecule such as maltose or isomaltose, dextransucrase redirects its catalytic activity towards the production of oligosaccharides, producing significantly less dextran in the process (Sharma et al., 2019).

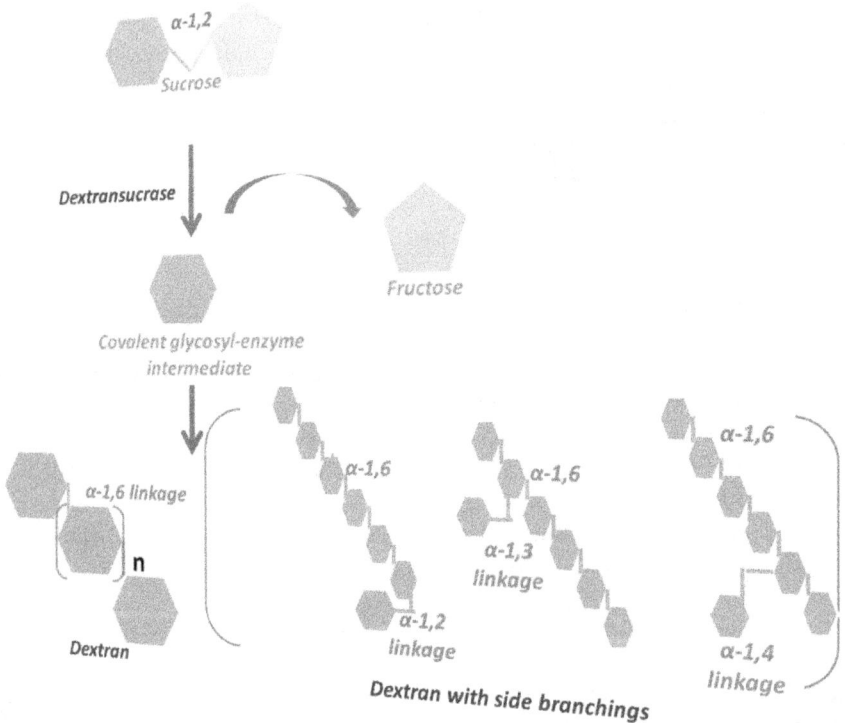

FIGURE 4.3   Catalytic function of dextransucrase towards sucrose and dextran synthesis

4.4 LEVAN

Levan is a fructose homopolysaccharide of commercial significance. It has multifarious uses in the healthcare, cosmetic, pharmaceutical, and food industries. Even with such a wide range of applications, its availability is limited due to the expensive process of production. Levan is comprised of multiple fructose units linked via β-2, 6 glycosidic linkages, with β-2, 1 branching point (de Siqueira et al., 2023). The molecular weight and degree of branching of levans may vary depending on the microbial strains. However, due to the numerous branches, bacterial levans have substantially greater molecular weights than levans of plant origin. The chemical synthesis of levans is possible, however, commercially it is produced via microbial fermentation or with the aid of microbial enzymes. Chemical synthesis of levan is neither economical nor environment friendly, whereas fermentation broth extraction gives poor yields and difficult separation (Srikanth et al., 2015). Industrially, the enzymatic and fermentation approaches are more promising for levan production. Levansucrase (EC 2.4.1.10) polymerizes fructose obtained from sucrose via a transglycosylation reaction. The polymerization is affected by sucrose availability, pH, temperature, metal ions, and the kind of reactor (Li et al., 2015).

Bacterial strains from a variety of species have been characterized for levan production in the extracellular matrix, for example, *Streptococcus Acetobacter, Bacillus, Azotobacter, Erwinia, Gluconobacter, Mycobacterium, Pseudomonas, Corynebacterium, Zymomonas*, and *Aerobacter* (Sarilmiser et al., 2015). Novel levansucrase genes can be identified by exploring diverse habitats at the genomic level. A levansucrase gene from a soil bacterium (*B. subtilis*) was found strictly regulated by a sucrose-inducible antitermination mechanism (Crutz et al., 1990). Overall, sucrose is necessary to activate the transcription levansucrase gene in the bacterium, including *Bacillus, Paenibacillus*, and *Geobacillus*. The genes encoding levansucrase can either be found in an operon region or at different loci in a number of bacteria. The stringent regulation of levanase transcription ensures levan biosynthesis over levan degradation. Levansucrase is adhered to the surface of the cell or discharged into the surrounding media. The majority of gram-negatives bacteria secretes levansucrase by a signal-peptide-independent pathway while gram-positive bacteria secrete levansucrase by involving cleavage of precursor molecules that include signal peptides (Öner et al., 2016). Levansucrase works on sucrose catalyzing simultaneous transfructosylation and hydrolysis reactions. In a study, levansucrase gene (MH684490) from *Leu. mesenteroides* MTCC 10508 was cloned and expressed in a heterologous host, and the protein was characterized for levan biosynthesis (Jadaun et al., 2019). Other genes had also been characterized for levan biosynthesis from sucrose by levansucrase catalytic activity, for example, *Leu. mesenteroides* NRRL B-512 F (AAY19523.1) and *Leu. Mesenteroides* Lm 17(ALF07532.2). However, the size of the levan produced by levansucrases might vary. Bacterial levans with a high degree of polymerization are synthesized via a processive mechanism in which the fructan chain remains bound to the enzyme (Ozimek et al., 2006). In case the substrate, sucrose, is exhausted, the fructan polymer acts as the fructosyl donor. Levansucrase

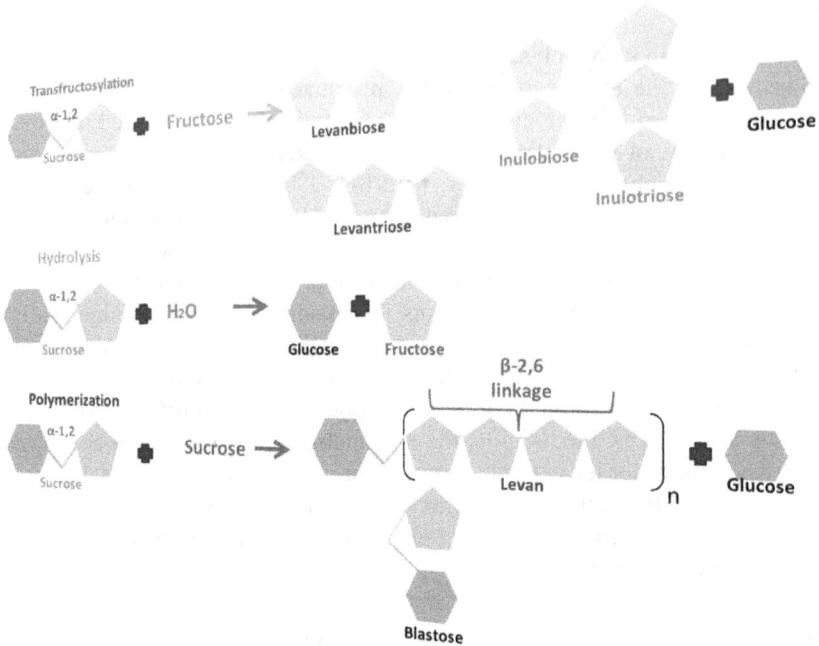

FIGURE 4.4 Levansucrase catalysis for the synthesis of levan and fructooligosaccharides

catalyzes the cleavage of β-(2, 6) links in exo-type fashion, triggering the release of the terminal fructose one at a time until the reaction ceases at the branching point (Öner et al., 2016). In this way, the branched levan appears to be protected from extensive hydrolysis by levansucrase. Levan polymer has been demonstrated to be useful in the preparation of films for wrapping and packaging applications (Agarwal et al., 2023).

Levansucrase is the enzyme that facilitates three catalytic reactions, polymerization by using a fructan molecule as acceptor, transfructosylation by utilizing mono- or di- or oligo-saccharide as an acceptor, and hydrolysis when a water molecule is used as an acceptor (Figure 4.4). Hydrolysis tends to take over when sucrose starts to deplete in the process. The incorporation of organic solvents and NaCl creates a water-restricted condition that favors transfructosylation over hydrolysis. The ideal temperature for sucrose hydrolysis is around 50–60°C; however fructans, that is, FOS or levan generation are preferred at lower temperatures (10–40°C) (Siqueira et al., 2023).

4.5 PULLULAN

Pullulan is a branched polymer of maltotriose units joined by α-1, 6-glycosidic linkages. The generation of pullulan by the black yeast-like fungus, *Aureobasidium* spp,

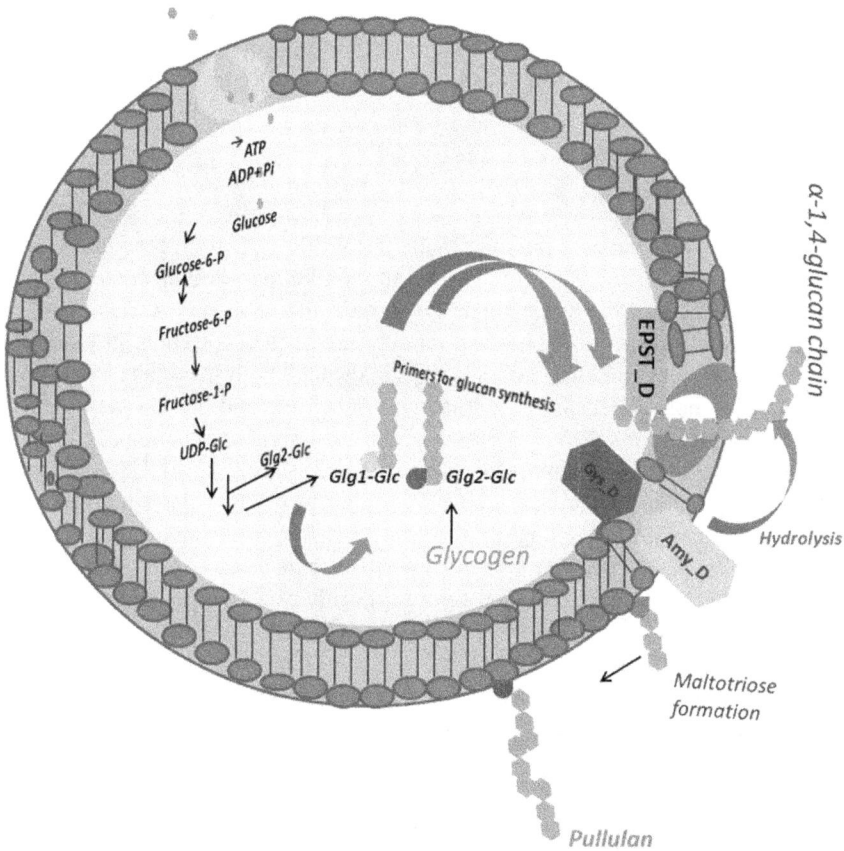

FIGURE 4.5 Metabolic pathway for pullulan production

is well-recognized, and it has a lot of promise to be employed in several biomedical research fields (Wei et al., 2021). Several strains of *Aureobasidium* spp. have been recognized that utilize a variety of substrates to produce pullulan, for example, glucose, sucrose, inulin, xylose, and agro-industrial wastes. Pullulan is currently used in the food and pharmaceutical sectors. Pullulan is non-mutagenic, non-carcinogenic, non-immunogenic, and nontoxic. The biomedical applications of pullulan include targeted drug therapy, perfusion, tissue engineering, wound healing, and lymph node target-specific imaging. It is also used in diagnostic applications, such as vascular compartment imaging (Prajapati et al., 2013).

Chen et al. (2020) have characterized a multidomain α-glucan synthetase 2 (AmAgs2) from *A. melanogenum* P16 for pullanan synthesis. Molecular analysis detected amylase domain (Amy_D), glycogen synthetase domain (Gys_D), and exopolysaccharide transporter domain (EPST_D) in AmAgs2 protein. Gys_D is a key domain in the construction of pullulan precursors, that is, long α-1,4-glucosyl chains (Figure 4.5). The precursors are transported to the transmembrane region by the action of EPST_D. Amy_D catalyzes the hydrolysis of endo-α-1,4-linkages in the precursor molecules, releasing maltotriose followed by its utilization in pullulan biosynthesis via α-1,6 glucosidic linkages among maltotriose units (Chen et al., 2018).

4.6 CURDLAN GUM

Curdlan is a water-insoluble polymer made up of 300 to 500 glucose units linked by β -1, 3-glycosidic linkages. It is soluble in an alkaline solution. It has a crucial function in cell signalling and cellular synchronization among bacterial communities (Zhang and Edgar, 2014). It has a unique three-helix molecular structure that allows macrophages and other immune cells to detect it with specificity, resulting in an immune response. Because of its distinct viscoelastic and thermo-gelling qualities, it is frequently employed in culinary and medicinal applications (Jindal and Khattar, 2018). This FDA approved polymer is widely used in the US and Europe as a food ingredient (Chen et al., 2010). Curdlan production has been characterized by microbial strains belonging to *A. faecalis, Rhizobium* sp., *Bacillus cereus, Gluconacetobacter xylinus, Agrobacterium* sp. (Verma et al., 2020). In reaction to stressors in the atmosphere, bacterial strains typically generate curdlan (Zhang et al., 2010, 2011). Nitrogen depletion serves as one of the most crucial environmental cues for curdlan biosynthesis.

Curdlan is biosynthesized by the polymerization of UDP-glucose, involving several enzymes. The three main components in curdlan biosynthesis are precursor synthesis, energy generation, and polymerization (Zhan et al., 2012). The primary component of curdlan, glucose, is phosphorylated to create glucose-6-phosphate (Glc-6-P). Phosphomutase partially transforms glucose-6-phosphate (Glc-6-P) into glucose-1- phosphate (Glc-1-P) as it passes through the TCA cycle and respiratory chain to be used in the production of ATP (EC 5.4.2.2). Uridine triphosphate (UTP) and Glc-1-P are then combined by UDP-Glc pyrophosphorylase (EC 2.7.7.9) to create UDP-Glc, which is then transported to the cell membrane by phospholipids (Liang et al., 2017). The curdlan synthetase catalyzes aggregation of activated glucose moieties from UDP-Glc to curdlan via β-1, 3-glucosidic linkages (Figure 4.6). Curdlan synthetase are massive transmembrane proteins, the biosynthesis of which requires a significant amount of energy and reducing power (Zhang et al., 2020). Thus, curdlan biosynthesis is influenced by the availability of nutrients (Karnezis et al., 2003). The ATP transporter-dependent pathway, the Wzx/Wzy-dependent pathway, the sucrose synthase extracellular synthesis method, and the synthase-dependent pathway are common processes for carbohydrate biosynthesis (Schmid and Sieber, 2015). The biosynthesis of heteropolysaccharides uses the Wzx/Wzy-dependent pathway (Schilling et al., 2020). In the outer cell membrane, the sucrose synthase pathway polymerizes monomers to create homopolymers or oligosaccharides. Homopolymers like curdlan, which have a polymeric transport channel that spans the entire cell membrane, are frequently created using synthetase-dependent mechanisms (Schmid, 2018). The genetic engineering efforts to increase the chain length and yield of curdlan include silencing of β-1,3-glucanase genes (*exoK, and exsH*) in *Agrobacterium* sp. CGMCC

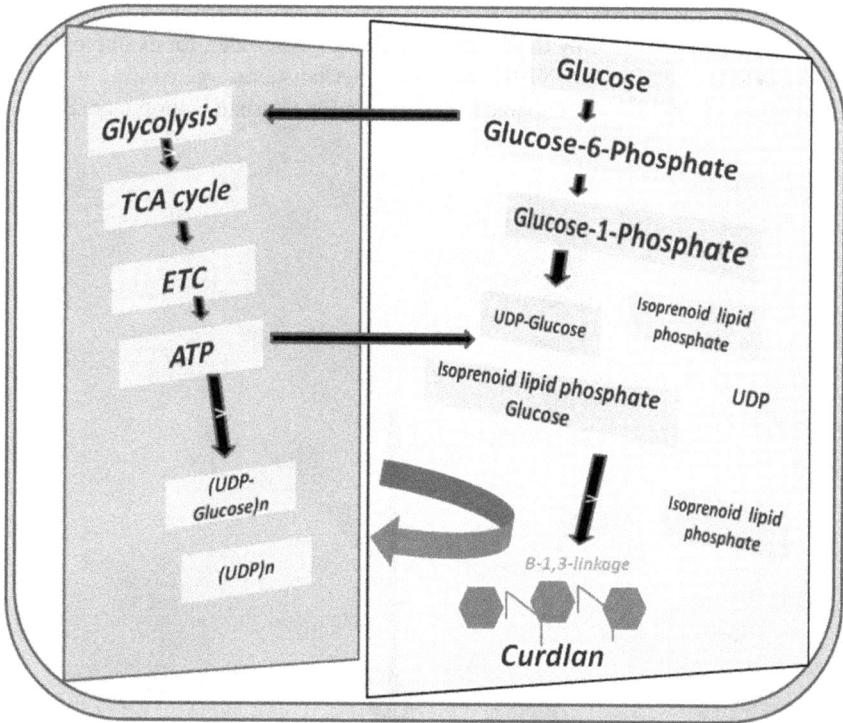

FIGURE 4.6 Metabolic pathway for curdlan production

11546 (Gao et al., 2020). Furthermore, the flux of precursors can be increased by overexpression of genes related to UDP-Glc synthesis pathways, such as *galU*, *exoC*, and *glK*, and UDP synthesis pathway (Yuan et al., 2021).

4.7 ALTERNAN

Alternan is a water-insoluble polymer containing alternate α-1, 3 and α-1, 6 linked d-glucose residues, with 46% and 54% abundance, respectively. A portion of the additional 3, 6-di-substituted d-glucosyl residues is also reported in its structure, which accounts for 7 to 11% abundance in the polymer. The unusual structural attributes of alternan provide it with the characteristics of low viscosity, good solubility, and resistance against microbial and mammalian enzymes (Saadat et al., 2019). In food and cosmetics, it can be used as a low-viscosity bulking agent and extender. It may also be useful as a non-caloric, carbohydrate-based dietary ingredient in foods that have been artificially sweetened. Alternan may also be used in place of malto-dextrins, polydextrose, and gum arabic as bulking agents, soluble fibres bulking agents, and binders, in food as well as inks, cosmetic creams, adhesives, and ointments (Monsan et al., 2001; Park and Khan, 2009).

Alternansucrase is the key enzyme for the biosynthesis of alternan. Alternansucrase is reported to be produced by three strains of *L. mesenteroides*, for example, NRRL B-1355, NRRL B-1501, and NRRL B-1498. Alternansucrase-generating wild-type *Leuconostoc* strains are also capable of generating the alternan biopolymer (Bounaix

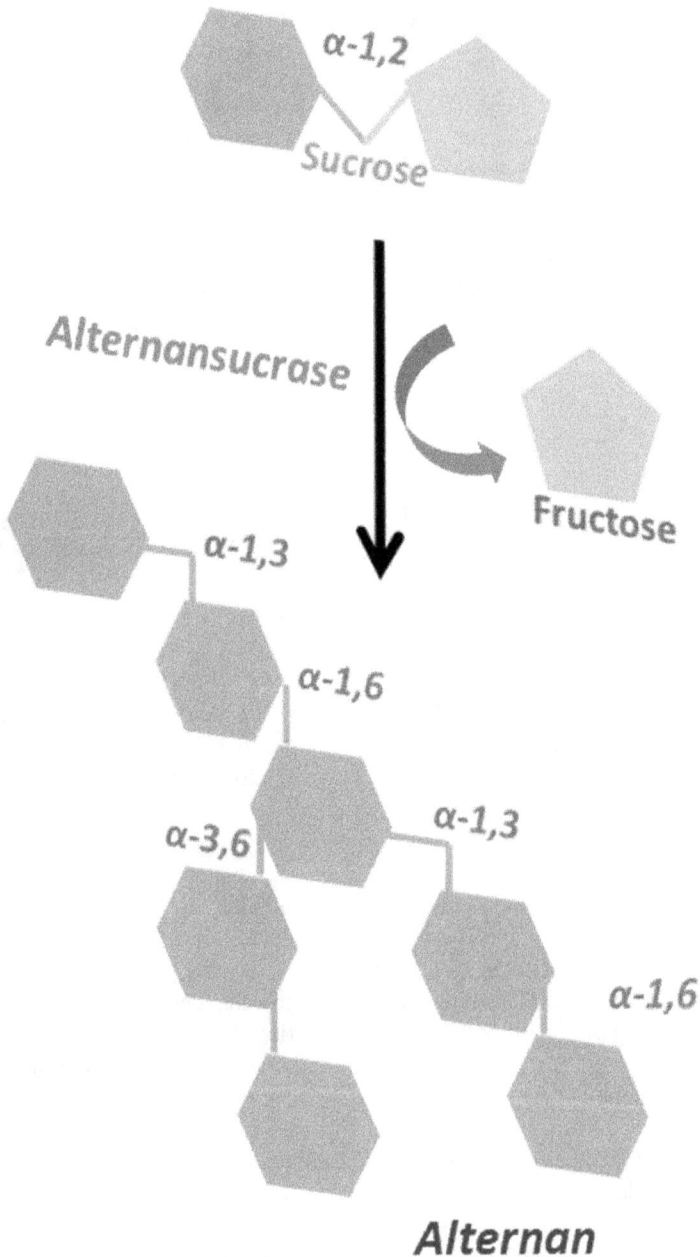

FIGURE 4.7 Catalytic biosynthesis of alternan by alternansucrase

et al., 2009, 2010). The conversion of D-glucopyronasyl residue from sucrose to alternan is catalyzed by alternansucrase, and fructosyl moiety is liberated (Figure 4.7). The analysis of the crystal structure of alternansucrase revealed that alternansucrase has a more complex active site than glucansucrases and dextransucrase. Moreover, alternansucrase has two acceptor binding sites responsible for the synthesis of polymer with alternately linked glucan (Wangpaiboon et al., 2021).

4.8 CONCLUSION

Systems-based approaches are required to manage and optimize EPS production in order to increase the yield. The majority of bacterial EPS of distinctive features have high production costs, and economic challenges related to yield and quality need to be solved. The fundamental knowledge about EPS biosynthesis is crucial to designing molecular strategies aims at the biosynthesis of high-quality polymers, with high yield and reduced cost of production. In-depth genomic information related to the biosynthetic pathway is crucial for designing innovative strains in the manufacture of microbial EPS. The engineered strains have economic significance since the fermentation medium may be simply supplied with the appropriate substrate to conveniently manage the product's spectrum. The EPS' qualities by changing their structure and chain length could be made possible by biocatalyst designing and recruitment of improved enzyme systems. A comprehensive understanding of the genomes of the bacteria producing EPS is critical in metabolic engineering for achieving a better yield. Artificial intelligence and machine learning offer a greater scope of gene editing and genome engineering for precise strain improvement.

ACKNOWLEDGMENTS

Authors acknowledge the Department of Biotechnology (DBT), Govt. of India, for all support.

REFERENCES

Agarwal, N., Jyoti, T.M., Mishra, M.M., Singh, S.P., 2023. Preparation and characterization of biodegradable films based on levan polysaccharide blended with gellan gum. Environ. Tech. Innov. 31, 103231.

Ates, O., 2015. Systems biology of microbial exopolysaccharides production. Front. Bioeng. Biotechnol. 3, 200.

Aziz, R.K., Bartels, D., Best, A.A., DeJongh, M., Disz, T., Edwards, R.A., Formsma, K., Gerdes, S., Glass, E.M., Kubal, M., Meyer, F., 2008. The RAST server: Rapid annotations using subsystems technology. BMC Genom. 9(1), 1–15.

Barrett, C.L., Kim, T.Y., Kim, H.U., Palsson, B.Ø., Lee, S.Y., 2006. Systems biology as a foundation for genome-scale synthetic biology. Curr. Opin. Biotechnol. 17(5), 488–492.

Boels, I.C., van Kranenburg, R., Hugenholtz, J., Kleerebezem, M., De Vos, W.M., 2001. Sugar catabolism and its impact on the biosynthesis and engineering of exopolysaccharide production in lactic acid bacteria. Int. Dairy J. 11(9), 723–732.

Bounaix, M.S., Gabriel, V., Morel, S., Robert, H., Rabier, P., Remaud-Simeon, M., Gabriel, B., Fontagne-Faucher, C., 2009. Biodiversity of exopolysaccharides produced from sucrose by sourdough lactic acid bacteria. J. Agric. Food Chem. 57(22), 10889–10897.

Bounaix, M.S., Gabriel, V., Robert, H., Morel, S., Remaud-Siméon, M., Gabriel, B., Fontagné-Faucher, C., 2010. Characterization of glucan-producing *Leuconostoc* strains isolated from sourdough. Int. J. Food Microbiol. 144(1), 1–9.

Bounaix, M.S., Robert, H., Gabriel, V., Morel, S., Remaud-Siméon, M., Gabriel, B., Fontagné-Faucher, C., 2010. Characterization of dextran-producing *Weissella* strains isolated from sourdoughs and evidence of constitutive dextransucrase expression. FEMS Microbial. Lett. 311(1), 18–26.

Cescutti, P., Foschiatti, M., Furlanis, L., Lagatolla, C., Rizzo, R., 2010. Isolation and characterisation of the biological repeating unit of cepacian, the exopolysaccharide produced by bacteria of the *Burkholderia cepacia* complex. Carbohydr. Res. 345(10), 1455–1460.

Chen, C., Wang, R., Sun, G., Fang, H., Ma, D., Yi, S., 2010. Effects of high pressure level and holding time on properties of duck muscle gels containing 1% curdlan. Innov. Food Sci. Emerg. Technol. 11(4), 538–542.

Chen, T.J., Chi, Z., Jiang, H., Liu, G.L., Hu, Z., Chi, Z.M., 2018. Cell wall integrity is required for pullulan biosynthesis and glycogen accumulation in *Aureobasidium melanogenum* P16. Biochim. Biophys. Acta Gen. Subj. 1862(6), 1516–1526.

Chen, T.J., Liu, G.L., Wei, X., Wang, K., Hu, Z., Chi, Z., Chi, Z.M., 2020. A multidomain α-glucan synthetase 2 (AmAgs2) is the key enzyme for pullulan biosynthesis in *Aureobasidium melanogenum* P16. Int. J. Biol. Macromol. 150, 1037–1045.

Crutz, A.M., Steinmetz, M., Aymerich, S., Richter, R., Le Coq, D., 1990. Induction of levansucrase in *Bacillus subtilis*: An antitermination mechanism negatively controlled by the phosphotransferase system. J. Bacterial. 172(2), 1043–1050.

de Belder, D., 1996. *Polysaccharides in Medicinal Applications*. Marcel Dekker, pp. 505–524.

de Siqueira, E.C., Öner, E.T., 2023. Co-production of levan with other high-value bioproducts: A review. Int. J. Biol. Macromol. 123800.

Freitas, F., Alves, V.D., Reis, M.A., 2011. Advances in bacterial exopolysaccharides: From production to biotechnological applications. Trends Biotechnol. 29(8), 388–398.

Gao, H., Xie, F., Zhang, W., Tian, J., Zou, C., Jia, C., Jin, M., Huang, J., Chang, Z., Yang, X., Jiang, D., 2020. Characterization and improvement of curdlan produced by a high-yield mutant of *Agrobacterium* sp. ATCC 31749 based on whole-genome analysis. Carbohydr. Polym. 245, 116486.

Jadaun, J.S., Narnoliya, L.K., Agarwal, N., Singh, S.P., 2019. Catalytic biosynthesis of levan and short-chain fructooligosaccharides from sucrose-containing feedstocks by employing the levansucrase from *Leuconostoc mesenteroides* MTCC10508. Int. J. Biol. Macromol. 127, 486–495.

Jindal, N., Khattar, J.S., 2018. Microbial polysaccharides in food industry. In: *Biopolymers for Food Design*. Academic Press, pp. 95–123.

Karnezis, T., Epa, V.C., Stone, B.A., Stanisich, V.A., 2003. Topological characterization of an inner membrane $(1 \rightarrow 3)$-β-D-glucan (curdlan) synthase from *Agrobacterium* sp. strain ATCC31749. Glycobiology. 13(10), 693–706.

Kaur, V., Bera, M.B., Panesar, P.S., Kumar, H., Kennedy, J.F., 2014. Welan gum: Microbial production, characterization, and applications. Int. J. Biol. Macromol. 65, 454–461.

Kavitake, D., Devi, P.B., Delattre, C., Reddy, G.B., Shetty, P.H., 2022. Exopolysaccharides produced by Enterococcus genus—An overview. Int. J. Biol. Macromol. 226, 111–120.

Keshavarz, T., Roy, I., 2010. Polyhydroxyalkanoates: Bioplastics with a green agenda. Curr. Opin. Microbial. 13(3), 321–326.

Kırtel, O., Avşar, G., Erkorkmaz, B.A., Öner, E.T., 2017. Microbial polysaccharides as food ingredients. In: Alexandru Grumezescu, Alina Maria Holban (Eds.), *Microbial Production of Food Ingredients and Additives.* Academic Press, pp. 347–383.

Kumar, S.A., Mody, K., Jha, B., 2007. Bacterial exopolysaccharides—a perception. J. Basic Microbial. 47(2), 103–117.

Lerman, J.A., Hyduke, D.R., Latif, H., Portnoy, V.A., Lewis, N.E., Orth, J.D., Schrimpe-Rutledge, A.C., Smith, R.D., Adkins, J.N., Zengler, K., Palsson, B.O., 2012. In silico method for modelling metabolism and gene product expression at genome scale. Nat. Commun. 3(1), 929.

Li, W., Yu, S., Zhang, T., Jiang, B., Mu, W., 2015. Recent novel applications of levansucrases. Appl. Microbial. Biotechnol. 99, 6959–6969.

Liang, Y., Zhu, L., Ding, H., Gao, M., Zheng, Z., Wu, J., Zhan, X., 2017. Enhanced production of curdlan by coupled fermentation system of *Agrobacterium* sp. ATCC 31749 and Trichoderma harzianum GIM 3.442. Carbohydr. Polym. 157, 1687–1694.

Monsan, P., Bozonnet, S., Albenne, C., Joucla, G., Willemot, R.M., Remaud-Siméon, M., 2001. Homopolysaccharides from lactic acid bacteria. Int. Dairy J. 11(9), 675–685.

Naessens, M., Cerdobbel, A.N., Soetaert, W., Vandamme, E.J., 2005. Leuconostoc dextransucrase and dextran: Production, properties and applications. J. Chem. Technol. Biotechnol Int. Res. Proc. Environ. Clean Technol. 80(8), 845–860.

Öner, E.T., Hernández, L., Combie, J., 2016. Review of levan polysaccharide: From a century of past experiences to future prospects. Biotechnol. Adv. 34(5), 827–844.

Ozimek, L.K., Kralj, S., Van der Maarel, M.J., Dijkhuizen, L., 2006. The levansucrase and inulosucrase enzymes of *Lactobacillus reuteri* 121 catalyse processive and non-processive transglycosylation reactions. Microbiology. 152(4), 1187–1196.

Park, J.K., Khan, T., 2009. Other microbial polysaccharides: Pullulan, scleroglucan, elsinan, levan, alternant, dextran. In: *Handbook of Hydrocolloids.* Woodhead Publishing, pp. 592–614.

Prajapati, V.D., Jani, G.K., Khanda, S.M., 2013. Pullulan: An exopolysaccharide and its various applications. Carbohydr. Polym. 95(1), 540–549.

Quevillon, E., Silventoinen, V., Pillai, S., Harte, N., Mulder, N., Apweiler, R., Lopez, R., 2005. InterProScan: Protein domains identifier. Nucleic Acids Res. 33(suppl_2), W116–W120.

Robyt, J.F., Yoon, S.H., Mukerjea, R., 2008. Dextransucrase and the mechanism for dextran biosynthesis. Carbohydr. Res. 343(18), 3039–3048.

Saadat, Y.R., Khosroushahi, A.Y., Gargari, B.P., 2019. A comprehensive review of anticancer, immunomodulatory and health beneficial effects of the lactic acid bacteria exopolysaccharides. Carbohydr. Polym. 217, 79–89.

Sarilmiser, H.K., Ates, O., Ozdemir, G., Arga, K.Y., Oner, E.T., 2015. Effective stimulating factors for microbial levan production by *Halomonas smyrnensis* AAD6T. J. Biosci. Bioeng. 119(4), 455–463.

Schilling, C., Badri, A., Sieber, V., Koffas, M., Schmid, J., 2020. Metabolic engineering for production of functional polysaccharides. Curr. Opin. Biotechnol. 66, 44–51.

Schmid, J., 2018. Recent insights in microbial exopolysaccharide biosynthesis and engineering strategies. Curr. Opin. Biotechnol. 53, 130–136.

Schmid, J., Sieber, V., 2015. Enzymatic transformations involved in the biosynthesis of microbial exo-polysaccharides based on the assembly of repeat units. Chembiochem. 16(8), 1141–1147.

Sharma, M., Patel, S.N., Lata, K., Singh, U., Krishania, M., Sangwan, R.S., Singh, S.P., 2016. A novel approach of integrated bioprocessing of cane molasses for production of prebiotic and functional bioproducts. Bioresour. Technol. 219, 311–318.

Sharma, M., Jadaun, J.S., Upadhyay, S.K., Singh, S.P., 2021. *Bioprospecting of Microorganism-Based Industrial Molecules.* Wiley, pp. 254–277. [Enzymatic Biosynthesis of Carbohydrate Biopolymers and Uses Thereof.]

Sharma, M., Sangwan, R.S., Khatkar, B.S., Singh, S.P., 2019. Alginate–pectin co-encapsulation of dextransucrase and dextranase for oligosaccharide production from sucrose feedstocks. Bioproc. Biosyst. Eng. 42, 1681–1693.

Soumya, M.P., Nampoothiri, K.M., 2021. An overview of functional genomics and relevance of glycosyltransferases in exopolysaccharide production by lactic acid bacteria. Int. J. Biol. Macromol. 184, 1014–1025.

Srikanth, R., Reddy, C.H.S., Siddartha, G., Ramaiah, M.J., Uppuluri, K.B., 2015. Review on production, characterization and applications of microbial levan. Carbohydr. Polym. 120, 102–114.

Takagi, K., Ioroi, R., Uchimura, T., Kozaki, M., Komagata, K., 1994. Purification and some properties of dextransucrase from *Streptococcus bovis* 148. J. Ferment. Bioeng. 77, 551–553.

Tao, F., Wang, X., Ma, C., Yang, C., Tang, H., Gai, Z., Xu, P., 2012. Genome sequence of *Xanthomonas campestris* JX, an industrially productive strain for Xanthan gum. J. Bacteriol. 194(17), 4755–4756.

Thorne, L., Mikolajczak, M.J., Armentrout, R.W., Pollock, T.J., 2000. Increasing the yield and viscosity of exopolysaccharides secreted by *Sphingomonas* by augmentation of chromosomal genes with multiple copies of cloned biosynthetic genes. J. Ind. Microbiol. Biotechnol. 25, 49–57.

van Hijum, S.A., Kralj, S., Ozimek, L.K., Dijkhuizen, L., van Geel-Schutten, I.G., 2006. Structure-function relationships of glucansucrase and fructansucrase enzymes from lactic acid bacteria. Microbiol. Mol. Biol. Rev. 70(1), 157–176.

Verma, D.K., Niamah, A.K., Patel, A.R., Thakur, M., Sandhu, K.S., Chávez-González, M.L., Shah, N., Aguilar, C.N., 2020. Chemistry and microbial sources of curdlan with potential application and safety regulations as prebiotic in food and health. Food Res. Int. 133, 109136.

Wang, X., Tao, F., Gai, Z., Tang, H., Xu, P., 2012. Genome sequence of the welan gum-producing strain *Sphingomonas* sp. ATCC 31555. J. Bacteriol. 194(21), 5989–5990.

Wangpaiboon, K., Sitthiyotha, T., Chunsrivirot, S., Charoenwongpaiboon, T., Pichyangkura, R., 2021. Unravelling regioselectivity of *Leuconostoc citreum* ABK-1 alternansucrase by acceptor site engineering. Inter. Mol. Sci. 22(6), 3229.

Wei, X., Liu, G.L., Jia, S.L., Chi, Z., Hu, Z., Chi, Z.M., 2021. Pullulan biosynthesis and its regulation in *Aureobasidium* spp. Carbohydr. Polym. 251, 1170.

Wu, J., Zhang, Y., Ye, L., Wang, C., 2021. The anti-cancer effects and mechanisms of lactic acid bacteria exopolysaccharides in vitro: A review. Carbohydr. Polym. 253, 117308.

Yu, L., Qian, Z., Ge, J., Du, R., 2022. Glucansucrase produced by lactic acid bacteria: Structure, properties, and applications. Ferment. 8(11), 629.

Yuan, M., Fu, G., Sun, Y., Zhang, D., 2021. Biosynthesis and applications of curdlan. Carbohydr. Polym. 273, 118597.

Zhan, X.B., Lin, C.C., Zhang, H.T., 2012. Recent advances in curdlan biosynthesis, biotechnological production, and applications. Appl. Microbiol. Biotechnol. 93, 525–531.

Zhang, H.T., Zhan, X.B., Zheng, Z.Y., Wu, J.R., Chen, D., 2010, June. New strategy for enhancement curdlan biosynthesis in alcaligenes faecalis by activating gene expression. In: *2010 4th International Conference on Bioinformatics and Biomedical Engineering.* IEEE, pp. 1–5.

Zhang, H.T., Zhan, X.B., Zheng, Z.Y., Wu, J.R., Yu, X.B., Jiang, Y., Lin, C.C., 2011. Sequence and transcriptional analysis of the genes responsible for curdlan biosynthesis in *Agrobacterium* sp. ATCC 31749 under simulated dissolved oxygen gradients conditions. App. Microbial. Biotechnol. 91, 163–175.

Zhang, R., Edgar, K.J., 2014. Properties, chemistry, and applications of the bioactive polysaccharide curdlan. Biomacromolecules. 15(4), 1079–1096.

Zhang, W., Gao, H., Huang, Y., Wu, S., Tian, J., Niu, Y., Zou, C., Jia, C., Jin, M., Huang, J., Chang, Z., 2020. Glutamine synthetase gene glnA plays a vital role in curdlan biosynthesis of Agrobacterium sp. CGMCC 11546. Int. J. Biol. Macromol. 165, 222–230.

5 Bioreactor design for the production of microbial polysaccharides

Neha Namdeo, Bikash Kumar, and Harit Jha

5.1 INTRODUCTION

Humans have historically exploited bacteria's capacity to manufacture multifaceted metabolic products for our advantage, particularly those metabolites with unique characteristics and no substitute sources (Adrio and Demain, 2006; Birch et al., 2019; Fazenda et al., 2008; Sanchez and Demain, 2008). Any novel microbial activity must compete favorably with existing metabolites with parallel qualities that have been isolated in the past from other natural sources. To do this, they must be manufactured at high production levels and in a way that makes recovery and purification easy and affordable (Pham et al., 2019). With polysaccharides from particularly viscous fermentation media, or even frequently from more conventional sources, this phase in the process is not always simple (Giavasis et al., 2000; Singh et al., 2017). It is nowadays getting easier to metabolically engineer EPS-producing strains or even to transfer the entire EPS synthesis genetic code into a different host microorganism (Cheng et al., 2022).

To produce goods on a large scale for commercial purposes, it is typically necessary to scale up the fermentation and use alternative methods for cell growth, including submerged and solid state under optimized and standardized conditions. This ensures that a constant product can be produced, whose composition and yield are not affected by seasonal or climatic factors. Additionally, these growing conditions make it easier to optimize yield and potentially operate the chemical (and consequently physical and biological) features of the EPS (Hsieh et al., 2006; Orr et al., 2009).

A major difficulty encountered during scale-up of EPS production is replicating the environmental and physical conditions required for the optimal growth of the organism (Garcia-Ochoa and Gomez, 2009). It is thus essential to review and assess the types of bioreactors being used for lab-scale and industrial-scale production of microbial exopolysaccharides. The minor modification in fermentation conditions can impact the overall yield and productivity of the microorganisms producing exopolysaccharides.

5.2 PROPERTIES OF MICROBIAL EXOPOLYSACCHARIDES

The major structural components of the cell wall are polysaccharides. Exopolysaccharides, also known as extracellular polysaccharides (EPS), are either covalently

 DOI: 10.1201/9781003342687-5

linked to the surface of cells in the form of capsules or are loosely bound in the form of slime (Sutherland, 1998, 1996). Studies suggested that EPS can be produced from several bacteria and fungi (Osemwegie et al., 2020). A few microbes have the great potential to exude EPS in significant amounts (>40 gL) under stressful conditions, however, the majority of microorganisms tend to exudate EPS without any stress into the extracellular environment (Lin and Casida Jr, 1984; Osemwegie et al., 2020; Papinutti, 2010; Ravella et al., 2010; Steluti et al., 2004). The main obstacle to EPS manufacturing, despite the industrially compelling benefits of microbial EPS, is the production cost (Hong et al., 2022). The cost is often associated with the substrate needed for microbial growth and the bioreactors used to grow large quantities of microorganisms (Ates, 2015; Béjar et al., 1998; Chi et al., 2007).

Homopolysaccharides and heteropolysaccharides are the two main categories of polysaccharides. Contrary to heteropolysaccharides, which have two or more sugar components, homopolysaccharides only have one sugar component (Abarquero et al., 2022; Mollet, 1996). Microbial polysaccharides are typically linear molecules, while some forms may also have side chains of varying lengths. Some of the examples of bacterial polysaccharides are xanthan gum (*Xanthomonas campestris*), gellan gum (*Sphingomonas elodea*), dextran (*Leuconostocmes enteroides*), and succinoglycan (*Rhizobium* sp.) (Schmid et al., 2015). Similarly, some of fungal EPS are pullulan (*Aureobasidium pullulans*), schizophyllan (*Schizophyllum commune*), chitin, and chitosan (*Candida albican, Zygosaccharomyces rouxii*, Filamentous fungi) (Osemwegie et al., 2020). A summarized table of bacterial and fungal polysaccharides and their source organisms, fermentation strategy, and productivity has been summarized in Table 5.1.

5.3 TYPES OF BIOREACTORS

A bioreactor for fermentation, often known as a "fermenter" is an instrument that guarantees all the ideal circumstances for the growth of microbial cells while remaining adaptable and versatile enough to be utilized for various tasks (Lattermann and Büchs, 2016; Mandenius, 2016). The majority of bioreactors for submerged fermentation are standard apparatuses with a variety of internal components and add-ons to increase oxygen mass transfer, provide culture broth homogenization, eliminate metabolic heat, prevent contamination with undesirable organisms, and maintain necessary pH, oxygen concentration values, and temperature, among other operating parameters (Hong et al., 2022). There are numerous commercial companies on the market that provide typical bioreactors to cultivate macromycetes, plant cells, and animal cells, in addition to microbes (Chegwin-Angarita et al., 2021; Ganeshan et al., 2021; Ruffoni et al., 2010; Wang et al., 2005). While designing and building bioreactors for submerged fermentation, agitation and aeration are crucial factors to be taken into account (da Rosa et al., 2019; Wang and Zhong, 2007; Zhou et al., 2018). Additionally, whether automated or not, bioreactor control is essential for the fermentation sector to succeed. The cost of the various process control devices (probes, transducers, controllers, and actuators) and the software can increase the cost of the bioreactor by up to 80% (without control devices) (Castillo et al., 2015). But some fermentation procedures for making polysaccharides need to be tightly

TABLE 5.1

Polysaccharides produced by bacteria for their commercial applications

Polysaccharide	Source	Fermentation strategy and reactor type	Productivity	Reference
Bacterial				
Levan	*Zymomonas mobilis*	Batch and continuous fermentations packed bed	299.1 g/L	(Silbir et al., 2014)
Dextran	*Leuconostoc mesenteroides*	Batch and Fed batch fermentation, 5-L Biostat B. plus bioreactor	37 and 35 g/L	(Nuwan et al., 2016)
Succinoglycan	*Agrobacterium tumefaciens*	Solid-state fermentation, prototype horizontal bioreactor	42 g/L	(Stredansky et al., 1998)
Xanthan gum	*Xanthamonas* sp. (*Xanthomonas campestris*)	Batch fermentation	8.93 g/L	(Rottava et al., 2009)
Alginate	*Pseudomonas putida*	Batch fermentation	262 mg/L	(Celik et al., 2008)
	Pseudomonas aeruginosa	Batch fermentation	368 mg/L	
Glucan	*Leuconostoc dextranicum*	Batch fermentation	1063 mg/L	(Majumder et al., 2009)
Gellan	*Sphingomonas paucimobilis*	Batch fermentation	35.70 g/l	(Gonçalves et al., 2009)
Cellulose	*Gluconacetobacter xylinum*	Batch fermentation	16.9 g/l	(Huang et al., 2010)
Fungal				
Botryosphaeran	*Botryosphaeria* sp.	Submerged culture, batch fermentation	1000 mg/L	(Steluti et al., 2004)
Pullulan	*Aureobasidium pullulans*	Batch Fermentation	51.4 ± 0.50 g/L	(Haghighatpanah et al., 2020)
Schizophyllan	*Schizophyllum commune*	Shake flask culture	64.99 mg/g extract	(Saetang et al., 2022)
Pleuran	*Pleurotus ostreatus*	Shake flask (250 mL) followed by 16 L Stirred tank bioreactor	Volumetric: 1980 mg/L Specific: 445 mg/g	(Maftoun et al., 2013)
Chitosans	*Benjaminiella poitrasii*	Submerged culture, batch fermentation	60.89 ± 2.30 mg/g of dry mycelial biomass	(Mane et al., 2022)
Chitosans oligosaccharides	*Trichoderma harzianum*	Submerged culture, batch fermentation	7–238 mg/mL	(Dario Rafael et al., 2019)

controlled for pH and temperature; consequently, it is important to assess the viability of each process to determine whether the level of sophistication of a particular bioreactor justifies its use in light of the cost and market for the intended product (Mitra and Murthy, 2022). It also needs to be assessed, whether the desired outcome is still present in the culture broth (exopolysaccharide) or within the cells (intrapolysaccharide (Sánchez Óscar and Montoya, 2021). For the purification and separation of these polysaccharides without losing their biological activity, multiunit processes are needed. These processes are intricate and challenging to design (Buksa et al., 2021; Chi et al., 2007; Li et al., 2021; Shi, 2016; Xue et al., 2022).

5.4 BIOREACTORS DESIGNING FOR MICROBIAL POLYSACCHARIDE PRODUCTION

During scale-up, it must be chosen whether to employ low-shear airlift fermenters, continuously stirred tank reactors (CSTRs) with either high or low-shear impellers, or other new configurations for EPS generation. All of the bioreactors used to produce EPS are specialized variants of reactors used for other purposes (Zampieri et al., 2022). Optimal mixing efficiencies require large diameter single bladed low shear axial flow impellers operating at low agitation speeds, and configurations like CSTRs represent compromises between obtaining efficient oxygen dispersal systems obtained with small diameter, high-speed multi-blade impeller systems, like high shear Rushton turbines (da Rosa et al., 2019; Garcia-Ochoa and Gomez, 2009; Gibbs and Seviour, 1998; Karthikeyan Akash and Joseph, 2022; McNeil and Harvey, 1993; Solomons, 1980; Zhong, 2011). It is also vital to assess the contamination and the potential of wild-type cells to be replaced by mutants there by hampering the productivity of EPS in continuous culture systems (Fazenda et al., 2008; Rau, 2004; Schuster et al., 1993; Seviour et al., 2011). The fermentation approach and application of biocatalyst will also rely on whether EPS production is related to biomass growth or not, as well as how long it takes the microbe to produce its maximum amount of EPS (Efremenko et al., 2022; Öner, 2013).

At both the scientific and industrial scales, stirred tank reactors (STRs) are the most frequently utilized fermenters. The continuous STR (CSTR) and the airlift reactor are the two fermenter systems for microbial cultivation that are most frequently utilized (ALR). In CSTR (Figure 5.1), radial flow impellers, such as Rushton turbine impellers, produce perturbation that ensures uniform mixing of the broth, facilitating the mass transfer of nutrients such as oxygen and nutrients to the cells. However, when the agitation grows, the shear pressure grows as well, which may affect the product's quality. ALR (Figure 5.2) is employed, in which air is injected through a sparger located at the bottom of the reactor, to function in low-shear circumstances. The broth rises to the top of the vessel, where its density increases, and then it drops through a downcomer or external loop (Castillo et al., 2015).

A variety of fermenter configurations have been reported to produce EPS from the diversity of microorganisms. As an illustration, continuous levan synthesis in a packed-bed bioreactor (Figure 5.3). The reactor contained Ca-alginate immobilized *Zymomonas mobilis* with high productivity collating with cheap manufacturing and input expenses (Silbir et al., 2014).

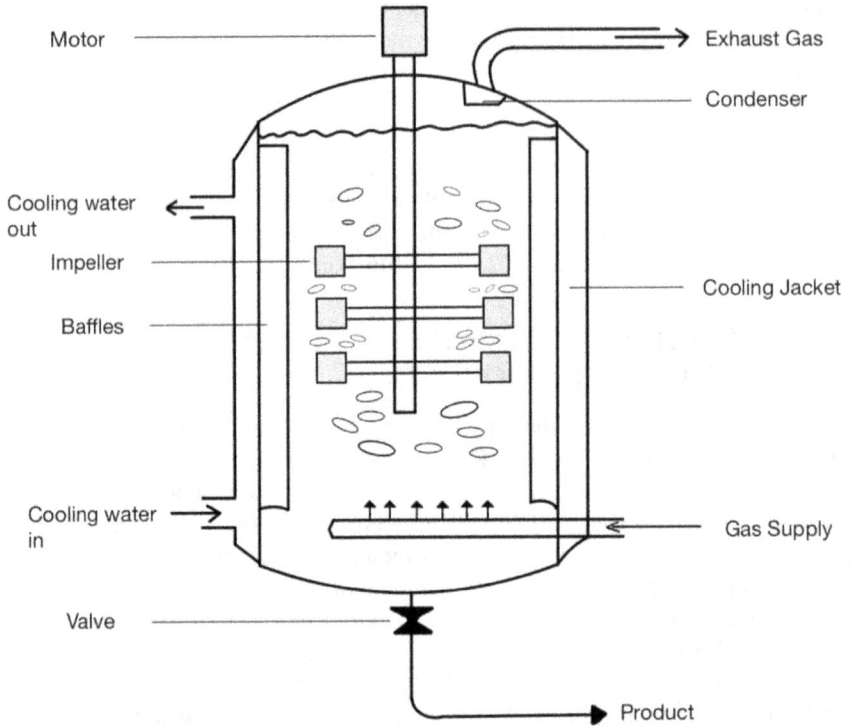

FIGURE 5.1   Continuous stirred tank reactor (CSTR)

The development of bioreactors for EPS production by thermophiles has certain unique challenges. Given that oxygen dissolves less readily at higher temperatures, adequate aeration and agitation should be supplied (Radchenkova et al., 2020, 2014; Sabin, 1955). Additionally, significant outlays are needed for heat exchangers, air compressors, and reactors that are particularly constructed to resist the high temperatures. The use of thermophiles has several benefits, including quick cultivation times, reduced chance of contamination, a better mass transfer rate, and broth with decreased viscosity due to their fast cell growth rates at high temperatures (Zhu et al., 2020). A major advantage of using thermophilic microorganisms in EPS production is their unique physicobiological properties suitable for industrial-scale bioprocesses (Kambourova et al., 2016).

For the effectiveness of the polysaccharide fermentation processes, constructing the industrial bioreactor is a crucial engineering task (Neway, 1989; Seviour et al., 2011). For microorganisms to produce with the greatest efficiency, ideal fermentation conditions are required. As a result, it's essential to closely monitor and control the fermentation parameters with the sensors to take into account scaling up laboratory methods for industrial applications (García et al., 2022a; Konstantinov and Yoshida, 1992). Bioreactors are currently divided into two

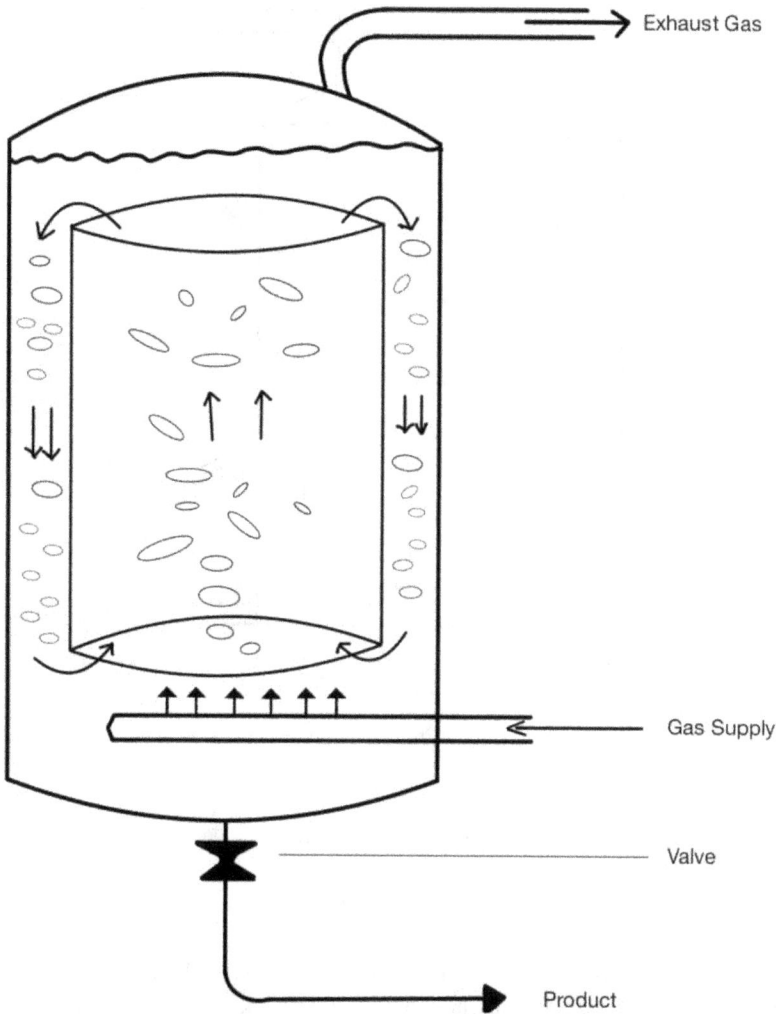

FIGURE 5.2   Airlift reactor (ALR)

categories: culture reactors and culture vessels that use particular biocatalysts like bacteria or tissues from plants and animals. Several studies showed that microorganisms in the bioreactor may be immobilized or suspended (Bhatia and Bera, 2015; Dianliang Wang et al., 2005; Thomas et al., 2013). The membranes, fibrous bed, or packed bed systems are the primary big scales of the bioreactors containing the immobilized microorganism (Laroche and Michaud, 2007; Singh et al., 2020). Generally speaking, there are three types of fermentations: batch, fed batch, and continuous operations. The fed-batch operation is a culture method in which additional nutrients are gradually or abruptly introduced to the starting medium after the cultivation has begun or at a point in the middle

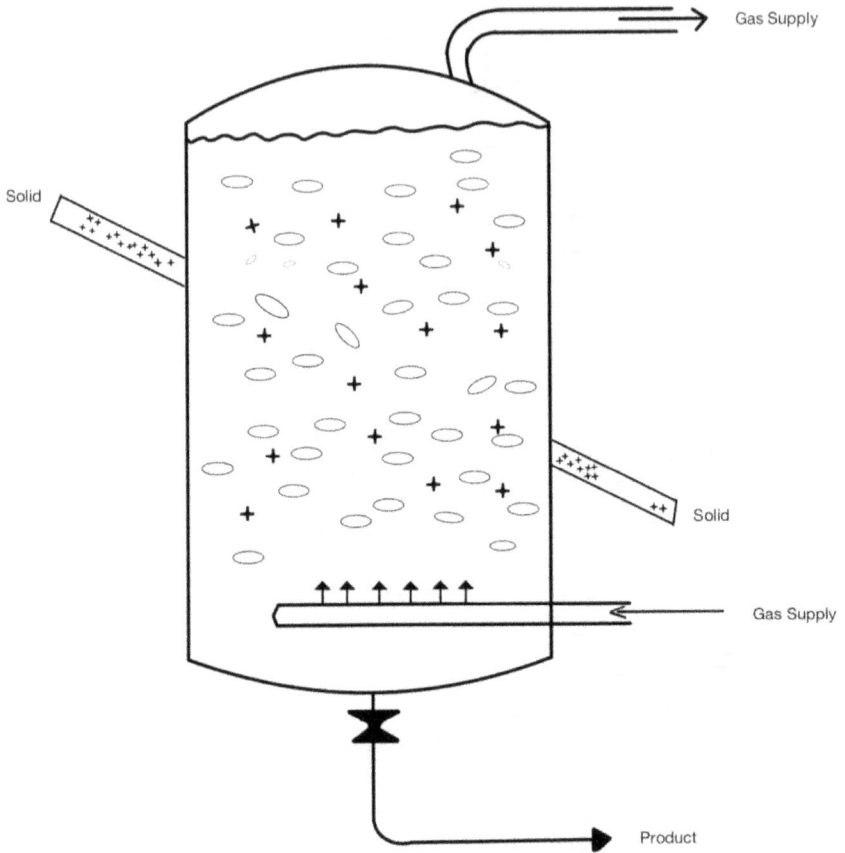

FIGURE 5.3 Continuous flow packed bed reactor (CFPBR)

of the batch procedure. The fed-batch process is frequently segregated into two models, the first of which does not include feedback control and the second of which does. The fed-batch techniques were created to prevent substrate inhibition, the glucose effect, and catabolite suppression. The fed-batch culture affords the opportunity to regulate response rates and get beyond bioreactor technology constraints in the vast majority of instances (Zampieri et al., 2022). The fundamental characteristics of these processes are the set total mass of each vessel batch owing to the closed system and the same residence period for the fluid components; nonetheless, the batch bioreactors became popular for the polysaccharides fermentation due to their high flexibility of application. Prior to cell cultivation, the fermentation medium is supplemented with all the nutrients required for the augmentation of the microbes and the production of metabolites during batch fermentation. Aeration provides growth with the oxygen it needs, and in most situations, the growth stops when the substrate's availability in the medium reaches a certain limit. The batch and fed-batch systems have been extensively researched and developed for the industrial production of the polysaccharides for all these reasons and benefits (García et al., 2022b; Mandal et al., 2007; Shabtai, 1990; Thirumal et al., 2018).

5.5 DIFFERENT PARTS AND PARAMETERS

5.5.1 AGITATION AND AERATION

To enhance the fermentation process, it is important to establish the aeration levels. For bacteria having an aerobic respiratory metabolism, the oxidation of organic molecules requires oxygen present in the air. These microbes grow quickly, consume oxygen at fast rates, and produce a substantial quantity of heat. The mass transport of oxygen to the liquid medium imposes restrictions on aerobic fermenters (Cappello, 2020). To accelerate the solubility of oxygen in the water, many bioreactors are equipped with a stirrer with one or more impellers. A sparger is also positioned under the lower impeller and serves similar functions (Tabuchi et al., 2021). Since the formation of foam in this gas-liquid system has the potential to negatively impact fermentation performance, the fermenter is equipped with antifoam materials or devices (Fazenda et al., 2008; Gong et al., 2021). Important elements impacting the fermentation include the impact of agitation speed and aeration rate on the formation of EPS. The type and physiology of the microorganisms utilized are strongly related to the ideal temperature and aeration for fermentation (Radchenkova et al., 2014). For instance, even though the bulk of polysaccharides are created aerobically, the lactic acid bacteria (LAB) are stimulated to make more polysaccharides when the oxygen level drops (De Vuyst et al., 2001; Sørensen et al., 2022). For the efficient generation of EPS, pH is a crucial factor. In fact, research on the submerged fermentation of *Ganoderma lucidum* has demonstrated that EPS synthesis requires a higher pH than cell development (Papinutti, 2010). Currently, it is possible to increase the production of biomass and EPS using a specialized approach, and particularly using a two-stage pH control strategy. The pH management and monitoring in the two-stage fermentation culture can have a significant impact on the yield, cell shape, chemical makeup, and molecular weight of the EPS produced (Delattre et al., 2008; Kim et al., 2006).

5.5.2 TRANSFER OF HEAT

The metabolic heat is generated by microbial cells as they multiply and expand in the culture medium. If this heat is not expelled, the system's temperature rises to levels that are not ideal for the biosynthesis or growth rate of the desired products. Even cell death can result at extreme temperatures (da Rosa et al., 2019). Bioreactors are equipped with a variety of heat exchangers for optimal heat transfer. Some of the commonly used heat exchange systems include heating jackets, heating filaments, and cold coils and indirect heat exchangers (Dhor et al., 2022; Mohamed, 2020; Shanmugam et al., 2022). These systems contain cooling water tubes that dissipate heat by conduction and convection mediated through metal coils. Since the heat transport in cylindrical tanks has been thoroughly investigated, there aren't many challenges in using industrial fermenters to produce polysaccharides (Calazans et al., 2000; Medina-Ramirez et al., 2020).

5.5.3 MIXING DEVICES AND FLUID FLOW

Impellers can be divided into two main categories as fluid flow inducers: axial flow and radial flow (Morrison et al., 1999; Radchenkova et al., 2014). Diverse-designed impellers induce different patterns of fluid flow.

1. Pumping efficiency and reduced shear using axial-flow impellers (movement).
2. Radial-flow impellers, which produce significant shear with minimal pumping.

Due to its effectiveness in gas-liquid contacting, the flat-bladed radial-flow impeller, also known as the "Rushton turbine," is frequently used in fermenters. Since Rushton turbines are typically provided as standard fittings by manufacturers of fermentation equipment, the term "standard" setup for a STR relates to their utilization.

Among other crucial factors are as follows (Morrison et al., 1999):

1. The power density per unit volume.
2. The placement of the impeller, including the shaft spacing.
3. For each type of impeller, there is normally an ideal value for the Di/Dt ratio.
4. Employing baffles.

Radchenkova et al. (2014) demonstrated enhanced exopolysaccharide production from *Aeribacillus palidus* 418 via controlling aeration and agitation speed using Narcissus impellers. The optimum condition was observed as Narcissus impeller set up at 900 rpm agitation speed, along with 0.5 vvm of aeration rate. The oxygen mass transfer coefficient (KLa) and bioprocess oxygen uptake rate (OUR) were evaluated for a better understanding of bioprocessing parameters.

5.5.4 IMPELLER PUMPING, FLOW INTENSITY, AND IMPELLER TURBULENT SHEAR STRESS

The impeller's ability to mix well creates a uniform or homogeneous environment throughout the full volume of culture broth, which is one of the impeller's key functions. Shear flow and circulation flow are the two different types of mixing (Calazans et al., 2000). There can be one of three sorts of flow regions in an STR:

1. The "micromixing area" surrounding the impeller, for example.
2. A dormant ("dead") zone with very low mixing.
3. The "macromixing region" is predominated by the circulating flow (pumping).

To prevent "dead zones," the tank's geometry and the circulation flow (pumping) generated by the agitator(s) must be adequate. The cells (together with the accompanying exopolymer) dynamically reside in the micromixing zone because the impeller's pumping action moves them away from the intense shear zone (Calazans et al., 2000). Pereira et al. (2011) contend that a quantitative indicator of this dynamic equilibrium is the shear-to-flow ratio, denoted as N/Di. Interzone mixing, often known as "blend times," is crucial for viscous fluids.

5.6 CHALLENGES IN EPS FERMENTATIONS

The key challenges in EPS production are similar to any bioproduct, including quality control, standardization of process, constant monitoring, and control. Adherence to regulatory guidelines is also an important aspect of large-scale production. The monitoring of bioprocess through inline, online, and real-time methods may be selected based on scale, cost, and control requirements. However, other than through offline analysis techniques, it is difficult to link critical process variables in EPS fermentations to the FDA's Critical to Quality (CtQ) qualities (Giavasis et al., 2003). Cultures that produce EPS need oxygen; hence, it must be provided to the culture. Gassing with sterile filtered air and mild mechanical agitation are typically used for this to spread tiny gas bubbles throughout the liquid. The aeration requirement may vary depending on the type of fermenter used (Figure 5.4). The viscosity is higher in the system containing system with yeast culture, consequently requiring enhanced aeration rate and stirrer speed to maintain cell density (Figure 5.4A) (Matthews, 2008). In EPS-producing cultures, the broth exhibits shear thinning behavior contrary to uniform density in other cultures (Figure 5.4B). As a result, it has a low

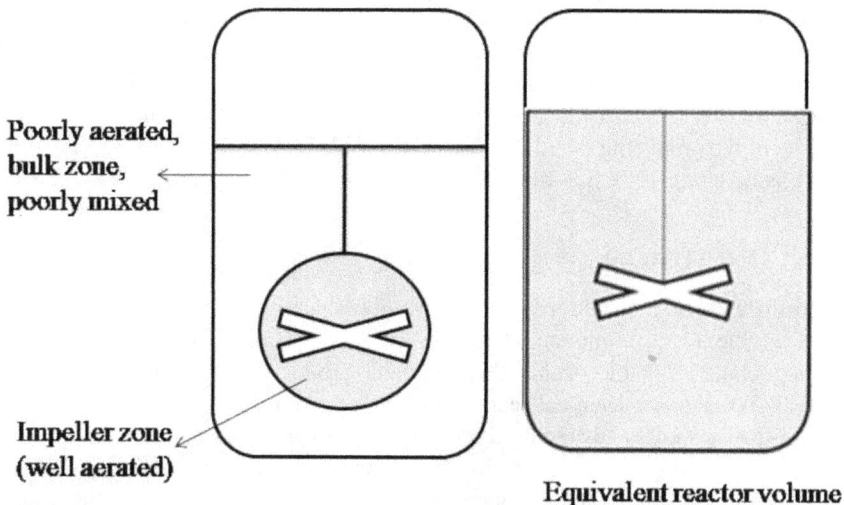

FIGURE 5.4 Rushton turbine—mechanical agitation (mixing in STR fermenters) for (A) Yeast culture and (B) Culture producing EPS

viscosity close to the impeller (high shear zone) and yields stress (no movement at all) or high viscosity in peripheral regions (Nienow, 1990).

This suggests that a sample obtained from the impeller zone might not be typical of the entire reactor media; further, these are very heterogeneous systems (de Jesus Assis et al., 2014; Q. Li et al., 2021; McNeil and Harvey, 1993; Odeleye et al., 2014). This situation has significant ramifications for our capacity to watch over and manage EPS procedures. Exopolysaccharide-producing bacterial processes are frequently quite aerobic.

In contrast to the reported promotion of EPS production by low DO in several fungal bioprocesses (Mahapatra and Banerjee, 2013; Seviour et al., 2011), xanthan production, for instance, increases significantly at higher oxygen transfer rates (Flores et al., 1994; Mohd Sauid et al., 2013; Peters et al., 1989; Suh et al., 1990). Since the CSTR culture occupies the majority of the periphery zone, which is poorly mixed, and just a tiny portion of the well-mixed, aerated impeller zone, sub-optimal xanthan production conditions prevail in the reactor as a whole. Raising gas flow rates to increase oxygen transport rates in such a system would fail because the gas would pass through the well mixed with little of the fermenter's apparent viscosity zone and just the impeller with water (Cappello et al., 2021; Matthews, 2008). Therefore, the overall enhanced aeration would have a relatively favorable impact despite a high shear fluid thinning rate; a cost penalty from the increased power inputs is significantly required for blending. Equally difficult problems surround such crucial processing elements as culture pH. Consequently, if culture pH decreases due to acid formation when ingesting substances such as ammonium salts, some will be alkalized. Due to these issues caused by the highly viscous nature of culture media containing EPS, there will be restrictions on mass, momentum, and heat transfer inside the fermentation vessels, which may significantly lower reactor productivity (Mohd Sauid et al., 2013; Rosalam and England, 2006; Seviour et al., 2011). All fermentation processes involve trade-offs, but EPS fermentation processes are particularly hampered by the physical constraints imposed by the altered behavior of the process fluid, the fermenter vessel's capacity to meet the needs of the producing organisms (oxygen transfer rates, stirrer speeds), practical considerations, and the physiological makeup of the producing strain (Freitas et al., 2017, 2011; McNeil and Harvey, 1993; Mohd Sauid et al., 2013; Schmidt, 2005; Seletzky, 2007).

5.7 CONCLUSION

The primary biological actions of bioactive polysaccharides are antibiotic, antimutagenic, antioxidant, immunomodulatory, anticoagulant, anticancer, hypoglycemic, and hypocholesterolemic—make them valuable products with significant uses, particularly in the pharmaceutical sector. Finding new sources and uses for these polysaccharides as well as advancing the technology for their mass production at scale are both required to meet the need for biopolymers on a global scale, particularly in industrialized nations. This chapter aims at providing an overview of the designing of a bioreactor for the production of microbial exopolysaccharides. As well as different parts and parameters required for designing a fermenter for improved EPS production is summarized.

REFERENCES

Abarquero, D., Renes, E., Fresno, J.M., Tornadijo, M.E., 2022. Study of exopolysaccharides from lactic acid bacteria and their industrial applications: A review. Int. J. Food Sci. Technol. 57, 16–26, https://doi.org/10.1111/ijfs.15227

Adrio, J.L., Demain, A.L., 2006. Genetic improvement of processes yielding microbial products. FEMS Microbiol. Rev. 30, 187–214. https://doi.org/10.1111/j.1574-6976.2005.00009.x

Ates, O., 2015. Systems biology of microbial exopolysaccharides production. Front. Bioeng. Biotechnol. 3, 200. https://doi.org/10.3389/fbioe.2015.00200

Béjar, V., Llamas, I., Calvo, C., Quesada, E., 1998. Characterization of exopolysaccharides produced by 19 halophilic strains of the species *Halomonas eurihalina*. J. Biotechnol. 61, 135–141. https://doi.org/10.1016/S0168-1656(98)00024-8

Bhatia, S., Bera, T., 2015. Chapter 7 — Classical and nonclassical techniques for secondary metabolite production in plant cell culture. In: S. Bhatia, K. Sharma, R. Dahiya, T. Bera (Eds.), *Modern Applications of Plant Biotechnology in Pharmaceutical Sciences*. Academic Press, pp. 231–291. https://doi.org/10.1016/B978-0-12-802221-4.00007-8

Birch, J., van Calsteren, M.-R., Pérez, S., Svensson, B., 2019. The exopolysaccharide properties and structures database: EPS-DB. Application to bacterial exopolysaccharides. Carbohydr. Polym. 205, 565–570. https://doi.org/10.1016/j.carbpol.2018.10.063

Buksa, K., Kowalczyk, M., Boreczek, J., 2021. Extraction, purification and characterisation of exopolysaccharides produced by newly isolated lactic acid bacteria strains and the examination of their influence on resistant starch formation. Food Chem. 362, 130221. https://doi.org/10.1016/j.foodchem.2021.130221

Calazans, G.M.T., Lima, R.C., de França, F.P., Lopes, C.E., 2000. Molecular weight and antitumour activity of *Zymomonas mobilis* levans. Int. J. Biol. Macromol. 27, 245–247.

Cappello, V., 2020. *Scale-Up of Gas-Liquid Agitated Bioreactors Three-Dimensional Modeling of Hydrodynamics, Mass Transfer and Kinetics*. Université Clermont Auvergne [2017–2020].

Cappello, V., Plais, C., Vial, C., Augier, F., 2021. Scale-up of aerated bioreactors: CFD validation and application to the enzyme production by *Trichoderma reesei*. Chem. Eng. Sci. 229, 116033. https://doi.org/10.1016/j.ces.2020.116033

Castillo, N.A., Valdez, A.L., Fariña, J.I., 2015. Microbial production of scleroglucan and downstream processing. Front. Microbiol. 6, 1106. https://doi.org/10.3389/fmicb.2015.01106

Celik, G.Y., Aslim, B., Beyatli, Y., 2008. Characterization and production of the exopolysaccharide (EPS) from *Pseudomonas aeruginosa* G1 and *Pseudomonas putida* G12 strains. Carbohydr. Polym. 73(1), 178–182. https://doi.org/10.1016/j.carbpol.2007.11.021

Chegwin-Angarita, C., Bello-Forero, R.A., Serrato-Bermúdez, J.C., 2021. *Optimization of Culture Conditions in a Bioreactor for the Production of Biomass and Metabolites of the Macromycete Lentinula Edodes*. https://doi.org/10.21203/rs.3.rs-610654/v1

Cheng, Y., Tian, K., Xie, P., Ren, X., Li, Y., Kou, Y., Chon, K., Hwang, M.-H., Ko, M.-H., 2022. Insights into the minimization of excess sludge production in micro-aerobic reactors coupled with a membrane bioreactor: Characteristics of extracellular polymeric substances. Chemosphere. 292, 133434. https://doi.org/10.1016/j.chemosphere.2021.133434

Chi, Z., Pyle, D., Wen, Z., Frear, C., Chen, S., 2007. A laboratory study of producing docosahexaenoic acid from biodiesel-waste glycerol by microalgal fermentation. Process Biochem. 42, 1537–1545. https://doi.org/10.1016/j.procbio.2007.08.008

Dario Rafael, O.H., Luis Fernándo, Z.G., Abraham, P.T., Pedro Alberto, V.L., Guadalupe, G.S., Pablo, P.J., 2019. Production of chitosan-oligosaccharides by the chitin-hydrolytic system of *Trichoderma harzianum* and their antimicrobial and anticancer effects. Carbohydr. Res. 486, 107836. https://doi.org/10.1016/j.carres.2019.107836

da Rosa, L.M., Koerich, D.M., Della Giustina, S.V., 2019a. Bioreactors operating conditions. In: A. Berenjian (Ed.), *Essentials in Fermentation Technology*. Springer International Publishing, pp. 169–212. https://doi.org/10.1007/978-3-030-16230-6_6

de Jesus Assis, D., Brandão, L.V., de Sousa Costa, L.A., Figueiredo, T.V.B., Sousa, L.S., Padilha, F.F., Druzian, J.I., 2014. A study of the effects of aeration and agitation on the properties and production of xanthan gum from crude glycerin derived from biodiesel using the response surface methodology. Appl. Biochem. Biotechnol. 172, 2769–2785. https://doi.org/10.1007/s12010-014-0723-7

Delattre, C., Laroche, C., Michaud, P., 2008. Production of bacterial and fungal polysaccharides. In: C. Larroche, A. Pandey, C.G. Dussap (Eds.), *Advances in Fermentation. Technology*, vol. 1. Asiatech Publisher, Inc., pp. 291–320.

De Vuyst, L., De Vin, F., Vaningelgem, F., Degeest, B., 2001. Recent developments in the biosynthesis and applications of heteropolysaccharides from lactic acid bacteria. Int. Dairy J. 11, 687–707. https://doi.org/10.1016/S0958-6946(01)00114-5

Dhor, N.R., Badri, D., Mohapatra, N., 2022. Bioreactors-technology and design analysis. Int. J. Adv. Res. Sci. Commun. Technol (IJARSCT). 2(2), 724–729. https://doi.org/10.48175/568

Efremenko, E., Senko, O., Maslova, O., Stepanov, N., Aslanli, A., Lyagin, I., 2022. Biocatalysts in synthesis of microbial polysaccharides: Properties and development trends. Catalysts. 12(11), 1377. https://doi.org/10.3390/catal12111377

Fazenda, M.L., Seviour, R., McNeil, B., Harvey, L.M., 2008. Submerged culture fermentation of "higher fungi": The macrofungi. In: *Advances in Applied Microbiology*, 63. 33–103. https://doi.org/10.1016/S0065-2164(07)00002-0

Flores, F., Torres, L.G., Galindo, E., 1994. Effect of the dissolved oxygen tension during cultivation of *X. campestris* on the production and quality of xanthan gum. J. Biotechnol. 34, 165–173. https://doi.org/10.1016/0168-1656(94)90086-8

Freitas, F., Alves, V.D., Reis, M.A.M., 2011. Advances in bacterial exopolysaccharides: From production to biotechnological applications. Trends Biotechnol. 29(8), 388–398. https://doi.org/10.1016/j.tibtech.2011.03.008

Freitas, F., Torres, C.A.V., Reis, M.A.M., 2017. Engineering aspects of microbial exopolysaccharide production. Bioresour. Technol. 245, 1674–1683. https://doi.org/10.1016/j.biortech.2017.05.092

Ganeshan, S., Kim, S.H., Vujanovic, V., 2021. Scaling-up production of plant endophytes in bioreactors: Concepts, challenges and perspectives. Bioresour. Bioprocess 8, 63. https://doi.org/10.1186/s40643-021-00417-y

García, A., Fernández-Sandoval, M.T., Morales-Guzmán, D., Martínez-Morales, F., Trejo-Hernández, M.R., 2022a. Advances in exopolysaccharide production from marine bacteria. J. Chem. Technol. Biotechnol. 97, 2694–2705. https://doi.org/10.1002/jctb.7156

García, A., Fernández-Sandoval, M.T., Morales-Guzmán, D., Martínez-Morales, F., Trejo-Hernández, M.R., 2022b. Advances in exopolysaccharide production from marine bacteria. J. Chem. Technol. Biotechnol. 97(10), 2694–2705. https://doi.org/10.1002/jctb.7156

Garcia-Ochoa, F., Gomez, E., 2009. Bioreactor scale-up and oxygen transfer rate in microbial processes: An overview. Biotechnol. Adv. 27(2), 153–176. https://doi.org/10.1016/j.biotechadv.2008.10.006

Giavasis, I., Harvey, L.M., McNeil, B., 2000. Gellan gum. Crit. Rev. Biotechnol. 20(3), 177–211. https://doi.org/10.1080/07388550008984169

Giavasis, I., Robertson, I., McNeil, B., Harvey, L.M., 2003. Simultaneous and rapid monitoring of biomass and biopolymer production *by Sphingomonas paucimobilis* using Fourier transform-near infrared spectroscopy. Biotechnol. Lett. 25, 975–979. https://doi.org/10.1023/A:1024040420799

Gibbs, P.A., Seviour, R.J., 1998. The production of exopolysaccharides by *Aureobasidium pullulans* in fermenters with low-shear configurations. Appl. Microbiol. Biotechnol. 49, 168–174. https://doi.org/10.1007/s002530051154

Gonçalves, V.M.F., Reis, A., Domingues, M.R.M., Lopes-da-Silva, J.A., Fialho, A.M., Moreira, L.M., Sá-Correia, I., Coimbra, M.A., 2009. Structural analysis of gellans produced by *Sphingomonas elodea* strains by electrospray tandem mass spectrometry. Carbohydr. Polym. 77(1), 10–19. https://doi.org/10.1016/j.carbpol.2008.11.035

Gong, Z., Yang, G., Che, C., Liu, J., Si, M., He, Q., 2021. Foaming of rhamnolipids fermentation: Impact factors and fermentation strategies. Microb. Cell Fact. 20, 77. https://doi.org/10.1186/s12934-021-01516-3

Haghighatpanah, N., Mirzaee, H., Khodaiyan, F., Kennedy, J.F., Aghakhani, A., Hosseini, S.S., Jahanbin, K., 2020. Optimization and characterization of pullulan produced by a newly identified strain of *Aureobasidium pullulans*. Int. J. Biol. Macromol. 152, 305–313. https://doi.org/10.1016/j.ijbiomac.2020.02.226

Hong, P.-N., Matsuura, N., Noguchi, M., Yamamoto-Ikemoto, R., Honda, R., 2022. Change of extracellular polymeric substances and microbial community in biofouling mitigation by continuous vanillin dose in membrane bioreactor. J. Water Process. Eng. 47, 102644. https://doi.org/10.1016/j.jwpe.2022.102644

Hsieh, C., Tseng, M.-H., Liu, C.-J., 2006. Production of polysaccharides from *Ganoderma lucidum* (CCRC 36041) under limitations of nutrients. Enzyme Microb. Technol. 38(1–2), 109–117. https://doi.org/10.1016/j.enzmictec.2005.05.004

Huang, H.-C., Chen, L.-C., Lin, S.-B., Hsu, C.-P., Chen, H.-H., 2010. In situ modification of bacterial cellulose network structure by adding interfering substances during fermentation. Bioresour. Technol. 101(15), 6084–6091. https://doi.org/10.1016/j.biortech.2010.03.031

Kambourova, M., Radchenkova, N., Tomova, I., Bojadjieva, I., 2016. Thermophiles as a promising source of exopolysaccharides with interesting properties. In: P. Rampelotto (Ed.), *Biotechnology of Extremophiles: Grand Challenges in Biology and Biotechnology*. Springer, pp. 117–139. https://doi.org/10.1007/978-3-319-13521-2_4

Karthikeyan, A., Joseph, A.S.R., 2022. Fermenter design. In: P. Verma (Ed.), *Industrial Microbiology and Biotechnology*. Springer, pp. 129–167. https://doi.org/10.1007/978-981-16-5214-1_5

Kim, H.M., Park, M.K., Won Yun, J., 2006. Culture pH affects exopolysaccharide production in submerged mycelial culture of *Ganoderma lucidum*. Appl. Biochem. Biotechnol. 134(3), 249–262. https://doi.org/10.1385/ABAB:134:3:249

Konstantinov, K.B., Yoshida, T., 1992. Knowledge-based control of fermentation processes. Biotechnol. Bioeng. 39, 479–486. https://doi.org/10.1002/bit.260390502

Laroche, C., Michaud, P., 2007. New developments and prospective applications for β (1, 3) glucans. Recent. Pat. Biotechnol. 1(1), 59–73. https://doi.org/10.2174/187220807779813938

Lattermann, C., Büchs, J., 2016. Design and operation of microbioreactor systems for screening and process development. In: C.F. Mandenius (Ed.), *Bioreactors*. John Wiley & Sons, Ltd, pp. 35–76. https://doi.org/10.1002/9783527683369.ch2

Li, Q., Ma, S., Shen, X., Li, M., Zou, Z., 2021. Effects of impeller rotational speed and immersion depth on flow pattern, mixing and interface characteristics for kanbara reactors using VOF-SMM simulations. Metals (Basel). 11(10), 1596. https://doi.org/10.3390/met11101596

Li, X., Chen, C., Leng, A., Qu, J., 2021. Advances in the extraction, purification, structural characteristics and biological activities of *Eleutherococcus senticosus* polysaccharides: A promising medicinal and edible resource with development value. Front. Pharmacol. 12, 753007. https://doi.org/10.3389/fphar.2021.753007

Lin, C.C., Casida Jr, L.E., 1984. GELRITE as a gelling agent in media for the growth of thermophilic microorganisms. Appl. Environ. Microbiol. 47(2), 427–429. https://doi.org/10.1128/aem.47.2.427-429.1984

Maftoun, P., Malek, R., Abdel-Sadek, M., Aziz, R., Enshasy, H., et al., 2013. Bioprocess for semi-industrial production of immunomodulator polysaccharide Pleuran by *Pleurotus ostreatus* in submerged culture. J. Sci. Ind. Res. 72, 655–662. http://nopr.niscpr.res.in/handle/123456789/22636

Mahapatra, S., Banerjee, D., 2013. Fungal exopolysaccharide: Production, composition and applications. Microbiol. Insights. 6, 1–16. https://doi.org/10.4137/mbi.s10957

Majumder, A., Singh, A., Goyal, A., 2009. Application of response surface methodology for glucan production from *Leuconostoc dextranicum* and its structural characterization. Carbohydr. Polym. 75(1), 150–156. https://doi.org/10.1016/j.carbpol.2008.07.014

Mandal, S.M., Ray, B., Dey, S., Pati, B.R., 2007. Production and composition of extracellular polysaccharide synthesized by a Rhizobium isolate of *Vigna mungo* (L.) Hepper. Biotechnol. Lett. 29, 1271–1275. https://doi.org/10.1007/s10529-007-9388-4

Mandenius, C.-F., 2016. Challenges for bioreactor design and operation. In: C-F. Mandenius (Ed.), *Bioreactors*. John Wiley & Sons, Ltd, pp. 1–34. https://doi.org/10.1002/9783527683369.ch1

Mane, S., Pathan, E., Tupe, S., Deshmukh, S., Kale, D., Ghormade, V., Chaudhari, B., Deshpande, M., 2022. Isolation and characterization of chitosans from different fungi with special emphasis on zygomycetous dimorphic fungus *Benjaminiella poitrasii*: Evaluation of its chitosan nanoparticles for the inhibition of human pathogenic fungi. Biomacromolecules. 23(3), 808–815. https://doi.org/10.1021/acs.biomac.1c01248

Matthews, G., 2008. Fermentation equipment selection: Laboratory scale bioreactor design considerations. In: B. McNeil, L.M. Harvey (Eds.), *Practical Fermentation Technology*. Wiley and Sons, Ltd, pp. 3–36. https://doi.org/10.1002/9780470725306.ch2

McNeil, B., Harvey, L.M., 1993. Viscous fermentation products. Crit. Rev. Biotechnol. 13, 275–304. https://doi.org/10.3109/07388559309075699

Medina-Ramirez, C.F., Castañeda-Guel, M.T., Fernanda Alvarez-Gonzalez, M., Montesinos-Castellanos, A., Morones-Ramirez, J.R., López-Guajardo, E.A., Gómez-Loredo, A., 2020. Application of extractive fermentation on the recuperation of exopolysaccharide from *Rhodotorula mucilaginosa* UANL-001L. Fermentation. 6(4), 108. https://doi.org/10.3390/fermentation6040108

Mitra, S., Murthy, G.S., 2022. Bioreactor control systems in the biopharmaceutical industry: A critical perspective. Syst. Microbiol. Biomanufacturing. 2, 91–112. https://doi.org/10.1007/s43393-021-00048-6

Mohamed, M., 2020. *Numerical Simulation of Fluid Flow and Heat Transfer Characteristics in Dimpled Jacket*. PhD thesis. Pennsylvania State University.

Mohd Sauid, S., Krishnan, J., Huey Ling, T., Veluri, M.V.P.S., 2013. Enhancement of oxygen mass transfer and gas holdup using palm oil in stirred tank bioreactors with xanthan solutions as simulated viscous fermentation broths. Biomed. Res. Int. 2013, 409675. https://doi.org/10.1155/2013/409675

Mollet, B., 1996. New technologies in fermented milk. Cerevisia 21(1), 63–65.

Morrison, N.A., Clark, R.C., Chen, Y.L., Talashek, T., Sworn, G., 1999. Gelatin alternatives for the food industry. In: K. Nishinari (Ed.), *Physical Chemistry and Industrial Application of Gellan Gum*. Springer, pp. 127–131. https://doi.org/10.1007/3-540-48349-7_19

Neway, J.O., 1989. *Fermentation Process Development of Industrial Organisms*. M. Dekker. Inc.

Nienow, A.W., 1990. Agitators for mycelial fermentations. Trends Biotechnol. 8, 224–233. https://doi.org/10.1016/0167-7799(90)90180-6

Nuwan, P., Piwpan, P., Jaturapiree, A., Jaturapiree, P., 2016. Production of dextran by *Leuconostoc mesenteroides* TISTR 053 in fed batch fermentation. In: *The 6 Th International Conference on Fermentation Technology for Value Added Agricultural Products July 29 Th—31st*, 2015 Centara Hotel & Convention Centre, p. 113. https://doi.org/10.14456/kkurj.2016.45

Odeleye, A.O.O., Marsh, D.T.J., Osborne, M.D., Lye, G.J., Micheletti, M., 2014. On the fluid dynamics of a laboratory scale single-use stirred bioreactor. Chem. Eng. Sci. 111, 299–312. https://doi.org/10.1016/j.ces.2014.02.032

Öner, E.T., 2013. Microbial production of extracellular polysaccharides from biomass. In: Z. Fang (Ed.), *Pretreatment Techniques for Biofuels and Biorefineries*. Springer, pp. 35–56. https://doi.org/10.1007/978-3-642-32735-3_2

Orr, D., Zheng, W., Campbell, B.S., McDougall, B.M., Seviour, R.J., 2009. Culture conditions affect the chemical composition of the exopolysaccharide synthesized by the fungus *Aureobasidium pullulans*. J. Appl. Microbiol. 107, 691–698. https://doi.org/10.1111/j.1365-2672.2009.04247.x

Osemwegie, O.O., Adetunji, C.O., Ayeni, E.A., Adejobi, O.I., Arise, R.O., Nwonuma, C.O., Oghenekaro, A.O., 2020. Exopolysaccharides from bacteria and fungi: Current status and perspectives in Africa. Heliyon. 6(6), e04205. https://doi.org/10.1016/j.heliyon.2020.e04205

Papinutti, L., 2010a. Effects of nutrients, pH and water potential on exopolysaccharides production by a fungal strain belonging to *Ganoderma lucidum* complex. Bioresour. Technol. 101(6), 1941–1946. https://doi.org/10.1016/j.biortech.2009.09.07

Pereira, L.T., Pereira, L.T., Teixeira, R.S., Bon, E.P., Freitas, S.P., 2011. Sugarcane bagasse enzymatic hydrolysis: Rheological data as criteria for impeller selection. J. Ind. Microbiol. Biotechnol. 38(8), 901–907. https://doi.org/10.1007/s10295-010-0857-8

Peters, H., Herbst, H., Hesselink, P.G.M., Lünsdorf, H., Schumpe, A., Deckwer, W., 1989. The influence of agitation rate on xanthan production by *Xanthomonas campestris*. Biotechnol. Bioeng. 34, 1393–1397. https://doi.org/10.1002/bit.260341108

Pham, J.V., Yilma, M.A., Feliz, A., Majid, M.T., Maffetone, N., Walker, J.R., Kim, E., Cho, H.J., Reynolds, J.M., Song, M.C., Park, S.R., Yoon, Y.J., 2019. A review of the microbial production of bioactive natural products and biologics. Front. Microbiol. 10, 1404. https://doi.org/10.3389/fmicb.2019.01404

Radchenkova, N., Boyadzhieva, I., Hasköylü, M.E., Atanasova, N., Yıldız, S.Y., Kuncheva, M.J., Panchev, I., Kisov, H., Vassilev, S., Oner, E.T., Kambourova, M.S., 2020. High bioreactor production and emulsifying activity of an unusual exopolymer by *Chromohalobacter canadensis* 28. Eng. Life Sci. 20, 357–367. https://doi.org/10.1002/elsc.202000012

Radchenkova, N., Vassilev, S., Martinov, M., Kuncheva, M., Panchev, I., Vlaev, S., Kambourova, M., 2014. Optimization of the aeration and agitation speed of *Aeribacillus palidus* 418 exopolysaccharide production and the emulsifying properties of the product. Process Biochem. 49, 576–582. https://doi.org/10.1016/j.procbio.2014.01.010

Rau, U., 2004. Glucans secreted by fungi. Turk. Electron. J. Biotechnol. 2, 30–36.

Ravella, S.R., Quiñones, T.S., Retter, A., Heiermann, M., Amon, T., Hobbs, P.J., 2010. Extracellular polysaccharide (EPS) production by a novel strain of yeast-like fungus *Aureobasidium pullulans*. Carbohydr. Polym. 82(3), 728–732. https://doi.org/10.1016/j.carbpol.2010.05.039

Rosalam, S., England, R., 2006. Review of xanthan gum production from unmodified starches by *Xanthomonas comprestris* sp. Enzyme. Microb. Technol. 39(2), 197–207.

Rottava, I., Batesini, G., Silva, M.F., Lerin, L., de Oliveira, D., Padilha, F.F., Toniazzo, G., Mossi, A., Cansian, R.L., Di Luccio, M., 2009. Xanthan gum production and rheological behavior using different strains of *Xanthomonas* sp. Carbohydr. Polym. 77(i), 65–71. https://doi.org/10.1016/j.carbpol.2008.12.001

Ruffoni, B., Pistelli, L., Bertoli, A., Pistelli, L., 2010. Plant cell cultures: Bioreactors for industrial production. In: G.M. Teresa, G. Rea and B. Berra (Eds.), *Bio-Farms for Nutraceuticals: Functional Food and Safety Control by Biosensors.* Springer, pp. 203–221. https://doi.org/10.1007/978-1-4419-7347-4_15

Sabin, A.B., 1955. Characteristics and genetic potentialities of experimentally produced and naturally occurring variants of poliomyelitis virus. Ann. N. Y. Acad. Sci. 61, 924–939. https://doi.org/10.1111/j.1749-6632.1955.tb42551.x

Saetang, N., Ramaraj, R., Unpaprom, Y., 2022. Optimization of ethanol precipitation of schizophyllan from *Schizophyllum commune* by applied statistical modelling. Biomass. Convers. Biorefin. https://doi.org/10.1007/s13399-022-02384-6

Sánchez, Ó.J., Montoya, S.V.L.M., 2021. Polysaccharide production by submerged fermentation. In: K.G. Ramawat, J.-M. Mérillon (Eds.), *Polysaccharides: Bioactivity and Biotechnology.* Springer International Publishing, pp. 1–19. https://doi.org/10.1007/978-3-319-03751-6_39-1

Sanchez, S., Demain, A.L., 2008. Metabolic regulation and overproduction of primary metabolites. Microb. Biotechnol. 1(4), 283–319. https://doi.org/10.1111/j.1751-7915.2007.00015.x

Schmid, J., Sieber, V., Rehm, B., 2015. Bacterial exopolysaccharides: Biosynthesis pathways and engineering strategies. Front. Microbiol. 6, 496. https://doi.org/10.3389/fmicb.2015.00496

Schmidt, F.R., 2005. Optimization and scale up of industrial fermentation processes. Appl. Microbiol. Biotechnol. 68, 425–435. https://doi.org/10.1007/s00253-005-0003-0

Schuster, R., Wenzig, E., Mersmann, A., 1993. Production of the fungal exopolysaccharide pullulan by batch-wise and continuous fermentation. Appl. Microbiol. Biotechnol. 39, 155–158. https://doi.org/10.1007/BF00228599

Seletzky, J.M., 2007. *Process Development and Scale-Up from Shake Flask to Fermenter of Suspended and Immobilized Aerobic Microorganisms.* PhD thesis. Biochemical Engineering, RWTH Aachen University.

Seviour, R.J., McNeil, B., Fazenda, M.L., Harvey, L.M., 2011. Operating bioreactors for microbial exopolysaccharide production. Crit. Rev. Biotechnol. 31(2), 170–185. https://doi.org/10.3109/07388551.2010.505909

Shabtai, Y., 1990. Production of exopolysaccharides by *Acinetobacter* strains in a controlled fed-batch fermentation process using soap stock oil (SSO) as carbon source. Int. J. Biol. Macromol. 12(2), 145–152. https://doi.org/10.1016/0141-8130(90)90066-J

Shanmugam, M.K., Mandari, V., Devarai, S.K., Gummadi, S.N., 2022. Types of bioreactors and important design considerations. In: R. Sirohi, A. Pandey, M.J. Taherzadeh, C. Larroche (Eds.), *Current Developments in Biotechnology and Bioengineering.* Elsevier, pp. 3–30. https://doi.org/10.1016/B978-0-323-91167-2.00008-3

Shi, L., 2016. Bioactivities, isolation and purification methods of polysaccharides from natural products: A review. Int. J. Biol. Macromol. 92, 37–48. https://doi.org/10.1016/j.ijbiomac.2016.06.100

Silbir, S., Dagbagli, S., Yegin, S., Baysal, T., Goksungur, Y., 2014. Levan production by *Zymomonas mobilis* in batch and continuous fermentation systems. Carbohydr. Polym. 99, 454–461. https://doi.org/10.1016/j.carbpol.2013.08.031

Singh, L., Yousuf, A., Mahapatra, D.M., 2020. *Bioreactors: Sustainable Design and Industrial Applications in Mitigation of Ghg Emissions,* 1st ed. Academic Press. https://doi.org/10.1016/C2019-0-02785-9

Singh, V., Haque, S., Niwas, R., Srivastava, A., Pasupuleti, M., Tripathi, C.K.M., 2017. Strategies for fermentation medium optimization: An in-depth review. Front. Microbiol. 7, 2087. https://doi.org/10.3389/fmicb.2016.02087

Solomons, G.L., 1980. Fermenter design and fungal growth. In: E. Smith, D.R. Berry, B. Kristiansen (Eds.), *Fungal Biotechnology*. Academic Press, pp. 55–80.

Sørensen, H.M., Rochfort, K.D., Maye, S., MacLeod, G., Brabazon, D., Loscher, C., Freeland, B., 2022. Exopolysaccharides of lactic acid bacteria: Production, purification and health benefits towards functional food. Nutrients. 14(14), 2938. https://doi.org/10.3390/nu14142938

Steluti, R.M., Giese, E.C., Piggato, M.M., Sumiya, A.F.G., Covizzi, L.G., Job, A.E., Cardoso, M.S., de Lourdes Corradi Da Silva, M., Dekker, R.F.H., Barbosa, A.M., 2004. Comparison of Botryosphaeran production by the ascomyceteous fungus *Botryosphaeria* sp., grown on different carbohydrate carbon sources, and their partial structural features. J. Basic. Microbiol. 44, 480–486. https://doi.org/10.1002/jobm.200410415

Stredansky, M., Conti, E., Bertocchi, C., Matulova, M., Zanetti, F., 1998. Succinoglycan production by *Agrobacterium tumefaciens*. J. Ferment. Bioeng. 85(4), 398–403. https://doi.org/10.1016/S0922-338X(98)80083-4

Suh, I.-S., Herbst, H., Schumpe, A., Deckwer, W.-D., 1990. The molecular weight of xanthan polysaccharide produced under oxygen limitation. Biotechnol. Lett. 12, 201–206. https://doi.org/10.1007/BF01026799

Sutherland, I.W., 1996. Extracellular polysaccharides. In: H-J. Rehm, G. Reed (Eds.), *Biotechnology: Products of Primary Metabolism*. pp. 613–657. https://doi.org/10.1002/9783527620999.ch16f

Sutherland, I.W., 1998. Novel and established applications of microbial polysaccharides. Trends Biotechnol. 16(1), 41–46. https://doi.org/10.1016/S0167-7799(97)01139-6

Tabuchi, S.C.T., Martiniano, S.E., Cunha, M.A.A., Barbosa-Dekker, A.M., Dekker, R.F.H., Prata, A.M.R., 2021. Kinetic study of lasiodiplodan production by *Lasiodiplodia theobromae* MMPI in a low-shear aerated and agitated bioreactor. J. Polym. Environ. 29, 89–102. https://doi.org/10.1007/s10924-020-01857-x

Thirumal, V., Chistoserdov, A., Bajpai, R., Bader, J., Popovic, M.K., Subramaniam, R., 2018. Effect of developed low cost minimal medium on lipid and exopolysaccharide production by *Lipomyces starkeyi* under repeated fed-batch and continuous cultivation. Chem. Biochem. Eng. Q. 32, 473–481. https://doi.org/10.15255/CABEQ.2018.1389

Thomas, L., Larroche, C., Pandey, A., 2013. Current developments in solid-state fermentation. Biochem. Eng. J. 81, 146–161. https://doi.org/10.1016/j.bej.2013.10.013

Wang, D., Liu, W., Han, B., Xu, R., 2005. The bioreactor: A powerful tool for large-scale culture of animal cells. Curr. Pharm. Biotechnol. 6(5), 397–403. https://doi.org/10.2174/138920105774370580

Wang, S.-J., Zhong, J.-J., 2007. Bioreactor engineering. In: S.-T. Yang (Ed.), *Bioprocessing for Value-Added Products from Renewable Resources*. Elsevier, pp. 131–161. https://doi.org/10.1016/B978-044452114-9/50007-4

Xue, H., Li, P., Bian, J., Gao, Y., Sang, Y., Tan, J., 2022. Extraction, purification, structure, modification, and biological activity of traditional Chinese medicine polysaccharides: A review. Front. Nutr. 9, 1–24. https://doi.org/10.3389/fnut.2022.1005181

Zampieri, R.M., Adessi, A., Caldara, F., De Philippis, R., Dalla Valle, L., La Rocca, N., 2022. In vivo anti-inflammatory and antioxidant effects of microbial polysaccharides extracted from Euganean therapeutic muds. Int. J. Biol. Macromol. (Part B) 209, 1710–1719. https://doi.org/10.1016/j.ijbiomac.2022.04.123

Zhong, J.-J., 2011. 2.21 — Bioreactor engineering. In: M. Moo-Young (Ed.), *Comprehensive Biotechnology*, 3rd ed. Pergamon, pp. 257–269. https://doi.org/10.1016/B978-0-444-64046-8.00077-X

Zhou, Y., Han, L.R., He, H.W., Sang, B., Yu, D.L., Feng, J.T., Zhang, X., 2018. Effects of agitation, aeration and temperature on production of a novel glycoprotein GP-1 by *Streptomyces kanasenisi* zx01 and scale-up based on volumetric oxygen transfer coefficient. Molecules. 23(1), 125. https://doi.org/10.3390/molecules23010125

Zhu, D., Adebisi, W.A., Ahmad, F., Sethupathy, S., Danso, B., Sun, J., 2020. Recent development of extremophilic bacteria and their application in biorefinery. Front. Bioeng. Biotechnol. 8, 1–18. https://doi.org/10.3389/fbioe.2020.00483

6 Utilization of agro-industrial by-products for exopolysaccharides production

Brahmeet Kaur, Gaurav Panesar,
Parmjit Singh Panesar, and Shashi Kant Bhatia

6.1 INTRODUCTION

One-third of the food produced worldwide is lost along the food chain. Due to this, several researchers have focused on evaluating the application of food by-products generated from food processing industries (Corrado et al., 2019). Conversion of agro-industrial by-products into valuable products may not only help to reduce the deleterious impact of waste disposal on the environment but also improve the functionality of the final developed product (Das et al., 2022; Kaur et al., 2022). Various by-products are generated from agro-industries in the form of seeds, skin, peels, pomace, and pericarp after the processing of fruits and vegetables like carrots, grapes, apples, guava, and potato. Similarly, other by-products like molasses from sugar-cane-processing industries, whey from the dairy industry, and bran from the cereal industry are also generated in large quantities (Panesar et al., 2015; Panesar et al., 2016; Kumar et al., 2017). High-added value compounds from these by-products have potential to be utilized in various applications of the food industry.

The demand for these natural exopolysaccharides has been found to increase in the past few decades due to their commercial application scope in various sectors of food, pharmaceutical, agriculture, and medicine (Kanimozhi et al., 2018). Polysaccharides produced through biotechnological pathway represent the currently existing market. Exopolysaccharides, in particular, have been produced by several microbial classes, including bacteria, yeast, and molds (Barcelos et al., 2020). Exopolysaccharides can be classified into two major groups, namely, homopolysaccharides (like dextran) and heteropolysaccharides (like gellan and xanthan). The majority of bacterial exopolysaccharides fall under the category of heteropolysaccharides (Aranda-selverio et al., 2010). The adoption of the biotechnological pathway for the production of EPS may offer several advantages over plant-derived products such as less consumption of time, independency from season and location, as well as a chance to utilize agro-industrial waste as fermentative substrates. However, the cost of production is a major limitation and causes hindrance owing to the requirements needed in the process (Rütering et al., 2016). An upsurge in the interest to develop culture media based on other sources of nutrients has been observed so as to decrease the cost of EPS production. The by-products generated from agro-industries are a rich source of sugars, including sucrose, fructose, and glucose, as well as contain appreciable quantities of nitrogen and vitamins which are useful for exopolysaccharide synthesis. Agro-waste are considered a natural, renewable, and inexpensive resource that can be employed for the cost-effective production of EPS. This may not only help utilize waste as raw materials and help in the waste management process but also provide economic advantages.

6.2   SOURCES AND COMPOSITION OF MAJOR AGRO-INDUSTRIAL BY-PRODUCTS

An enormous quantity of waste is generated by agro-industries, particularly food processing industries such as breweries, cereal, sugarcane, dairy, fruit, and vegetables (Panesar et al., 2016). These generate wastes (solid, liquid, and gaseous) that emerge not only from processing operations but also from their disposal and treatment. Biomass resources can be generally categorized into two physical forms, liquid resources which include molasses, juice, syrups, olive mill wastewater, and whey as well as solid resources like pomace and lignocellulosic biomass. The employment of complex media for microbial growth is economically not feasible due to the high amount of necessary expensive nutrients like peptone, yeast extract, and salts. Thus, the selection of agro-industrial by-products for the production of exopolysaccharides may lower the cost involved in its production. Some of the agro-industrial wastes commonly utilized for microbial production of exopolysaccharides have been depicted in Figure 6.1.

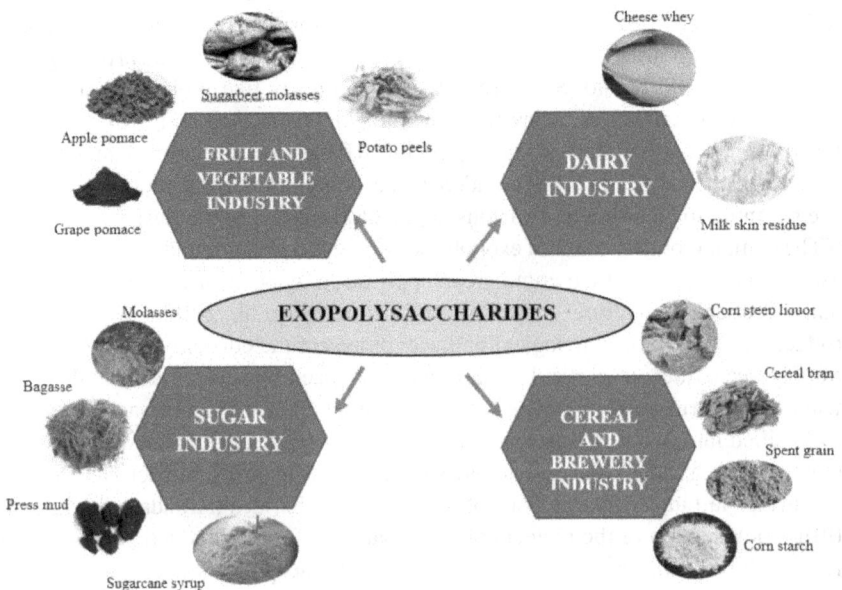

FIGURE 6.1   Some commonly utilized agro-industrial wastes for microbial production of exopolysaccharides

6.2.1 SYRUP AND MOLASSES

Molasses obtained from sugar crystallization constitutes one of the chief by-products generated from the sugar industry. It is a viscous substance similar to honey with about 9% fructose, 7% glucose, 35% sucrose, 3% reducing sugars, 20% water, 4% carbohydrates, 12% ash, 5% non-nitrogenous acids, 4.5% nitrogenous compounds, and 5% other substances (Teclu et al., 2009). Sugarcane molasses is composed of several components including total sugars (447–587 g/kg), sucrose (157–469 g/kg), sugar reducing agents (97–300 g/kg), nitrogen (0.25–1.5 g/kg), potassium (19–54 g/kg), phosphorus (0.3–0.7 g/kg), calcium (6–12 g/kg), and magnesium (4–11 g/kg). It has a pH range of 5–7 with ash and mineral content between 8–13% and salt content of 2–8% (Amorim et al., 2003; Clarke, 2003). It is also considered as an important dietary supplement as unlike refined sugars, it possesses a significantly high content of vitamin B6 as well as minerals (iron, manganese, and magnesium). The fermentative production of commercial exopolysaccharides such as xanthan, levan, pullulan, gellan, dextran, etc. may be effectively carried out using syrups and molasses as substrates (Özcan and Öner, 2015). Molasses obtained from sugarcane and sugar beet has been identified as suitable carbon sources for the production of xanthan gum, pullulan, gellan gum, welan gum, and pantoan by *Aureobasidium pullulans*, *Sphingomonas* sp., *Xanthomonas campestris*, *Alcaligens* sp., and *Pantoea* sp. (Gudiña et al., 2023). The presence of high concentrations of ions like Na^+, K^+, Fe^{+2}, and Zn^{+2} in syrups and molasses can serve as triggering stress factors for EPS formation of (Abdel-Aziz et al., 2012).

6.2.2 POMACE

Pomace generated as a by-product from the fruit and vegetable industry is a rich source of crude fiber, pectin, and minerals like Mg, K, Mn, and Fe (Shalini and Gupta, 2010). The fruit peels and pulp are considered a rich source of dietary fiber, and the insoluble fraction constitutes the major portion of the pomace of most fruits and vegetables (Russo et al., 2014). Apple pomace accounts for 1 million tons and grape pomace (stem, seeds, and, skin) constitutes about 10 million tons of total production every year (Shalini and Gupta, 2010; Öner, 2013). Cellulose (43%) and hemicellulose (20–32%) are found abundantly in apple pomace (Rabetafika et al., 2014). A significant level of calcium and magnesium were found in the apple peels along with the high level of iron, zinc, and copper. Grape pomace is generated as a major biowaste of the winemaking industry and accounts for 15–20% of the total grape weight (Machado and Domínguez-Perles, 2017). It is considered as an excellent source of flavonoids and anthocyanins (Sehm et al., 2011). The pomace left after grape processing is rich in dietary fiber (cellulose, hemicellulose, and little pectin) and polyphenolic compounds (flavonols, phenolic acids, and anthocyanins) (Brenes et al., 2016; Kammerer et al., 2005). Potato peels, which form a major part of the waste generated from potato starch plants, have a high content of dietary fibers (mainly rhamnogalacturonan I) (Byg et al., 2012). It has been reported to contain a crude fiber content of 61–125 g/kg (dry basis) and 27–35 g/100 g fiber content with phenolic content of about 50% (Afifi, 2011; Sharoba et al., 2013; Friedman, 1997). By-products generated from the fruit and vegetable industry such as apple pomace, lemon and orange peels, and coconut husk are considered highly potential substrates for solid state fermentation due to their higher capacity of water absorption. This makes it valuable for the growth of microorganisms (Goldmeyer et al., 2014). The peels, sediment, and filter cake of pomegranate were found to be a rich source of phenolic compounds and thus possess high antioxidant activity (Surek and Nilufer-Erdil, 2016). Pineapple peels and crown, generated as a major by-product from pineapple processing industries, account for 29–40% of the total pineapple weight. Pineapple peels possess a significantly high content of bioactive compounds like lutein, alpha-carotene, beta-carotene, and vitamin A. Flour of pineapple peel waste can serve as a potential carbon source for fermentative production of probiotic bacteria, *Pediococcus pentosaceus* (Imandi et al., 2008).

6.2.3 CHEESE WHEY

Whey is considered the major dairy industry by-product obtained after the processing of cheese. Million tons of by-products are produced annually by the dairy industry, out of which, cheese whey constitutes the net fraction remaining after milk coagulation (Panesar and Kennedy, 2012). The composition of whey includes water (95%), lactose (70%), minerals (20%), proteins (10%), and solids (5%) (Rocha and Guerra, 2020). The production of 1 kg of cheese results in the production of 9–10 L of whey, and if discarded without treatment, it may create serious environmental issues (Castelli and Du Vale, 2013). Cheese whey possesses a high chemical oxygen demand (COD) and biochemical oxygen demand (BOD). The high COD value of cheese whey may be attributed to the rich content of lactose (35–50 g/L) in it (Carvalho et al., 2013). It has a yellow-greenish color due to the presence of riboflavin with about 65 g of total solids per liter. The whey composition is dependent on the origin of milk, the type of cheese manufactured (acid coagulation or rennet), as well as on other factors which influence milk composition (feed, breed, seasonal cycles, and lactation phase) (Pires et al., 2021). On the basis of the processing technique used, it can be categorized into two types of acid and sweet whey. The pH of acid whey varies from 4.5 to 5.8 (Argenta and Scheer, 2019) and is composed of fat (5–6 g/L), protein (6–8 g/L), lactose (44–46 g/L), and minerals (4.3–7.2 g/L) (Guo and Wang, 2019). On the other hand, the pH of sweet bovine whey varies from 6–7 with fat content of 5–6 g/L, protein 6–10 g/L, lactose 46–52 g/L, and minerals 2.5–4 g/L. It is formed as a result of the production of most types of cheese as well as from some casein products (Ryan and Walsh, 2016; Guo and Wang, 2019). The rich pool of nutrients and growth factors present in whey increases its potential to serve as a stimulating substance for the growth of microorganisms.

6.2.4 OLIVE MILL WASTEWATER

Olive mill wastewater (OMWW) is generated as the major liquid effluent after olive oil extraction. It is generally recognized as dark color juice which is composed of water from the olive, machinery cooling waters, fruit washings, and the remainder of the fruit (Öner, 2013). OMWW is considered highly toxic as it possesses high content of phenolic compounds and marked acidity (Domingues et al., 2018). Its organic load is extremely high along with recalcitrant organic compounds like polyphenols having antimicrobial properties. The chemical composition of OMWW is dependent on the region and origin of the olive tree, the type and maturity of olives, and the climatic conditions. The composition may also be influenced by the method of oil extraction. OMWW has been reported to contain 83–94% water, 4–16% inorganic fraction, and 0.4–2.5% organic fraction (w/w) (Davies et al., 2004). It is also composed of valuable metals such as magnesium and calcium. The composition of organic portion of OMWW consists of proteins, lipids, and carbohydrates as well as a number of other organic compounds including polyaromatic and monoaromatic molecules (Aguilera et al., 2008). Apart from these, it also possesses copper, iron, and zinc in different amounts (Anne, 2011). The phenolic fraction of OMWW is divided into high and low molecular weight components (Obied et al., 2007).

6.2.5 CORN STEEP LIQUOR

The main by-product generated after corn starch production is in the form of corn steep liquor (CSL). It is a soluble solid which constitutes 40–50% of the dry weight of corn and has a rich composition of reducing sugars and various amino acids (Cornejo-Villegas et al., 2018; Martinez-Arcos et al., 2021). The presence of reducing sugars, total sugars, sulfites, nitrogen, and free amino acids makes CSL an important substrate for fermentation, even though its chemical and physical functions may be affected by the corn varieties (Chovatiya et al., 2011). CSL may be potentially utilized as ruminant feed due to its low fiber content, high crude protein, and high energy (Nisa et al., 2004; Obayori et al., 2010). It is also rich in vitamin B complex, minerals, and phenolic antioxidants (Chovatiya et al., 2011; Niwa et al., 2001). Since it is considered a cheap source of nitrogen, carbon, and vitamins, it may be utilized in the production of antibiotics, lactic acid, glutamate, and other biotechnological products (Hull et al., 1996).

The chemical composition of some major agro-industrial by-products utilized for exopolysaccharide production has been presented in Table 6.1.

TABLE 6.1
Composition of some major agro-industrial by-product utilized for exopolysaccharide production

| Agro-industrial by-product | Components | | | | | | References |
	Lactose/ Carbohydrates	Protein	Water	pH	Fat	Mineral	
Corn steep liquor	5.80#	24#	–	3.86	1.0#	8.8#	(Achal et al., 2010)
Apple pomace	2.5–22.7*	1.2–6.91*	3.97–9.75	-	0.26–8.49*	1.65*	(Antonic et al., 2020; Zeraik et al., 2010)
Passion fruit peel	6.78*	0.67*	87.64*	-	0.01*	-	
Sweet whey	46–52#	6.0–10.0#	93–94#	6.2–6.4	-	-	(Jelen, 2000; Olusola et al., 2012)
Acid whey	44–46^	6.0–8.0#	93–94#	4.6–5.0	-	-	
Olive mill wastewater	2–8*	-	83–94*	4.7–5.7	0.03–1.1*	-	(Rahmanian et al., 2014)
Cane molasses	50.9#	-	12.0#	6.07	-	-	(Yatim et al., 2017)

'*'—g/100 g, '#'—%, '^'—g/L, '@'—g/cm³

6.3   MICROBIAL SOURCES

6.3.1   BACTERIA

Bacterial and archaebacterial are one of the chief sources of exopolysaccharides. Some of the major bacterial species producing exopolysaccharides include *Streptococcus, Xanthomonas*, and *Acetobacter*. In thermophilic and halophilic bacteria like *Archaeglobus fulgidus, Thermococcus litoralis*, and *Thermococcus sulfolobus*, it occurs in the form of surface biofilm (Poli et al., 2011). A microbial group called myxobacteria is responsible for the production of an enormous variety of polysaccharides, which plays a role in locomotion, and protection against dehydration and phagotrophs (Hogg, 2013). Xanthan, gellan gum, and dextran are the most common exopolysaccharides derived from these prokaryotic species. Dextran, a water-soluble glucan was the first bacterial EPS commercially explored and produced by bacterial species of the genera *Lactobacillus, Streptococcus, Leuconostoc, Pediococcus*, and *Weisella*. These are generally regarded as safe (GRAS) and commonly utilized in the manufacturing of different varieties of food products (Patel and Prajapati, 2013). Curdlan is a water insoluble-gel polymer and is produced by the bacterial species *Agrobacterium* and *Alcaligens*. Both curdlan and dextran are FDA (Food and Drug Administration) approved and commonly utilized in food, cosmetic, and medical applications (Yang et al., 2016). Cellulose, which is a water-insoluble biopolymer has high water absorption capacity and mechanical strength produced by species of the genera *Rhizobium, Acetobacter*, and *Glucanoacetobacter* (Gomes et al., 2013). Xanthan, a heteropolysaccharide that has a glucose backbone with side chains of trisaccharide, is commercially derived from a bacterial species, *Xanthomonas campestris* (Silva et al., 2009). Both xanthan and dextran are considered as vector molecules (Corrêa et al., 2016), which delivers drugs to specific target or tissue and plays a significant role as natural anionic polyelectrolytes (Luo and Wang, 2014). A bacterial species, *Streptococcus zooepidemicus* are responsible for the production of hyaluronic acid (HA), which is a linear polymer composed of repeating disaccharide units of N-acetylglucosamine and glucuronic acid (Jia et al., 2013). A branched homofructan, Levan is produced from sucrose by the action of Levansucrase, which is an enzyme secreted by a wide variety of bacterial species, namely, *Halomonas, Zymomonas, Bacillus,* and *Pseudomonas* (Öner et al., 2016). Some of the characteristics of Levan, including water and oil solubility, compatibility with salts and surfactants, film forming, emulsifying capacity, and adhesive capacity make it suitable for use in a number of food applications (Silbir et al., 2014). Several other types of EPS with distinctive structural and novel characteristics are secreted by numerous other bacteria like *Pseudomonas oleovorans, Klebsiella pneumoniae, Enterobacter A47, Escherichia coli*, etc. (Patel and Prajapati, 2013). EPS finds applications in a number of applications pertaining to food manufacturing, quality, processing, and preservation. EPSs like dextran helps improve moisture retention in confectionary products and increases sugar crystallization and viscosity, while xanthan plays an important role as an emulsifier, thickening agent, and stabilizer in several food industries. The supplementation of EPS in foods changes its visco-electrolyte and rheological properties of the water (Madhuri and Prabhakar, 2014). EPS obtained from LAB are commonly utilized in dairy and fermented foods like yoghurt, sour cream, cheese, etc. to enhance the taste, flavor, texture as well as shelf life of fermented foods (Roca et al., 2015).

6.3.2 FUNGI AND YEAST

A wide array of fungi of the genera *Candida, Cryptococcus, Aureobasidium*, etc. has been associated with the production of EPS. Among these, homopolysaccharides are more common, but heteropolysaccharides are also produced by several fungal species. Pullulan, a linear alpha-glucan with high mechanical strength and adhesiveness is secreted by the black yeast *A. pullulans* (Prajapati et al., 2013). A water soluble beta-glucan named scleroglucan is secreted by a plant pathogen, *Sclerotium* sp. (Schmid et al., 2011). Scleroglucan has been utilized in several applications in food, pharmaceutical, and cosmetic industries (Castillo et al., 2015). Many higher fungi of the genera *Tremella, Poria*, and *Ganoderma* have been reported to secrete EPS with bioactive properties like antioxidant, antimicrobial, antitumor, and immunostimulating properties (Osińska-Jaroszuk et al., 2015). In the past, endophytic fungi and mushroom have been utilized for the production of useful exopolysaccharides (Sood et al., 2013). In addition to this, a vast array of fungi including *Fusarium* sp., *Pleurotus* sp., *Phelinus linteus, Ganoderma lucidium*, and *Inonotus obliquus* have also been exploited as a potential source of exopolysaccharides (Guo et al., 2013). Yeast isolates such as *Candida utilis, Saccharimyces ludwigii, S. cerevisiae*, and *S. fibuligera* have commonly been applied for the production of beta-glucan (Bzducha-Wróbel et al., 2019). A fungus of the genera *Diaporthe* sp. has been reported to produce an exopolysaccharide that has 8% protein and 91% carbohydrate indicative of high-quality EPS (Orlandelli et al., 2016). EPSs with high molecular weight have been observed to possess effective biological activity in comparison to that of low molecular weight. The variation in molecular weight as well as sugar composition of fungal EPSs are determined by the physical conditions during the fermentation process as well as composition of culture media.

EPS produced by fungi holds applications in food and cosmetic industries. Exopolysaccharides produced from fungal strains like glucan, lentinan, and schizophyllan are commonly considered as biological response modifiers due to their medicinal properties. These compounds have the capacity of triggering a nonspecific reaction against tumor cells, bacterial and viral infections, as well as inflammation (Giavasis, 2014). EPS secreted by a fungal strain, *Ganoderma neojaponicum*, can play an important role in immunomodulatory activity as it may stimulate the immune system to fight against infectious diseases (Nik Ubaidillah et al., 2015).

Some of the common microbial sources of exopolysaccharides with their structure and molecular weight have been presented in Table 6.2.

TABLE 6.2

Microbial sources of exopolysaccharides with their molecular weight

Type of EPS	EPS name	Molecular weight (D)	Microbial source	References
Homopolysaccharide	Xanthan	3×10^6	*Xanthomonas* spp.	Angelin and Kavitha (2020), Rana and Upadhyay (2020)
Homopolysaccharide	Pullulan	362×10^3-480×10^3	*Rhodotorula bacarum, Rhodosporidium paludigenum, Cyttaria darwinii, Cyttaria harioti, Cryphonectria parasitica, Aspergillus japonicus, Teloschistes flavicans, Tremella mesenterica, Micrococcus leuteus*	Sugumaran et al. (2017), Mishra et al. (2018)
Homopolysaccharide	Dextran	10^3-10^7	*Leuconostoc, Streptococcus,* and *Acetobacter*	Mensink et al. (2015)
Homopolysaccharide	Levan	10^4-10^8	*Zymomonas mobilis Pseudomonas fluorescens, Serratia Levanicum, Microbacterium laevaniformans, Lactobacillus* spp., *Bacillus stearothermophilus*	Inthanavong et al. (2013), Mensink et al. (2015), Ferreira et al. (2022)
Homopolysaccharide	Curdlan	2×10^6	*Alcaligenes faecalis* var. *myxogenes,* some *Rhizobium* strains, and *Cellulomonas* spp.	Laroche and Michaud (2007), Nishinari (2009), McIntosh et al. (2005)
Homopolysaccharide	Scleroglucan	6×10^6	*Sclerotium rolfsii*	Li et al. (2020)
Heteropolysaccharide	Gellan	5×10^5	*Pseudomona elodea* and *Sphingomonas* spp.	Prajapati et al. (2013), Dev et al. (2022)
Heteropolysaccharide	Alginate	33×10^3-400×10^3	*Pseudomonas aeruginosa, Azotobacter vinelandii*	Iwaki et al. (2012), Rana and Upadhyay (2020)

6.4 METHODS FOR THE PRODUCTION OF EXOPOLYSACCHARIDES

6.4.1 SOLID STATE FERMENTATION

Solid state fermentation has been considered an appropriate tool for the biotransformation of agro-industrial wastes into high-value products such as bioplastics, biofuels, and bioactive compounds (Sharma et al., 2021). It is an emerging technology in which microbial growth and product formation occur on the surface of solid materials. The moisture is absorbed on to the solid substrate for supporting microbial growth and metabolism (Pirota et al., 2014). Substrates used most commonly in SSF include lignocellulosic materials like straws, wood shavings or sawdust, cereal grains (corn, barley, wheat, and rice), wheat bran, legume seeds, and a wide range of plant and animal materials. These are low cost and easily obtained substrates composed of polymeric compounds which remain sparingly soluble or insoluble in water offering a source of nutrients for growth of microorganisms (Sadh et al., 2018). The lower amount of water or the absence of water in SSF is linked with several advantages such as low production processing cost, reduced downstream processing, smaller fermenter-size, easy product recovery, and also reduced requirements of energy for sterilization and stirring (Pandey, 2003).

The use of agro-industrial by-products as substrates for SSF is an eco-friendly method of waste management as its disposal adds extra cost to the processors and its direct disposal into landfills or soil may lead to serious environmental issues. The fermentation media may contribute to almost 50% of the total cost in the microbial fermentation process (Küçükaşik et al., 2011). Employment of complex media may be economically unattractive due to the high cost of nutrients like peptone, yeast extract, salts, etc. Thus, selecting agro wastes as raw materials for the production of exopolysaccharides may help reduce the production cost of the overall process. The EPS production by *Bacillus subtilis* was found to increase when sucrose was replaced with cane molasses and rice bran at same concentration (20 g/L), resulting in 4.86 g/L of EPS (Razack et al., 2013). Similarly, apple pomace had been utilized to produce inulinase using *Mucor circinelloides* via SSF (Singh et al., 2020).

Solid state fermentation involves the utilization of water-insoluble materials for the growth of microorganisms and their metabolism (Noomhorm et al., 1992) in the absence or near absence of water. The technology is beneficial in terms of the direct utilization of solid substrates for the production of microbial biomass products under aerobic conditions. The solid substrates work as the nutrient supply to the growing microbes present on it. An increase in the nutrient level through microbial synthesis may be observed during the process with negligible loss of nutrients (Wee, 1991). Microorganisms multiply and produce protein in large quantities by utilization of organic acids and sugars present in the substrates (Stabnikova et al., 2005). Solid state fermentation offers high quality and activity of the extracts. The absence of organic solvents, lower capital, operational expenses, and downstream refining makes it an overall excellent process for industrial use (Cano y Postigo et al., 2021). The development of a reliable SSF process involves the usage of particular kinds of bioreactors such as rotating disc, tray, column, rotating drum, fixed bed, fluidized bed, airlift, air pressure pulsation, and immersion bioreactors. Tray bioreactors, the most appropriate and traditional bioreactors have a simple design and ease of use for scaling up of SSF (Selo et al., 2021). A wide range of microorganisms have been employed to check its potential in SSF. Results have indicated fungi, specifically filamentous fungi, as the most compliant candidate for SSF. Although SSF process involves the extraction of materials of industrial importance, it has a few disadvantages, including difficulty in maneuvering yield, turnaround time of the process, optimization, and purification of the products. Apart from this, SSF is a labor-intensive process with substrate selection influencing the yield of the target product and optimization (Ihayere et al., 2017).

6.4.2 SUBMERGED FERMENTATION

Submerged fermentation involves the utilization of liquid substrates at both laboratory and industrial scales. Agro-industrial residues including molasses, industrial wastewater, fruit residue juices, and solid residues extract from fermentations are among the commonly employed substrates used for SmF (Mohapatra et al., 2017). The method is favorably apt for the microorganisms which require moisture content in the range higher than SSF, which occurs in absence of free water. The ease of manipulation of growth factors (Papaspyridi et al., 2010) and requirement of nutrients (Xiao et al., 2010) for small and medium-scale production as well as longer storage of mycelia without any changes in genetic integrity accounted for the increased focus on SmF (Corrêa et al., 2016). The effectiveness of the fermentation process is dependent on the metabolic capacity of the microorganisms, growth nutrients, and physical conditions like temperature and pH. The research in SmF is focused on the determination of production economics of the process through enhancement in productivity and product yields (Castilho et al., 2000). In SMF or SSF process, the bioreactor creates a suitable environment for the growth of microorganisms and biological activity (Krishna, 2005). Effective bioreactors are generally associated with biologically high reaction rates, including substrate consumption, product formation, growth performance, and synthesis rate of by-products (Bastos et al., 2016).

A number of useful products of commercial importance can be obtained through SmF using agro-industrial by-products such as seeds, stones, peels from fruits and vegetables as they possess valuable biochemical constituents (Sagar et al., 2018). Growth of microbial cells on agro-residues as a substrate is gaining interest with fungal mycelia being of prime importance as it utilizes complex organic compounds in agro-wastes through fermentation to form useful biomolecules which act as nutritional supplements (Gupta et al., 2019). Plant biomass waste is, therefore, considered a desirable substitute of raw material for the production of various functional products as it is readily available and bio-renewable. A fungal species of the genus *Pleurotus* has been reported to utilize different agro-residual waste to produce polysaccharides (Baeva et al., 2019). Fungal exopolysaccharides are found to have many therapeutic effects such as anticancer, antioxidant, and antimicrobial properties (Osemwegie et al., 2020). Submerged culture of *Pleurotus pulmonarius* has been utilized to produce exopolysaccharides using peels of pineapple, plantain, groundnut shell, mango, walnut husk. and coconut coir, as substrates (Ogidi et al., 2020). It has been reported that most of the fungi responsible for the production of EPS are either facultative anaerobic or aerobic, and oxygen limitation did not support the production of EPS (Sandford, 1979). The constituents of broth and its behavior at very agitation speeds, low shear rate, and firm airflow access play a critical role in the synthesis process during submerged fermentation which results in ideal growth conditions for microorganisms (Reddy et al., 2021). The production of EPS under submerged fermentation conditions can lead to the irregular distribution of oxygen and higher viscosity which can negatively influence the EPS quality and synthesis. Such problems can be overcome by using different rates of agitation depending on the type of microorganism used (Kumar et al., 2007). The evaluation of EPS production by *Aspergillus parasiticus* in static and agitated submerged conditions indicated more EPSs (0.41 g/L) formation using the agitated culture technique than static culture (0.18 g/L) (Ruperez and Leal, 1981).

6.5 PRODUCTION OF SELECTED MICROBIAL EXOPOLYSACCHARIDES

6.5.1 XANTHAN GUM

Xanthan gum is one of the most important commercial hetero-polysaccharide which is secreted mainly by a bacterium, *Xanthomonas campestris* (Habibi and Khosravi-Darani, 2017). The use of sucrose and glucose (expensive substrates) as carbon sources as well as the cost of downstream processing (around 50% of the total cost) makes xanthan gum an expensive microbial polysaccharide since a high level of purity is essential before its use in the food industry (Li et al., 2016).

The optimization of the biotechnological process of xanthan gum production using low cost-substrates such as agro-industrial by-products is a reliable alternative to high-cost substrates, which increases the processing cost (Benny et al., 2014). Some of the alternative sources which have been utilized for the production of xanthan gum include cheese whey, cocoa residue, juice residue, sugar cane broth, treated tapioca pulp, shrimp cell, waste sugar beet, green coconut shell, and glycerin derived from the biodiesel production (dos Santos et al., 2016). The presence of other carbohydrates such as pectin, lactose, and other nutrients (vitamins, proteins, minerals) plays an important role in increasing the yield of xanthan (Menezes et al., 2012). Four types of citrus waste fractions, including whole citrus waste from lemons, oranges, and grape fruits, hemicellulosic extracts, pectic extracts, and cellulosic extracts have been utilized for the production of xanthan gum using *Xanthomonas campestris* ATCC 13951 (Bilanovic et al., 1994). Low substrate consumption and high yield was obtained when xanthan gum was produced by the bacterium *Xanthomonas campestris* using green coconut shells (50% and 5.5 g/L) and passion fruit (54.7% and 6.7 g/L) (dos Santos et al., 2016). When *Xanthomonas citri* was used for the production of xanthan under solid state fermentation using potato peels as substrate, a higher yield of 2.90 g/50 g was obtained. This was followed by *X. campestris* (2.87 g/50 g), *X. oryzae* (1.50 g/50 g), and *X. musacearum* (0.50 g/50 g), respectively (Vidhyalakshmi et al., 2012). Production of xanthan gum has also been successfully performed using dairy waste (cheese whey) as substrate in a three phase semifluidized bed biofilm bioreactor using aerobic microbes such as *X. campestris* (Narayana and Narayan, 2019).

A number of alternative substrates have been reported recently for the low-cost xanthan gum production including sugarcane broth, rice straw, green coconut shells, cheese whey, etc. Some of the companies involved in large-scale commercialization of xanthan gum include Sanofi-EIF and Mero Rousselot-Santia (France), Jungbunzlauer (Austria), Merck and Pfizer (USA), Saidy Chemica (China) (Habibi and Khosravi-Darani, 2017).

6.5.2 Gellan gum

Gellan gum is an anionic linear heteropolysaccharide that consists of repeating units of alpha-rhamnose, two residues of beta-D glucuronate and beta-D-glucose possessing high thermal stability and viscosity (Moscovici, 2015). It is produced by aerobic submerged fermentation using a nonpathogenic genus *Sphingomonas paucimobilis*. Variation in extraction conditions leads to the formation of a wide range of gellan gums with varied degrees of esterification. Sucrose as a carbon source and yeast extract as a nitrogen source results in the maximum production of gellan gum (Zhang et al., 2015).

The raw materials cost has a significant effect on gellan gum production. Due to the high production cost associated with the fermentation process, gellan gum has a higher market price than xanthan gum. Out of the total cost of gellan gum, 30% of it is contributed by media formulation and fermentation (Huang et al., 2020). Some of the cost-effective substrates containing agro-industrial wastes and by-products such as glycerol (Freitas et al., 2009), molasses, and cheese whey (Wang et al., 2020) have been utilized to enhance the exopolysaccharides production such as gellan gum. The microbial culture of *Sphingomonas paucimobilis* ATCC31461 has been employed to grow in a production medium containing lactose or cheese whey for the production of highly viscous gellan gum (Fialho et al., 1999). Grape pomace, which constitutes 15–25% of the processed grapes, is obtained as a result of the compression and compaction of grapes and stems during the production of wine, grape juice, and molasses. The high nutritional value of fresh grape pomace constituted by 15–20% of soluble (sucrose, fructose, glucose, etc.) and fermentable carbohydrates is significant for the production of gellan gum (Kamer et al., 2021). Carbon source is considered as the most important media component involved in gellan gum production as it directly affects the composition, structure, technological properties, and production yield (Bajaj et al., 2007; Pollock, 2005). It was observed that using grape pomace at 4% inoculum volume, 50 g/L carbon ratio, and 300 rpm agitation rate resulted in a maximum production yield of gellan gum (12.08 g/L). Since grape pomace is an economical substrate, contains high sugar content, and does not require any pre-treatment or enzymatic hydrolysis, it may serve as an important carbon source for gellan production (Kamer et al., 2021). Production of gellan gum where milk skin residue was used as a carbon source in the medium formulation resulted in maximum yield (8.98 g/L) (Viswanathan et al., 2020).

Gellan gum is considered to be one of the widely utilized exopolysaccharides as being thermoreversible. It possesses good acid and thermal stability as well as has the ability to form a transparent gel in the presence of multivalent cations. Some of the commercialized gellan products include Kelcogel and Gelrite being a potent replacer for gelatin and agar (Madhuri and Prabhakar, 2014).

6.5.3 Pullulan

Pullulan is a linear homopolysaccharide that consists of alpha-(1–4) glycosidic bonds connecting maltotriose units and alpha-(1–6) units connecting maltotrioses. It is produced by yeast-like fungi *A. pullulans* with 362–480 kDa of average molecular weight (Mishra et al., 2011). Properties such as low viscosity, water solubility, good oxygen barrier, excellent adhesive properties, capacity to mimic petrochemical-derived polymers, and film-forming properties (Barcelos et al., 2020). Addition to the expenses of production is attributed to the high cost of nutrients required in the synthesis of pullulan (Mishra and Varjani, 2019). However, waste generated by many food processing industries is rich in organic and inorganic compounds essential for pullulan production. A multitude of by-products from food industries has been reported as suitable to be used in pullulan production (Wani et al., 2021; Abdeshahian et al., 2021).

Despite the several advantages offered by pullulan, it is less exploited than other biopolymers due to its high cost (Leathers, 2003). Replacement of conventional media components like yeast extract, dextrose, peptone, etc. utilized in biopolymer production with low-cost substrates may be economically advantageous (Sharma et al., 2013). Carbon and nitrogen sources are the prerequisites for the fermentation production of pullulan. However, the cost of pullulan production may increase with these valuable substrates (Sugumaran et al., 2013). Agro-industrial wastes are considered rich alternative sources of organic and inorganic constituents (Singh and Kaur, 2019), which makes them cost-effective substrates for pullulan production (Wang et al., 2014). Some of the agroindustrial wastes like soybean hydrolysate, coconut by-product, potato starch waste, cassava starch, grape skin pulp extract, eggplant peel, soybean meal hydrolysate, corn steep liquor, jack fruit seed, jaggery, and rice hull hydrolysate has been utilized in the production of pullulan through fermentation (Akdeniz Oktay et al., 2022). Utilization of dried hazelnut husk hydrolysate and sugarcane molasses residue as substrate with *Aureobasidium pullulans* AZ-6 resulted in pullulan yield of 30.02 g/L and 33.59 g/L, respectively. Corn starch and whole corn crop hydrolysate have been utilized as the carbon source by *A. pullulans* EV6 strain for efficient production of pullulan (He et al., 2023). Replacement of agro-industrial residues (soybean oil cake, rice bran oil cake, mustard seed oil cake, corn steep liquor, and cottonseed oil cake) as nutrients in place of conventional media components (yeast extract, peptone) has been carried out for the production of pullulan. Corn-steep liquor along with glucose (15%, w/v) as a carbon source was found to give the maximum yield (88.59 g/L) of pullulan and thus found to be an effective replacement of conventional media (Sharma et al., 2013). Agricultural wastes utilized as starting point of carbon generation in fermentation media may result in an efficient production of pullulan and, hence, considered as a promising substrate from an economic and ecological point of view (Dailin et al., 2019; Akdeniz Oktay et al., 2022).

A number of applications of pullulan have been documented in food industries. It has proven useful in making edible coatings, as substitute of starch in low calorie food recipes, as a flavoring and spice in microencapsulated seasoning agents as well as a plant nutrient, seed coat, and tobacco binder (Priyadarshi et al., 2021). The oxygen resistance capacity of pullulan makes it suitable to be utilized as a film for lipids and vitamins in food (Abdeshahian et al., 2021). Enzymatic hydrolysis of pullulan and utilization of debranching enzyme pullulanase may lead to formation of maltotriose syrup (Priyadarshi et al., 2021).

6.5.4 LEVAN

Levan is a nontoxic high molecular weight homofructan found in bacteria (*Bacillus, Erwinia, Zymomonas, Streptococcus, Aerobacter, Rahnella, Microbacterium, Pseudomonas,* and *Halomonas*) few strains of yeast, fungi, and in plants (Barcelos et al., 2020). Levan is a 2,6 linked beta-D-fructose polymer having a molecular weight of 107 kDa, which corresponds to approximately 60,000 fructose units.

The fermentation medium employed in the production of the exopolysaccharide can contribute around 50% of the total microbial fermentation cost (Küçükaşik et al., 2011). The employment of complex media for microbial growth is not economically attractive due to the high amount of expensive nutrients such as peptone, yeast extract, and salts. Therefore, the selection of agro-wastes as raw material for exopolysaccharide production helps in the reduction of production costs. Waste from the orange juice industry has been utilized as an alternative culture medium with *B. atrophaeus* for production of levan (3 g/L for 30 h) (González-Garcinuño et al., 2017). The production cost of media is influenced by the amount of carbon, nitrogen, micro, and macronutrients. Therefore, increased demand of low cost and high yielding raw material for production of levan has been ongoing (Yezza et al., 2006). In this context, agro-industrial waste has been focused as suitable low-cost carbon source for the production of levan. Some of these agro-industrial waste includes sugarbeet molasses, sugarcane molasses, beet molasses, date syrup, and starch molasses (Moosavi-Nasab et al., 2010; Küçükaşik et al., 2011). Chicken feather peptone (CFP) derived from poultry waste has been utilized as a low-cost complex supplemental nutrient source for microbial production of levan. CFP (2 g/L) added to the medium resulted in maximum levan yield (0.26 ± 0.04 g/g sucrose) comparable to that obtained from commercial medium (0.31 ± 0.02 g/g sucrose) (Veerapandian et al., 2020). Species of *Halomonas* AAD6 cells cultivated on pretreated beet molasses (30 g/L) resulted in a considerable amount of levan yield (12.4 g/L) (Küçükaşik et al., 2011). Fermentation of sucrose by the bacteria *Zymomonas mobilis* resulted in the production of 2% levan as one of the products (Panesar et al., 2006). Currently, levan from *Bacillus subtilis* (Natural Polymers Inc., Bainbridge, GA, USA), *Z. mobilis* (Real Biotech Co., Chungnam, Korea), and *Streptococcus salivarius* (Advance Co., Tokyo, Japan) are commercialized in the market with registered trademark (Barcelos et al., 2020).

6.5.5 MISCELLANEOUS

Dextran is one of the exopolysaccharides made up of 1,6 alpha-D glucopyranose units and commonly secreted by bacterial strain, *Leuconostoc mesenteroides* NRRL B-512 F. Another newly isolated strains like *L. mesenteroides* PCSIR-9 and PCSIR-4 have also been documented to produce dextran of another quality. Sugarbeet molasses has been utilized for the production of dextran (50 g/L) under submerged fermentative conditions using *L. mesenteroides* (Vedyashkina et al., 2005). A combination of cheese whey and molasses under submerged fermentation conditions resulted in a dextran yield of 9.51 g/L (Moosavi-Nasab et al., 2010). Carob extract obtained from carob pod residues were employed for the production of dextran (8.56 g/L) under a fermentation period of 12 h by using *L. mesenteroides* NRRL B512 (Santos et al., 2005). In another study, date seeds were utilized for the production of dextran under SSF by using *Saccharomyces cerevisiae*, which resulted in the highest yield of 6 g/flask (Seesuriyachan et al., 2010). Succinoglycan is a branched EPS with galactose and glucose units in the main chain and tetra-sachharides composed of modified sugar residues in the main chain (Kanimozhi et al., 2018). Several soil bacteria like *Rhizobium, Pseudomonas, Alcaligens,* and *Agrobacterium* have been employed for the production of succinoglycan (Glenn et al., 2007). Various agro-industrial waste such as spent malt grains, grated carrots, etc. has been used as substrates for the production of succinoglycan by *Agrobacterium tumefaciens* under SSF (Stredansky and Conti, 1999). Rice husk hydrolysate has been utilized as a fermentative substrate for the production of suc-cinoglycan (69 g/L) using *Rhizobium radiobacter* (Pedroso et al., 2019). Alginate is an anionic polymer which comprise of mannuronic acid and glucuronic acid resi-due. It is widely produced by species of *Azotobacter,* and *Pseudomonas.* Alginate has been produced by an agro-industrial waste, whey broth by using *Azotobacter chroococcum* species (Khanafari and Sepahei, 2007). Similarly, another EPS cur-dlan has been produced with date palm juice and by-products as substrate by using *Rhizobium radiobacter* (Salah et al., 2011).

The commonly utilized agro-industrial wastes utilized for the fermentative pro-duction of microbial exopolysaccharides have been presented in Table 6.3.

TABLE 6.3
Agro-industrial wastes utilized for fermentative production of microbial exopolysaccharides

EPS name	Agro-industrial waste	Microbial source	Yield (%)	References
Xanthan	Passion fruit peel, green coconut shell, corn cobs	*Xanthomonas campestris*	6.7 g/L (using passion fruit peel)	dos Santos et al. (2016)
Xanthan	Potato peels	*Xanthomonas* sp.	2.9 g/50 g peel (6 days)	Vidhyalakshmi et al. (2012)
Xanthan	Tapioca pulp	*X. campestris* NCIM 2954	7.1 g/L	Gunasekar et al. (2014)
Pullulan	Jatropha seed cake and corn steep liquor	Aureobasidium pullulans RBF 4A3	66.25 g/L	Mehta et al. (2014)
Pullulan	Citrus peels, grape pomace, hydrolysates of hazelnuts and chestnut shells, sugarcane molasses residue, dried and fresh hazelnut husks, pumpkin peel	*Aureobasidium pullulans* AZ-6	33.59 g/L (using sugarcane molasses)	Akdeniz Oktay et al. (2022)
Pullulan	Corn cob and straw hydrolysates	*Aureobasidium pullulans* CCTCC M 2012259	20.25 g/L (using corn straw)	He et al. (2023)

(Continued)

TABLE 6.3 (*Continued*)
Agro-industrial wastes utilized for fermentative production of microbial exopolysaccharides

EPS name	Agro-industrial waste	Microbial source	Yield (%)	References
Pullulan	Soybean oil cake, rice bran oil cake, cottonseed oil cake, mustard oil cake, corn steep liquor	*Aureobasidium pullulans* RBF 4A3	88.59 g/L	Sharma et al. (2013)
Pullulan	De-oiled jatropha seed cake	*Aureobasidium pullulans* RBF 4A3	-	Choudhury et al. (2012)
Gellan	Milk skin residue	*Sphingomonas paucimobilis* MTCC2308	8.98 g/L, 16 h	Viswanathan et al. (2020)
Gellan	Grape pomace	*Sphingomonas paucimobilis* (ATCC 31461)	12.08 g/L	Kamer et al. (2021)
Gellan	Sugar beet molasses	*S. paucimobilis* ATCC-31461	13.81 g/L, 48 h	Banik and Santhiagu (2006)
Levan	Orange juice pulp	*Bacillus atrophaeus*	3 g/L, 30 h	González-Garcinuño et al. (2017)
Levan	Chicken feather peptone	*Bacillus subtilis* MTCC 441	0.26 ± 0.04 g/g, 20 h	Veerapandian et al. (2020)
Levan	Starch molasses	*Halomonas* sp. AAD6	12.4 g/L, 210 h	Küçükaşik et al. (2011)

6.6 CONCLUSION AND FUTURE PERSPECTIVES

Increasing environmental concerns and a greater focus on the use of biotechnological strategies in diverse fields have prompted microbes as a renewable source for the production of exopolysaccharides. Moreover, the rapid growth of microorganisms, and high productivity without safety concerns of the produced EPS have enabled the use of these generated compounds as an inexpensive ingredient in the improvement of organoleptic and nutritional properties of food products. However, despite their potential, many concerns regarding the high cost of the substrate, lower yields, and environmental issues involved with the fermentative process of EPS production have been of prime focus among the research community. The use of agro-industrial waste to circumvent these problems can be of great interest in terms of sustainable and economical production of EPS. Further studies can be focused on the use of metabolic engineering and immobilization technology to increase the yield of exopolysaccharides. Immobilization technology can be applied for the effective separation and regeneration of extracellular polysaccharides, whereas, the construction of new microbial strains can be achieved through metabolic engineering for increased production of exopolysaccharides.

REFERENCES

Abdel-Aziz, S.M., Hamed, H.A., Mouafi, F.E., 2012. Acidic exopolysaccharide flocculant produced by the fungus *Mucor rouxii* using beet-molasses. Res. Biotechnol. 3(6), 1–13.

Abdeshahian, P., Ascencio, J.J., Philippini, R.R., Antunes, F.A.F., de Carvalho, A.S., Abdeshahian, M., da Silva, S.S., 2021. Valorization of lignocellulosic biomass and agri-food processing wastes for production of glucan polymer. Waste Biomass Valorization. 12, 2915–2931.

Achal, V., Mukherjee, A., Reddy, M.S., 2010. Biocalcification by *Sporosarcina pasteurii* using corn steep liquor as the nutrient source. Ind. Biotechnol. 6(3), 170–174.

Afifi, M.M., 2011. Enhancement of lactic acid production by utilizing liquid potato wastes, Int. J. Biotechno. Biochem. 5(2), 91–102.

Aguilera, M., Quesada, M.T., Aguila, V.G., Morillo, J.A., Rivadeneyra, M.A., Romos-Cormenzana, A., Monteoliva-Sanchez, M., 2008. Characterization of *Paenibacillus jamilae* strains that produce exopolysaccharide during growth on and detoxification of olive mill wastewaters. Bioresour. Technol. 99, 5640–5644.

Akdeniz Oktay, B., Bozdemir, M.T., Özbaş, Z.Y., 2022. Evaluation of some agro-industrial wastes as fermentation medium for pullulan production by *Aureobasidium pullulans* AZ-6. Curr. Microbiol. 79, 93.

Amorim, H.V., Baso, L.C., Lopes, M.L., 2003. Sugar cane juice and molasses, beet molasses and sweet sorghum: Composition and usage. In: Ingledew W.M., Kelsall D.R., Austin G.D., Kluhspies C. (Eds.), *Alcohol Textbook*. Nothingham University Press, pp. 39–46.

Angelin, J., Kavitha, M., 2020. Exopolysaccharides from probiotic bacteria and their health potential. Int. J. Biol. Macromol. 162, 853–865.

Anne, K., 2011. Magnesium and calcium in drinking water and heart diseases. In: J. Nriagu (Ed.), *Encyclopedia of Environmental Health*. Elsevier, pp. 535–544.

Antonic, B., Jancikova, S., Dordevic, D., Tremlova, B., 2020. Apple pomace as food fortification ingredient: A systematic review and meta-analysis. J. Food Sci. 85(10), 2977–2985.

Aranda-Selverio, G., Penna, A.L.B., Campos-Sás, L.F., Santos Junior, O.D., Vasconcelos, A.F.D., Silva, M.D.L.C.D., Silveira, J.L.M., 2010. Rheological properties and effect of the salt addition on the exopolysaccharides viscosity produced by bacteria of *Rhizobium* genus. Quim. Nova. 33, 895–899.

Argenta, A.B., Scheer, A.D.P., 2019. Membrane separation processes applied to whey: A review. Food Rev. Int. 36, 1–30.

Baeva, E., Bleha, R., Lavrova, E., Sushytskyi, L., Copíkova, J., Jablonsky, I., Kloucek, P., Synytsya, A., 2019. Polysaccharides from Basidiocarps of cultivating mushroom *Pleurotus ostreatus*: Isolation and structural characterization. Molecules. 24(15), 2740.

Bajaj, I.B., Survase, S.A., Saudagar, P.S., Singhal, R.S., 2007. Gellan gum: Fermentative production, downstream processing and applications. Food Technol. Biotechnol. 45(4), 341–354.

Banik, R.M., Santhiagu, A., 2006. Improvement in production and quality of gellan gum by *Sphingomonas paucimobilis* under high dissolved oxygen tension levels. Biotechnol. Lett. 28, 1347–1350. https://doi.org/10.1007/s10529-006-9098-3

Barcelos, M.C., Vespermann, K.A., Pelissari, F.M., Molina, G., 2020. Current status of biotechnological production and applications of microbial exopolysaccharides. Crit. Rev. Food Sci. Nutr. 60(9), 1475–1495.

Bastos, R.G., Motta, F.L., Santana, M.H.A., 2016. Oxygen transfer in the solid-state cultivation of *D. monoceras* on polyurethane foam as an inert support. Braz. J. Chem. Eng. 33, 793–799.

Benny, I.S., Gunasekar, V., Ponnusami, V., 2014. Review on application of xanthan gum in drug delivery. Int. J. Pharmtech Res. 6(40), 1322–1326.

Bilanovic, D., Shelef, G., Green, M., 1994. Xanthan fermentation of citrus waste. Biores. Technol. 48(2), 169–172.

Brenes, A., Viveros, A., Chamorro, S., Arija, I., 2016. Use of polyphenol-rich grape by-products in monogastric nutrition. A review. Anim. Feed Sci. Technol. 211, 1–17.

Byg, I., Diaz, J., Øgendal, L.H., Harholt, J., Jørgensen, B., Rolin, C., Ulvskov, P., 2012. Large-scale extraction of rhamnogalacturonan I from industrial potato waste. Food Chem. 131(4), 1207–1216.

Bzducha-Wróbel, A., Koczoń, P., Błażejak, S., Kozera, J., Kieliszek, M., 2019. Valorization of deproteinated potato juice water into β-glucan preparation of *C. utilis* origin: Comparative study of preparations obtained by two isolation methods. Waste Biomass Valor. 11, 3257–3271.

Cano y Postigo, L.O., Jacobo-Velázquez, D.A., Guajardo-Flores, D., Garcia Amezquita, L.E., García-Cayuela, T., 2021. Solid-state fermentation for enhancing the nutraceutical content of agrifood by-products: Recent advances and its industrial feasibility. Food Biosci. 41, 100926.

Carvalho, F., Prazeres, A.R., Rivas, J., 2013. Cheese whey wastewater: Characterization and treatment. Sci. Total. Environ. 445, 385–396.

Castelli, H., Du Vale, L., 2013. *Handbook on Cheese: Production, Chemistry and Sensory Properties.* Nova Science Publishers, Inc.

Castilho, L.R., Polato, C.M.S., Baruque, E.A., Sant'Anna jr, G.L., Freire, D.M.G., 2000. Economic analysis of lipase production by *Penicillium restrictum* in solid-state and submerged fermentations. Biochem. Eng. J. 4, 239–247.

Castillo, N.A., Valdez, A.L., Fariña, J.I., 2015. Microbial production of scleroglucan and downstream processing. Front. Microbiol. 6, 1–19.

Choudhury, A.R., Bhattacharyya, M.S., Prasad, G.S., 2012. Application of response surface methodology to understand the interaction of media components during pullulan production by *Aureobasidium pullulans* RBF-4A3. Biocatal. Agric. Biotechnol. 1(3), 232–237.

Chovatiya, S.G., Bhatt, S.S., Shah, A.R., 2011. Evaluation of corn steep liquor as a supplementary feed for *Labeo rohita* (Ham.) fingerlings. Aquac. Int. 19(1), 1–12.

Cornejo-Villegas, M., Rincón-Londoño, N., Real-López, D., Rodríguez-García, M.E., 2018. The effect of Ca^{2+} ions on the pasting, morphological, structural, vibrational, and mechanical properties of corn starch-water system. J. Cereal. Sci. 79, 174–182.

Corrado, S., Caldeira, C., Eriksson, M., Hanssen, O.J., Hauser, H.E., van Holsteijn, F., Sala, S., 2019. Food waste accounting methodologies: Challenges, opportunities, and further advancements. Glob. Food Sec. 20, 93–100.

Corrêa, R.C.G., Brugnari, T., Bracht, A., Peralta, R.M., Ferreira, I.C., 2016. Biotechnological, nutritional and therapeutic uses of *Pleurotus* spp. (Oyster mushroom) related with its chemical composition: A review on the past decade findings. Trends Food Sci. Technol. 50, 103–117.

Dailin, D.J., Low, L.Z.M.I., Kumar, K., Abd Malek, R., Natasya, K.H., Keat, H.C., 2019. Agro-industrial waste: A potential feedstock for pullulan production. Biosci. Biotechnol. Res. Asia. 16(2), 229–250.

Das, D., Panesar, P.S., Panesar, G., Timilsena, Y., 2022. Sources, composition, and characterization of agro-industrial byproducts. In: A.K. Anal, P.S. Panesar (Eds.), *Valorization of Agro-Industrial Byproducts*. CRC Press, pp. 11–30.

Davies, L.C., Vilhena, A.M., Novais, J.M., Martins-Dias, S., 2004. Olive mill wastewater characteristics: Modelling and statistical analysis. Grasas y Aceites. 55(3), 233–241.

Dev, M.J., Warke, R.G., Warke, G.M., Mahajan, G.B., Patil, T.A., Singhal, R.S., 2022. Advances in fermentative production, purification, characterization and applications of gellan gum. Bioresour. Technol. 127498.

Domingues, E., Gomes, J., Quina, M.J., Quinta-Ferreira, R.M., Martins, R.C., 2018. Detoxification of olive mill wastewaters by Fenton's process. Catalysts. 8(12), 662.

Dos Santos, F.P., Jra, A.M.O., Nunesa, T.P., de Farias Silvab, C.E., de Souza Abud, A.K., 2016. Bioconversion of agro-industrial wastes into xanthan gum. Chem. Eng. 49, 145–150.

Ferreira, J.D.A.S., Sampaio, I.C.F., da Cruz Hora, C.E., Matos, J.B.T.L., de Almeida, P.F., Chinalia, F.A., 2022. Culturing strategy for producing levan by upcycling oil produced water effluent as base medium for *Zymomonas mobilis*. Process Biochem. 115, 49–56.

Fialho, A.M., Martins, L.O., Donval, M.L., Leitão, J.H., Ridout, M.J., Jay, A.J., Sá-Correia, I., 1999. Structures and properties of gellan polymers produced by *Sphingomonas paucimobilis* ATCC 31461 from lactose compared with those produced from glucose and from cheese whey. Appl. Environ. Microbiol. 65(6), 2485–2491.

Freitas, F., Alves, V.D., Pais, J., Costa, N., Oliveira, C., Mafra, L., Reis, M.A., 2009. Characterization of an extracellular polysaccharide produced by a *Pseudomonas* strain grown on glycerol. Bioresour. Technol. 100, 859–865.

Friedman, M., 1997. Chemistry, biochemistry, and dietary role of potato polyphenols. A review. J. Agric. Food Chem. 45(5), 1523–1540.

Giavasis, I., 2014. Bioactive fungal polysaccharides as potential functional ingredients in food and nutraceuticals. Curr. Opin. Biotechnol. 26, 162–173.

Glenn, S.A., Gurich, N., Feeney, M.A., González, J.E., 2007. The ExpR/sin quorum-sensing system controls succinoglycan production in *Sinorhizobium meliloti*. J. Bacteriol. 189, 7077–7088.

Goldmeyer, B., Pena, N.G., Melo, A., da Rosa, C.S., 2014. Physicochemical characteristics and technological functional properties of fermented blueberry pomace and their flours. Rev. Bras. Frutic. 36(4), 980–987.

Gomes, F.P., Silva, N.H.C.S., Trovatti, E., Serafim, L.S., Duarte, M.F., Silvestre, A.J.D., Neto Freire, C.S.R., 2013. Production of bacterial cellulose by *Gluconacetobacter sacchari* using dry olive mil residue. Biomass Bioenerg. 55, 205–211.

González-Garcinuño, Á., Tabernero, A., Sánchez-Álvarez, J.M., Galán, M.A., Martin del Valle, E.M., 2017. Effect of bacteria type and sucrose concentration on levan yield and its molecular weight. Microb. Cell Factories. 16(1), 1–11.

Gudiña, E.J., Couto, M.R., Silva, S.P., Coelho, E., Coimbra, M.A., Teixeira, J.A., Rodrigues, L.R., 2023. Sustainable exopolysaccharide production by *Rhizobium viscosum* CECT908 using corn steep liquor and sugarcane molasses as sole substrates. Polym. J. 15, 20.

Gunasekar, V., Reshma, K.R., Treesa, G., Gowdhaman, D., Ponnusami, V., 2014. Xanthan from sulphuric acid treated tapioca pulp: Influence of acid concentration on xanthan fermentation. Carbohydr. Polym. 102, 669–673.

Guo, M., Wang, G., 2019. History of whey production and whey protein manufacturing. In: M. Guo (Eds.), *Whey Protein Production, Chemistry, Functionality, and Applications*. John Wiley & Sons, pp. 1–12.

Guo, S., Mao, W., Li, Y., Tian, J., Xu, J., 2013. Structural elucidation of the exopolysaccharide produced by fungus *Fusarium oxysporum* Y24-2. Carbohydr. Res. 365, 9–13.

Gupta, S., Summuna, B., Gupta, M., Annepu, S.K., 2019. Edible mushrooms: Cultivation, bioactive molecules, and health benefits. In: J.M. Merillon, K. Ramawat (Eds.), *Bioactive Molecules in Food. Reference Series in Phytochemistry*. Springer, pp. 1–33.

Habibi, H., Khosravi-Darani, K., 2017. Effective variables on production and structure of xanthan gum and its food applications: A review. Biocatal. Agric. Biotechnol. 10, 130–140.

He, C., Zhang, X., Zhang, Z., Wang, C., Wang, D., Wei, G., 2023. Whole-crop biorefinery of corn biomass for pullulan production by *Aureobasidium pullulans*. Bioresour. Technol. 370, 128517.

Hogg, S., 2013. Essential Microbiology, 2nd Edn., John Wiley and Sons, West Sussex.

Huang, J., Zhu, S., Li, C., Zhang, C., Ji, Y., 2020. Cost-effective optimization of gellan gum production by *Sphingomonas paucimobilis* using corn steep liquor. Prep. Biochem. Biotechnol. 50, 191–197.

Hull, S.R., Yang, B.Y., Venzke, D., Kulhavy, K., Montgomery, R., 1996. Composition of corn steep water during steeping. J. Agric. Food Chem. 44(7), 1857–1863.

Ihayere, C.A., Okhuoya, J.A., Osemwegie, O.O., 2017. Cultivation of Ganodermalucidum (W.Curtis:Fr.) P. Karst on sawdust of brachystagianigerica Hoyle & A.P.D. Jones, Agric. Food Sci. J. Ghana. 10(1), 852–862.

Imandi, S.B., Bandaru, V.V.R., Somalanka, S.R., Bandaru, S.R., Garapati, H.R., 2008. Application of statistical experimental designs for the optimization of medium constituents for the production of citric acid from pineapple waste. Bioresour. Technol. 99(10), 4445–4450.

Inthanavong, L., Tian, F., Khodadadi, M., Karboune, S., 2013. Properties of *Geobacillus stearothermophilus* levansucrase as potential biocatalyst for the synthesis of levan and fructooligosaccharides. Biotechnol. Prog. 29(6), 405–415.

Iwaki, Y.O., Escalona, M.H., Briones, J.R., Pawlicka, A., 2012. Sodium alginate-based ionic conducting membranes. Mol. Cryst. Liq. 554(1), 221–231.

Jelen, P., 2000. Whey: Composition, properties, processing, and uses. In: F.J. Francis (Ed.), *Encyclopedia of Food Science and Technology*. John Wiley, pp. 2652–2661.

Jia, Y., Zhu, J., Chen, X., Tang, D., Su, D., Yao, W., Gao, X., 2013. Metabolic engineering of *Bacillus subtilis* for the efficient biosynthesis of uniform hyaluronic acid with controlled molecular weights. Bioresour. Technol. 132, 427–431.

Kamer, D.D.A., Gumus, T., Palabiyik, I., Demirci, A.S., Oksuz, O., 2021. Grape pomace as a promising source for gellan gum production. Food Hydrocoll. 114, 106584.

Kammerer, D., Claus, A., Schieber, A., Carle, R., 2005. A novel process for the recovery of polyphenols from grape (*Vitis vinifera* L.) pomace. J. Food Sci. 70(2), C157–C163.

Kanimozhi, J., Sivasubramanian, V., Achary, A., Vasanthi, M., Vinson, S.P., Sivashankar, R., 2018. Bioprocessing of agrofood industrial wastes for the production of bacterial exopolysaccharide. In: *Bioprocess Engineering for a Green Environment*, 1st ed. CRC Press, pp. 67–98.

Kaur, B., Panesar, P.S., Anal, A.K., Chu-Ky, S., 2022. Recent trends in the management of mango by-products. Food Rev. Int. 1–21.

Khan, S., Lu, F., Kashif, M., Shen, P., 2021. Multiple effects of different nickel concentrations on the stability of anaerobic digestion of molasses. Sustainability. 13(9), 4971.

Khanafari, A., Sepahei, A.A., 2007. Alginate biopolymer production by *Azotobacter chroococcum* from whey degradation. Int. J. Env. Sci. Technol. 4(4), 427–432.

Krishna, C., 2005. Solid-state fermentation systems—an overview. Crit. Rev. Biotechnol. 25(1–2), 1–30.

Küçükaşik, F., Kazak, H., Güney, D., Finore, I., Poli, A., Yeniqün, O., Nicolaus, B., Oner, E.T., 2011. Molasses as fermentation substrate for levan production by *Halomonas* sp. Appl. Microbiol. Biotechnol. 89(6), 1729–1740.

Küçükaşik, F., Kazak, H., Güney, D., Finore, I., Poli, A., Yenigün, O., Öner, E.T., 2011. Molasses as fermentation substrate for levan production by *Halomonas* sp. Appl. Microbiol. Biotechnol. 89, 1729–1740.

Kumar, K., Yadav, A.N., Kumar, V., Vyas, P., Dhaliwal, H.S., 2017. Food waste: A potential bioresource for extraction of nutraceuticals and bioactive compounds. Bioresour. Bioprocess. 4, 1–14.

Kumar, S.A., Mody, K., Jha, B., 2007. Bacterial exopolysaccharides—a perception. J. Basic Microbiol. 47, 103–117. https://doi.org/10.1002/jobm.200610203.

Laroche, C., Michaud, P., 2007. New developments and prospective applications for beta (1,3) glucans. Recent Pat. Biotechnol. 1(1), 59–73.

Leathers, T.D., 2003. Bioconversions of maize residues to value-added coproducts using yeast-like fungi. FEMS Yeast Res. 3, 133–140.

Li, P., Li, T., Zeng, Y., Li, X., Jiang, X., Wang, Y., Zhang, Y., 2016. Biosynthesis of xanthan gum by Xanthomonas campestris LRELP-1 using kitchen waste as the sole substrate. Carbohydr. Polym. 151, 684–691.

Li, X., Lu, Y., Adams, G.G., Zobel, H., Ballance, S., Wolf, B., Harding, S.E., 2020. Characterisation of the molecular properties of scleroglucan as an alternative rigid rod molecule to xanthan gum for oropharyngeal dysphagia. Food Hydrocoll. 101, 105446.

Luo, Y., Wang, Q., 2014. Recent development of chitosan-based polyelectrolyte complexes with natural polysaccharides for drug delivery. Int. J. Biol. Macromol. 64, 353–367.

Machado, N.F., Domínguez-Perles, R., 2017. Addressing facts and gaps in the phenolics chemistry of winery by-products. Molecules. 22(2), 286.

Madhuri, K.V., Prabhakar, K.V., 2014. Microbial exopolysaccharides: Biosynthesis and potential applications. Orient. J. Chem. 30, 1401–1410.

Martínez-Arcos, A., Moldes, A.B., Vecino, X., 2021. Adding value to secondary streams of corn wet milling industry. CyTA-J. Food. 19(1), 675–681.

McIntosh, M., Stone, B.A., Stanisich, V.A., 2005. Curdlan and other bacterial (1->3)-beta-D-glucans. Appl. Microbiol. Biotechnol. 68(2), 163–173.

Mehta, A., Prasad, G.S., Choudhury, A.R., 2014. Cost effective production of pullulan from agri-industrial residues using response surface methodology. Int. J. Biol. Macromol. 64, 252–256.

Menezes, J.D.S., Druzian, J.I., Padilha, F.F., Souza, R.R., 2012. Biotechnological production of xanthan gum in some agro-industrial waste, characterization and applications. Rev. Eletrônica Gest. Educ. Tecnol. Ambient. 8(8), 1761–1776 (in Portuguese).

Mensink, M.A., Frijlink, H.W., van der Voort Maarschalk, K., Hinrichs, W.L.J., 2015. Inulin, a flexible oligosaccharide I: Review of its physicochemical characteristics. Carbohydr. Polym. 130, 405–419.

Mishra, B., Varjani, S., 2019. Evaluation of pullulan production by a newly isolated *Micrococcus luteus*. Indian J. Exp. Biol. 57, 813–820.

Mishra, B., Vuppu, S., Rath, K., 2011. The role of microbial pullulan, a biopolymer in pharmaceutical approaches: A review. J. Appl. Pharm. Sci. 1, 45–50.

Mishra, B., Zamare, D., Manikanta, A., 2018. Selection and utilization of agro-industrial waste for biosynthesis and hyper-production of pullulan: A review. In: *Biosynthetic Technology and Environmental Challenges*. Springer, pp. 89–103.

Mohapatra, S., Sarkar, B., Samantaray, D.P., Daware, A., Maity, S., Pattnaik, S., Bhattacharjee, S., 2017. Bioconversion of fish solid waste into PHB using *Bacillus subtilis* based submerged fermentation process. Environ. Technol. 38(24), 3201–3208.

Moosavi-Nasab, M., Layegh, B., Aminlari, L., Hashemi, M.B., 2010. Microbial production of levan using date syrup and investigation of its properties. World Acad. Eng. Technol. 44, 1248–1254.

Moscovici, M., 2015. Present and future medical applications of microbial exopolysaccharides. Front. Microbiol. 6, 1012.

Narayanan, C.M., Narayan, V., 2019. Biological wastewater treatment and bioreactor design: A review. Sustain. Environ. Res. 29(1), 1–17.

Nik Ubaidillah, N.H., Abdullah, N., Sabaratnam, V., 2015. Isolation of the intracellular and extracellular polysaccharides of *Ganoderma neojaponicum* (Imazeki) and characterization of their immunomodulatory properties. Electronic J. Biotechnol. 18(3), 188–195.

Nisa, M., Sarwar, M., Khan, M.A., 2004. Influence of ad libitum feeding of urea treated wheat straw with or without corn steep liquor on intake, in situ digestion kinetics, nitrogen metabolism, and nutrient digestion in Nili-Ravi buffalo bulls. Aust. J. Agric. Res. 55(2), 229–233.

Nishinari, K., Zhang, H., Funami, T., 2009. Curdlan. In: G.O. Phillips, P.A. Williams (Eds.), *Handbook of Hydrocolloids*. CRC Press, pp. 567–591.

Niwa, T., Doi, U., Kato, Y., Osawa, T., 2001. Antioxidative properties of phenolic antioxidants isolated from corn steep liquor. J. Sci. Food Agric. 49(1), 177–182.

Noomhorm, A., Ilangantileke, S., Bautista, M.B., 1992. Factors in the protein enrichment of cassava by solid state fermentation. J. Sci. Food Agric. 58(1), 117–123.

Obayori, O.S., Ilori, M.O., Adebusoye, S.A., Oyetibo, G.O., Omotayo, A.E., Amund, O.O., 2010. Effects of corn steep liquor on growth rate and pyrene degradation by *Pseudomonas* strains. Curr. Microbiol. 60(6), 407–411.

Obied, H.K., Bedgood Jr, D.R., Prenzler, P.D., Robards, K., 2007. Bioscreening of Australian olive mill waste extracts: Biophenol content, antioxidant, antimicrobial and molluscicidal activities. Food Chem. Toxicol. 45(7), 1238–1248.

Ogidi, C.O., Ubaru, A.M., Ladi-Lawal, T., Thonda, O.A., Aladejana, O.M., Malomo, O., 2020. Bioactivity assessment of exopolysaccharides produced by *Pleurotus pulmonarius* in submerged culture with different agro-waste residues. Heliyon. 6(12), e05685.

Olusola, O.J., Macdosnald, I.O., Mathew, M.O., 2012. Proximate composition of whey from South West Nigeria. Adv. Biores. 3, 14–16.

Öner, E.T., 2013. Microbial production of extracellular polysaccharides from biomass. In: Z. Feng (Ed.), *Pretreatment Techniques for Biofuels and Biorefineries*. Springer, pp. 35–56.

Öner, E.T., Hernàndez, L., Combie, J., 2016. Review of levan polysaccharide: From a century of past experiences to future prospects. Biotechnol. Adv. 34(5), 827–844.

Orlandelli, R.C., Vasconcelos, A F.D., Azevedo, J.L., da Silva, M.D.L.C., Pamphile, J.A., 2016. Screening of endophytic sources of exopolysaccharides: Preliminary characterization of crude exopolysaccharide produced by submerged culture of *Diaporthe* sp. JF766998 under different cultivation time. Biochim. 2, 33–40.

Osemwegie, O.O., Adetunji, C.O., Ayeni, E.A., Adejobi, O.I., Arise, R.O., Nwonuma, C.O., Oghenekaro, A.O., 2020. Exopolysaccharides from bacteria and fungi: Current status and perspectives in Africa. Heliyon. 6(6), e04205.

Osińska-Jaroszuk, M., Jarosz-Wilkołazka, A., Jaroszuk-Ściseł, J., Szałapata, K., Nowak, A., Jaszek, M., Ozimek, E., Majewska, M., 2015. Extracellular polysaccharides from Ascomycota and Basidiomycota: Production conditions, biochemical characteristics, and biological properties. World J. Microbiol. Biotechnol. 31, 1823–1844.

Özcan, E., Öner, E.T., 2015. Microbial production of extracellular polysaccharides from biomass sources. In: K.G. Ramawat, J.M. Merillon (Eds.), *Polysaccharides*. Springer, pp. 161–184.

Pandey, A., 2003. Solid state fermentation. Biochem. Eng. J. 13, 81–84.

Panesar, P.S., Kaur, R., Gisha, Sangwan, R.S., 2016. Bio-processing of agro-industrial wastes for the production of food grade enzymes: Progress and prospects. Appl. Food Biotechnol. 3(4), 208–227.

Panesar, P.S., Kennedy, J.F., 2012. Biotechnological approaches for the value addition of whey. Crit. Rev. Biotechnol. 32(4), 327–348.

Panesar, P.S., Marwaha, S.S., Kennedy, J.F., 2006. *Zymomonas mobilis*: An alternative ethanol producer. J. Chem. Technol. Biotechnol. 81(4), 623–635.

Panesar, R., Kaur, S., Panesar, P.S., 2015. Production of microbial pigments utilizing agroindustrial waste: A review. Curr. Opin. Food Sci. 1, 70–76.

Papaspyridi, L.M., Katapodis, P., Gonou Zagou, Z., Kapsanaki Gotsi, E., Christakopoulos, P., 2010. Optimization of biomass production with enhanced glucan and dietary fibres content by *Pleurotus ostreatus* ATHUM 4438 under submerged culture. Biochem. Eng. J. 50(3), 131–138.

Patel, A., Prajapati, J.B., 2013. Food and health applications of exopolysaccharides produced by lactic acid bacteria. J. Adv. Dairy Res. 1–8.

Pedroso, G.B., Silva, L.O., Araujo, R.B., Saldanha, L.F., Denardi, L., Martins, A.F., 2019. An innovative approach for the biotechnological production of succinoglycan from rice husks. Ind. Crops Prod. 137, 615–627.

Pires, A.F., Marnotes, N.G., Bella, A., Viegas, J., Gomes, D.M., Henriques, M.H., Pereira, C.J., 2021. Use of ultrafiltrated cow's whey for the production of whey cheese with kefir or probiotics. J. Sci. Food Agric. 101(2), 555–563.

Pirota, R.D., Delabona, P.S., Farinas, C.S., 2014. Simplification of the biomass to ethanol conversion process by using the whole medium of filamentous fungi cultivated under solid-state fermentation. Bioenergy Res. 7, 744–752.

Poli, A., Di Donato, P., Abbamondi, G.R., Nicolaus, B., 2011. Synthesis, production, and biotechnological applications of exopolysaccharides and polyhydroxyalkanoates by archaea. *Archaea*. 2011(693253). Hindawi Publishing Corporation Archaea.

Pollock, T.J., 2005. Sphingan group of exopolysaccharides (EPS). In: A. Steinbuchel (Ed.), *Biopolymers*. Wiley Publications.

Prajapati, V.D., Jani, G.K., Khanda, S.M., 2013. Pullulan: An exopolysaccharide and its various applications. Carbohydr. Polym. 95(1), 540–549.

Priyadarshi, R., Kim, S.M., Rhim, J.W., 2021. Pectin/pullulan blend films for food packaging: Effect of blending ratio. Food Chem. 347, 129022.

Rabetafika, H.N., Bchir, B., Blecker, C., Paquot, M., Wathelet, B., 2014. Comparative study of alkaline extraction process of hemicelluloses from pear pomace. Biomass Bioenergy. 61, 254–264.

Rahmanian, N., Jafari, S.M., Galanakis, C.M., 2014. Recovery and removal of phenolic compounds from olive mill wastewater. J. Am. Oil Chem. Soc. 91, 1–18.

Rana, S., Upadhyay, L.S.B., 2020. Microbial exopolysaccharides: Synthesis pathways, types and their commercial applications. Int. J. Biol. Macromol. 157, 577–583.

Razack, S.A., Velayutham, V., Thangavelu, V., 2013. Medium optimization for the production of exopolysaccharide by *Bacillus subtilis* using synthetic sources and agro wastes. Turk. J. Biol. 37(3), 280–288.

Reddy, C.C., Khilji, I.A., Gupta, A., Bhuyar, P., Mahmood, S., Saeed AL-Japairai, K.A., Chua, G.K., 2021. Valorization of keratin waste biomass and its potential applications. J. Water Process Eng. 40, 101707.

Roca, C., Alves, V.D., Freitas, F., Reis, M.A., 2015. Exopolysaccharides enriched in rare sugars: Bacterial sources, production, and applications. Front. Microbiol. 6, 288.

Rocha, J.M., Guerra, A., 2020. On the valorization of lactose and its derivatives from cheese whey as a dairy industry by-product: An overview. Eur. Food Res. Technol. 246(11), 2161–2174.

Ruperez, P., Leal, J.A., 1981. Extracellular galactosaminogalactan from *Aspergillus parasiticus*. Trans. Br. Mycol. Soc. 77(3), 621–625.

Russo, M., Bonaccorsi, I., Torre, G., Saro, M., Dugo, P., Mondello, L., 2014. Underestimated sources of flavonoids, limonoids and dietary fibre: Availability in lemon's by-products. J. Funct. Foods. 9, 18e26.

Rütering, M., Schmid, J., Rühmann, B., Schilling, M., Sieber, V., 2016. Controlled production of polysaccharides—exploiting nutrient supply for levan and heteropolysaccharide formation in *Paenibacillus* sp. Carbohydr. Polym. 148, 326–334.

Ryan, M.P., Walsh, G., 2016. The biotechnological potential of whey. Rev. Environ. Sci. Bio. Technol. 15, 479–498.

Sadh, P.K., Duhan, S., Duhan, J.S., 2018. Agro-industrial wastes and their utilization using solid state fermentation: A review. Bioresour. Bioprocess. 5(1), 1–15.

Sagar, N.A., Pareek, S., Sharma, S., Yahia, E.M., Lobo, M.G., 2018. Fruit and vegetable waste: Bioactive compounds, their extraction, and possible utilization. Compr. Rev. Food Sci. Food Saf. 17(3), 512–531.

Salah, R.B., Jaouadi, B., Bouaziz, A., Chaari, K., Blecker, C., Derrouane, C., Besbes, S., 2011. Fermentation of date palm juice by curdlan gum production from *Rhizobium radiobacter* ATCC 6466™: Purification, rheological and physico-chemical characterization. LWT. 44(4), 1026–1034.

Sandford, P.A., 1979. Exocellular microbial polysaccharides. Adv. Carbohyd. Chem. Biochem. 36, 265–313.

Santos, M., Rodrigues, A., Teixeira, J.A., 2005. Production of dextran and fructose from carob pod extract and cheese whey by *Leuconostoc mesenteroides* NRRL B512. Biochem. Eng. J. 25, 1–6.

Schmid, J., Meyer, V., Sieber, V., 2011. Scleroglucan: Biosynthesis, production and application of a versatile hydrocolloid. Appl. Microbiol. Biotechnol. 91, 937–947.

Seesuriyachan, P., Techapun, C., Shinkawa, H., Sasaki, K., 2010. Solid state fermentation for extracellular polysaccharide production by *Lactobacillus confusus* with coconut water and sugarcane juice as renewable wastes. Biosci. Biotechnol. Biochem. 74, 423–426.

Sehm, J., Treutter, D., Lindermayer, H., Meyer, H.H.D., Pfaffl, M.W., 2011. The influence of apple- or red-grape pomace enriched piglet diet on blood parameters, bacterial colonisation, and marker gene expression in piglet white blood cells. Food Nutr. Sci. 2, 366–376.

Selo, G., Planinić, M., Tišma, M., Tomas, S., Koceva Komlenić, D., Bucić-Kojić, A., 2021. A comprehensive review on valorization of agro-food industrial residues by solid-state fermentation. Foods. 10, 927.

Shalini, R., Gupta, D.K., 2010. Utilization of pomace from apple processing industries: A review. J. Food Sci. Technol. 47(4), 365–371.

Sharma, N., Prasad, G.S., Choudhury, A.R., 2013. Utilization of corn steep liquor for biosynthesis of pullulan, an important exopolysaccharide. Carbohyd. Polym. 93(1), 95–101.

Sharma, P., Gaur, V.K., Sirohi, R., Varjani, S., Kim, S.H., Wong, J.W., 2021. Sustainable processing of food waste for production of bio-based products for circular bioeconomy. Bioresour. Technol. 325, 124684.

Sharoba, A.M., Farrag, M.A., El-Salam, A., 2013. Utilization of some fruits and vegetables wastes as a source of dietary fibers in cake making. J. Food Dairy Sci. 4(9), 433–453.

Silbir, S., Dagbagli, S., Yegin, S., Baysal, T., Goksungur, Y., 2014. Levan production by *Zymomonas mobilis* in batch and continuous fermentation. Carbohydr. Polym. 99, 454–461.

Silva, M.F., Fornari, R.C.G., Mazutti, M.A., Oliveira, D., Padilha, F.F., Cichoski, A.J., Cansian, R.L., Luccio, M.D., Treichel, H., 2009. Production and characterization of xanthan gum by *Xanthomonas campestris* using cheese whey as sole carbon source. J. Food Eng. 90, 119–123.

Singh, R.S., Chauhan, K., Kaur, K., Pandey, A., 2020. Statistical optimization of solid state fermentation for the production of fungal inulinase from apple pomace. Bioresour. Technol. Rep. 9. https://doi.org/10.1016/J.BITEB.2019.100364

Singh, R.S., Kaur, N., 2019. Understanding response surface optimization of medium composition for pullulan production from de-oiled rice bran by *Aureobasidium pullulans*. Food Sci. Biotechnol. 28, 1507–1520.

Sood, G., Sharma, S., Kapoor, S., Khanna, P.K., 2013. Optimization of extraction and characterization of polysaccharides from medicinal mushroom *Ganoderma lucidum* using response surface methodology. J. Med. Plants Res. 7, 2323–2329.

Stabnikova, O., Wang, J.Y., Ding, H.B., 2005. Biotransformation of vegetable and fruit processing wastes into yeast biomass enriched with selenium. Bioresour. Technol. 96(6), 747–751.

Stredansky, M., Conti, E., 1999. Xanthan production by solid state fermentation. Process Biochem. 34, 581–587.

Sugumaran, K.R., Gowthami, E., Swathi, B., Elakkiya, S., Srivastava, S.N., Ravikumar, R., Ponnusami, V., 2013. Production of pullulan by *Aureobasidium pullulans* from Asian palm kernel: A novel substrate. Carbohyd. Polym. 92(1), 697–703.

Sugumaran, K.R., Ponnusami, V., 2017. Conventional optimization of aqueous extraction of pullulan in solid-state fermentation of cassava bagasse and Asian palm kernel. Biocatal. Agric. Biotechnol. 10, 204–208.

Surek, E., Nilufer-Erdil, D., 2016. Phenolic contents, antioxidant activities and potential bioaccessibilities of industrial pomegranate nectar processing wastes. Int. J. Food Sci. Technol. 51(1), 231–239.

Teclu, D., Tivchev, G., Laing, M., Wallis, M., 2009. Determination of the elemental composition of molasses and its suitability as carbon source for growth of sulphate-reducing bacteria. J. Hazard. Mater. 161, 1157–1165.

Vedyashkina, T.A., Revin, V.V., Gogotov, I.N., 2005. Optimizing the conditions of dextran synthesis by the bacterium *Leuconostoc mesenteroides* grown in a molasses-containing medium. Appl. Biochem. Microbiol. 41, 361–364.

Veerapandian, B., Shanmugam, S.R., Varadhan, S., Sarwareddy, K.K., Mani, K.P., Ponnusami, V., 2020. Levan production from sucrose using chicken feather peptone as a low-cost supplemental nutrient source. Carbohyd. Polym. 227, 115361.

Vidhyalakshmi, R., Vallinachiyar, C., Radhika, R., 2012. Production of xanthan from agro-industrial waste. J. Adv. Sci. Res. 3, 56–59.

Viswanathan, H.S., Rahman, S.S.A., Venkatachalam, P., Karuppiah, S., 2020. Production of gellan gum using milk skin residue (MSR)—A tea shop waste: statistical optimization and downstream processing. Biomass Convers. Biorefin. 1–15.

Wang, D., Ju, X., Zhou, D., Wei, G., 2014. Efficient production of pullulan using rice hull hydrolysate by adaptive laboratory evolution of *Aureobasidium pullulans*. Bioresour. Technol. 164, 12–19.

Wang, D., Kim, H., Lee, S., Kim, D.H., Joe, M.H., 2020. Improved gellan gum production by a newly-isolated *Sphingomonas azotifigens* GL-1 in a cheese whey and molasses based medium. Process Biochem. 95, 269–278.

Wani, S.M., Mir, S.A., Khanday, F.A., Masoodi, F.A., 2021. Advances in pullulan production from agro-based wastes by *Aureobasidium pullulans* and its applications. Innov. Food Sci. Emerg. 74, 102846.

Wee, K.L., 1991. Use of non-conventional feedstuff of plant origin as fish feeds is it practical and economically feasible. Asian Fish. Soc. 5, 205.

Xiao, J.H., Xiao, D.M., Xiong, Q., Liang, Z.Q., Zhong, J.J., 2010. Nutritional requirements for the hyperproduction of bioactive exopolysaccharides by submerged fermentation of the edible medicinal fungus *Cordyceps taii*. Biochem. Eng. J. 49(2), 241–249.

Yang, M., Zhu, Y., Li, Y., Bao, J., Fan, X., Qu, Y., Wang, Y., Hu, Z., Li, Q., 2016. Production and optimization of curdlan produced by *Pseudomonas* sp. QL212. Int. J. Biol. Macromol. 89, 25–34.

Yatim, A.F.M., Syafiq, I.M., Huong, K.H., Amirul, A.A., Effendy, A.W.M., Bhubalan, K., 2017. Bioconversion of novel and renewable agro-industry by-products into a biodegradable poly (3-hydroxybutyrate) by marine *Bacillus megaterium* UMTKB-1 strain. BioTechnologia. 98(2).

Yezza, A., Tyagi, R.D., Valéro, J.R., Surampalli, R.Y., 2006. Bioconversion of industrial wastewater and wastewater sludge into *Bacillus thuringiensis* based biopesticides in pilot fermentor. Bioresour. Technol. 97, 1850–1857.

Zeraik, M.L., Pereira, C.A.M., Zuin, V.G., Yariwake, J.H., 2010. Passion fruit: A functional food? Rev. Bras. Farmacogn. 20(3), 459–471.

Zhang, J., Dong, Y.C., Fan, L.L., Jiao, Z.H., Chen, Q.H., 2015. Optimization of culture medium compositions for gellan gum production by a halobacterium *Sphingomonas paucimobilis*. Carbohydr. Polym. 115, 694–700.

7 Exopolysaccharides from extremophiles and their potential applications

Loreni Chiring Phukon, Rounak Chourasia,
Md Minhajul Abedin, Swati Sharma,
Sudhir P. Singh, and Amit Kumar Rai

7.1 INTRODUCTION

Microorganisms are ubiquitous and are known to thrive in environments with extreme conditions such as high or low temperature, high salinity, high/low pressure, acidic and basic environments. Extremophiles are those group of microorganisms that not only tolerate extreme environments but requires extreme conditions to survive (Madigan and Oren, 1999). According to the environment, extremophiles are psychrophiles (optimum subzero to 20°C), psychrotolerant (optimum >20°C with the ability to grow at low temperatures) (Kumari et al., 2021; Phukon et al., 2022; Chiring Phukon et al., 2020), thermophiles (optimum 45–80°C), hyperthermophiles (80->100°C), halophiles (high salt concentration), alkaliphiles (alkaline), acidophiles (acidic), metallophiles (high metal resistance), and piezophiles (high pressure) (López-Ortega et al., 2021; Madigan and Oren, 1999). Extremophiles have found importance in the current industrial system due to their ability to produce a wide range of extremophilic enzymes, proteins, metabolites, and extracellular polysaccharides or exopolysaccharides (EPS) of industrial value. EPS are carbohydrate-based high molecular-weight polymers secreted into the surrounding environment as slime or as individual capsular polysaccharides by microorganisms (López-Ortega et al., 2021; Wang et al., 2019b; Chourasia et al., 2022). Microbial EPS can be either homopolysaccharide or heteropolysaccharide depending on the monosaccharide composition of the EPS and can be bound with or without other molecules such as amino acids, proteins, acids, ions, and sulfate residues (Angelin and Kavitha, 2020; López-Ortega et al., 2021; Abedin et al., 2023). The monosaccharide units usually found in most of the EPS include glucose, galactose, mannose, xylose, ribose, arabinose, fructose, fucose, and rhamnose (López-Ortega et al., 2021). EPS have gained interest in recent times due to the extensive demand for natural polymers in the industrial and biotechnology sector such as in the food industry, pharmaceuticals, cosmetics, and bioremediation of the environment (Poli et al., 2010; Vinothkanna et al., 2022; Chourasia et al., 2021). Microbial EPS currently used on an industrial scale include dextran, xanthan, and gellan gum, which are being extensively used

DOI: 10.1201/9781003342687-7

in the food industry replacing plant and synthetic polysaccharides. EPS from the microbial source is easy to produce in a short period and has shown no known environmental disadvantage as compared to synthetic and plant polysaccharides.

Extremophiles have adapted unique features to survive extreme environments, for instance, thermophiles produce thermostable enzymes (DNA polymerase), which has proved to be a boon to humankind in the biotechnological process. Similarly, the production of EPS is one of the survival mechanisms used by extremophiles which can be exploited for the greater good of the human race (Wang et al., 2019b). EPS produces an environment around the cells providing cell adhesion, concentrating nutrients, and retaining water, the water-holding property of EPS helps the thermophiles and psychrophiles from dehydration (Nicolaus et al., 2010). Although the majority of the reports on the production of microbial EPS are from mesophiles due to the ease of cultivation, the production of EPS by extremophiles has gained importance as they are less likely to get contaminated by mesophiles, high growth rate, a unique composition such as over-sulfation, physiological properties such as thermostability, cryo-stable, pseudo-plasticity, and biological properties such as antioxidative, anticancer, antiviral, antimicrobial, and bio-remedial properties (Wang et al., 2021; Yasar Yildiz et al., 2014). EPS from extremophiles have been reported to have potential applications in various sectors such as health, food, cosmetics, and most importantly in the bioremediation of the environment. An increase in the number of reports on the production of EPS from extremophiles has shown the potentiality of EPS produced by the microorganisms thriving in extreme niches of the world, however more extensive research on the production of EPS by extremophiles is required to fully utilize the unique resources of the nature. An overview on the isolation of extremophiles from extreme habitats and the potential application of EPS produced by extremophilic microorganisms is presented in Figure 7.1. This chapter focuses on recent progress in the production of EPS from different groups of extremophiles and their potential application in the biomedical and bioremediation sector.

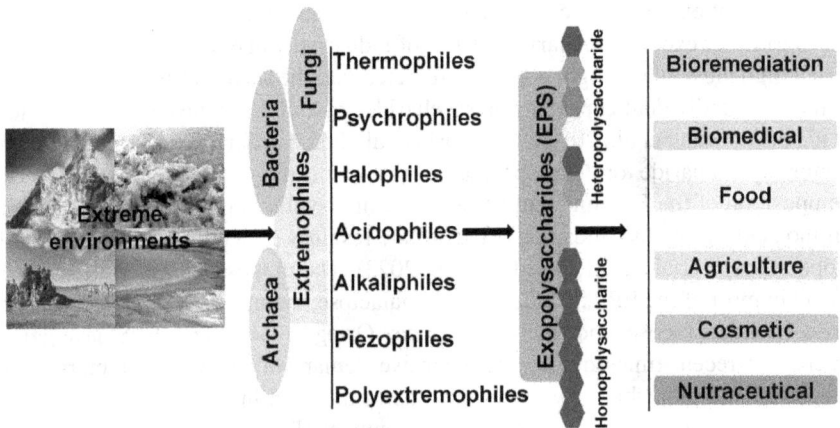

FIGURE 7.1 Exopolysaccharides (EPS) from extremophilic microorganisms and potential industrial applications

7.2 PRODUCTION OF EXOPOLYSACCHARIDE BY EXTREMOPHILES

7.2.1 EXOPOLYSACCHARIDE FROM THERMOPHILES

Microorganisms inhabiting regions with elevated temperatures are known to produce thermostable proteins, enzymes, and extracellular polymeric substances to withstand the adverse effect of high temperatures. EPS contributes to providing a suitably hydrated environment for thermophilic microorganisms to thrive by retaining water and concentrating nutrients (Wang et al., 2019a; Sharma et al., 2020). EPS production by thermophiles is believed to be an adaptation strategy for surviving extreme temperatures (Wang et al., 2019). At the industrial scale, EPS is produced mainly from mesophiles, as EPS production in thermophiles is lower than in most mesophiles. However, in recent past years, EPS produced from thermophilic bacteria and archaea have gained significance industrially over EPSs produced from mesophiles due to the relatively low/lack of pathogenicity, low chances of contaminations, and efficient production in a relatively short period (Wang et al., 2021; Yasar et al., 2014). The beneficial attributes of thermophiles for EPS production over mesophiles have motivated researchers around the globe to study EPS-producing thermophilic microbes.

Thermophilic microorganisms from deep-sea hydrothermal vents, volcanic eruption sites, industrial waste dumping sites, and marine and terrestrial geothermal hot springs have been studied worldwide for the production of EPS. Carbon sources have been found to have a significant impact on the yield and chemical composition of the EPS produced by thermophiles. As thermophiles are known to use a variety of carbon sources such as glucose, xylose, fructose, mannose, galactose, and disaccharides such as lactose, sucrose, and trehalose (López-Ortega et al., 2021; Nicolaus et al., 2002). Apart from a carbon source, nitrogen/amino acid supplementation also impacts the growth and production of EPS in some thermophilic microbes. Two hyperthermophiles *Thermococcus litoralis* (archaeon) and *Thermotoga maritima* (bacterium) showed similar growth metrices in the presence of maltose. However, it was observed that EPS production was impacted by the nitrogen source. The archaeon *T. litoralis* was able to produce EPS in media supplemented with amino acids both in the presence/absence of maltose; on other hand, the bacterium *T. maritima* was unable to utilize amino acids while it was able to produce EPS in the presence of maltose and ammonium chloride (Rinker and Kelly, 2000).

Geobacillus strains are among the most dominant microflora found in extremely hot environments worldwide, and several studies on the production of EPS have been done worldwide using *Geobacillus* strains. A thermostable EPS with a thermal degradation of 280°C was produced by a thermophilic strain *Geobacillus tepidamans* isolated from a geothermal hot spring, Bulgaria. Maltose was found to be the most appropriate carbon source for EPS production by *G. tepidamans* with the highest yield of 111 mg/L (Kambourova et al., 2009). Two thermophilic strains *Geobacillus thermodenitrificans* ArzA-6 and *Geobacillus toebii* ArzA-8 showed a maximum EPS yield of 0.27 g/g dry cells and 0.22 g/g dry cells in the presence of fructose (Panosyan et al., 2018). The major monomeric constituents of the EPS produced by *G. thermodenitrificans* ArzA-6 and *G. toebii* ArzA-8 were found to be mannose, while glucose was the main constituent of the thermostable EPS produced by *G.*

tepidamans. Apart from *Geobacillus* strains, thermophilic/thermotolerant *Bacillus* strains were also studied for the production of thermostable EPS. *Bacillus thermode-nitrificans* DSM 465 produced two EPSs with glucose and mannose as the main constituents (Nicolaus et al., 2000); in another study *Bacillus licheniformis* strain B3–15 also produced a glucomannan EPS, which has potential applicability in pharmaceutical industries (Caccamo et al., 2020). Similarly, *Geobacillus* sp. strain WSUCF1 produced two EPS using glucose as the sole carbon source. The major constituents of one of the EPS were found to be glucomannan, while the other EPS was composed of mannose only (Wang et al., 2021). The distinguishable attributes of EPS such as thermostability produced by the thermophiles, as well as the advantageous attributes of the thermophiles over mesophiles can be harnessed and lead the EPS industries toward a cost-effective economy.

7.2.2 EXOPOLYSACCHARIDE FROM PSYCHROPHILES

About one-fifth of the total surface of our planet is comprised of regions with temperatures below freezing point. Microorganisms from cold environments including terrestrial land, aquatic environments, as well as underwater have evolved distinctive features to adapt to cold temperatures (Phukon et al., 2022). Psychrophilic microorganisms have potential biotechnological and industrial significance due to their capability to produce important macromolecules such as enzymes and EPS (Kumari et al., 2021; Mukhopadhyay et al., 2014). The psychrophilic microbial communities inhabiting these cold habitats secrete high molecular weight EPS that act as a cryoprotective layer to avoid cell damage and help regulate metabolic processes (López-Ortega et al., 2021). The cryoprotective properties of the EPS produced by psychrophiles are contributed by its increased water-holding capacity, preventing crystallization of water, and antioxidative properties towards superoxide ions (Ali et al., 2020; Casillo et al., 2017; Chatterjee et al., 2018; López-Ortega et al., 2021).

EPS production has been demonstrated as a well-defined surviving mechanism adapted by cold-loving microorganisms in several studies. *Colwellia psychrery-thraea* strain 34H, a marine psychrophilic bacteria, showed tenfold higher EPS production when grown at −8°C than at −4°C (Casillo et al., 2017). Likewise, another marine bacterium, *Pseudoalteromonas* sp. CAM025, also showed 30-fold higher EPS production at −2°C and 10°C than at 20°C (Nichols et al., 2005a). Similarly, the cryoprotective nature of EPS can be demonstrated by calculating the number of viable cells under repeated freeze-thaw experiments. *Pseudomonas* sp. 1DI demonstrated the cryoprotective properties of EPS by maintaining a cell viability count of 93% and 75% at −80°C and −20°C after repeated freeze-thaw cycles (Carrión et al., 2015). Liu et al. (2013) also reported similar results, *Pseudoalteromonas* sp. Strain SM20310 showed a seven times higher viable cell count in the presence of EPS (Liu et al., 2013). Likewise, the EPS produced by psychrophilic organisms can also be used as a universal cryoprotectant to protect non-EPS-producing cells. Several studies based on the production of EPS by psychrophiles demonstrated the cryoprotective properties of EPS on non-EPS-producing *Escherichia coli* cells. *E. coli* viable cell count of 100.5–91.6% and 64.13% was obtained with the addition of 1% (w/v) EPS from *Pseudoalteromonas arctica* KOPRI 21653 and 10% EPS from *Pseudomonas* sp. ID1 (Carrión et al., 2015; Sung and Joung, 2007), while an 18-times higher vial *E. coli* cell count was obtained in a study conducted in the presence of 30 mg/mL EPS from *Pseudoalteromonas* sp. Strain SM20310 as compared to control the group that was tested in the absence of EPS (Liu et al., 2013). The survivability of probiotic bacteria during cold storage (4°C) and in vitro gastric digestion using 5 g/L EPS produced by a probiotic fungal strain *Cordyceps sinensis* showed 50% and 70% cell survival (Song et al., 2019).

Similar to other groups of microorganisms, the production of EPS by psychrophiles is affected by several factors such as carbon source, inoculum load, incubation period, and other growth parameters (López-Ortega et al., 2021). Although psychrophiles can use different sugar and non-sugar carbon source for the production of EPS, glucose and sucrose are the most commonly utilized carbons sources for EPS production. Psychrophilic filamentous fungus *Thelebolus* sp. IITKGP-BT12 produced 1.94 g/L basal salt medium containing 50 g glucose (Mukhopadhyay et al., 2014). *Pseudomonas* sp. BGI-2, a cold-adapted bacterium isolated from a glacier showed EPS production in a wide variety of carbon sources, including glucose, glycerol, mannose, mannitol, and molasses; however, the highest EPS production of 2 g/L was obtained in an optimal medium supplemented with 100 g/L glucose (Ali et al., 2020). A similar study with four antarctic bacteria with growth temperatures between 4–30°C showed EPS production in a medium supplemented with glucose and sucrose. Sucrose supplementation was found to be the best carbon source for EPS production in *Shewanella* sp. CAL606, *Colwellia* sp. GW185, and *Winogradskyella* sp. CAL396 as it significantly increased the growth and EPS production, while the growth and EPS production were significantly increased by glucose supplementation for *Winogradskyella* sp. CAL384. The authors also reported that incubation at low temperatures promoted EPS production in three bacterial strains, *Shewanella* sp. CAL606, *Winogradskyella* sp. CAL384, and *Winogradskyella* sp. CAL396 (Caruso et al., 2018a). Incubating temperature has been demonstrated in several studies as one of the important parameters affecting EPS production in psychrophilic microorganisms. Several studies have reported that EPS production was higher when the strains were grown at temperatures below the optimal growth temperature. For instance, *C. psychrerythraea* strain 34H showed tenfold higher EPS production at −8°C as compared to −4°C, *Pseudoalteromonas* sp. CAM025 showed a 30-fold increase in EPS production at −2° and 10°C than at 20°C, while *Pseudomonas* sp. BGI-2 showed higher EPS production at 4–15°C as compared to its optimal growth temperatures of 20°C and 30°C (Ali et al., 2020; Casillo et al., 2017; Nichols et al., 2005). The EPS produced by psychrophilic microorganisms has been studied mainly for its cryoprotective nature; however, other functions, as an emulsifying agent and therapeutic activities such as antioxidative and antitumor activities, have also been documented.

7.2.3 Exopolysaccharides from Halophiles, Alkalophiles, and Acidophiles

Organisms inhabiting regions with high salt concentrations that exceed the seawater salt concentration are known as halophiles or halophilic organisms. Some of the famous hypersaline sites worldwide include the Great Salt lake, the Dead Sea, the Tuz lake, the Gaet'ale Pond, the Don Juan Pond lake, and the Uyuni salt flat. Halophiles can be classified as extremely, moderately, and slightly halophilic based on the NaCl concentration requirement for optimum growth. Extremely halophilic microorganisms have optimum growth with NaCl concentration ranging from 2.5–5.2 M, moderately halophiles range from 0.5–2.5 M, while slightly halophiles have optimum growth at NaCl ranging from 0.3–0.5 M (López-Ortega et al., 2021). Halophiles have adapted/evolved various strategies to withstand the adverse effects of high salinity such as accumulation of compatible solutes, potassium and chloride ions, and production of polymeric substances such as lectins, EPS, and polyhydroxyalkanoates (Chambi et al., 2021; López-Ortega et al., 2021; Margesin and Schinner, 2001). Secretion of EPS by halophiles in the environment avoids the dessication of cells due to salt stress, as well as helps retain water, keeping the cells hydrated. EPS produced by halophiles have been used as emulsifying agents, heavy metal binding, biological activity, and gelling agents (Wang et al., 2019).

Several archaeal and bacterial species isolated from hypersaline sites have been reported from EPS production. A xanthan-like sulfated trisaccharide heteropolysaccharide EPS consisting of repeated units of one mannose and two moieties of 2-acetamido-2-deoxy glucuronic acid produced by the halophilic archaeon *Haloferax mediterranei* were the first reported and studied archaeal EPS (Antón et al., 1988; Koller et al., 2015). Sulfated EPS production has been reported from halophilic bacterial species also, two novel halophilic bacteria *Halomonas ventosae* and *Halomonas anticariensis* produced sulfated EPS containing glucose, galactose, and mannose as the main monosaccharide units (Juan Antonio Mata et al., 2006). EPS produced by mesophiles lack sulfate, while sulfated EPS and the presence of uronic acid in EPS are two distinguishing characteristics of EPS produced by halophilic organisms (Wang et al., 2019). Media compositions such as carbon, nitrogen, and salt concentrations affect the production of EPS by halophiles. Different studies observed an increase/decrease in EPS production by halophilic microorganisms with increases in salinity, and EPS production by extremely halophilic archaeon *H. mediterranei* decreased with an increase in salt concentration from 75 g/L to 250 g/L (Cui et al., 2017). While an increase in EPS production was observed in *Halomonas eurihalina* ATHM 37, *Arthrobacter luteolus* ATHM, *Bacillus aerius* ATHM35, *Staphylococcus saprophyticus*, and *Oceanobacillus profundus* with an increase in NaCl concentration (Aisha Waheed Qurashi, 2011; Moshabaki et al., 2018). Carbon sources play an important role in the production of EPS by extremophiles, and a wide variety of carbon sources have been observed to have both positive as well as negative responses to individual microorganisms. *Halomonas smyrnensis* AAD6[T] was reported as the first *Halomonas* sp. to produce levan. In different studies, *H. smyrnensis* AAD6[T] produced levan using different carbon sources such as sucrose, mannitol, and molasses (Ates, 2015; Poli et al., 2009). Some archaeon species such as *Halomicrobium* strains, *Haloferax prahovense* (DSM 18310), and *Natronococcus jeotgali* (DSM 18795) have also been reported to produce levan and inulin like fructan-based EPS (López-Ortega et al., 2021).

Acidophilic and alkaliphilic microorganisms are those extremophiles that inhabit regions with extreme pH. Acidophiles usually have optimum growth at pH 3 or below, while alkaliphiles grow best above or at pH 9. The growth of both the pH-dependent extremophiles is either very slow or no growth is possible in environments with neutral pH (Wang et al., 2019b). Only a few studies on the production of EPS by acidophiles and alkaliphiles have been reported so far. EPS secretion in acidophiles is generally considered as an accumulation of by-products during another bioprocess such as bioleaching. EPS containing fatty acids, polysaccharides, proteins, and ferric ions was produced as a by-product during bioleaching by *Acidithiobacillus caldus* and *Leptospirillum ferriphilum* dominated acidophilic mixed culture (Zeng et al., 2010). Similar production of EPS with varied monomeric saccharides was observed during the bioleaching process using *Thiobacillus ferrooxidans*. EPS production by *T. ferrooxidans* was found to be beneficial during the bioleaching process. The EPS helped in the attachment of the cells to pyrite and sulfur (Gehrke et al., 1998). Alkaliphilic microorganisms have been isolated worldwide for the production of various alkaliphilic enzymes such as lipases, cellulases, amylases, and proteases. Production of EPS by alkaliphiles isolated from high pH sites such as soda lakes, volcanic eruption sites, and hypersaline sites has been reported; however, the in-depth study on the production of EPS such as fermentation conditions, molecular and genetic studies are limited (Wang et al., 2019). A very potential EPS with anticancerous and antibacterial properties was produced by a haloalkalitolerant bacteria *Alkalibacillus* sp. w3. Glucose and yeast extract was the most effective carbon and nitrogen source for the production of EPS by *Alkalibacillus* sp. w3 (Arayes et al., 2022). Similarly, an EPS showing 75% similarity to dextran was produced at pH 10 by a haloalkaliphilic bacteria *Vagococcus carniphilus* isolated from alkaline Lonar lake, Maharashtra, India (Joshi and Kanekar, 2011). The adaptive and coping mechanism of the Acidophilic and alkaliphilic microorganisms in harsh environments can be harnessed for potential use as EPS producers at an industrial scale. Apart from being able to produce EPS for various applications, the EPS can be used as a protective agent in other bioprocesses involving extreme pH such as bioleaching.

7.3　APPLICATION OF EXOPOLYSACCHARIDE PRODUCED BY EXTREMOPHILES

Extremophilic microorganisms inhabiting extreme niches withstand the extreme environment by producing enzymes, polysaccharides, and proteins capable of having advantageous physiochemical characteristics as compared to mesophiles (Nicolaus et al., 2010). Although the production of extracellular polysaccharides by extremophilic microorganisms has been a distinguishing adaptive feature to survive extreme conditions, the production of EPS using extremophiles on an industrial scale is scarce (Banerjee et al., 2021; Nicolaus et al., 2010). Mesophilic microorganisms are the dominant EPS producer on an industrial scale, however diverse studies over the past decades have shown the potential of using extremophilic microorganisms as EPS producers. Several studies have shown the potential applicability of the EPS produced by extremophiles in the food, pharmaceutical, bioremediation, and biotechnological sector. An overview of the potential application of EPS produced by extremophiles is presented in Table 7.1. The health benefits and the remedial properties of the exopolysaccharides produced by extremophiles are the most researched properties for demonstrating the potentiality and applicability of extremophile-produced exopolysaccharides.

7.3.1 HEALTH-PROMOTING PROPERTIES

EPS has found immense application in different sectors, however, biological active property aiding/promoting good health is among the most potential utilities of EPS. Several studies on EPS from extremophiles have been reported to show health-promoting activities, including anticancer, antioxidative, antiviral, immunoregulatory, and prebiotic. Cancer is one of the leading causes of death worldwide resulting in extensive research for a cure. EPS produced by extremophilic microorganisms has been reported to show promising in vitro anticancer activities. A high molecular weight EPS produced by an Antarctic bacterium *Pseudoaltermonas* sp. S-5 exhibited cytotoxic activity inhibiting the growth of human leukemia K562 cells, suggesting potential application as an anticancer drug capable of targeting cancerous cells for apoptosis (Chen et al., 2015). Similar results of growth inhibition of K562 cells were observed in the presence of the acidic capsular EPS produced by *Psychrobacter submarinus* KMM 225 isolated from the deep sea (Kokoulin et al., 2021). The EPS produced by the thermophilic bacterium *Geobacillus* sp. TS3–9 isolated from radioactive radon hot-spring showed antiproliferative activity against hepatoma carcinoma cells (Liang, 2017); similarly, antitumoural activity against T cell lines from acute lymphoblastic leukemia was observed when the heteropolysaccharide (B100S) produced by *Halomonas stenophila* B100 was over sulfated. The B100S has been demonstrated as the very first bacterial EPS to exert an apoptotic effect on cancer T cells whilst showing no effect on primary T cells (Ruiz-ruiz et al., 2011). Bacteria are among the most commonly found microorganisms thriving in extreme environments; however, few single-celled yeasts as well as multi-celled fungi capable of producing EPS are also reported to thrive in extreme environments. Recently both micro and macro eukaryotic organisms have gained immense interest from industrial as well as biotechnological points of view as an efficient and economic source for the development of healthcare products. The EPS produced by fungi has shown potential application in the healthcare sector in several studies. EPS produced by the cold-adapted yeast *Rhodotorula mucilaginosa* sp. GUMS16 showed reduced viable K562 cells at a dosage of 1500 μg/mL. The EPS considerably reduced the viability of the chronic myeloid leukemia K562 cell line without showing any toxic effect on normal non-cancerous cells (Kheyrandish et al., 2022). A newly identified psychrophilic Antarctic fungus *Thelebolus* sp. IITKGP-BT12 produced Thelebolan, a bioactive EPS that showed promising apoptotic inducing and antiproliferative activities against mouse skin carcinoma cell lines B16-F0. Thelebolan showed 50% antiproliferative activity against B16-F0 cells at a dosage of 275.42 μg/mL (Mukhopadhyay et al., 2014).

EPS produced by extremophiles has also been reported to have other promising activities such as immunomodulatory, antioxidative, and biocompatibility. Reactive oxygen species (ROS) has been associated with the development of chronic and degenerative diseases by causing oxidative stress to cells. Biomolecules from a biological source having antioxidant properties are favoured over synthetic molecules

TABLE 7.1

Exopolysaccharide from extremophiles and their biological properties

Microorganism (Source)	Nature and growth condition of extremophiles	Optimum condition for EPS production (Sugar/nitrogen/ etc.)	Yield	Molecular weight	Chemical composition of the EPS	Biological properties	Reference
Aeribacillus pallidus 418 (Rupi hot spring, Bulgaria)	Thermophilic (35–72°C, pH 62.9, 2.5% NaCl)	Maltose/Ammonium chloride and yeast extract/55°C, pH 7.0	53 mg/L	**EPS1** 700 kDa **EPS2** 1000 kDa	**EPS1**-Man, Gal, Glc, GalN, GlcN, Rib **EPS2**-Man, Gal, Glc, GalN, GlcN, Rib, Ara	Emulsifying and foaming activity	(Radchenkova et al., 2013)
Bacillus licheniformis B3-15 (Vulcano Island hot spring, Italy)	Thermotolerant (45°C, pH- 7, 2% NaCl)	Glucose/45°C	165 mg/L	600 kDa	Man, Glc/Gal	Antiviral activity Immunomodulatory activity	(Arena et al., 2006; Maugeri et al., 2002)
Aeribacillus pallidus YM-1 (Yumen Oilfield, China)	Thermophilic (55°C, pH 7, 10% NaCl)	Glucose/yeast extract/55°C/pH 7.5/1% NaCl	4.24 g/ml	540 kDa	Glc, Alt, Man, Gal, proteins, lipids	Emulsifying activity	(Zheng et al., 2012)
Geobacillus tepidamans V264 (Velingrad hot spring, Bulgaria)	Thermophilic (60°C, pH- 7, 1% NaCl)	Maltose/diammonium hydrogen phosphate/60°C/pH 7	47.5 mg/L	>1000 kDa	Glc, Gal, Fuc, Fru	Anti-cytotoxicity activity Biocompatibility	(Kambourova et al., 2009)
Geobacillus sp. WSUCF1 (Compost facility, Washington State University, US)	Thermophilic (60°C, pH- 7)	Glucose/yeast extract/60°C/ pH- 7/3% NaCl	525.7 mg/L	**EP1** ~1000 kDa **EP2** ~1000 kDa	**EP1**-Man, Glc **EP2**-Man	Anti-cytotoxicity activity Biocompatibility activity Antioxidative activity	(Wang et al., 2021)
Geobacillus thermodenitrificans ArzA-6 (Arzakan geothermal spring, Armenia)	Thermophilic (65°C, pH 7.0)	Fructose and glucose/65°C/pH 7.0	76 mg/L	500 kDa	Man, Gal, Ara, Fru, Glc	NT	(Panosyan et al., 2018)

(*Continued*)

TABLE 7.1 (Continued)

Exopolysaccharide from extremophiles and their biological properties

Microorganism (Source)	Nature and growth condition of extremophiles	Optimum condition for EPS production (Sugar/nitrogen/ etc.)	Yield	Molecular weight	Chemical composition of the EPS	Biological properties	Reference
Geobacillus toebii ArzA-8 (Arzakan geothermal spring, Armenia)	Thermophilic (65°C, pH 7.0)	Fructose and glucose/65°C/pH 7.0	80 mg/L	600 kDa	Man, Gal, Glc, Ara	NT	(Panosyan et al., 2018)
Rhodothermus marinus DSM4252	Thermophilic (65°C)	Lactose/65°C	3.3 g/L	73.5 kDa	Xyl, Ara, Glc	NT	(Sardari et al., 2017)
R. marinus MAT493	Thermophilic (65°C)	Lactose/65°C	460 mg/L	85.5 kDa	Glc, Ara, Xyl, Man	NT	(Sardari et al., 2017)
Geobacillus sp. TS3–9 (Taishun radon hot Spring, China)	Thermophilic (60°C, pH 8)	Lactose/55°C/pH 8	87 mg/L	3200 kDa	Man, Glc, Rha, acetyls, and uronic acid	Antioxidative activity Anticancer activity	(Liang, 2017)
Anoxybacillus sp. R4-33 (Radon hot spring, China)	Radiation resistant and thermophilic (60°C, pH 7.5)	Glucose/yeast extract and tryptone/55°C/pH 8/5% NaCl	10.83 mg/ml	>1000 kDa	Man, Glc, and acetyls	Biosorption of heavy metal	(Zhao et al., 2014)
Rhodotorula mucilaginosa sp. GUMS16 (Leaf debris, Iran)	Psychrotolerant (4–30°C, 5% NaCl)	Sucrose/25°C/pH 6	28.5 mg/ml	84 kDa	Gcu, Man	Antiproliferative activity Biocompatibility Antioxidative activity	(Hamidi et al., 2020)
Pseudoalteromonas sp. CAM036 (Melted ice, Antarctic)	Psychrophilic (20°C, pH 7.0)	Glucose/peptone and yeast extract/20°C/pH 7.0	NR	1700 kDa	Ara, Man, Gal, Glc, GalA, GalNAc, GlcNAc	Cryoprotective activity	(Nichols et al., 2005)
Pseudoalteromonas sp. CAM064 (Melted ice, Antarctic)	Psychrophilic (20°C, pH 7.0)	Glucose/peptone and yeast extract/20°C/pH 7.0	NR	100 kDa	Ara, Man, Gal, Glc, GlcA, GalNAc, GlcNAc	Cryoprotective activity	(Nichols et al., 2005)

Organism (source)	Type	Growth conditions	Yield	Molecular weight	Monosaccharide composition	Activity/application	Reference
Shewanella livingstonensis CAM090 (Melted ice, Antarctic)	Psychrophilic (20°C, pH 7.0)	Glucose/peptone and yeast extract/20°C/pH 7.0	NR	80 kDa	Ara, Rha, Xyl, Man, Gal, Glc, GlcA, GalNAc	Cryoprotective activity	(Nichols et al., 2005)
Pseudoalteromonas sp. CAM023 (Melted ice, Antarctic)	Psychrophilic (20°C, pH 7.0)	Glucose/32°C/7.5% sea salt	NR	1800 kDa	Ara, Man, Gal, Glc, GalA, GlcA, GalNAc,	Cryoprotective activity	(Nichols et al., 2005)
Mucilaginibacter sp. ERMR7:07 (Rathong glacier, Sikkim, India)	Psychrotrophic (20°C, pH 7.4)	Starch and sucrose/peptone/20°C/pH 7.4	26.4 mg/ml	3.05 kDa	Glc	Cryoprotective activity; Biosorption of heavy metals	(Kumar et al., 2022)
Polaribacter sp. SM1127 (Arctic alga)	Psychrotropic (15°C/pH 7/30% sea salt)	Glucose/peptone and yeast extract/15°C/pH 7/30% sea salt	2.1 g/L	220 kDa	GlcN, Man, GlcA, Gal, Fuc, Glc, Rha	Antioxidative activity; Cryoprotective effect on human dermal fibroblast; High moisture retention	(Sun et al., 2015)
Pseudoalteromonas sp. CAM003 (Melted ice, Antarctic)	Psychrophilic (20°C, pH 7.0)	Glucose/peptone and yeast extract/20°C/pH 7.0	NR	1800 kDa	Ara, Rib, Rha, Fuc, Man, Glc, GlcA, GalNAc, GlcNAc	Cryoprotective activity	(Nichols et al., 2005)
Pseudoalteromonas sp. CAM015 (Melted ice, Antarctic)	Psychrophilic (20°C, pH 7.0)	Glucose/peptone and yeast extract/20°C/pH 7.0	NR	2800 kDa	Ara, Rha, Xyl, Man, Gal, Glc, GlcA, GalNAc	Cryoprotective activity	(Nichols et al., 2005)
Pseudomonas sp. BGI-2 (Batura glacier, Pakistan)	Psychrotropic (25°C, pH 7)	Glucose/yeast extract/15°C/pH 6.0/10% NaCl	20.01 mg/ml	NR	Glc, GlcN, Gal	Cryoprotective	(Ali et al., 2020)
Colwellia psychrerythraea 34H (Marine sediments, Artic)	Psychrophilic (8°C, pH 7.6, 19.5% NaCl)	Yeast extract and peptone/8°C/pH 7.2	Not reported	>1500 kDa	N-acetylquinovosamine, GalA	Cryoprotective; Anti-biofilm activity; Antifreeze activity	(Carillo et al., 2015; Casillo et al., 2017; Marx et al., 2009)

(Continued)

TABLE 7.1 (Continued)
Exopolysaccharide from extremophiles and their biological properties

Microorganism (Source)	Nature and growth condition of extremophiles	Optimum condition for EPS production (Sugar/nitrogen/ etc.)	Yield	Molecular weight	Chemical composition of the EPS	Biological properties	Reference
Flavobacterium frigidarium CAM005 (Melted ice, Antarctic)	Psychrophilic (20°C, pH 7.0)	Glucose/peptone and yeast extract/20°C/pH 7.0	NR	1810 kDa	Ara, Man, Gal, Glc, GlcA, GlcNAc	Cryoprotective activity	(Nichols et al., 2005)
Pseudoalteromonas sp. MER144 (Terra Nova Bay, Antarctic)	Psychrophile	Sucrose/4°C/pH 7.0/3% NaCl	318.26 mg/L	250 kDa	Glu, Man, GlCNAc, Ara, GlcA, GalA, Gal	Cryoprotective activity; Biosorption of heavy metals	(Caruso et al., 2018)
Myroides odoratus CAM030 (Melted ice, Antarctic)	Psychrophilic (20°C, pH 7.0)	Glucose/peptone and yeast extract/20°C/pH 7.0	NR	190 kDa	Ara, Rha, Xyl, Man, Gal, Glc, GalA, GlcA, GalNAc, GlcNAc	Cryoprotective activity	(Nichols et al., 2005)
Polaribacter irgensii CAM006 (Melted ice, Antarctic)	Psychrophilic (20°C, pH 7.0)	Glucose/peptone and yeast extract/20°C/pH 7.0	NR	2100 kDa	Ara, Fuc, Man, Gal, Glc, GlcA, GalNAc, GlcNAc	Cryoprotective activity	(Nichols et al., 2005)
Halomonas maura S-30 (Saline soil, Asilah, Morocco)	Halophilic (32°C, pH 7.0, 7.5% sea salts, 5.1% NaCl)	Beet molasses and Glucose/yeast extract and protease peptone/32°C, pH 7.0, 2.5% sea salts	3 g/ml	4700 kDa	Man, Gal, Glc, GlcA, sulfate, phosphate, acetyls, and uronic acid	Biosorption of lead	(Arias et al., 2003)
Alkalibacillus sp. w3 (Al Hamra lake, Egypt)	Haloalkalitolerant (35°C, pH- 9, 15% NaCl)	Glucose/yeast extract and peptone/ pH-10/25% NaCl	3 g/ml	NR	Carbohydrate, proteins, and lipids	Anticancer activity; Antimicrobial activity	(Arayes et al., 2022)

Organism (source)	Characteristics	Medium/conditions	Yield	Molecular weight	Monosaccharide composition	Applications	References
Bacillus licheniformis T14 (Hydrothermal vent, Italy)	Haloalkaliphilic, thermophilic (50°C, pH- 8, 5% NaCl)	Sucrose/yeast extract/50° C/ pH- 8/5% NaCl	366 mg/L	1000 kDa	Man	Anti-cytotoxic activity Antiviral activity Immunomodulatory activity	(Gugliandolo et al., 2014; Spanò et al., 2013)
Cobetia marina DSMZ 4741 (Coastal seawater, USA)	Halophilic (25°C, pH 7.6)	Glucose/yeast extract and peptone/20°C/pH 7.6	0.4 g/L	270 kDa	Rib, Pyr, KDO	NT	(Lelchat et al., 2015)
Halomonas nitroreducens WB1 (Bakreshwar hot springs, India)	Halothermophilic (60°C, pH 7.5, 5% NaCl)	Glucose (60°C, pH 7.5, 5% NaCl)	~1.2 mg/ml	EPS1 5200 kDa EPS2 30 kDa EPS3 1.3 kDa	EPS1-Glc, Man, Gal EPS2-Glc, Man, Rha, Ara EPS3-Glc, Man, Gal, GalA	Biocompatibility Antioxidant activity Emulsifying activity Biosorption of heavy metals	(Chikkanna et al., 2018)
Halomonas ventosa A112 (Saline soils, Fuente de Piedra, Spain)	Halophilic (32°C, pH 7, 7.5% sea salt)	Glucose/peptone and yeast extract/20°C/pH 7.0	283.5 mg/L	53 kDa	Glc, Man, Gal, Xyl, Ara, GalA	Heavy metal chelation Antiviral activity Anticoagulation activity	(Mata et al., 2006)
Kocuria rosea ZJUQH CCTCC M2016754 (Chaka Salt Lake, Tibet)	Halophilic (25°C, pH 6.8–7.2, 10% NaCl)	Yeast extract and peptone/30°C	70.64 g/L	56.59 kDa	Glc	Bioremediation of hypersaline environment	(Gu et al., 2017)
Halomonas anticariensis FP35 (Saline soils, Fuente de Piedra, Spain)	Halophilic (32°C, pH 7, 7.5% sea salt)	Glucose/32°C/7.5% sea salt	296.5 mg/L	20 kDa	Glc, Man, GalA	Heavy metal chelation Antiviral activity Anticoagulation activity	(Mata et al., 2006)

(Continued)

TABLE 7.1 (Continued)
Exopolysaccharide from extremophiles and their biological properties

Microorganism (Source)	Nature and growth condition of extremophiles	Optimum condition for EPS production (Sugar/nitrogen/ etc.)	Yield	Molecular weight	Chemical composition of the EPS	Biological properties	Reference
Salipiger mucosus A3 (Hypersaline soil of solar saltern, Calbanche)	Halophilic (32°C, 7.5% salt)	Glucose/32°C	1.35 g/L	250 kDa	Glc, Man, Gal, Fuc, acetyls, proteins, sulfates, phosphates, pyruvic acid, and hexoamines	Emulsifying activity Chelation of heavy metals	(Llamas et al., 2010)
Halomonas almeriensis M8 (Saltern, Spain)	Halophilic (32°C, 7.5% salt)	Glucose/32°C	1.7 g/L	15 kDa	Man, Glc, Rha	Pseudoplasticity Heavy metal chelation Emulsifying activity	(Llamas et al., 2012)
Halomonas anticariensis FP36 (Saline soils, Fuente de Piedra, Spain)	Halophilic (32°C, pH 7, 7.5% sea salt)	Glucose/32°C/7.5% sea salt	499.5 mg/L	46 kDa	Glc, Man, GalA	Heavy metal chelation Antiviral activity Anticoagulation activity	(Mata et al., 2006)
Idiomarina fontislapidosi F23 (Hypersaline habitats, Spain)	Halophilic (32°C, 7.5% total salt)	Glucose/32°C/7.5% sea salt	14.5 g/ml	**EPS1** 1500 kDa **EPS2** 15 kDa	Glc, Man, Gal, Xyl	Emulsifying activity and in the chelation of heavy metals	(Mata et al., 2008)
I. ramblicola R22 (Hypersaline habitats, Spain)	Halophilic (32°C, 7.5% total salt)	Glucose/32°C/7.5% sea salt	15 g/ml	**EPS1** 550 kDa **EPS2** 20 kDa	EPS1-Glc, Man, Rha EPS2-Glc, Man, Rha, Xyl, GalA	Emulsifying activity and in the chelation of heavy metals	(Mata et al., 2008)

Organism	Type/Conditions	Carbon source/Conditions	Yield	Molecular weight	Monosaccharide composition	Applications	References
Alteromonas hispanica F32 (Hypersaline habitats, Spain)	Halophilic (32°C, 7.5% total salt)	Galactose/32°C/7.5% sea salt	12.5 g/ml	19000 kDa	Glc, Man, Rha, Xyl	Emulsifying activity and in the chelation of heavy metals	(Mata et al., 2008)
Chromohalobacter canadensis 28 (Pomorie salterns, Bulgaria)	Halophilic (30°C, pH 7.5, 10% NaCl)	Lactose/yeast extract and peptone/30°C, pH 8.5, 15% NaCl	172 mg/L	>1000 kDa	GlcN, Glc, Rha, Xyl, other unknown sugars, and proteins	In the cosmetic industry as an emulsifier, foaming agent, and water-binding agent	(Radchenkova et al., 2018)
Halomonas smyrnensis AAD6 (Çamalti Saltern, Turkey)	Halophilic (37°C, pH 7, 13.7% NaCl)	Sucrose/peptone and ammonium sulfate/37°C/pH 7/13.7% NaCl	8.84 g/L	>1000 kDa	Fru	Anticancer activity Biocompatibility	(Kazak Sarilmiser et al., 2015; Kazak Sarilmiser and Toksoy Oner, 2014)
Halomonas elongata S6 (Sehline Sebkha Salt Lake, Tunisia)	Halophilic (35°C, pH 7, 5% salt)	Glucose/yeast extract and peptone/35°C/pH 7/5% NaCl	NR	270 kDa	GlcN, Man, Rha, Glc	Anti-Biofilm activity Antioxidative activity Bio-flocculation Emulsifying activity	(Joulak et al., 2020a)

NR—not reported, N—not tested, Man—mannose, Gal—galactose, Ara—arabinose, Fru—fructose, Glc—glucose, Alt—altrose, Fuc—fucose, Xyl—xylose, GlcN—glucosamine, GalN—galactosamine, GlcNAc—N-acetylglucosamine, GalNAc—N-acetylgalactosamine, Ri—ribose, ManN—mannosamine, GalA—Galacturonic acid, GlcA—glucuronic acid, KDO-3-deoxy-d-manno-oct-2-ulosonic acid, Pyr—pyruvate, Rha—Rhamnose

for scavenging ROS as a measure against ROS related diseases. The EPS produced by *Geobacillus* sp. TS3–9 was reported to be multifunctional as it not only has anti-cancer properties but also showed high antioxidative activity (Liang, 2017). The exopolysaccharides EPS-1 produced by a thermotolerant *B. licheniformis* B3–15 showed immunomodulatory and antiviral effects. The replication of HSV type 2 was impaired in the presence of EPS-1, which induced the production of IL-12, IFN-γ, TNF-α, IL-18, and IFN-α (Arena et al., 2006). Similar antiviral activity on HSV-2 and the immunomodulatory effect of EPS produced by haloalkaliphilic, and thermophilic *B. licheniformis* T14 was reported (Gugliandolo et al., 2014). Both the EPS produced by the *Bacillus* sp. showed non-cytotoxicity to normal cells, capable of use as a biocompatible immunomodulatory agent. The same authors in a different study also reported partial restoration of immunological disorder caused by infection of HSV-2 upon treatment with EPS produced by *G. thermodenitrificans* B3–72 (Arena et al., 2009). The EPS produced by *G. thermodenitrificans* B3–72 hindered the replication of HSV-2 in human peripheral blood mononuclear cells and showed a high level of cytokine production. Biocompatibility/non-cytotoxicity to normal healthy cells is a bonus addition to the biological activity of the molecule and has a higher chance of applicability in the pharmaceutical and nutraceutical industries. For instance, the EPS produced by *G. tepidamans* V264 inhibited the cytotoxic effect of toxic sesquiterpene hydroquinone (avarol) on brine shrimp larvae by increasing the lethal dose (LD_{50}) by 12-fold from 0.18 µg/mL to 2.24 µg/mL (Kambourova et al., 2009). Similar results were reported for the EPSs produced by *Geobacillus* sp. WSUCF1 showed high antioxidant activity, thermostability, and non-cytotoxicity, making both the EPSs potential drug delivery carriers in the biomedical industry (Wang et al., 2021). The cold-adapted yeast *Rhodotorula mucilaginosa* sp. GUMS16 isolated from leaf debris also showed higher antioxidative activity while showing zero toxicity towards the healthy normal human cell, suggesting potential application in the pharmaceutical industries as a biocompatible antioxidative agent (Hamidi et al., 2020).

7.3.2 BIOREMEDIATION

We live in an era of rapid industrialization that has led to the generation of a huge amount of hazardous solid and liquid wastes. Various conventional strategies such as chlorination and filtration of wastewater, landfilling, and incineration of solid waste, etc. are currently applied for the treatment of hazardous waste materials. However, such conventional treatments are non-eco-friendly and expensive. Biological treatments through the use of living organisms or biological molecules are preferred over conventional treatments as they are eco-friendly, sustainable, and do not produce toxic by-products (Banerjee et al., 2021). Application of EPS for bioremediation is mostly based on its absorption, adsorption, and degradation properties. Most EPS-mediated remediation strategies follow the use of EPS as (a) filtering agents in biofilters in wastewater treatment reactor beds, (b) coagulant in the coagulation/flocculation and as emulsifiers, and (c) bioaccumulation of heavy materials (Banerjee et al., 2021; López-Ortega et al., 2021).

The increase in population size has increased the demand for industrial products such as paper, leathers, and textiles, which are a popular source of release of wastewater with high amount of toxic dyes and accounts for about 20% of total water pollution worldwide (Banerjee et al., 2021). EPS-mediated flocculation of toxic dye was first reported from a psychrotolerant antarctic bacteria *Pseudoalteromonas* sp. Bsi20310. The EPS produced by *P.* sp. Bsi20310 enhanced the coagulation activity of ferric chloride of synthetic brilliant RX-3B dye by 90%. A dosage of 150 mg/L EPS with 55 mg/L ferric chloride, pH 10 was the best proportion to achieve the highest flocculation by Fe(III) (Zhou et al., 2010). Similar flocculation activity of EPS was reported for the EPS produced by *Pseudoalteromonas* sp. SM9913, EPS SM9913 was able to flocculate a large number of suspended solid materials such as soil, MgO, kaolin clay, Al_2O_3, and activated carbon (Li et al., 2008). The flocculation ability of EPS SM9913 was enhanced in the presence of calcium and ferric ions and was also found to have better flocculation ability than aluminum sulfate (a known flocculation agent) at high salinity and low temperatures (Li et al., 2008). EPS-mediated coagulation treatment of wastewater is based on the electrostatic and bridging effect of the EPS used in the treatment solution, as well as the composition of the EPS (Banerjee et al., 2021; Shi et al., 2017). The presence of proteins, acids, and ions in EPS helps in reducing the electrostatic force and increasing the hydrophobicity of the cell surface (Poli et al., 2010; Shi et al., 2017). More than 90% of the decolorization of congo red wastewater (a synthetic azo dye) was achieved by the marine bacterium *Aliiglaciecola lipolytica* by adsorbing onto cells with its EPS (Wang et al., 2020). About 54% of the dye was adsorbed on biomass, while 46% of the dye was degraded by the enzymes azoreductase and laccase, and through co-metabolism with glucose. *A. lipolytica* secreted EPS that were tightly bound with proteins and humic acid that can decrease the electrostatic force between individual cells resulting in increased hydrophobicity of the cell surface, thereby improving adsorption and self-flocculation (Wang et al., 2020). EPS produced by extremophiles is mostly reported to be tightly bound with other molecules such as proteins, inorganic ions, and sulfate residues that enhance the functionality, physiochemical properties, and potential applicability of these naturally produced EPS in the bioremediation of industrial wastewater.

Heavy metal pollution of land and water bodies is another pressing threat worldwide. Ore mining and industries dealing with the production of pesticides, petrochemical, and chemical are the largest contributors to heavy metals contamination. Bioaccumulation of heavy metals in the ecological food chain has a serious impact on human health. The efficiency of microbial EPS in the bioremediation of heavy metals through biosorption, bioleaching, and bioaccumulation has been reported by several studies (Banerjee et al., 2021). Biosorption of metal using EPS is based on the polyanionic properties of the EPS through an ion exchange mechanism or by direct interaction with the metal ions (Banerjee et al., 2021; Joulak et al., 2020). EPS from microbes of extreme origin has been demonstrated to have high functionality than EPS from mesophiles through various studies. It is mainly due to the monosaccharide composition and presence of bound molecules. For instance, the thermophilic bacteria *Anoxybacillus* sp. R4–33 isolated from a hot spring produced an acidic EPS that can remove approximately 2 mg of zinc and 1.4 mg of cadmium per gram of EPS (Zhao et al., 2014). Similarly, a marine halophilic cyanobacterium *Synechococcus elongatus* BDU130911 bio-adsorbed 75% uranium and emulsified 51% petroleum hydrocarbon (Rashmi et al., 2021). In another example, biosorption of lead was reported for the EPS produced by a metallophilic bacterium *Pseudomonas* sp. W6 isolated from Karbi Anglong hot spring in Assam, India, the EPS removed about 65% lead from a synthetic groundwater medium (Kalita and Joshi, 2017). Similar biosorption of lead and cobalt was also reported for the EPSs produced by *Salipiger mucosus* A3 and *Halomonas almeriensis*, the possible mechanism of chelation of lead was attributed to the anionic property of the EPS that formed possible interactions with the cationic lead ions (Llamas et al., 2010, 2012).

Bioleaching of metal using microbes has been associated with the production of important EPS that helps in the leaching process. The use of biological cells for the extraction of valuable metals has fewer environmental threats as compared to leaching using chemicals. *T. ferrooxidans* EPS production during the leaching process was found to be helpful in the adhesion of the cells to the metal ores (Gehrke et al., 1998); similar functionality was observed for the EPS produced by the acidophile *Acidithiobacillus thiooxidans* (Díaz et al., 2018). Due to this property, the bacterium *A. thiooxidans* was used in the industrial mining site for the treatment of wastewater. Bioremediation of wastewater is among the major functionality of EPS from extremophiles. Apart from hazardous contaminants from industrial wastewater, petroleum pollution in the water system is also a major ecological threat to the world. Petroleum hydrocarbons are the major source of energy used worldwide to run industries and transport. Contamination of water bodies as a result of oil spillage through fishing vessels, supply tankers, cruise ships, and water-based industries such as oil and gas offshore drilling has become a major environmental threat, especially in the Antarctic region (Nagar et al., 2021). Due to extreme environmental conditions in the Antarctic region, natural biodegradation of petroleum pollution is slow; however, several studies have reported the isolation of various biofilm-producing hydrocarbon-degrading microbes from extreme environments and oil-contaminated regions (Abdulrasheed et al., 2020; Gentile et al., 2016; Nagar et al., 2021; Roslee et al., 2020; Vandana and Das, 2021). Environmental pollution is an ever-rising issue in our today's world. Development is linked with an increase in the number of industries and usage of petroleum hydrocarbon, which in turn leads to an increase in the amount of release of hazardous substances in land and water bodies. The unique properties of the extremophilic EPS, such as the presence of tightly bound macro and micro molecules, acids, ions, and sulfation, can be applied at the field level for the bioremediation of the environment.

7.4 CONCLUSION

It is well-considered that microorganisms inhabiting extreme environmental conditions are an invaluable source of promising biotechnological and industrial resources. The EPS produced by the extremophiles has been observed to have unique physio-chemical properties that can be endured for the benefit of the people. Extremophilic EPS have found application in various industrial sector such as food, cosmetics, pharmaceuticals, nutraceuticals, and in the bioremediation of the environment. More and more reports are being documented on the production of EPS worldwide, however, a strong insight into the chemical structure and in-depth study on the potential application of these EPS in various fields will be helpful in the possible commercialization of this precious resource of nature.

REFERENCES

Abdulrasheed, M., Zakaria, N.N., Roslee, A.F.A., Shukor, M.Y., Zulkharnain, A., Napis, S., Convey, P., Alias, S.A., Gonzalez-Rocha, G., Ahmad, S.A., 2020. Biodegradation of diesel oil by cold-adapted bacterial strains of *Arthrobacter* spp. from Antarctica. Antarctic Sci. 32(5), 341–353.

Abedin, M.M., Chourasia, R., Phukon, L.C., Sarkar, P., Ray, R.C., Singh, S.P., Rai, A.K. 2023. Lactic acid bacteria in the functional food industry: biotechnological properties and potential applications. Crit. Rev. Food Sci. Nutr. 0(0), 1–19. https://doi.org/10.1080/10408398.2023.2227896

Aisha, W.Q., 2011. Osmoadaptation and plant growth promotion by salt tolerant bacteria under salt stress. African J. Microbiol. Res. 5(21), 3546–3554.

Ali, P., Shah, A.A., Hasan, F., Hertkorn, N., Gonsior, M., Sajjad, W., Chen, F., 2020. A glacier bacterium produces high yield of cryoprotective exopolysaccharide. Front. Microbiol. 10, 3096.

Angelin, J., Kavitha, M., 2020. Exopolysaccharides from probiotic bacteria and their health potential. Int. J. Biol. Macromol. 162, 853–865.

Antón, J., Meseguer, I., Rodrífguez-Valera, F., 1988. Production of an extracellular polysaccharide by haloferax mediterranei. Appl. Environ. Microbiol. 54(10), 2381.

Arayes, M.A., Mabrouk, M.E.M., Sabry, S.A., Abdella, B., 2022. Exopolysaccharide production from *Alkalibacillus* sp. w3: Statistical optimization and biological activity. Biologia. 1, 1–12.

Arena, A., Gugliandolo, C., Stassi, G., Pavone, B., Iannello, D., Bisignano, G., Maugeri, T.L., 2009. An exopolysaccharide produced by *Geobacillus thermodenitrificans* strain B3–72: Antiviral activity on immunocompetent cells. Immun. Lett. 123(2), 132–137. https://doi.org/10.1016/j.imlet.2009.03.001

Arena, A., Maugeri, T.L., Pavone, B., Iannello, D., Gugliandolo, C., Bisignano, G., 2006. Antiviral and immunoregulatory effect of a novel exopolysaccharide from a marine thermotolerant *Bacillus licheniformis*. Int. Immunopharmacol. 6(1), 8–13.

Arias, S., Del Moral, A., Ferrer, M.R., Tallon, R., Quesada, E., Béjar, V., 2003. Mauran, a exopolysaccharide produced by the halophilic bacterium Halomonas maura, with a novel composition and interesting properties for biotechnology. Extremophiles. 7(4), 319–326.

Ates, O., 2015. Systems biology of microbial exopolysaccharides production. Front. Bioengin. Biotech. 3, 200.

Banerjee, A., Sarkar, S., Govil, T., González-Faune, P., Cabrera-Barjas, G., Bandopadhyay, R., Salem, D.R., Sani, R.K., 2021. Extremophilic exopolysaccharides: Biotechnologies and wastewater remediation. Front. Microbiol. 12.

Caccamo, M.T., Gugliandolo, C., Zammuto, V., Magazù, S., 2020. Thermal properties of an exopolysaccharide produced by a marine thermotolerant *Bacillus licheniformis* by ATR-FTIR spectroscopy. Int. J. Biol. Macromol. 145, 77–83.

Carillo, S., Casillo, A., Pieretti, G., Parrilli, E., Sannino, F., Bayer-Giraldi, M., Cosconati, S., Novellino, E., Ewert, M., Deming, J.W., Lanzetta, R., Marino, G., Parrilli, M., Randazzo, A., Tutino, M.L., Corsaro, M.M., 2015. A unique capsular polysaccharide structure from the psychrophilic marine bacterium *Colwellia psychrerythraea* 34H that mimics antifreeze (glyco)proteins. J. Am. Chem. Soc. 137(1), 179–189.

Carrión, O., Delgado, L., Mercade, E., 2015. New emulsifying and cryoprotective exopolysaccharide from *Antarctic Pseudomonas* sp. ID1. Carbohydr. Polym. 117, 1028.

Caruso, C., Rizzo, C., Mangano, S., Poli, A., di Donato, P., Finore, I., Nicolaus, B., di Marco, G., Michaud, L., Lo Giudice, A., 2018a. Production and biotechnological potential of extracellular polymeric substances from sponge-associated Antarctic bacteria. Appl. Environ. Microbiol. 84(4).

Caruso, C., Rizzo, C., Mangano, S., Poli, A., Di Donato, P., Nicolaus, B., Di Marco, G., Michaud, L., Lo Giudice, A., 2018b. Extracellular polymeric substances with metal adsorption capacity produced by *Pseudoalteromonas* sp. MER144 from Antarctic seawater. Environ. Sci. Poll. Res. 25(5), 4667–4677.

Casillo, A., Parrilli, E., Sannino, F., Mitchell, D.E., Gibson, M.I., Marino, G., Lanzetta, R., Parrilli, M., Cosconati, S., Novellino, E., Randazzo, A., Tutino, M.L., Corsaro, M.M., 2017. Structure-activity relationship of the exopolysaccharide from a psychrophilic bacterium: A strategy for cryoprotection. Carbohydr. Polym. 156, 364–371.

Chambi, D., Romero-Soto, L., Villca, R., Orozco-Gutiérrez, F., Vega-Baudrit, J., Quillaguamán, J., Hatti-Kaul, R., Martín, C., Carrasco, C., 2021. Exopolysaccharides production by cultivating a bacterial isolate from the hypersaline environment of salar de uyuni (Bolivia) in pretreatment liquids of steam-exploded quinoa stalks and enzymatic hydrolysates of Curupaú sawdust. Fermentation. 7(1), 1–16.

Chatterjee, S., Mukhopadhyay, S.K., Gauri, S.S., Dey, S., 2018. Sphingobactan, a new α-mannan exopolysaccharide from Arctic *Sphingobacterium* sp. IITKGP-BTPF3 capable of biological response modification. Int. Immunopharmacol. 60, 84–95.

Chen, G., Qian, W., Li, J., Xu, Y., Chen, K., 2015. Exopolysaccharide of Antarctic bacterium *Pseudoalteromonas* sp. S-5 induces apoptosis in K562 cells. Carbohydr. Polym. 121, 107–114.

Chikkanna, A., Ghosh, D., Kishore, A., 2018. Expression and characterization of a potential exopolysaccharide from a newly isolated halophilic thermotolerant bacteria Halomonas nitroreducens strain WB1. Peer J. 4, e4684.

Chiring Phukon, L., Minhajul Abedin, M., Chourasia, R., Sahoo, D., Parameswaran, B., Kumar Rai, A., 2020. Production of xylanase under submerged fermentation from *Bacillus firmus* HS11 isolated from Sikkim Himalayan region. Indian J. Exp. Biol. 58.

Chourasia, R., Abedin, M.M., Chiring Phukon, L., Sahoo, D., Singh, S.P., Rai, A.K., 2021. Biotechnological approaches for the production of designer cheese with improved functionality. Compr. Rev. Food Sci. Food Sa. 20(1), 960–979. https://doi.org/10.1111/1541-4337.12680

Chourasia, R., Phukon, L.C., Abedin, M.M., Sahoo, D., Rai, A.K., 2022. Microbial transformation during gut fermentation. In: Amit Kumar Rai, Anu Appaiah K.A. (Eds.), *Bioactive Compounds in Fermented Foods: Health Aspects.*

Cui, Y.W., Gong, X.Y., Shi, Y.P., Wang, Z., 2017. Salinity effect on production of PHA and EPS by *Haloferax mediterranei*. RSC Adv. 7(84), 53587–53595.

Díaz, M., Castro, M., Copaja, S., Guiliani, N., 2018. Biofilm formation by the acidophile bacterium *Acidithiobacillus thiooxidans* involves c-di-GMP pathway and Pel exopolysaccharide. Genes. 9, 113.

Gehrke, T., Telegdi, J., Thierry, D., Sand, W., 1998. Importance of extracellular polymeric substances from *Thiobacillus ferrooxidans* for bioleaching. Appl. Enviro. Microbiol. 64(7), 2743–2747.

Gentile, G., Bonsignore, M., Santisi, S., Catalfamo, M., Giuliano, L., Genovese, L., Yakimov, M.M., Denaro, R., Genovese, M., Cappello, S., 2016. Biodegradation potentiality of psychrophilic bacterial strain *Oleispira antarctica* RB-8T. Mar. Poll. Bull. 105(1), 125–130.

Gu, D., Jiao, Y., Wu, J., Liu, Z., Chen, Q., 2017. Optimization of EPS production and characterization by a halophilic bacterium, *Kocuria rosea* ZJUQH from Chaka Salt Lake with response surface methodology. Molecules. 22(5).

Gugliandolo, C., Spanò, A., Lentini, V., Arena, A., Maugeri, T.L., 2014. Antiviral and immunomodulatory effects of a novel bacterial exopolysaccharide of shallow marine vent origin. J. Appl. Microbiol. 116(4).

Hamidi, M., Gholipour, A.R., Delattre, C., Sesdighi, F., Mirzaei Seveiri, R., Pasdaran, A., Kheirandish, S., Pierre, G., Safarzadeh Kozani, P., Safarzadeh Kozani, P., Karimitabar, F., 2020. Production, characterization and biological activities of exopolysaccharides from a new cold-adapted yeast: *Rhodotorula mucilaginosa* sp. GUMS16. Int. J. Biol. Macromol. 151, 268–277.

Joshi, A.A., Kanekar, P.P., 2011. Production of exopolysaccharide by *Vagococcus carniphilus* MCM B-1018 isolated from alkaline Lonar Lake, India. Ann. Microbiol. 61(4), 733–740.

Joulak, I., Azabou, S., Finore, I., Poli, A., Nicolaus, B., Donato, P.D.I., Bkhairia, I., Dumas, E., Gharsallaoui, A., Immirzi, B., Attia, H., 2020a. Structural characterization and functional properties of novel exopolysaccharide from the extremely halotolerant *Halomonas elongata* S6. Int. J. Biol. Macromol. 164, 95–104.

Joulak, I., Finore, I., Poli, A., Abid, Y., Bkhairia, I., Nicolaus, B., Di Donato, P., Dal Poggetto, G., Gharsallaoui, A., Attia, H., Azabou, S., 2020b. Hetero-exopolysaccharide from the extremely halophilic *Halomonas smyrnensis* K2: Production, characterization and functional properties in vitro. 3 Biotech. 10(9), 1–12.

Kalita, D., Joshi, S.R., 2017. Study on bioremediation of Lead by exopolysaccharide producing metallophilic bacterium isolated from extreme habitat. Biotechnol. Rep. 16, 48–57.

Kambourova, M., Mandeva, R., Dimova, D., Poli, A., Nicolaus, B., Tommonaro, G., 2009. Production and characterization of a microbial glucan, synthesized by *Geobacillus tepidamans* V264 isolated from Bulgarian hot spring. Carbohydr. Polym. 77(2), 338–343.

Kazak Sarilmiser, H., Ates, O., Ozdemir, G., Arga, K.Y., Toksoy Oner, E., 2015. Effective stimulating factors for microbial levan production by *Halomonas smyrnensis* AAD6T. J. Biosci. Bioengin. 119(4), 455–463.

Kazak Sarilmiser, H., Toksoy Oner, E., 2014. Investigation of anti-cancer activity of linear and aldehyde-activated levan from *Halomonas smyrnensis* AAD6T. Biochem. Engin. J. 92, 28–34.

Kheyrandish, S., Rastgar, A., Hamidi, M., Sajjadi, S.M., Sarab, G.A., 2022. Evaluation of anti-tumor effect of the exopolysaccharide from new cold-adapted yeast, *Rhodotorula mucilaginosa* sp. GUMS16 on chronic myeloid leukemia K562 cell line. Int. J. Biol. Macromol. 206, 21–28.

Kokoulin, M.S., Kuzmich, A.S., Romanenko, L.A., Chikalovets, I.V., 2021. Structure and in vitro antiproliferative activity of the acidic capsular polysaccharide from the deep-sea bacterium *Psychrobacter submarinus* KMM 225T. Carbohydr. Polym. 262, 117941.

Koller, M., Chiellini, E., Braunegg, G., 2015. Study on the production and re-use of poly(3-hydroxybutyrate-co-3-hydroxyvalerate) and extracellular polysaccharide by the archaeon *Haloferax mediterranei* strain DSM 1411. Chem. Biochem. Engin. Quar. 29(2), 87–98.

Kumar, A., Mukhia, S., Kumar, R., 2022. Production, characterisation, and application of exopolysaccharide extracted from a glacier bacterium *Mucilaginibacter* sp. ERMR7:07. Process Biochem. 113, 27–36.

Kumari, M., Padhi, S., Sharma, S., Phukon, L.C., Singh, S.P., Rai, A.K., 2021. Biotechnological potential of psychrophilic microorganisms as the source of cold-active enzymes in food processing applications. 3 Biotech. 11, 1–18.

Lelchat, F., Cérantola, S., Brandily, C., Colliec-Jouault, S., Baudoux, A.C., Ojima, T., Boisset, C., 2015. The marine bacteria *Cobetia marina* DSMZ 4741 synthesizes an unexpected K-antigen-like exopolysaccharide. Carbohydr. Polym. 124, 347–356.

Li, W.W., Zhou, W.Z., Zhang, Y.Z., Wang, J., Zhu, X.B., 2008. Flocculation behavior and mechanism of an exopolysaccharide from the deep-sea psychrophilic bacterium *Pseudoalteromonas* sp. SM9913. Biores. Technol. 99(15), 6893–6899.

Liang, X., 2017. Structural characterization and bioactivity of exopolysaccharide synthesized by *Geobacillus* sp. TS3–9 isolated from Radioactive Radon Hot Spring. Adv. Biotechnol. Microbiol. 4(2).

Liu, S.B., Chen, X.L., He, H.L., Zhang, X.Y., Xie, B. Bin, Yu, Y., Chen, B., Zhou, B.C., Zhang, Y.Z., 2013. Structure and ecological roles of a novel exopolysaccharide from the arctic sea ice bacterium *Pseudoalteromonas* sp. Strain SM20310. Appl. Environ. Microbiol. 79(1), 224–230.

Llamas, I., Amjres, H., Mata, J.A., Quesada, E., Béjar, V., 2012. The potential biotechnological applications of the exopolysaccharide produced by the halophilic bacterium *Halomonas almeriensis*. Molecules. 17, 7103–7120.

Llamas, I., Mata, J.A., Tallon, R., Bressollier, P., Urdaci, M.C., Quesada, E., Béjar, V., 2010. Characterization of the exopolysaccharide produced by *Salipiger mucosus* A3T, a halophilic species belonging to the Alphaproteobacteria, isolated on the Spanish mediterranean seaboard. Mar. Drugs. 8(8), 2240–2251.

López-Ortega, M.A., Chavarría-Hernández, N., López-Cuellar, M. del R., Rodríguez-Hernández, A.I., 2021. A review of extracellular polysaccharides from extreme niches: An emerging natural source for the biotechnology. From the adverse to diverse! Int. J. Biol. Macromol. 177, 559–577.

Madigan, M.T., Oren, A., 1999. Thermophilic and halophilic extremophiles. Curr. Opin. Microbiol. 2(3), 265–269.

Margesin, R., Schinner, F., 2001. Potential of halotolerant and halophilic microorganisms for biotechnology. Extremophiles. 5(2), 73–83.

Marx, J.G., Carpenter, S.D., Deming, J.W., 2009. Production of cryoprotectant extracellular polysaccharide substances (EPS) by the marine psychrophilic bacterium *Colwellia psychrerythraea* strain 34H under extreme conditions. Can. J. Microbiol. 55(1), 63–72.

Mata, J.A., Béjar, V., Bressollier, P., Tallon, R., Urdaci, M.C., Quesada, E., Llamas, I., 2008. Characterization of exopolysaccharides produced by three moderately halophilic bacteria belonging to the family Alteromonadaceae. J. Appl. Microbiol. 105(2), 521–528.

Mata, J.A., Béjar, V., Llamas, I., Arias, S., Bressollier, P., Tallon, R., Urdaci, M.C., Quesada, E., 2006. Exopolysaccharides produced by the recently described halophilic bacteria *Halomonas ventosae* and *Halomonas anticariensis*. Res. Microbiol. 157(9), 827–835. https://doi.org/10.1016/J.RESMIC.2006.06.004

Maugeri, T.L., Gugliandolo, C., Caccamo, D., Panico, A., Lama, L., Gambacorta, A., Nicolaus, B., 2002. A halophilic thermotolerant *Bacillus* isolated from a marine hot spring able to produce a new exopolysaccharide. Biotechnol. Lett. 24(7), 515–519.

Moshabaki Isfahani, F., Tahmourespour, A., Hoodaji, M., Ataabadi, M., Mohammadi, A., 2018. Characterizing the new bacterial isolates of high yielding exopolysaccharides under hypersaline conditions. J. Clean. Prod. 185, 922–928.

Mukhopadhyay, S.K., Chatterjee, S., Gauri, S.S., Das, S.S., Mishra, A., Patra, M., Ghosh, A.K., Das, A.K., Singh, S.M., Dey, S., 2014. Isolation and characterization of extracellular polysaccharide Thelebolan produced by a newly isolated psychrophilic Antarctic fungus Thelebolus. Carbohydr. Polym. 104(1), 204–212.

Nagar, S., Antony, R., Thamban, M., 2021. Extracellular polymeric substances in Antarctic environments: A review of their ecological roles and impact on glacier biogeochemical cycles. Polar. Sci. 30, 100686.

Nichols, C.M., Bowman, J.P., Guezennec, J., 2005a. Effects of incubation temperature on growth and production of exopolysaccharides by an antarctic sea ice bacterium grown in batch culture. Appl. Environ. Microbiol. 71(7), 3519–3523.

Nichols, C.M., Lardière, S.G., Bowman, J.P., Nichols, P.D., Gibson, J.A.E., Guézennec, J., 2005b. Chemical characterization of exopolysaccharides from Antarctic marine bacteria. Microb. Ecol. 49(4), 578–589.

Nicolaus, B., Kambourova, M., Oner, E.T., 2010. Exopolysaccharides from extremophiles: From fundamentals to biotechnology. Environ. Technol. 31(10), 1145–1158.

Nicolaus, B., Lama, L., Panico, A., Moriello, V.S., Romano, I., Gambacorta, A., 2002. Production and characterization of exopolysaccharides excreted by thermophilic bacteria from shallow, marine hydrothermal vents of flegrean ares (Italy). Syst. Appl. Microbiol. 25(3), 319–325.

Nicolaus, B., Panico, A., Manca, M.C., Lama, L., Gambacorta, A., Maugeri, T., Gugliandolo, C., Caccamo, D., 2000. A thermophilic Bacillus isolated from an Eolian shallow hydrothermal vent, able to produce exopolysaccharides. Syst. Appl. Microbiol. 23, 426–432.

Panosyan, H., Di Donato, P., Poli, A., Nicolaus, B., 2018. Production and characterization of exopolysaccharides by *Geobacillus thermodenitrificans* ArzA-6 and Geobacillus toebii ArzA-8 strains isolated from an Armenian geothermal spring. Extremophiles. 22(5), 725–737.

Phukon, L.C., Chourasia, R., Padhi, S., Abedin, M.M., Godan, T.K., Parameswaran, B., Singh, S.P., Rai, A.K., 2022. Cold-adaptive traits identified by comparative genomic analysis of a lipase-producing *Pseudomonas* sp. HS6 isolated from snow-covered soil of Sikkim Himalaya and molecular simulation of lipase for wide substrate specificity. Curr. Genet. 68(3–4), 375–391.

Poli, A., Anzelmo, G., Nicolaus, B., 2010. Bacterial exopolysaccharides from extreme marine habitats: Production, characterization and biological activities. Mar. Drugs. 8, 1779–1802.

Poli, A., Kazak, H., Gürleyendağ, B., Tommonaro, G., Pieretti, G., Öner, E.T., Nicolaus, B., 2009. High level synthesis of levan by a novel Halomonas species growing on defined media. Carbohydr. Polym. 78(4), 651–657.

Radchenkova, N., Boyadzhieva, I., Atanasova, N., Poli, A., Finore, I., Di Donato, P., Nicolaus, B., Panchev, I., Kuncheva, M., Kambourova, M., 2018. Extracellular polymer substance synthesized by a halophilic bacterium *Chromohalobacter canadensis* 28. Appl. Microbiol. Biotechnol. 102(11), 4937–4949.

Radchenkova, N., Vassilev, S., Panchev, I., Anzelmo, G., Tomova, I., Nicolaus, B., Kuncheva, M., Petrov, K., Kambourova, M., 2013. Production and properties of two novel exopolysaccharides synthesized by a thermophilic bacterium *Aeribacillus pallidus* 418. Appl. Biochem. Biotechnol. 171(1), 31–43.

Rashmi, V., Darshana, A., Bhuvaneshwari, T., Saha, S.K., Uma, L., Prabaharan, D., 2021. Uranium adsorption and oil emulsification by extracellular polysaccharide (EPS) of a halophilic unicellular marine cyanobacterium *Synechococcus elongatus* BDU130911. Curr. Res. Green Sustain. Chem. 4, 100051.

Rinker, K.D., Kelly, R.M., 2000. Effect of carbon and nitrogen sources on growth dynamics and exopolysaccharide production for the hyperthermophilic archaeon Thermococcus litoralis and bacterium Thermotoga maritima. Biotechnol. Bioengin. 69(5), 537–547.

Roslee, A.F.A., Zakaria, N.N., Convey, P., Zulkharnain, A., Lee, G.L. Y., Gomez-Fuentes, C., Ahmad, S.A., 2020. Statistical optimisation of growth conditions and diesel degradation by the Antarctic bacterium, *Rhodococcus* sp. strain AQ5–07. Extremophiles. 24(2), 277–291.

Ruiz-ruiz, C., Srivastava, G.K., Carranza, D., Mata, J.A., Llamas, I., Santamaría, M., Quesada, E., Molina, I.J., 2011. An exopolysaccharide produced by the novel halophilic bacterium *Halomonas stenophila* strain B100 selectively induces apoptosis in human T leukaemia cells. Appl. Microbiol. Biotechnol. 89, 345–355.

Sardari, R.R.R., Kulcinskaja, E., Ron, E.Y.C., Björnsdóttir, S., Friðjónsson, Ó.H., Hreggviðsson, G.Ó., Karlsson, E.N., 2017. Evaluation of the production of exopoly-saccharides by two strains of the thermophilic bacterium *Rhodothermus marinus*. Carbohydr. Polym. 156, 1–8.

Sharma, N., Kumar, J., Abedin, M.M., Sahoo, D., Pandey, A., Rai, A.K., Singh, S.P., 2020. Metagenomics revealing molecular profiling of community structure and metabolic pathways in natural hot springs of the Sikkim Himalaya. BMC Microbiol. 20(1), 1–17. https://doi.org/10.1186/S12866-020-01923-3/FIGURES/6

Shi, Y., Huang, J., Zeng, G., Gu, Y., Chen, Y., Hu, Y., Tang, B., Zhou, J., Yang, Y., Shi, L., 2017. Exploiting extracellular polymeric substances (EPS) controlling strategies for perfor-mance enhancement of biological wastewater treatments: An overview. Chemosphere. 180, 396–411.

Song, A.X., Mao, Y.H., Siu, K.C., Tai, W.C.S., Wu, J.Y., 2019. Protective effects of exopoly-saccharide of a medicinal fungus on probiotic bacteria during cold storage and simu-lated gastrointestinal conditions. Int. J. Biol. Macromol. 133, 957–963.

Spanò, A., Gugliandolo, C., Lentini, V., Maugeri, T.L., Anzelmo, G., Poli, A., Nicolaus, B., 2013. A novel EPS-producing strain of *Bacillus licheniformis* isolated from a shallow vent off Panarea Island (Italy). Curr. Microbiol. 67(1), 21–29.

Sun, M.L., Zhao, F., Shi, M., Zhang, X.Y., Zhou, B.C., Zhang, Y.Z., Chen, X.L., 2015. Characterization and biotechnological potential analysis of a new exopolysaccharide from the Arctic marine bacterium *Polaribacter* sp. SM1127. Sci. Rep. 5, 1–12.

Sung, J.K., Joung, H.Y., 2007. Cryoprotective properties of exopolysaccharide (P-21653) produced by the Antarctic bacterium, *Pseudoalteromonas* arctica KOPRI 21653. J. Microbiol. 45(6), 510–514.

Vandana, Das, S., 2021. Structural and mechanical characterization of biofilm-associated bacterial polymer in the emulsification of petroleum hydrocarbon. 3 Biotech. 11(5), 1–15.

Vinothkanna, A., Sathiyanarayanan, G., Rai, A.K., Mathivanan, K., Saravanan, K., Sudharsan, K., Kalimuthu, P., Ma, Y., Sekar, S., 2022. Exopolysaccharide produced by probiotic *Bacillus albus* DM-15 isolated from ayurvedic fermented *Dasamoolarishta*: characterization, antioxidant, and anticancer activities. Front. Microbiol. 13, 832109.

Wang, J., Goh, K.M., Salem, D.R., Sani, R.K., 2019a. Genome analysis of a thermophilic exo-polysaccharide-producing bacterium—*Geobacillus* sp. WSUCF1. Sci. Rep. 9(1), 1–12.

Wang, J., Salem, D.R., Sani, R.K., 2019b. Extremophilic exopolysaccharides: A review and new perspectives on engineering strategies and applications. Carbohydr. Polym. 205, 8–26.

Wang, J., Salem, D.R., Sani, R.K., 2021. Two new exopolysaccharides from a thermo-philic bacterium *Geobacillus* sp. WSUCF1: Characterization and bioactivities. New Biotechnol. 61, 29–39.

Wang, Y., Jiang, L., Shang, H., Li, Q., Zhou, W., 2020. Treatment of azo dye wastewater by the self-flocculating marine bacterium *Aliiglaciecola lipolytica*. Environ. Technol. Innov. 19, 100810.

Yasar Yildiz, S., Anzelmo, G., Ozer, T., Radchenkova, N., Genc, S., Di Donato, P., Nicolaus, B., Toksoy Oner, E., Kambourova, M., 2014. *Brevibacillus themoruber*: A promis-ing microbial cell factory for exopolysaccharide production. J. Appl. Microbiol. 116, 314–324.

Zeng, W., Qiu, G., Zhou, H., Liu, X., Chen, M., Chao, W., Zhang, C., Peng, J., 2010. Characterization of extracellular polymeric substances extracted during the bioleaching of chalcopyrite concentrate. Hydrometall. 100, 177–180.

Zhao, S., Cao, F., Zhang, H., Zhang, L., Zhang, F., Liang, X., 2014. Structural characterization and biosorption of exopolysaccharides from *Anoxybacillus* sp. R4–33 isolated from radioactive radon hot spring. Appl. Biochem. Biotechnol. 172, 2732–2746.

Zheng, C., Li, Z., Su, J., Zhang, R., Liu, C., Zhao, M., 2012. Characterization and emulsifying property of a novel bioemulsifier by *Aeribacillus pallidus* YM-1. J. Appl. Microbiol. 113(1), 44–51.

Zhou, W., Shen, B., Meng, F., Liu, S., Zhang, Y., 2010. Coagulation enhancement of exopolysaccharide secreted by an Antarctic sea-ice bacterium on dye wastewater. Sep. Purif. Technol. 76(2), 215–221.

8 Exopolysaccharides from marine microbial resources

Rupinder Pal Singh, Harinderjeet Kaur,
Saurabh Gupta, Amrit Singh, Yadvinder Singh,
Harpal Singh, Mansimran Kaur Randhawa,
and Parmjit Singh Panesar

8.1 INTRODUCTION

Polysaccharides from microbes are generated as exopolysaccharides (EPS) and capsular polysaccharides (CPS). Sutherland (1972) coined the term "exopolysaccharide" to refer to carbohydrates of high molecular weight polymers made by numerous marine microorganisms. Since then, exopolysaccharides have been employed to denote extracellular polymeric compounds with a broader definition (Nichols et al., 2005). Exopolysaccharides are carbohydrate polymers with a high molecular weight that are secreted by bacteria throughout the development and metabolic processes. Since the beginning of time, humanity has placed a high value on marine environments. Polysaccharides were traditionally extracted from plants, algae, etc. The first studies of their production by certain microbes were recorded in 1880s. Exopolysaccharides (EPS), also referred to as extracellular polymeric compounds, are polysaccharide-containing macromolecule mixtures secreted into the extracellular matrix by micro-algae during the entire lifecycle (Xiao et al., 2016). Exopolysaccharides can be linked to lipids, proteins, chemical and inorganic substances, DNA, and metal ions. They are composed of repeating units of sugar moieties coupled to a carrier lipid. Exopolysaccharides (EPSs) build up on the surface of microbial cells and shield the cells from harsh environment by regulating the structure of the membrane and are also used as energy and carbon reserves. In addition to shielding cells from toxins and dehydration; serving as energy and carbon sinks in response to nutrient challenges, exopolysaccharides are crucial for the environment. Exopolysaccharides are regularly released from the membrane into the environment, appearing to be unattached, renewable, free, and simple to extract from orgasmic bio-masses by specific fermentation procedures (Dave et al., 2020). Exopolysaccharides have been discovered from various sources. Microbial-origin exopolysaccharides are abundant in nature, have distinct characteristics, and can be extracted from fresh water, marine water, extreme conditions, and the soil ecosystem. In an environment with high salt content, high pressure, cold and oligotrophic conditions, marine microorganisms

DOI: 10.1201/9781003342687-8

can create a variety of specialized active compounds. Due to their many functional domains as marine nutraceuticals, marine bioremediation, marine bioenergy, marine natural products for health, these exopolysaccharides have explored their prospective uses in the pharma, food, cosmetics, as well as in environmental clean-up and agriculture (Bhatt et al., 2022). Additionally, marine creatures can develop distinctive bioactive compounds that can be used by industries. Exopolysaccharides from marine microbes are useful in several sectors, in addition to bioactive compounds.

Exopolysaccharides are further classified based on structural composition, as demonstrated in Figure 8.1. Homo-exopolysaccharides contain only one kind of monosaccharide as fructans, α-D-glucans, β-D-glucans, and poly-galactans. However, Hetero-exopolysaccharide contains different kinds of monosaccharides mostly D-glucose, L-rhamnose, D-galactose, and derivatives (Delbarre et al., 2014). On the basis of fundamental properties of the molecules, like the branching patterns, molecular weights, and main chain bond patterns, the difference arises among the homopolysaccharides (Harutoshi, 2013). Additionally, EPS were divided into seven groups based on their functions (Flemming and Wingender, 2010), such as surface active, nutritional groups, instructive, sorptive, structural or constructive, redox active, and active.

Exopolysaccharides are chains of linear and branched recurring group of sugars or sugar derivatives. In varying proportions, these sugars are mostly composed of glucose, mannose, galactose, N-acetylgalactosamine, N-acetylglucosamine, and

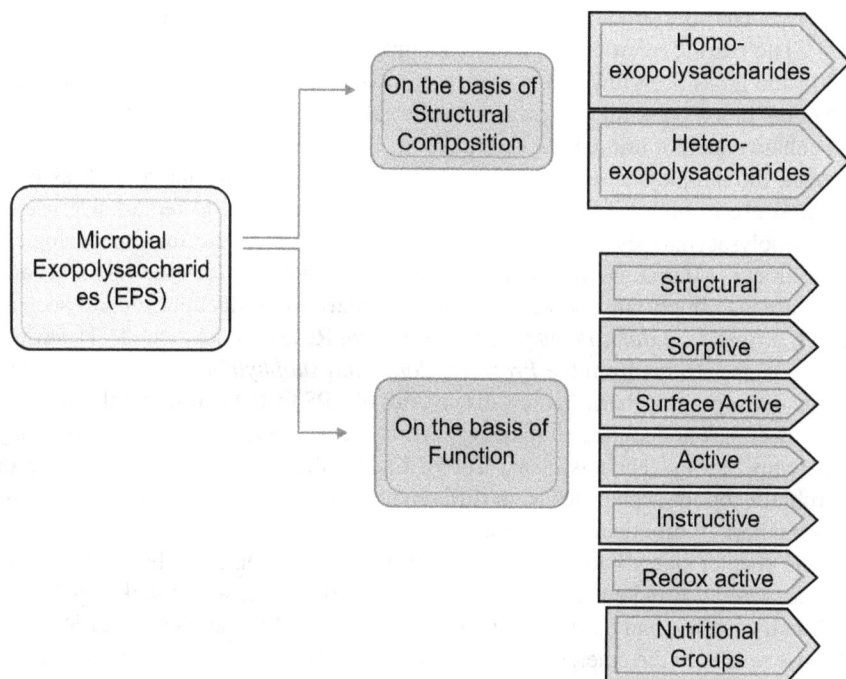

FIGURE 8.1 Classification of microbial exopolysaccharide

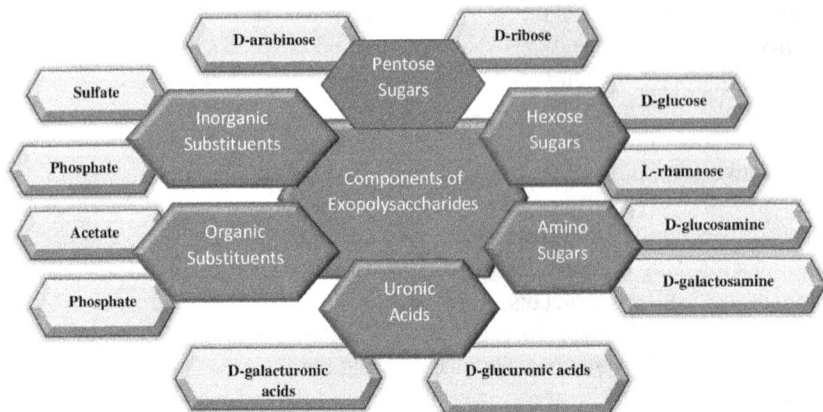

FIGURE 8.2 Components of bacterial exopolysaccharides

rhamnose (Patel et al., 2012). Figure 8.2 lists the usual bacterial exopolysaccharides composition. Marine microbes are considered to be significant natural reservoir of exopolysaccharide producers, and various marine microorganisms have been extracted from the marine environment for exopolysaccharide synthesis. Marine microorganisms, notably *Alteromonas, Halomonas, Bacillus, Enterobacter, Pseudoalteromonas, Planococcus, Vibrio, and Rhodococcus*, etc., are the main exopolysaccharides producers and have been thoroughly investigated (Finore et al., 2014). The marine microbial EPS has a complex and diverse structure (Wang et al., 2020), and it has been linked to a variety of biological actions, including cholesterol reduction, blood pressure lowering, antibacterial, antioxidant, anticancer, antifreeze, and enhancement of immunological function.

Also, the unique characteristics of the marine microbial habitat as well as their potential use in bioremediation and carbon sequestration led to remarkable use of the exopolysaccharides generated by marine microbes in the marine ecological environment. Microalgae and cyanobacteria along with the marine bacteria, and fungi, are also important sources of exopolysaccharides production, sources such as *Porphyridium sordidum, Porphyridium marinum*. Research done on the 11 variety of microalgae strains from the *Proteorhodophytina subphylum* focused on the generation of exopolysaccharides. Few discoveries of EPS from hydrothermal vent habitats in deep-sea are connected to heavy metals, tremendous pressure, and extremely high temperatures (Nichols et al., 2005). Lately, the commercial significance of microbial exopolysaccharides from different environment was recognized, proposing a plethora of promising industrial applications.

This chapter intends to provide in depth information on microbial exopolysaccharide. Researchers have drawn their interests on the seas, which make up 71% of the Earth's surface and have an average temperature of 2°C and pressure of 38 MPa, all these results in the emergence of a variety of extreme environment in the marine ecosystem, such as deep sea, hot water vent, marine salterns, where these exopolysaccharides thrive. Their origins, chemical structural properties, and prospective uses in bioremediation and biological activity, implies that the marine microbial exopolysaccharides can be considered as a novel and intriguing source.

8.2 SOURCES OF MARINE EXOPOLYSACCHARIDES

The use of polysaccharides in various products as stabilizers, thickeners, and gel-forming agents in food industries, and as antioxidant, anticancer, or prebiotic substance in the pharmaceutical industry has been studied (Dave et al., 2020). They are usually obtained from the different origins as plants, fungi, algae, and bacteria. However, oceanic and terrestrial sub surfaces have been found to contain the highest percentage of microbial diversity. The marine environment is very dynamic, a habitat to a wide variety of marine microorganisms, and also thought to generate chemicals and bioactive substances (Shearer et al., 2007). Due to the significance in marine food chains, their part in the dynamics of nutrient cycling, and biogeochemical processes, marine microorganisms such as bacteria, fungus, and viruses are environmentally varied populations. In these habitats, bacteria are among the most significant and prevalent populations. Marine bacteria or their products like pigments, exopolymers, biosurfactants, enzymes, and antibacterial components have a broad scope in environmental remediation as well as in food, pharmaceutical, and textile industry. Polysaccharides produced by microbes can be classified into three categories as cell wall polysaccharides, extracellular polysaccharides, and intracellular polysaccharides. Several microorganisms produce exopolysaccharides while intracellular and cell wall polysaccharides are essential components of the cell. To determine the link between EPS-generating strains surviving in a variety of hypersaline environments, taxonomic research was published by Martínez-Cánovas et al. (2004) in which nearly 134 strains of bacteria were isolated from several saline habitats such as marine salterns, inland salterns, and also soils. Numerous marine microbes were extracted from the sea for the generation of exopolysaccharides as they are considered to be the natural exopolysaccharide-producing factories (Satpute et al., 2010). The Mediterranean Sea is the source of a halophilic archaea bacteria, *Haloferax mediterranei*, that produce exopolysaccharides. Mata et al. (2008) identified fairly common EPS-producing halophilic bacteria, a member of the *Alteromonadaceae* family, isolated from hypersaline environments in Spain. Many thermophilic strains were obtained near the Lucrino area (gulf of Pozzuoli, Naples, Italy) and Ischia Island (Flegrean areas, Italy), from shallow hydrothermal vents for the examination of exopolysaccharides production. The majority of the exopolysaccharides-producing bacteria, such as *Pseudomonas, Alteromonas, Vibrio,* and *Acinetobacter,* are gram-negative in nature. Microbial origin exopolysaccharides have significant benefits above polysaccharides that are derived from various sources such as algae, plants, and animals (Banerjee et al., 2021). These exopolysaccharides possess various characteristics. One is structural reproducibility, on the other hand, because the polysaccharide structures of plant and animal sources are reliant on different factors like climate, environmental conditions, and feeding requirements (Casillo et al., 2018). Because of the unique qualities possessed by the exopolysaccharides, which includes structural diversity, biodegradability, bioactivities, and biocompatibility, EPS derived from marine microorganisms have received attention and are being researched in the field of food industries, pharmacy, and cosmetics (Radchenkova et al., 2013). Moreover, by enhancing the growth conditions and by employing biotechnological methods, which have now become a significant and developing research area, it is feasible to achieve the maximum yield of exopolysaccharides. The marine environment is very dynamic, a habitat to a wide variety of marine microorganisms, and also thought to generate chemicals and bioactive substances (Shearer et al., 2007).

Marine microorganisms, among various kinds of aquatic microorganisms, are responsible for nearly 50% of all the Earth's organic matter. Till now, the majority of the exopolysaccharides producing marine strains, species such as *Pseudoalteromonas, Bacillus, Alteromonas*, and *Vibrio* (Guezennec et al., 2002; Zanchetta et al., 2003) were isolated. From those prevalent strains, a number of exopolysaccharides with distinct structural characteristics and biological activity have been discovered, enlisted in Table 8.1. An EPS was obtained from a marine source and generated by a probiotic, namely, *Pediococcus pentosaceus* M41 (Ayyash et al., 2020). Exopolysaccharide EPS273 derived from the marine microbe *P. stutzeri* 273 could shatter the preexisting biofilms of *P. aeruginosa* PAO1 and suppresses the biofilm generation, demonstrating that the exopolysaccharide EPS273 had such potential in the treatment of the bacteria biofilm-associated illness (Wu et al., 2016). RSK CAS4, a *Bacillus thuringiensis* bacterium, derived from the ascidian *Didemnum granulatum*, by applying the response surface approach, the conditions of the exopolysaccharide production have improved (Ramamoorthy et al., 2018). Microbacterium strain FSW-25, identified on the Rasthakaadu beach, Kanyakumari possess the capability to generate a significant number of exopolysaccharides (Sran et al., 2022). Also, a bacterial strain *Sphingobium yanoikuyae* BBL01, discovered from the eastern sea of South Korea, produced an EPS that was stable at high temperature up to 233°C and exhibited no effect of different carbon sources on the monomer composition of exopolysaccharide (Bhatia et al., 2021). Recently, a strain belonging to *Bacillus cereus* was discovered off the coast of Saudi Arabia in the Red Sea. The exopolysaccharide extracted from the marine strain, a significant portion of which was EPSR3, exhibited anticancer, antioxidant, and anti-inflammatory properties. The biological actions caused were due to the uronic acids in EPSR3 (Selim et al., 2022).

Furthermore, exopolysaccharides produced from the halophilic bacteria tolerate osmotic pressure (Bhatt et al., 2022). Therefore, the search for unique, physiologically active EPS from halophilic bacteria is still in its initial stages because of the extensive dispersion in saline environments. *Halolactibacillus miurensis* produced an EPS called HMEPS that was initially isolated and demonstrated strong antioxidant activity (Arun et al., 2017). *Halomonas saliphila* LCB169T, a unique species, produces a high-salt fermented broth in which a unique EPS, also known as hsEPS, was extracted successfully.

Marine fungus is also regarded as a significant source of ecologically major elements and comparatively unexplored bioresources for innovative natural products. However, there is little information about their biodiversity and classification (Tisthammer et al., 2016; Raghukumar, 2004). The coastlines, particularly riverine, mangrove, and estuarine ecosystems, have been reported to have the highest diversity of marine mushrooms indicating the influence of environmental disasters on the habitation of terrestrial fungi. Numerous variables can affect the abundance and distribution of fungi in a marine environment. As a result, marine fungi have few morphological characteristics that are similar to those of their terrestrial equivalents (Menezes et al., 2010). Obligate and facultative are two such groups of marine fungi. Marine Obligate fungi germinate and sporulate in the marine environments. While growing and sporulating in marine habitats, facultative marine fungus inhabits freshwater and terrestrial ecosystem (Raghukumar, 2004). Marine algal exopolysaccharides are widespread and abundant in the marine ecosystem, wherein they execute vital roles that improve survival and adaptability. Exopolysaccharides (EPS) from marine environments (prokaryotes and eukaryotes) have a wide range of applications that make them an intriguing alternative to their synthetic equivalents. Marine algal and microbial exopolysaccharides are primarily composed of both homopolymeric and heteropolymeric composition and molecular weight of about 10–30 kDa, exhibiting a wide range of chemical configurations (Singha, 2012). Various exopolysaccharides developed from marine microorganisms are enlisted in Table 8.1.

Various new marine microorganisms have already been discovered in the last ten years because of the enhancement of isolation and classification technologies. Unique exopolysaccharides have been found with distinct biological characteristics, by improving the growth parameters. Exopolysaccharides of microbial origin is a complex type of polymers that typically contain one single sugar (monosaccharides), such as amino sugars, uronic acids, pentoses, and hexoses. These exopolysaccharides usually bind with proteins, lipids, and non-carbohydrate molecules, as supplementary structural compounds, which include pyruvate, sulfate, acetate, succinate, and phosphate (Andrew and Jayaraman, 2020). Numerous monosaccharides are abundantly identified in exopolysaccharides derived from marine bacteria, rather than polysaccharides extracted from plants, which include fucose, ribose, uronic acid, and amino saccharides. The biological processes carried out by these marine bacterial exopolysaccharides are likewise intimately correlated with the presence of such unique monosaccharides. The composition of an exopolysaccharide identified from *Colwellia psychrerythraea* 34H consists of recurring units of two galacturonic acid residues both adorned and alanine amino acids, a N-acetyl-quinovosamine (QuiN), and a substantial cryoprotective activity. The pseudo-helicoidal composition of the exopolysaccharide could maintain the water molecules tetrahedral order in the initial hydration envelope and also prevent the crystal transformation of the ice, according to NMR and computational studies (Casillo et al., 2018). Marine researchers identified a novel exopolysaccharide FSW-25, derived from marine *Microbacterium aurantiacum*. FSW-25 is a heteropolysaccharide having the high molecular weight and the presence of abundant uronic acid.

Furthermore, the research is required to determine how the monosaccharide composition affects the heat stability of marine bacteria's exopolysaccharides. The marine bacterial exopolysaccharides differ from exopolysaccharides derived from marine fungi in a way that marine fungal EPS shows less significant diversity in the monosaccharide composition. These fungal-origin exopolysaccharides are primarily made up of neutral monosaccharides, comprising glucose, mannose, and galactose

TABLE 8.1

Exopolysaccharides derived from marine microorganisms

Types	Name	Source of the exopolysaccharide	Monosaccharides composition	References
Bacterial Exopolysaccharides	EPS-DR3A	*Pseudoalteromonas* sp. YU16-DR3A	Fucose, Erythrotetrose, Glucose, Ribose	Dhanya et al., 2022
	EPS Mi25	*Microbacterim* FSW-25	Glucose, Mannose, Fucose, Glucuronic acid	Sran et al., 2022
	EPS	*Sphingobium yanoikuyae* BBL01	α-D-mannopyranose, β-D-mannopyranose, α-D-xylopyranose, β-D-xylopyranose, β-D-glucopyranose, α-D-talopyranose, and β-d-galacturonic acid	Bhatia et al., 2021
	EPS5SH	*Bacillus* sp. H5	Mannose, Glucosamine, Glucose, Galactose	Wei et al., 2021
	SM1127EPS	*Polaribacter* sp.	Rhamnose, Fucose, Glucuronic acid, Mannose, Galactose, Glucose, N-Acetyl-D-glucosamine	Wang et al., 2021
	EPS	*Enterobacter* sp. ACD2	Glucose, Galactose, Fucose, Glucuronic acid	Almutairi and Helal, 2021
	Sphingobatan	*Sphingobacterium* sp. IITKGP-BTPF3	Mannose	Ali et al., 2020
	EPS-M41	*Pediococcus pentosaceus* M41	Arabinose, Mannose, Glucose, Galactose	Ayyash et al., 2020
	hsEPS	*Halomonas saliphila* LCB169T	Mannose, Glucose, Arabinose, Xylose, Galactose, Fucose	Gan et al., 2020
	EPS	*Bacillus cereus* KMS31	Mannose, Glucose, Xylose, Rhamnose	Krishnamurthy et al., 2020
	PEP	*Pseudoalteromonas* sp.	Glucose, Galactose, Mannose	Caruso et al., 2019
	EPS	*C. psychrerythraea* 34H	Quinovose, Galacturonic acid	Casillo et al., 2018

EPS RH-7	*Rhodobacter johrii CDR-SL 7Cii*	Glucose, Rhamnose, Glucuronic acid, Galactose	Chatterjee et al., 2018
EPS	*Bacillus thuringiensis RSK CAS4*	Fucose, Galactose, Xylose, Glucose, Rhamnose, Mannose	Ramamoorthy et al., 2018
EPS	*Halolactibacillus miurensis*	Galactose, Glucose	Arun et al., 2017
EPS	*Alteromonas sp. JL2810*	Galacturonic acid, Mannose, Rhamnose	Zhang et al., 2017
EPS	*Pseudoalteromonas sp. MD12–642*	Galacturonic Acid, Glucuronic Acid, Rhamnose, Glucosamine	Roca et al., 2016
EPS273	*P. stutzeri 273*	Glucosamine, Glucose, Rhamnose	Wu et al., 2016
EPS2E1	*Halomonas sp. 2E1*	Mannose, Glucose	Carrión et al., 2015
EPS	*Oceanobacillus iheyensis*	Mannose, Glucose, Arabinose	Kavita et al., 2014
A101	*Vibrio sp. QY101*	Glucuronic acid, Galacturonic acid, Glucosamine, Rhamnose	Jiang et al., 2011
SM20310	*Issachenkonii*	Rhamnose, Xylose, Mannose, Galactose, Glucose, N-Acetyl-D-glucosamine, N-Acetyl-D-galactosamine	Qin et al., 2007
Fungal Exopolysaccharides			
AUM-1	*Aureobasidium melanogenum SCAU-266*	Glucose, Mannose, Galactose	Lin et al., 2022
LCJ-5-4	*Aspergillus versicolor*	Glucose, Mannose	Chen et al., 2012
HPA	*Hansfordia sinuosae*	Mannose, Galactose, Glucose	Li et al., 2018
AS2–1	*Alternaria sp.*	Mannose, Glucose, Galactose	Chen et al., 2016
Fw-1	*Fusarium oxysporum*	Galactose, Glucose, Mannose	Chen et al., 2015
GW-12	*Penicillium solitum*	Mannose	Yan et al., 2014

(Continued)

TABLE 8.1 (*Continued*)
Exopolysaccharides derived from marine microorganisms

Types	Name	Source of the exopolysaccharide	Monosaccharides composition	References
	YSS	*Aspergillus terreus*	Glucose, Mannose	Chunyan et al., 2013
	AWP	*Aspergillus versicolor*	Glucose, Mannose	Chen et al., 2013
	AEPS-1	*Cordyceps sinensis*	Glucose, Galactose, Mannose, Fucose, Rhamnose, Xylose, Arabinose, Ribose	Leung et al., 2009
Algal Exopolysaccharides	EPS	*Leptolyngbya* sp.	Mannose, Arabinose, Glucose, Rhamnose, uronic acid	Gongi et al., 2022
	EPS	*Porphyridium sordidum*	Fucose, Rhamniose, Arabinose, Galactose, Glucose, Xylose, Glucosamine	Drira et al., 2021
	EPS-0C, EPS-2C, EPS-5C	*Porphyridium marinum*	Xylose, Galactose, Glucose, Fucose, Arabinose, Glucuronic acid	Gargouch et al., 2021
	EPS	*Tetraselmis suecica*	Arabinose, Ribose, Mannose, Galacturonic acid, Galactose, Glucose, Glucuronic acid	Parra-Riofrio et al., 2020
	EPS	*Nostoc* sp.	Uronic acid, Rhamnose, Fucose, Arabinose, Xylose, Mannose, Galactose, Glucose	Uhliariková et al., 2020
	Cyanoflan	*Cyanothece* sp. CCY 0110	Mannose, Glucose, uronic acid, Galactose, Xylose, Rhamnose, Fucose, Arabinose	Mota et al., 2019
	EPS	*Flintiella sanguinaria*	Xylose, Galactose, Glucuronic acid, Rhamnose, Glucose, Arabinose	Gaignard et al., 2018
	EPS	*Nostoc carneum*	Xylose, Glucose	Hussein et al., 2015
	EPS	*Chlamydonas reinhardtii*	Galacturonic acid, Ribose, Rhamnose, Arabinose, Galactose, Glucose, Xylose	Bafana, 2013
	EPS (sulfated)	*Phaeodactylum tricornutum*	Glucose, Xylose	Chen et al., 2013

with various molar ratios, according to the monosaccharide compositions. These exopolysaccharides typically possess powerful antioxidant properties.

8.3 BACTERIAL EXOPOLYSACCHARIDE

The biosynthesis of extracellular polysaccharides requires various enzymes, which makes it a complicated process. Numerous investigations have been conducted to determine the precise mechanisms of EPS biosynthesis in bacteria, although the different ways of EPS biosynthesis in marine bacteria are still unknown. Three potential EPS production mechanisms in marine bacteria have been proposed by numerous researchers and reviewers, notably the Wzx/Wzy-dependent system, the ATP-binding cassette (ABC) transporter-dependent pathway, and the synthase-dependent pathway. In gram-negative bacteria, the Wzx/Wzy-dependent pathway has received much attention, particularly for the formation of heteropolysaccharides (Delbarre et al., 2014)). A range of membrane-spanning proteins works together to support the Wzx/Wzy-dependent pathway in the cytoplasm (Rehm, 2010). Since several phosphoglycosyl transferases act in the cytoplasm to initiate extracellular polysaccharide biosynthesis, the ABC transporter-dependent pathway is fairly alike the Wzx/Wzy-dependent pathway (Rehm, 2010). The independent excretion of the whole polymer strands from flippase through cell wall and the membranes is a distinctive characteristic of the synthase-dependent process (Schmid et al., 2015).

Exopolysaccharides are generated inside bacterial cells or in the capsular material which tightly encircles them or have slime in the nearby region that isn't connected to any specific cell (Nichols et al., 2005). The majority of the bacteria found in microbial consortia of the natural surroundings have functional and structural stability dependent on the presence of an extracellular polymeric matrix. The approach of growing a single strain under controlled growth parameters is applied in order to study the synthesis of microbial exopolysaccharides and to develop a hypothesis regarding how these molecules behave in the surrounding environment (Kumar et al., 2007). Once the substrate gets into the cell unchanged or following phosphorylation, this is considered to be one of the initial steps in the biosynthesis of the exopolysaccharides (Dave et al., 2020). Activated substrates and carrier molecules in the cytoplasmic membrane are used in the synthesis. Several of the enzymes responsible for the generation of the nucleotides are membrane-bound. Thus, it is unclear if their products are generated nearby the enzymes required for the polymer formation or whether they exist in the cytoplasm (Sutherland, 1997).

The enzymes required for the synthesis of exopolysaccharides from microbes are present in the different regions of microbial cells. The intracellular enzymes as hexokinase, phosphoglucomutase enables the cell metabolic processes, and converts glucose to glucose-1-phosphate (Glc-1-P). The other intracellular enzyme uridine diphosphate-glucose pyrophosphorylase converts the Glc-1-P to uridine diphosphate glucose (UDP-Glc). UDP-Glc is the prime component in the synthesis of EPS. Further, the enzymes (glucosyl transferases) of cell periplasmic membrane transfers UDP-Glc to the repeating units linked to the glycosyl carrier lipid. The shifting of the required monosaccharides from sugar nucleotides to isoprenoid alcohol phosphate, a carrier lipid, is necessary for the formation of the repeating units. Several polymers with glycosyl repeating units outside the cell membrane, like teichoic acids, peptidoglycan, and lipopolysaccharides, likewise require this carrier lipid for the production of exopolysaccharides (Kumar et al., 2007). Following polymerization, a highly specialized enzyme may hydrolyze the polysaccharide chain from the isoprenoid carrier lipid to form loose slime or EPS capsule surrounding the microbial cell. The polysaccharide is delivered simultaneously through the membranes. The biosynthesis process of exopolysaccharides by utilization of glucose by bacteria is presented in Figure 8.3.

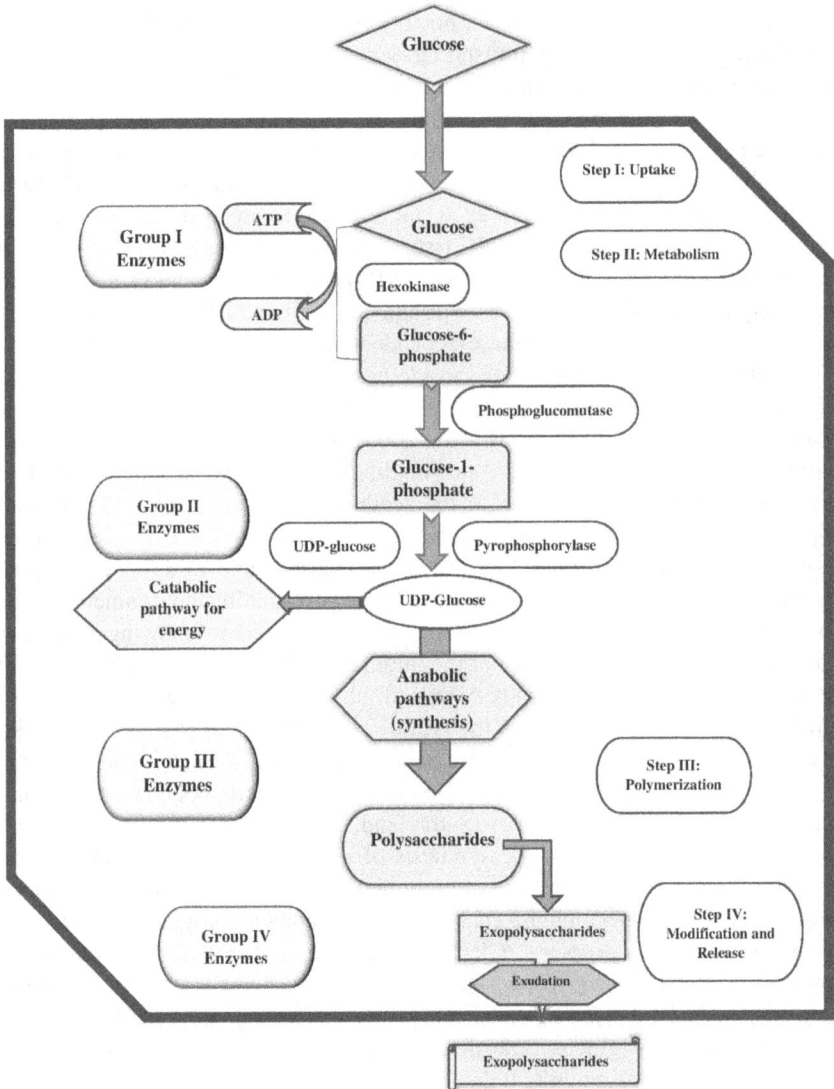

FIGURE 8.3 Biosynthesis process of microbial exopolysaccharides

A ligase process in strains that produce capsules may release the polymer chain from the carrier lipid and form covalent bond with the protein in the outer membrane or on the surface of the cell. Through S layers, noncovalently attached proteins, or glycoproteins can only be loosely bound to the layer of peptidoglycan and then released as amorphous slime. A number of factors, including the culture medium's composition, the mineral salts and trace elements, the sources of nitrogen, carbon, and precursor molecules, the kind of strain, and also the fermentation conditions, including temperature, oxygen concentration, pH, and agitation affect the composition, structure, and viscosity of EPSs. Depending upon the nature, composition, and structure of microbial extracellular polysaccharides, they could be used for a variety of purposes (Dave et al., 2020).

EPS is made up of only a few monosaccharides, and their structural variety dictates its potential applications. Because of their physical, rheological, and certain distinct features, exopolysaccharides are employed in the food, pharmaceutical, health, and bioremediation areas. Exopolysaccharides generated by marine *Vibrio* and *Pseudomonas*, as well as several other bacterial taxa, have anticancer, antiviral, and immunostimulant properties. *Alteromonas infernus*, extracted from deep-sea hydrothermal vents, produced a heparin-like exopolysaccharide with low molecular weight, and possesses anticoagulant properties (Colliec et al., 2001).

8.4   FUNGAL EXOPOLYSACCHARIDES

Exopolysaccharides with unique structural and chemical moieties, as well as those with a variety of biological functions, are frequently produced by marine fungi. Exopolysaccharides generated by marine fungus and bacteria throughout metabolic activities are multifunctional, high molecular weight biopolymers, according to a recent investigation (Prathyusha et al., 2018). In terms of structural and functional characteristics, molecular weight range, reactive groups, and chemical composition, exopolysaccharides are diverse in nature. It is worth noting that these polysaccharides can be easily adjusted to meet the demand and necessity, because of the occurrence of derivable groups on the molecular structures. Marine fungus, however, are thought to be a significant source of exopolysaccharides with future uses in the field of pharmaceuticals, culinary, cosmetics, and medical due to the tremendous biochemical and genetic diversity (Mahapatra and Banerjee, 2013).

These exopolysaccharides can be generated in much bigger quantities in a short period of time than polysaccharide production from plant or algae, and their upstream and downstream processing is simpler (Szewczuk-Karpisz et al., 2016). Exopolysaccharides investigation is presently booming, though, because of its intriguing physicochemical and rheological characteristics and additional functionality, which serve as a basic biomaterial with a variety of uses in increased microbial oil recovery, cosmetic, wastewater treatment (Sveistrup et al., 2016; Agrawal et al., 2018), as well as in pharmaceutical industry, etc. Exopolysaccharide synthesis and structural variations between marine and terrestrial fungus are greatly influenced by salinity, wide variation in mineral concentration, high pH, low temperature, high pressure, and specific light conditions (Bonugli-Santos et al., 2015).

Exopolysaccharides generated by marine fungus vary chemically in their molecular weight, sugar content, degree of polymerization, and type of bond. Exopolysaccharides are categorized according to their nature (acidic, neutral, and basic polysaccharides) and the type of sugar they consist (homopolysaccharides and heteropolysaccharides). A few numbers of marine fungal exopolysaccharides are homopolysaccharides with glucose as the predominant monomer, while most are acidic in nature and heteropolysaccharides. Although different marine fungus produces exopolysaccharides with different saccharide contents, the most common are glucose, xylose, galactose, rhamnose, mannose, and fucose sugars.

According to several researches, polysaccharides are typically colorless, nontoxic, odorless, amorphous, mainly soluble in water, and without a sweet flavor (Mahapatra and Banerjee, 2013; Osińska-Jaroszuk et al., 2015). There have been numerous studies on the significance of exopolysaccharide's molecular structure and biological function since the pattern of their structure may influence bioavailability. Interestingly, the chemical composition, branching rate, glycosidic linkages, molecular weight, polymerization, backbone length, and 3D confrontation of exopolysaccharides are all specifically correlated to the biological responses they elicit. To better understand the relationship between structural and functional features as well as their uses in nutrition and medicine, thorough characterizations of exopolysaccharide are necessary. Various researches on exopolysaccharides reported that a small amount of them tend to have immunomodulatory, antioxidant, and anticancer properties (Jing et al., 2014).

A yeast like fungus known as *Aureobasidium pullulans* was frequently observed in the terrestrial and aquatic ecosystem and was quite prevalent in both temperate and tropical regions. According to reports, these fungi generate "pullulan," an edible exopolysaccharide that is safe, eco-friendly, tasteless, odorless, non-ionic, and is water soluble (insoluble in organic solvents) having a Poly α (1–6) maltotriose repeating units.

8.5 MARINE ALGAL EXOPOLYSACCHARIDES

Deep-sea environments exhibit distinct characteristics such as high pressure, severe temperature, and heavy metals according to the latest studies in deep-sea hydrothermal vent surroundings (Nichols et al., 2005). The commercial significance of deep-sea exopolysaccharides has lately been known, and these exopolysaccharides show various applications in the field of industry. These exopolysaccharides become untouched and provide vast opportunities in biotechnology utilization. Exopolysaccharides from microalgae are known as innovative and attractive opportunities for both the pharmaceuticals and the food industries. They are significant marine bioactive constituents (Zhang et al., 2019). Due to the complicated interaction of several structural patterns, exopolysaccharides demonstrate a broad range of biological activities such as molecular weight, the constitution of the sugar deposit, forms of glycosidic linkages, and the composition of monosaccharide and also the occurrence of some non-sugar particles in the backbone of the polysaccharides (Berthon et al., 2017; Pierre et al., 2019). Due to the distinctive physical characteristics, such as high viscosity and outstanding rheological qualities, exopolysaccharides from microalgae were initially designed as thickeners and bio-lubricants in food applications. As more and more species of exopolysaccharides from microalgae are discovered, studies concerning their use are gradually diversifying. Exopolysaccharides have an excellent importance for producing food as a carbon pool, similar to biomass. Several research have revealed that exopolysaccharides derived from microalgae exhibits several biological activities, alike antineoplastic, anti-inflammatory, and antioxidant characteristics (Qi et al., 2022). As a result, it is anticipated to be a viable substrate for pharmaceuticals, dietary supplements, and cosmetics. Exopolysaccharides may be dissolved in the culture media as a by-product of the microalgae growth process, it is easier to prepare other than intracellular products, and seems to have more industrial benefits. Microalgal exopolysaccharides is now thought to have many potential applications. With the use of extensive cultivation, Frutarom Industries Ltd. (Israel) has effectively recovered the exopolysaccharide and transformed it into cosmetics. Apparently, no other microalgal exopolysaccharide species have been able to achieve commercial uses. When it comes to optimally transforming solar anergy and carbon dioxide into polysaccharides, cyanobacteria outperform plants and microorganisms (Gupta et al., 2013). The extracellular sulfated heteropolysaccharide known as emulcyan, produced by the benthic cyanobacterium *Phormidium* J-1, has been demonstrated to have emulsifying properties.

The *Trebouxiophyceae* (Chlorophyta) family, which includes the unicellular microalga *B. braunii*, is widely distributed in both fresh and brackish waters of temperate and tropical zones. Due to its ability to produce significant amounts of lipids and hydrocarbons, *B. braunii* has been widely recognized as a potential source of environmentally friendly bioenergy. *B. braunii* is a viable candidate for mass development and uses due to its high efficiency in producing lipids and hydrocarbons as well as the ability to secrete massive quantities of exopolysaccharides. *B. braunii* SCS-1905, a microalgae strain, was recently discovered in Guangdong Province, China. A variety of exopolysaccharides were produced after the medium was extracted, isolated, and purified, which includes crude exopolysaccharides (CEPS), deproteinized exopolysaccharides (DEPS), and portions of EPS-1, EPS-2, and EPS-3.

8.6 DIATOMS

The unicellular, colonial photoautotrophs known as diatoms (Bacillariophyta) are part of a larger class of heterokonts. Diatoms are thought to have evolved through supplementary endosymbiosis since heterokont chloroplasts resembling red algae. Exopolysaccharides from diatoms are bioactive substances dispersed in the ecosystem. Diatoms contribute significantly to marine phytoplankton and generate up to 25% of the annual primary organic matter. Photosynthetic fixation of the inorganic carbon results in the production of marine diatom's exopolysaccharides. They provide an important source of organic matter for the benthic populations and are a part of the marine coenoses food chain (Underwood et al., 2004). Diatom exudates play a crucial role in the development of the carbohydrate components of bottom deposits (Underwood et al., 2004; Bellinger et al., 2005).

Extracellular polysaccharides have a number of uses. The diatom cells are encased in exopolysaccharide mucous sheaths and envelop, creating a habitat that shields them from abruptly unfavorable changes in the environment (Ordain et al., 2003). The detoxification of heavy metals and the immobilization of hazardous chemicals are made possible by the carboxylic and sulfonic groups present in exopolysaccharide. Algal cells at low tide are kept from drying out by the watered polymer chains of exopolysaccharide.

Exopolysaccharide production is crucial for the movement of the diatoms (Underwood et al., 2004). Theories were proposed for the explanation of their capability to shift. The energy released by the solubility of mucus generated by diatoms serves as the driving force for moving cells, as per the theory of hydration (Gordon, 1987). The second explanation, put forth by Edgar and Pickett-Heaps (1982), is the one that is currently more frequently accepted. It is based on research showing that the cellular actin bundles are present in the frutescent region, and TEM analysis image shows that adhesively charged polysaccharides are excreted by flagellated diatoms through their raphe. This hypothesis postulates that the slime ejected from the raphe and various holes in the frustule becomes moist and expands, enabling it to adhere to the substrate. After treating with certain actin and myosin inhibitors, cell mobility stops (Underwood et al., 2004).

8.7 APPLICATIONS OF MARINE EXOPLOYSACCHARIDES

Due to the favorable marine microbial environment, microbial carbohydrate polymers are diverse in nature and structure. These polysaccharides have found extensive applicability in food as well as pharmaceutical industries as texturizing, encapsulating, emulsifying, cryoprotective, gelling, binding, and drug delivery agent as presented in Figure 8.4. Due to the anti-tumor and immunomodulatory activity, these have largely been considered in biochemical fields. However, in the last few years, the unique properties of exopolysaccharides synthesized by marine bacteria have intriguing prospects in the field of biotechnology and marine ecological environment as carbon sequestration, oil pollution remediation, bioflocculant, etc. In the food, pharmaceutical, clothing, cosmetic, paper, and petroleum industries, EPS from a *Pseudomonas* strain has the potential to be a cost-effective substitute for more expensive natural polysaccharides like guar gum due to its good flocculating, emulsifying, and film-forming qualities.

8.7.1 FOOD APPLICATIONS

Marine EPSs may be used in food products as thickeners, gelling, emulsifying, and stabilizing agents. EPSs from halotolerant bacteria, which have been crucial in food biotechnology for the development of fermented foods and food additives, are said to exhibit good emulsification property, which can also be helpful for medical applications (Raguènès et al., 2003). Rougeaux's team looked at EPSs from a variety of microbial isolates from deep-sea hydrothermal vents, for their intriguing characteristics, as gel formation or increase in viscosity of aqueous solution (Gutierrez, 2008).

Potential applications of marine derived EPS have been discussed under Table 8.2. The marine bacterium *Alteromonas* has the potential to be used as a thickening

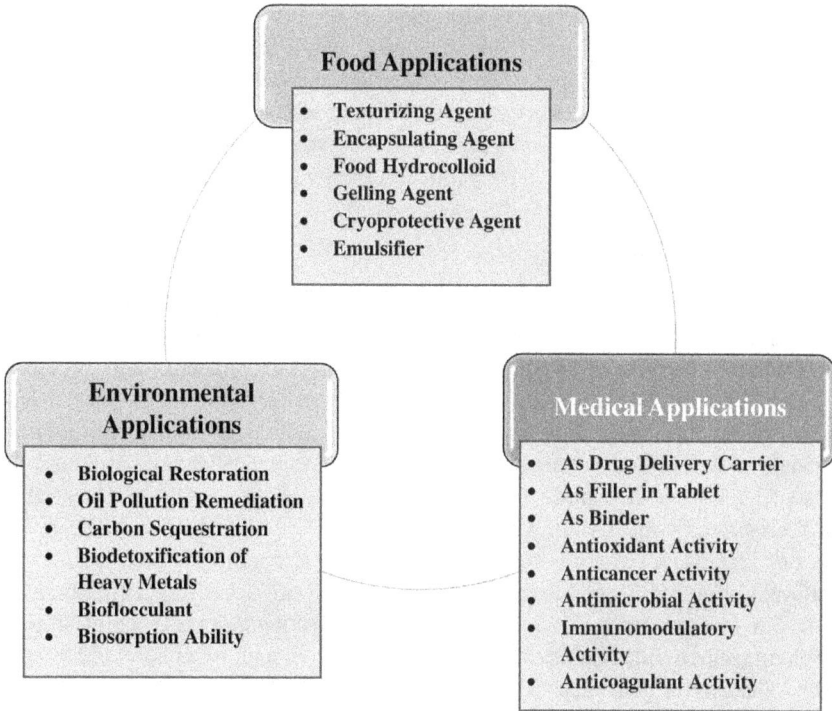

Food Applications

- Texturizing Agent
- Encapsulating Agent
- Food Hydrocolloid
- Gelling Agent
- Cryoprotective Agent
- Emulsifier

Environmental Applications

- Biological Restoration
- Oil Pollution Remediation
- Carbon Sequestration
- Biodetoxification of Heavy Metals
- Bioflocculant
- Biosorption Ability

Medical Applications

- As Drug Delivery Carrier
- As Filler in Tablet
- As Binder
- Antioxidant Activity
- Anticancer Activity
- Antimicrobial Activity
- Immunomodulatory Activity
- Anticoagulant Activity

FIGURE 8.4 Illustration of applications of marine exopolysaccharide

agent because its viscosity is comparable to that of xanthan (Rougeaux et al., 2001). An extracellular polysaccharide "PE12," isolated from a *Pseudoalteromonas* strain displays good emulsification ability against a variety of oil substrates. Cultures of an *Enterobacter* sp. have yielded an extracellular microbial polysaccharide that exhibits peculiar gelation characteristics with possible technological significance. The marine bacteria *Halomonas eurihalina* developed an exopolysaccharide that made acidic solutions viscous and acted as an emulsifier (Martinez-Checa et al., 2002). Also, halophilic bacterium *Enterobacter cloacae* developed an exopolysaccharide known as "EPS 71," that emulsified a number of hydrocarbons and vegetable oils. At 35°C, with the addition of traces of sodium chloride, hexane, and groundnut oil, emulsions remained stable for up to ten days between pH 2 and 10. The ability of "EPS 71" to stabilize emulsions was comparable with numerous commercial gums, including xanthan. The food sector may find the polysaccharide to be an intriguing ingredient as a viscosity-increasing agent, particularly in goods containing edible acids like citric acid and ascorbic acid (Iyer et al., 2006). Another, EPS made by *E. cloacae* emulsified groundnut, jojoba, coconut, sunflower, castor, and coconut seed oils. It could also emulsify benzene, hexane, paraffin oil, kerosene, and xylene. Also, at 35°C in the presence of 5 to 50 mg/ml sodium chloride, peanut oil and hexane emulsions remained stable for up to ten days. Even at high temperatures, this exopolysaccharide maintained good viscosity (Martinez-Checa et al., 2002).

Sulfation of hydroxyl groups of polysaccharides synthesized by filamentous fungi *Phomaherbarum* improved free-radical scavenging activity of the extracellular polysaccharide and protected erythrocytes from oxidative damage. Despite these discoveries, no practical applications have yet been investigated.

8.7.2 MEDICAL APPLICATIONS

The production of EPSs, from marine bacteria, fungi, and algae serves crucial biological and physiological purposes. In the biochemical and restorative domains, EPS are considered because of its immunomodulatory and anticancer effects. The basic structural variety and variability of bioactive polysaccharides and polysaccharide-bound proteins may play a role in their ability to regulate immunological functions. Marine EPS exhibits strong antioxidant activity due to their different structural features such as monosaccharide residues, glycosidic linkages, sulfate concentration, and their binding site (Andrew and Jayaraman, 2020). The marine bacteria EPS develops of biofilms to adapt to harsh environmental conditions such as low temperature, high osmotic pressure, and high salinity (Squillaci et al., 2016; Sun et al., 2014). For instance, the AEPS from the *Rhodella reticulata* are more effective at scavenging superoxide anions than the common antioxidant, tocopherol (Chen et al., 2009). The Arctic Sea bacterium *Polaribacter sp.* SM1127's isolated EPS exhibits strong antioxidant properties. The higher antioxidant activity of this EPS than Hyaluronic acid (HA) (a typical free ion scavenging compound in cosmetics), indicates that this extracellular carbohydrate polymer has possibilities of application in cosmetics antioxidant area. Additionally, SM1127 could eliminate too many reactive oxygen species created by inflammation and infection of the wound, hastening the healing process. Therefore, it is possible that this EPS will be used to quicken the recovery from burns, frostbite, and other wounds (Wang et al., 2021). Similarly, Exopolysaccharide generated by *P. stutzeri* 273 named EPS27 possesses good antioxidant activity and has significant application in food and healthcare fields (Wu et al., 2016). The biggest organ of the human body is the skin, and skin wound recovery is a significant medical issue (Groeber et al., 2011; Zhou et al., 2010). Natural products like extracellular polysaccharides have become more essential and are frequently advised as a substitute medicine for wound recovery because synthetic drugs possess high risk of side effects, such as allergies and drug resistance. EPS produced from marine fungus also exhibits good antioxidant activity. As a novel extracellular polysaccharide (YSS) having a molecular weight of 18.6, kDa was discovered by Wang et al. from the marine fungus *Aspergillus terreus* (Chen et al., 2013). YSS showed DPPH radical scavenging ability with EC50 of 2.8 mg/mL. The biological activity of several microbial exopolysaccharide is listed in Table 8.2.

Inducing apoptosis is a successful strategy for treating malignant growth, although this becomes complex by the development of multi-drug resistance (MDR) mechanisms. According to several studies (Wei et al., 2008; Kambourova et al., 2009), EPS can reduce P-glycoprotein and reverse MDR. Bacterial EPS are very biologically active and have roles in immunological modulation, cell division and separation

TABLE 8.2
Some potential applications of marine exopolysaccharides

Isolate	Functions	References
Aureobasidium melanogenum SCAU-266	Immunomodulatory Activity	Lin et al., 2022
Bacillus cereus	Antioxidant, Antitumor, Anti-inflammatory Activities	Selim et al., 2022
Sphingobium yanoikuyae BBL01	Antioxidant Activity, Metal Chelating Activity, Emulsifying, Flocculating Properties, and as a Stabilizer	Bhatia et al., 2021
Polaribacter sp.	Promotion of Wound Recovery, Prevention of Frostbite Injury, Free Radical Scavenging Activity	Wang et al., 2021
Sphingobacterium sp. IITKGPBTPF3	Immunomodulatory Activity	Ali et al., 2020
Pantoea sp. YU16-S3	Promotion of Wound Recovery	Sahana and Rekha, 2020
Bacillus cereus KMS3-1	Wastewater Treatment	Krishnamurthy et al., 2020
Marinobacter sp. W1–16	Emulsifying, Heavy Metal Binding Ability, and Cryoprotective Properties	Caruso et al., 2019
Rhodobacter johrii CDRSL 7Cii	Emulsifying Activity	Chatterjee et al., 2018
Alternaria sp., *Hansfordia sinuosae*, *Aspergillus versicolor*, *Aspergillus terreus*	Anticancer, Antioxidant	Li et al., 2018; Chen et al., 2016; Chunyan et al., 2013
Bacillus thuringiensis E4, *Brevundimonas subvibrioides* MSA1, *Advenella kashmirensis* NRC-7, *Pseudomonas fluorescens* SGA3, and *Bacillus amyloliquefaciens* MGA2	Antiproliferative Activities Against Hepatocellular Carcinoma Cells (HepG2)	Asker et al., 2018
Oceanobacillus iheyensis	Anti-biofilm	Kavita et al., 2014
Polaribacter sp. SM1127	High Viscosity, Antioxidant Activity, and Cryoprotective Activity	Dong et al., 2012
Alteromonas macleodii subsp. Fijiensis biovar deepsane HYD 657	For Cosmetics	Le Costaouec et al., 2012
Pseudoalteromonas ruthenica	Texturizer due to Pseudoplastic Nature, Good Shearing Property, and Better Stability at Higher pH Levels	Saravanan and Jayachandran, 2008
Issachenkonii	Antifreeze	Qin et al., 2007
Vibrio alginolyticus	Binders and Adhesives	Jayaraman and Seetharaman, 2003
Cyanobacteria, Desulfovibrio, Desulfobacter	Texturizer due to Good Shearing Properties	Rougeaux et al., 2001

control, as well as having anticancer, antioxidant, and antiviral effects (Zhang et al., 2003; Selim et al., 2018). Umezawa et al. (1983) tested marine microorganisms' extracellular polysaccharides for their ability to treat a sarcoma-180 sarcoma tumor in mice. Ruiz-Ruiz et al. (2011) reported that the new halophilic bacteria *Halomonas stenophila* strain B100 produces extracellular polysaccharides of high sulfate content. The over sulfated EPS exhibited antitumor activity on T cell lines derived from acute lymphoblastic leukemia (ALL). Only cancerous cells were susceptible to the apoptosis induced by the sulfated EPS (B100S), whereas vital immunity cells were protected. So EPS are currently receiving more attention from the medical community. Polysaccharides with antitumor property possess distinct structure, chemical configuration, and physical properties. Numerous glycans, ranging from homopolymers to heteropolymers, exhibit antitumor action (Ooi and Liu, 2000). The size of the particles, branching rate, shape, and dissolvability in water are all factors that affect biological activity of EPS. The anticancer and immunomodulatory effects of marine EPS are (1) prevention of the growth of several types of cancerous cells; (2) immunostimulating action against tumors in conjunction with chemotherapy; and (3) inhibiting the spread or migration of cancerous cells within the body.

The sulfated EPS (including fucoidan) reduced cytotoxicity when compared to other antiviral drugs used in clinical practice. Sulfated extracellular carbohydrate polymers inhibits the entrance of infections into host cells. Another exopolysaccharide (EPS-2) from a thermotolerant *Bacillus licheniformis* isolated from a narrow outlet of Vulcano Island has been shown to have antiviral and immunoregulatory effects (Arena et al., 2009). EPS improves the immunological examination of human peripheral blood mononuclear cells (PBMC) for HSV-2 (herpes simplex infection) illness caused by the production of Th1-type cytokines. Higher doses of EPS-2 (300 and 200 ng/ml) produce a greater amount of Th1 cytokines. As a result, the development of a cytokine is associated with an increase in protection against viral replication. Marine exopolysaccharides find wide application in antimicrobial activity. Nguyen et al. (2008) developed an EPS (cellulose) from bacteria *Gluconacetobacter xylinus* K3 containing nisin to control Listeria monocytogenes in food. In 2012, Orsod et al. isolated two EPS from marine bacteria and tested their efficacy against gram-positive bacteria *Lysinibacillus* and *Paenibacillus* sp. and gram-negative bacteria (*Pseudomonas* sp., *Escherichia coli*). Both EPS exhibited antibacterial activities against all the examined species.

Additionally, marine EPS can serve as good binders in making tablets using the wet granulation technique. Binders are added as a solution or a solid to the powder mixture (after which the granulating liquid, typically water is added). Most polysaccharides have high water holding capacity due to the high concentration of hydroxyl groups, which causes wet granulation (Paolucci et al., 2015). Bacterial marine EPS as chitin and chitosan can also be used as diluents/fillers and binder in tablets. Chitin and chitosan cause the greatest compression during filling and tablet pressure handling since they have the least mass and tap thickness. EPS also serves as drug delivery agents. EPS from marine bacteria are bio-perfect, nontoxic, and biodegradable (Dhanaraju et al., 2011), and delivery of the drug is controlled viably. Drug delivery systems can be created using a variety of methods and can take different forms, such as hydrogels, nanoparticles, and capsules, which are suitable for the delivery of bioactive agents, as proteins and nucleic acids. Also, by increasing the porosity, wettability, and swelling ability, marine EPS encourages the conversion of the tablet into granules upon digestion into the stomach (Thomas and Kim, 2013). According to Hoare and Kohane (2008), marine-derived bacteria may serve as a disintegration enhancer for poorly soluble drugs. These powders can weaken the adhesion forces that hold a tablet shape together and encourage dissociation into smaller granules, increasing the surface area that is potentially susceptible to disintegration.

8.7.3　Environmental Applications

Wastewater treatment is crucial given the advancement in industrialization. Mixing, precipitation, filtration, or chlorine disinfection are some common traditional wastewater treatment techniques; however, they are costly and vulnerable to secondary pollutant (Crini et al., 2019; Syafiuddin and Fulazzaky, 2021). Contrarily, biological treatments are considered to be environmentally friendly. The usage of marine microbial derived exopolysaccharides flocculation is one of the crucial techniques of biological system (Banerjee et al., 2021). The physical and chemical characteristics of microbial aggregates are significantly influenced by these exopolysaccharides (Salama et al., 2016), which promotes flocculation and the pollutants precipitation in sewage. The interaction of exopolysaccharides with the cells results in a vast network that shields them from toxins and dehydration. Additionally, these marine microbial derived exopolysaccharides are polyanionic with uronic acids, sulfated units, and phosphate groups. These chemical groups give exopolysaccharides a negative charge that enables them to bind to trace and hazardous metals, dissolved cations, and other molecules (López-Ortega et al., 2021). The extracellular polysaccharides produced from marine bacteria has been demonstrated to absorb several metal species from marine sediment, making EPS a possible metal-chelating agent. These polymers could eliminate harmful substances including zinc, cadmium, and lead. The maximal absorption capacities for zinc, cadmium, and lead were 77, 154, and 316 mg/g, respectively. Thus, these EPSs are a potential replacement for existing physical or chemical biosorbents in wastewater treatment because of their potent chelating capabilities. According to Zhou et al., *Pseudoalteromonas* SP, produces Bsi 20310, can significantly speed up the flocculation development in the ferric chloride coagulation RX-3B simulation of fuel wastes (Zhou et al., 2010). A similar positive association between LB-EPS and the rate of sludge flocculation was also reported by Yang and Li (2009). Similarly, microbial exopolysaccharides are helpful in addressing other environmental issues. A typical technique for removing heavy metals and other contaminants from the sewage called flocculation. According to Paritosh Parmar et al., a fish, an Indian mackerel, and an Indian squid were used to extract exopolysaccharides derived from luminescent bacteria, PBR1 and PBRL1, belongs to vibrio genera. The result of a β hemolysin test for PBR1 and PBRL 1 revealed that these bacterial strains are non-pathogenic, and exopolysaccharide derived from them can be useful in the removal of heavy metals from water, and due to their luminescence properties, these strains could be used to construct luminescent biosensors, despite the fact that the majority of the vibrio strains are pathogenic in nature (Parmar et al., 2020). Salama et al. suggested five effective strategies to comprehend exopolysaccharide in context of its significance in wastewater treatment: (1) exopolysaccharide extraction techniques, (2) setup of exopolysaccharide in situ analysis methods, (3) exopolysaccharides sub-fractions identification, (4) clarification of exopolysaccharide's primary functions, (5) determination of the primary influences on the production of exopolysaccharides (Salama et al., 2016).

Microbial EPS are used for the bioremediation process because of their capacity to bind to anion and cations (Saikia et al., 2013). Certain remediation techniques include EPS undergone chemical modifications such as phosphorylation, acetylation, sulfonylation, and methylation. The selectivity of metal binding is determined by EPS acetylation (Sutherland, 1997). The EPS's ability to bind metal significantly contributes to the removal of metals from wastewater. The immobilized EPS produced from *Paenibacillus polymyxa and Chryseomonas* demonstrated the elimination of copper, cobalt, lead, and cadmium from water (Acosta et al., 2005). The dead cell bound extracellular polysaccharide from *Bacillus pumilus, Bacillus cereus*, and *Pentoea* agglomerants 85.5–89% chromium (Sultan et al., 2012). The organism's ability to bond with minerals and hence extract metals from the sulfide ores is facilitated by *Acidithiobacillus ferrooxidans*' EPS (Yu et al., 2011).

According to Salehizadeh and Shojaosadati (2003), the EPS of *Bacillus firmus* was able to absorb lead (98.3%), copper (74.9%), and zinc (61.8%). The XU1 EPS from *Azotobacter chroococcum* showed the reduction of lead and mercury by 40.48% and 40.77%, respectively (Rasulov et al., 2013). Also, EPS from *Ensifer meliloti* revealed reduction of zinc, nickel, and lead ions by 66, 85, and 89%, respectively.

Marine exopolysaccharides have been utilized in analogous ways to address offshore crude oil leaks. According to Iwabuchi et al. (2002) *Monascus* derived exopolysaccharides can be utilized to cope with the coastal oil leak and hasten the breakdown of the crude oil. For instance, Ant-3b has the ability to significantly reduce the crude oil pollution in low temperature environments since it could emulsify n-hexadecane at low temperatures. Meanwhile, Ilori et al. demonstrated that the biosurfactants are far better than the artificial surfactants (Ilori et al., 2005).

In addition, the carbon cycle benefits have been found in the deep ocean by the production of exopolysaccharides from marine microbes. According to Joseph et al., exopolysaccharide produced by *Colwellia psychrerythraea* strain 34H, a marine psychrophilic bacterium, may have an impact on the deep sea's carbon cycle and nutrient transport (Marx et al., 2009). The exopolysaccharide stimulates the landscape level build-up of soil organic matter in depleted in Antarctic soil (Zayed et al., 2022; Mergelov et al., 2020). Every season, Antarctica green algae store 479,000 tons of carbon. The exopolysaccharide produced from the microbes in Antarctic environment forms biofilms, biological soil crusts, cell agglomeration, microbial mats, and other structures that affect the world's carbon cycle (Anesio et al., 2010).

8.8 CONCLUSION AND FUTURE PROSPECTS

Owing to the remarkable properties of marine EPS, these EPS may partially replace non-marine EPS in future. But presently, the application of EPS in biotechnological, pharmaceutical, and food processing areas is still under research, due to the small-scale production at high prices. The EPS's high production costs have a significant impact on the gross margin at the market level. Fermentation media is the primary cause of the high production costs, which typically account for around 30% of the expenses for the fermentation process, due to the use of expensive and specific nutrient sources. Though marine EPS has recently been investigated for its diverse industrial applications, still only few reports have focused on the synthesis and recovery of marine microbial EPS. So in-depth study in this area is required to fully comprehend the characteristics of EPS. The properties of marine bacterial strains can be enhanced through genetic engineering (use of mutagenic strains, gene alterations), and also EPS with specified features and structures can be created using genetic engineering. This will result in larger EPS yields. The current procedures for EPS structural determination are time-consuming and labor-intensive. In order to simplify the procedure, appropriate adjustments can be added to the current protocols, or new methods can be created. Additionally, it is necessary to modify the EPS extraction procedures in a cost-effective way, which would drastically reduce the overall cost of the downstream operations. Thence, substitution of non-marine EPS could only be possible by large scale production of marine extracellular polysaccharide at economic level.

REFERENCES

Acosta, M.P., Valdman, E., Leite, S.G., Battaglini, F., Ruzal, S.M., 2005. Biosorption of copper by *Paenibacillus polymyxa* cells and their exopolysaccharide. World J. Microbiol. Biotechnol. 21, 1157–1163.

Agrawal, S., Adholeya, A., Barrow, C.J., Deshmukh, S.K., 2018. Marine fungi: An untapped bio-resource for future cosmeceuticals. Phytochem. Lett. 23, 15–20.

Ali, P., Shah, A.A., Hasan, F., Hertkorn, N., Gonsior, M., Sajjad, W., Chen, F., 2020. A glacier bacterium produces high yield of cryoprotective exopolysaccharide. Front. Microbiol. 10, 3096.

Almutairi, M.H., Helal, M.M., 2021. Biological and microbiological activities of isolated *Enterobacter* sp. ACD2 exopolysaccharides from Tabuk region of Saudi Arabia. J. King. Saud. Univ. Sci. 33, 101328.

Andrew, M., Jayaraman, G., 2020. Structural features of microbial exopolysaccharides in relation to their antioxidant activity. Carbohydr. Res. 487.

Anesio, A.M., Sattler, B., Foreman, C., Telling, J., Hodson, A., Tranter, M., Psenner, R., 2010. Carbon fluxes through bacterial communities on glacier surfaces. Ann. Glaciol. 51, 32–40.

Arena, A., Gugliandolo, C., Stassi, G., Pavone, B., Iannello, D., Bisignano, G., Maugeri, T.L., 2009. An exopolysaccharide produced by *Geobacillus thermodenitrificans* strain B3–72: Antiviral activity on immunocompetent cells. Immunol. Lett. 123, 132–137.

Arun, J., Selvakumar, S., Sathishkumar, R., Moovendhan, M., Ananthan, G., Maruthiah, T., Palavesam, A., 2017. In vitro antioxidant activities of an exopolysaccharide from a salt pan bacterium. *Halolactibacillus miurensis*. Carbohydr. Polym. 155, 400406.

Asker, M.S., El, Sayed, O.H., Mahmoud, M.G., Yahya, S.M., Mohamed, S.S., Selim, M.S., El, Awady, M.S., Abdelnasser, S.M., Abo, Elsoud, M.M., 2018. Production of exopolysaccharides from novel marine bacteria and anticancer activity against hepatocellular carcinoma cells (HepG2). Bull. Natl. Res. Cent. 42, 30.

Ayyash, M., Abu-Jdayil, B., Olaimat, A., Esposito, G., Itsaranuwat, P., Osaili, T., Obaid, R., Kizhakkayil, J., Liu, S.Q., 2020. Physicochemical, bioactive and rheological properties of an exopolysaccharide produced by a probiotic *Pediococcus pentosaceus* M41. Carbohydr. Polym. 229, 115462.

Bafana, A., 2013. Characterization and optimization of production of exopolysaccharide from *Chlamydomonas reinhardtii*. Carbohydr. Polym. 95, 746–752.

Banerjee, A., Sarkar, S., Govil, T., González-Faune, P., Cabrera-Barjas, G., Bandopadhyay, R., Salem, D.R., Sani, R.K., 2021. Extremophilic exopolysaccharides: Biotechnologies and wastewater remediation. Front. Microbiol. 12, 721365.

Bellinger, J., Abdullahi, A.S., Gretz, M.R., Underwood, G.J.C., 2005. Biofilm polymers: Relationship between carbohydrate biopolymers from estuarine mudflats and unialgal cultures of benthic diatoms. Aquat. Microbiol. Ecol. 38, 169–180.

Berthon, J.Y., Nachat-Kappes, R., Bey, M., Cadoret, J.P., Renimel, I., Filaire, E., 2017. Marine algae as attractive source to skin care. Free. Radical. Res. 51, 555–567.

Bhatia, S.K., Gurav, R., Choi, Y.K., Choi, T.R., Kim, H.J., Song, H.S., Lee, S.M., Park, S.L., Lee, H.S., Kim, Y.G., Ahn, J., Yang, Y.H., 2021. Bioprospecting of exopolysaccharide from marine *Sphingobium yanoikuyae* BBL01: Production, characterization, and metal chelation activity. Bioresour. Technol. 324, 124674.

Bhatt, H.B., Baria, D.M., Raval, V.H., Singh, S.P., 2022. Multifunctional properties of polysaccharides produced by halophilic bacteria and their new applications in biotechnology. In: Raghvendra Pratap Singh, Geetanjali Manchanda, Kaushik Bhattacharjee, Hovik Panosyan (Eds.), *Microbial Syntrophy-mediated Eco-enterprising*. Academic Press, pp. 41–70.

Bonugli-Santos, R.C., dos, Santos, Vasconcelos, M.R., Passarini, M.R., Vieira, G.A., Lopes, V.C., Mainardi, P.H., 2015. Marine-derived fungi: Diversity of enzymes and biotechnological applications. Front. Microbiol. 6, 269.

Carrión, O., Delgado, L., Mercade, E., 2015. New emulsifying and cryoprotective exopolysaccharide from Antarctic *Pseudomonas* sp. ID1. Carbohydr. Polym. 117, 1028–1034.

Caruso, C., Rizzo, C., Mangano, S., Poli, A., Di, Donato, P., Nicolaus, B., Giudice, A.L., 2019. Isolation, characterization and optimization of EPSs produced by a cold-adapted *Marinobacter* isolate from Antarctic seawater. Antarctic. Sci. 31, 69–79.

Casillo, A., Lanzetta, R., Parrilli, M., Corsaro, M.M., 2018. Exopolysaccharides from marine and marine extremophilic bacteria: Structures, properties, ecological roles and applications. Mar. Drugs. 16, 69.

Chatterjee, S., Mukhopadhyay, S.K., Gauri, S.S., Dey, S., 2018. Sphingobactan, a new alphamannan exopolysaccharide from Arctic *Sphingobacterium* sp. IITKGP-BTPF3 capable of biological response modification. Int. Immuno. 60, 84–95.

Chen, B., You, W., Huang, J., Yu, Y., Chen, W., 2009. Isolation and antioxidant property of the extracellular polysaccharide from *Rhodella reticulata*. World J. Microbiol. Biotechnol. 26, 833–840.

Chen, Y., Mao, W., Gao, Y., Teng, X., Zhu, W., Chen, Y., Zhao, C., Li, N., Wang, C., Yan, M., et al., 2013. Structural elucidation of an extracellular polysaccharide produced by the marine fungus *Aspergillus versicolor*. Carbohydr. Polym. 93, 478–483.

Chen, Y.-L., Mao, W.J., Tao, H.-W., Zhu, W.-M., Yan, M.-X., Liu, X., Guo, T.-T., Guo, T., 2015. Preparation and characterization of a novel extracellular polysaccharide with antioxidant activity, from the mangrove-associated fungus *Fusarium oxysporum*. Mar. Biotechnol. 17, 219–228.

Chen, Y.-L., Mao, W.J., Yan, M.X., Liu, X., Wang, S.Y., Xia, Z., Xiao, B., Cao, S.J., Yang, B.Q., Li, J., 2016. Purification, chemical characterization, and bioactivity of an extracellular polysaccharide produced by the marine sponge endogenous fungus *Alternaria* sp. SP-32. Mar. Biotechnol. 18, 301–313.

Chen, Y.-L., Mao, W.J, Yang, Y., Teng, X., Zhu, W., Qi, X., Chen, Y., Zhao, C., Hou, Y., Wang, C., 2012. Structure and antioxidant activity of an extracellular polysaccharide from coral-associated fungus, *Aspergillus versicolor* LCJ-5-4. Carbohydr. Polym. 87, 218–226.

Chunyan, W., Wenjun, M., Zhengqian, C., Weiming, Z., Yanli, C., Chunqi, Z., Na, L., Mengxia, Y., Xue, L., Tiantian, G., 2013. Purification, structural characterization and antioxidant property of an extracellular polysaccharide from *Aspergillus terreus*. Process. Biochem. 48, 1395–1401.

Colliec, J.S., Chevolot, L., Helley, D., Ratiskol, J., Bros, A., Sinquin, C., Roger, O., Fischer, A.M., 2001. Characterization, chemical modifications and in vitro anticoagulant properties of an exopolysaccharide produced by *Alteromonas infernos*. Biochim. Biophys. Acta. 1528, 141–151.

Crini, G., Lichtfouse, E., 2019. Advantages and disadvantages of techniques used for wastewater treatment. Environ. Chem. Lett. 17, 145–155.

Dave, S.R., Upadhyay, K.H., Vaishnav, A.M., Tipre, D.R., 2020. Exopolysaccharides from marine bacteria: Production, recovery and applications. Environ. Sustain. 3, 139–154.

Delbarre-Ladrat, C., Sinquin, C., Lebellenger, L., Zykwinska, A., Colliec-Jouault, S., 2014. Exopolysaccharides produced by marine bacteria and their applications as glycosaminoglycans-like molecules. Front. Chem. 2, 85.

Dhanaraju, M.D., Elisabeth, S., Thirumurugan, G., 2011. Triamcinolone-loaded glutaraldehyde cross-linked chitosan microspheres: Prolonged release approach for the treatment of rheumatoid arthritis. Drug. Delivery. 18, 198–207.

Dhanya, B.E., Prabhu, A., Rekha, P.D., 2022. Extraction and characterization of an exopolysaccharide from a marine bacterium. Int. Microbiol. 25, 285–295.

Dong, S., Yang, J., Zhang, X.Y., Shi, M., Song, X.Y., Chen, X.L., Zhang, Y.Z., 2012. Cultivable alginate lyase-excreting bacteria associated with the arctic brown alga *Laminaria*. Mar. Drugs. 10, 2481–2491.

Drira, M., Elleuch, J., Ben, Hlima, H., Hentati, F., Gardarin, C., Rihouey, C., Le, Cerf, D., Michaud, P., Abdelkafi, S., Fendri, I., 2021. Optimization of exopolysaccharides production by *Porphyridium sordidum* and their potential to in-duce defense responses in *Arabidopsis thaliana* against *Fusarium oxysporum*. Biomolecules. 11, 282.

Edgar, L.A., Pickett-Heaps, I.D., 1982. Ultrastructural localisation of polysaccharides in the motile diatom *Navicula cuspidata*. Protoplasma. 113, 10–22.

Finore, I., Di Donato, P., Mastascusa, V., Nicolaus, B., Poli, A., 2014. Fermentation technologies for the optimization of marine microbial exopolysaccharide production. Mar. Drugs. 12, 3005–3024.

Flemming, H.C., Wingender, J., 2010. The biofilm matrix. Nut. Rev. Microbiol. 8, 623–633.

Gaignard, C., Macao, V., Gardarin, C., Rihouey, C., Picton, L., Michaud, P., Laroche, C., 2018. The red microalga *Flintiella sanguinaria* as a new exopolysaccharide producer. J. Appl. Phycol. 30, 2803–2814.

Gan, L., Li, X., Zhang, H., Zhang, R., Wang, H., Xu, Z., Peng, B., Tian, Y., 2020. Preparation, characterization and functional properties of a novel exopolysaccharide produced by the halophilic strain *Halomonas saliphila* LCB169T. Int. J. Biol. Macromol. 156, 372–380.

Gargouch, N., Elleuch, F., Karkouch, I., Tabbene, O., Pichon, C., Gardarin, C., Rihouey, C., Picton, L., Abdelkafi, S., Fendri, I., Laroche, C., 2021. Potential of exopolysaccharide-from *Porphyridium marinum* to contend with bacterial proliferation, biofilm formation, and breast cancer. Mar. Drugs. 19, 66.

Gongi, W., Cordeiro, N., Pinchetti, J.L.G., Ben, Ouada, H., 2022. Functional, rheological, and antioxidant properties of extracellular polymeric substances produced by a thermophilic cyanobacterium *Leptolyngbya* sp. J. Appl. Phycol. 34, 1423–1434.

Gordon, R., 1987. A retaliatory role for algal projectiles, with implications for the mechano-chemistry of diatom gliding motility. J. Theor. Biol. 126, 419–436.

Groeber, F., Holeiter, M., Hampel, M., Hinderer, S., Schenke-Layland, K., 2011. Skin tissue engineering—In vivo and in vitro applications. Adv. Drug. Deliv. Rev. 63, 352–366.

Guezennec, J., 2002. Deep-sea hydrothermal vents: A new source of innovative bacterial exopolysaccharides of biotechnological interest. J. Ind. Microbiol. Biot. 29, 204–208.

Gupta, V., Ratha, S.K., Sood, A., Chaudhary, V., Prasann, R., 2013. New insights into the biodiversity and applications of cyanobacteria (blue-green algae)—Prospects and chal-lenges. Algal. Res. 2, 79.

Gutierrez, T., 2008. Emulsifying and metal ion binding activity of a glycoprotein exopolymer produced by *Pseudoalteromonas* sp. strain TG12. Appl. Environ. Microbial. 74, 4867.

Harutoshi, T., 2013. Exopolysaccharides of lactic acid bacteria for food and colon health applications. Foods. 11(2), 156.

Hoare, T.R., Kohane, D.S., 2008. Hydrogels in drug delivery: Progress and challenges. Polym. 49, 1993–2007.

Hussein, M.H., Abou-ElWaf, G.S., Shaaban-De, S.A., Hassan, N.I., 2015. Characterization and antioxidant activity of exopolysaccharide secreted by *Nostoc carneum*. Int. J. Pharmacol. 11, 432–439.

Ilori, M.O., Amobi, C.J., Odocha, A.C., 2005. Factors affecting biosurfactant production by oil degrading *Aeromonas* spp. isolated from a tropical environment. Chemosphere. 61, 985–992.

Iwabuchi, N., Sunairi, M., Urai, M., Itoh, C., Anzai, H., Nakajima, M., Harayama, S., 2002. Extracellular polysaccharides of *Rhodococcus rhodochrous* S-2 stimulate the degrada-tion of aromatic components in crude oil by indigenous marine bacteria. Appl. Env. Microb. 68, 2337–2343.

Iyer, A., Mody, K., Jha, B., 2006. Emulsifying properties of a marine bacterial EPS. Enz. Microb. Technol. 38, 220.

Jayaraman, M., Seetharaman, J., 2003. Physiochemical analysis of the EPSs produced by a marine biofouling bacterium, *Vibrio alginolyticus*. Proc. Biochem. 38, 841.

Jiang, P., Li, J.B., Han, F., Duan, G.F., Lu, X.Z., Gu, Y.C., Yu, W.G., 2011. Antibiofilm activ-ity of an exopolysaccharide from marine bacterium *Vibrio* sp. QY101. PLoS One. 6, e18514.

Jing, Y., Cui, X., Chen, Z., Huang, L., Song, L., 2014. Elucidation and biological activities of a new polysaccharide from cultured *Cordyceps militaris*. Carbohydr. Polym. 102, 288–296.

Kambourova, M., Mandeva, R., Dimova, D., Poli, A., Nicolaus, B., Tommonaro, G., 2009. Production and characterization of a microbial glucan, synthesized by *Geobacillus tepidamans* V264 isolated from Bulgarian hot spring. Carbohydr. Polym. 77, 338–343.

Kavita, K., Singh., V.K., Mishra, A., Jha, B., 2014. Characterization and anti-biofilm activ-ity of extracellular polymeric substances from *Oceanobacillus iheyensis*. Carbohydr. Polym. 101, 29–35.

Krishnamurthy, M., Uthaya, C.J., Thangavel, M., AnnaduraI, V., Rajendran, R., Gurusamy, A., 2020. Optimization, compositional analysis, and characterization of exopolysac-charides produced by multimetal resistant *Bacillus cereus* KMS3-1. Carbohydr. Polym. 227, 115369.

Kumar, A.S., Mody, K., Jha, B., 2007. Bacterial exopolysaccharides—A perception. J. Basic. Microbio. 47, 103–117.

Le Costaouec, T., Cérantola, S., Ropart, D., Ratiskol, J., Sinquin, C., Colliec-Jouault, S., Boisset, C., 2012. Structural data on a bacterial exopolysaccharide produced by a deep-sea *Alteromonas macleodii* strain. Carbohyd. Polym. 90, 49–59.

Leung, P.H., Zhao, S., Ho, K.P., Wu, J.Y., 2009. Chemical properties and antioxidant activity of exopolysaccharides from mycelial culture of *Cordyceps sinensis* fungus Cs-HK1. Food. Chem. 114, 1251–1256.

Li, H., Cao, K., Cong, P., Liu, Y., Cui, H., Xue, C., 2018. Structure characterization and antitumor activity of the extracellular polysaccharide from the marine fungus *Hansfordia sinuosae*. Carbohydr. Polym. 190, 87–94.

Lin, Y., Yang, J., Luo, L., Zhang, X., Deng, S., Chen, X., Li. Y., Bekhit, A.E.A., Xu, B., Hung, R., 2022. Ferroptosis related immunomodulatory effect of a novel extracellular polysaccharides from marine fungus *Aureobasidium melanogenum*. Mar. Drugs. 20, 332.

López-Ortega, M.A., Chavarría-Hernández, N., del, Rocio, Lopez-Cuellar, M., Rodríguez-Hernández, A.I., 2021. A review of extracellular polysaccharides from extreme niches: An emerging natural source for the biotechnology. From the adverse to diverse! Int. J. Biol. Macromol. 177, 559–577.

Mahapatra, S., Banerjee, D., 2013. Fungal exopolysaccharide: Production, composition and applications. Microbiol. Insight. 6, 10957.

Martínez-Cánovas, M.J., Quesada, E., Llamas, I., Bejar, V., 2004. *Halomonas ventosae* sp. nov., a moderately halophilic, exopolysaccharide-producing bacterium. Int. J. Syst. Evol. Microbiol. 54, 733–737.

Martínez-Checa, F., Toledo, F.L., Vilchez, R., Quesada, E., Calvo, C., 2002. Yield production, chemical composition, and functional properties of emulsifier H28 synthesized by *Halomonas eurihalina* strain H-28 in media containing various hydrocarbons. Appl. Microbiol. Biotechnol. 58, 358–363.

Marx, J.G., Carpenter, S.D., Deming, J.W., 2009. Production of cryoprotectant extracellular polysaccharide substances (EPS) by the marine psychrophilic bacterium *Colwellia psychrerythraea* strain 34H under extreme conditions. Can. J. Microbiol. 55, 63–72.

Mata, J., Be´jar, V., Bressollier, P., Tallon, R., Urdaci, M., Quesada, E., Llamas, I., 2008. Characterization of exopolysaccharides produced by three moderately halophilic bacteria belonging to the family Alteromonadaceae. J. Appl. Microbiol. 105, 521–528.

Menezes, C.B.A., Bonugli-Santos, R.C., Miqueletto, P.B., Passarini, M.R.Z., Silva, C.H.D., Justo, M.R., Sette, L.D., 2010. Microbial diversity associated with algae, ascidians and sponges from the north coast of São Paulo state, Brazil. Microbiol. Res. 165, 466–482.

Mergelov, N., Dolgikh, A., Shorkunov, I., Zazovskaya, E., Soina, V., Yakushev, A., Fedorov-Davydov, D., Pryakhin, S., Dobryansky, A., 2020. Hypolithic communities shape soils and organic matter reservoirs in the ice-free landscapes of East Antarctica. Sci. Rep. 10, 1–19.

Mota, R., Vidal, R., Pandeirada, C., Flores, C., Adessi, A., De, Philippis, R., Nunes, C., Coimbra, M.A., Tamagnini, P., 2019. Cyanoflan: A cyanobacterial sulfated carbohydrate polymer with emulsifying properties. Carbohydr. Polym. 229, 115525.

Nguyen, V.T., Gidley, M.J., Dykes, G.A., 2008. Potential of a nisin-containing bacterial cellulose film to inhibit *Listeria monocytogenes* on processed meats. Food. Microbiol. 25, 471–478.

Nichols, C.A., Guezennec, J., Bowman, J.P., 2005. Bacterial exopolysaccharides from extreme marine environments with special consideration of the southern ocean, sea ice, and deep-sea hydrothermal vents: A review. Marine. Biotechnol. 7, 253–271.

Ooi, V.E.C., Liu, F., 2000. Immunomodulation and anticancer activity of polysaccharide-protein complexes. Curr. Med. Chem. 7, 715–729.

Ordain, F., Galois, R., Barnard, C., 2003. Carbohydrate production in relation to microphytobenthic biofilm development: An integrated approach in a tidal mesocosm. Microbiol. Ecol. 45, 237–251.

Orsod, M., Joseph, M., Huyop, F., 2012. Characterization of exopolysaccharides produced by *Bacillus cereus* and *Brachybacterium* sp. isolated from Asian Sea Bass (Lates calcarifer). Malays. J. Microbiol. 8, 170–174.

Osińska-Jaroszuk, M., Jarosz-Wilkołazka, A., Jaroszuk-Ściseł, J., Szałapata, K., Nowak, A., Jaszek, M., Majewska, M., 2015. Extracellular polysaccharides from Ascomycota and Basidiomycota: Production conditions, biochemical characteristics, and biological properties. World J. Microbiol. Biotechnol. 31, 1823–1844.

Paolucci, M., Fasulo, G., Volpe, M.G., 2015. Employment of marine polysaccharides to manufacture functional biocomposites for aquaculture feeding applications. Mar. Drugs. 13, 2680–2693.

Parmar, P., Shukla, A., Goswami, D., Gaur, S., Patel, B., Saraf, M., 2020. Comprehensive depiction of novel heavy metal tolerant and EPS producing bioluminescent *Vibrio alginolyticus* PBR1 and *V. rotiferianus* PBL1 confined from marine organisms. Microbiol. Res. 238, 126526.

Parra-Riofrío, G., García-Márquez, J., Casas-Arrojo, V., Uribe-Tapia, E., Abdala-Díaz, R., 2020. Antioxidant and cytotoxic effects on tumor cells of exopolysaccharides from *Tetraselmis suecica* (Kylin) butcher grown under autotrophic and heterotrophic conditions. Mar. Drugs. 18, 534.

Patel, S., Majumder, A., Goyal, A., 2012. Potentials of exopolysaccharides from lactic acid bacteria. Indian J. Microbiol. 52, 3–12.

Pierre, G., Delattre, C., Dubessay, P., Jubeau, S., Vialleix, C., Cadoret, J.P., Probert, I., Michaud, P., 2019. What is in store for EPS microalgae in the next decade? Molecules. 24, 1–25.

Prathyusha, A.M.V.N., Sheela, G., Bramhachari, P.V., 2018. Chemical characterization and antioxidant properties of exopolysaccharides from mangrove filamentous fungi *Fusarium equiseti* ANP2. Biotechnol. Rep. 19, e00277.

Qi, M., Zheng, C., Wu, W., Yu, G., Wang, P., 2022. Exopolysaccharides from marine microbes: Source, structure and application. Mar. Drugs. 20, 512.

Qin, G., Zhu, L., Chen, X., Wang, P.G., Zhang, Y., 2007. Structural characterization and ecological roles of a novel exopolysaccharide from the deep-sea psychrotolerant bacterium *Pseudoalteromonas* sp. Microbiol. 153, 1566–1572.

Radchenkova, N., Vassilev, S., Panchev, I., Anzelmo, G., Tomova, I., Nicolaus, B., Kuncheva, M., Petrov, K., Kambourova, M., 2013. Production and properties of two novel exopolysaccharides synthesized by a thermophilic bacterium *Aeribacillus pallidus*. Appl. Biochem. Biotech. 171, 31–43.

Raghukumar, C., 2004. Marine fungi and their enzymes for decolorization of colored effuents. In: *Marine Microbiology: Facets and Opportunities*. National Institute of Oceanography, pp. 145–158.

Raguénès, G., Cambon-Bonavita, M.A., Lohier, J.F., Boisset, C., Guezennec, J., 2003. A novel, highly viscous polysaccharide excreted by an *Alteromonas* isolated from a deep-sea hydrothermal vent shrimp. Curr. Microbiol. 46, 448–452.

Ramamoorthy, S., Gnanakan, A., Lakshmana, S.S., Meivelu, M., Jeganathan, A., 2018. Structural characterization and anti-cancer activity of extracellular polysaccharides from ascidian symbiotic bacterium *Bacillus thuringiensis*. Carbohydr. Polym. 190, 113–120.

Rasulov, B.A., Yili, A., Aisa, H.A., 2013. Biosorption of metal ions by exopolysaccharide produced by *Azotobacter chroococcum* XU1. J. Environ. Prot. 4, 989.

Rehm, B.H.A., 2010. Bacterial polymers: Biosynthesis, modifications and applications. Nat. Rev. Microbiol. 8, 578–592.

Exopolysaccharides from marine microbial resources **191**

Roca, C., Lehmann, M., Torres, C.A., Baptista, S., Gaudêncio, S.P., Freitas, F., Reis, M.A., 2016. Exopolysaccharide production by a marine *Pseudoalteromonas* sp. strain isolated from Madeira Archipelago ocean sediments. New. Biotechnol. 33, 460–466.

Rougeaux, H., Guezennec, M., Che, L.M., Payri, C., Deslandes, E., Guezennec, J., 2001. Microbial communities and exopolysaccharides from Polynesian mats. J. Mar. Biotechnol. 3, 181–187.

Ruiz-Ruiz, C., Srivastava, G.K., Carranza, D., Mata, J.A., Llamas, I., Santamaría, M., Quesada, E., Molina, I.J., 2011. An exopolysaccharide produced by the novel halophilic bacterium *Halomonas stenophila* strain B100 selectively induces apoptosis in human T leukaemia cells. Appl. Microbiol. Biotechnol. 89, 345–355.

Sahana, T.G., Rekha, P.D., 2020. A novel exopolysaccharide from marine bacterium *Pantoea* sp. YU16-S3 accelerates cutaneous wound healing through Wnt/beta-catenin pathway. Carbohydr. Polym. 238, 116191.

Saikia, U., Bharanidharan, R., Vendhan, E., Yadav, S., Siva, Shankar, S., 2013. A brief review on the science, mechanism and environmental constraints of microbial enhanced oil recovery (MEOR). Int. J. Chem. Technol. Res. 5, 1205–1212.

Salama, Y., Chennaoui, M., Sylla, A., Mountadar, M., Rihani, M., Assobhei, O., 2016. Characterization, structure, and function of extracellular polymeric substances (EPS) of microbial biofilm in biological wastewater treatment systems: A review. Desalin. Water. Treat. 57, 16220–16237.

Salehizadeh, H., Shojaosadati, S.A., 2003. Removal of metal ions from aqueous solution by polysaccharide produced from *Bacillus firmus*. Water. Res. 37, 4231–4423.

Saravanan, S., Jayachandran, P., 2008. Preliminary characterization of exopolysaccharides produced by a marine bioilm-forming bacterium *Pseudoalteromonas ruthenica* (SBT 033). Lett. Appl. Microbiol. 46, 1.

Satpute, S.K., Banat, I.M., Dhakephalkar, P.K., Banpurkar, A.G., Chopade, B.A., 2010. Biosurfactants, bioemulsifiers and exopolysaccharides from marine microorganisms. Biotechnol. Adv. 28, 436–450.

Schmid, J., Sieber, V., Rehm, B., 2015. Bacterial exopolysaccharides: Biosynthesis pathways and engineering strategies. Front. Microbiol. 6, 496.

Selim, M.S., Almuhayawi, M.S., Alharbi, M.T., Nagshabandi, M.K., Alanazi, A., Warrad, M., Hagagy, N., Ghareeb, A., Ali, A.S., 2022. In vitro assessment of antistaphylococci, antitumor, immunological and structural characterization of acidic bioactive exopoly-saccharides from marine *Bacillus cereus* isolated from Saudi Arabia. Metabolites. 12, 132.

Selim, M.S., Amer, S.K., Mohamed, S.S., Mounier, M.M., Rifaat, H.M., 2018. Production and characterisation of exopolysaccharide from *Streptomyces carpaticus* isolated from marine sediments in Egypt and its effect on breast and colon cell lines. J. Genetic. Eng. Biotechnol. 16, 23–28.

Shearer, C.A., Descals, E., Kohlmeyer, B., Kohlmeyer, J., Marvanová, L., Padgett, D., Voglymayr, H., 2007. Fungal biodiversity in aquatic habitats. Biodivers. Conserv. 16, 49–67.

Singha, T.K., 2012. Microbial extracellular polymeric substances: Production, isolation and applications. IOSR J. Pharm. 2, 276.

Squillaci, G., Finamore, R., Diana, P., Restaino, O.F., Schiraldi, C., Arbucci, S., Ionata, E., La, Cara, F., Morana, A., 2016. Production and properties of an exopolysaccharide synthesized by the extreme halophilic archaeon *Haloterrigena turkmenica*. Appl. Microbiol. Biot. 100, 613–623.

Sran, K.S., Bisht, B., Mayilraj, S., Choudhury, A.R., 2022. Structural characterization and antioxidant potential of a novel anionic exopolysaccharide produced by marine *Microbacterium aurantiacum* FSW-25. Int. J. Biol. Macromol. 131, 343–352.

Sultan, S., Mubashar, K., Faisal, M., 2012. Uptake of toxic Cr (VI) by biomass of exopolysaccharides producing bacterial strains. J. Microbiol. Res. 6, 3329–3336.

Sun, M.-L., Liu, S.-B., Qiao, L.-P., Chen, X.-L., Pang, X., Shi, M., Zhang, X.-Y., Qin, Q.-L., Zhou, B.-C., Zhang, Y.-Z., et al., 2014. A novel exopolysaccharide from deep-sea bacterium *Zunongwangia profunda* SM-A87: Low-cost fermentation, moisture retention, and antioxidant activities. Appl. Microbiol. Biotechnol. 98, 7437–7445.

Sutherland, I.W., 1972. Bacterial exopolysaccharides. Adv. Microbiol. Physiol. 8, 143–213.

Sutherland, I.W., 1997. Microbial exopolysaccharides—structural subtleties and their consequences. Pure. Appl. Chem. 69, 1911–1917.

Sveistrup, M., Mastrigt, F., Norrman, J., Picchioni, F., Paso, K., 2016. Viability of biopol enhanced oil recovery. J. Dispers. Sci. Technol. 37, 1160–1169.

Syafiuddin, A., Fulazzaky, M.A., 2021. Decolorization kinetics and mass transfer mechanisms of Remazol Brilliant Blue R dye mediated by different fungi. Biotechnol. Rep. 29, e00573.

Szewczuk-Karpisz, K., Wiśniewska, M., Pac-Sosińska, M., Choma, A., Komaniecka, I., 2016. Stability mechanism of the silica suspension in the *Sinorhizobium meliloti* 1021 exopolysaccharide presence. J. Ind. Eng. Chem. 35, 108–114.

Thomas, N.V., Kim, S.K., 2013. Beneficial effects of marine algal compounds in cosmeceuticals. Mar. Drugs. 11, 146–164.

Tisthammer, K.H., Cobian, G.M., Amend, A.S., 2016. Global biogeography of marine fungi is shaped by the environment. Fungal. Ecol. 19, 39–46.

Uhliariková, I., Matulová, M., Capek, P., 2020. Structural features of the bioactive cyanobacterium *Nostoc* sp. exopolysaccharide. Int. J. Biol. Macromol. 164, 2284–2292.

Umezawa, H., Okami, Y., Kurasawa, S., Ohnuki, T., Ishizuka, M., 1983. Marinactan, antitumor polysaccharide produced by marine bacteria. J. Antibiot. 36, 471–477.

Underwood, G.J.C., Boulcot, M., Raines, C.A., Waldron, K., 2004. Environmental effects on exopolymer production by marine benthic diatoms—dynamics, changes in composition and pathways of production. J. Phycol. 40, 293–304.

Wang, Q., Wei, M., Zhang, J., Yue, Y., Wu, N., Geng, L., Sun, C., Zhang, Q., Wang, J., 2021. Structural characteristics and immuneenhancing activity of an extracellular polysaccharide produced by marine *Halomonas* sp. 2E1. Int. J. Biol. Macromol. 183, 1660–1668.

Wang, Y.Z., Li, H., Dong, F.K., Yan, F., Cheng, M., Li, W.Z., Chang, Q., Song, T.Z., Liu, A.Y., Song, B., 2020. Therapeutic effect of calcipotriol pickering nanoemulsions prepared by exopolysaccharides produced by *Bacillus halotolerans* FYS strain on psoriasis. Int. J. Nanomed. 15, 10371–10384.

Wei, D.L., Bo, D.Z., Wei, R., Ji, L.L., Zhi, L.L., 2008. Reversal effect of *Ganoderma lucidum* polysaccharide on multi drug resistance in K562/ADM cell line. Acta. Pharmacol. Sin. 29, 620–627.

Wei, M., Geng, L., Wang, Q., Yue, Y., Wang, J., Wu, N., Wang, X., Sun, C., Zhang, Q., 2021. Purification, characterization and immunostimulatory activity of a novel exopolysaccharide from *Bacillus* sp. H5. Int. J. Biol. Macromol. 189, 649–656.

Wu, S., Liu, G., Jin, W., Xiu, P., Sun, C., 2016. Antibiofilm and anti-infection of a marine bacterial exopolysaccharide against *Pseudomonas aeruginosa*. Front. Microbiol. 7, 102.

Xiao, R., Zheng, Y., 2016. Overview of microalgal extracellular polymeric substances (EPS) and their applications. Biotechnol. Adv. 34, 1225–1244.

Yan, M., Mao, W., Chen, C., Kong, X., Gu, Q., Li, N., Liu, X., Wang, B., Wang, S., Xiao, B., 2014. Structural elucidation of the exopolysaccharide produced by the mangrove fungus *Penicilliums solitum*. Carbohydr. Polym. 111, 485–491.

Yang, S.-F., Li, X.-Y., 2009. Influences of extracellular polymeric substances (EPS) on the characteristics of activated sludge under nonsteady-state conditions. Process. Biochem. 44, 91–96.

Yu, R.L., Yang, O.U., Tan, J.X., Wu, F.D., Jing, S.U., Lei, M.I., Zhong, D.L., 2011. Effect of EPS on adhesion of *Acidithiobacillus ferrooxidans* on chalcopyrite and pyrite mineral surfaces. Trans. Nonferrous. Metal. Soc. 21, 407–412.

Zanchetta, P., Lagarde, N., Guezennec, J., 2003. A new bone-healing material: A hyaluronic acid like bacterial exopolysaccharide. Calcif. Tissue. Int. 72, 74–79.

Zayed, A., Mansour, M.K., Sedeek, M.S., Habib, M.H., Ulber, R., Farag, M.A., 2022. Rediscovering bacterial exopolysaccharides of terrestrial and marine origins: Novel insights on their distribution, biosynthesis, biotechnological production, and future perspectives. Crit. Rev. Biotechnol. 42, 597–617.

Zhang, H.N., He, J.H., Yuan, L., Lin, Z.B., 2003. In vitro and in vivo protective effect of *Ganoderma lucidum* polysaccharides on alloxan-induced pancreatic islets damage. Life Sci. 73(18), 2307–2319.

Zhang, J., Liu, L., Ren, Y., Chen, F., 2019. Characterization of exopolysaccharides produced by microalgae with antitumor activity on human colon cancer cells. Int. J. Biol. Macromol. 128, 761–767.

Zhang, Z., Cai, R., Zhang, W., Fu, Y., Jiao, N., 2017. A novel exopolysaccharide with metal adsorption capacity produced by a marine bacterium *Alteromonas* sp. JL2810. Mar. Drugs. 15, 175.

Zhou, W., Shen, B., Meng, F., Liu, S., Zhang, Y., 2010. Coagulation enhancement of exopolysaccharide secreted by an Antarctic seaice bacterium on dye wastewater. Sep. Purif. Technol. 76, 215–221.

9 Microbial exopolysaccharides for food application

Tien-Thanh Nguyen, Tuan-Anh Pham,
Lan-Huong Nguyen, Tuan Le, and Son Chu-Ky

9.1 INTRODUCTION

Exopolysaccharides (EPS), among different extracellular polymeric substances, produced and secreted by various microorganisms. These are naturally high molecular weight and biodegradable consisting of sugar and sugar-derivative units (Barcelos et al., 2020; Mohd Nadzir et al., 2021; Rana and Upadhyay, 2020). In general, most of the EPS are produced and secreted to form biofilm or capsulation in combination with other polymers such as protein, and lipids (Barcelos et al., 2020; Kumar et al., 2007). Exopolysaccharides can be recognized by the mucoid colonies on solid agar plate surfaces and the high viscosity of liquid medium culture of microorganisms (Barcelos et al., 2020; Mohd Nadzir et al., 2021; Rana and Upadhyay, 2020).

Even though the real functions of EPS for microbes are still not elucidated (Abdalla et al., 2021), the potential physiological roles of EPS have been reported as follows:

- To enhance interactions between microorganisms and their environment, to anchor the cell on the solid surface, as in the case of biofilm on the solid substratum.
- Role as protective surrounding matrix against harsh conditions such as the presence of toxic compounds (antibiotics, bile salts, metal ions), hydrolyzing enzymes (lysozyme, gastric and pancreatic enzymes), and stress factors (pH shock, temperature or osmolarity) (Jurášková et al., 2022; Lynch et al., 2018; Prete et al., 2021).
- To protect bacterial cells from phage attacks and predatory attacks from protozoa (Ciszek-Lenda, 2011).
- Can be used as nutrient storage for microbial cells (Barcelos et al., 2020).

EPS can be produced by yeasts, molds, and bacteria, thus varied in physical and chemical properties. The commonly reported EPS producers include lactic acid bacteria, *Xanthomonas campestris, Aureobasidum pullulans, Pseudomonas elodea, Alcaligenes faecalis, Sclerotinum rolfsii, Zymomonas mobilis, Acetobacter xylinum*

DOI: 10.1201/9781003342687-9

(Mohd Nadzir et al., 2021; Mollakhalili and Mohammadifar, 2015; Rana and Upadhyay, 2020; Yildiz and Karatas, 2018). Some halophilic archaea can produce EPS, and most of them belong to the genus *Halomonas* (Joulak et al., 2019; Mata et al., 2006). The produced EPS from different microbial sources are diverse in backbone structure, molecular weight, and solubility, hence the technological properties.

Researchers can classify EPS based on their linkage type (a -, b-bonds) in structure or functionality (sorptive, surface active) (Rana and Upadhyay, 2020). However, the chemical nature of EPS is the most used for the classification of EPS. Thereby, two groups of EPS including homopolysaccharides (HoPs) and heteropolysaccharides (HePs) are reported (Mohd Nadzir et al., 2021; Mollakhalili and Mohammadifar, 2015; Rana and Upadhyay, 2020; Yildiz and Karatas, 2018).

- Homopolysaccharides (HoPs) consists of only one kind of monosaccharide, for example, D-glucose or D-fructose. Depending on the glycosyl type, linkage variety, and carbon position join in the linkage, they can be further classified (Abdalla et al., 2021).
- Heteropolysaccharides (HePs) consist of at least two types of monosaccharides (such as rhamnose, fructose, galactose, or glucose). Furthermore, monosaccharides can be modified including pyruvylation, acetylation, and phosphorylation (Prete et al., 2021).

Literature reported four different EPS biosynthesis pathways, including the Wzx/Wzy-, the ABC transporter-, the synthase-, and the extracellular sucrase-dependent pathway (Angelin and Kavitha, 2020; Rana and Upadhyay, 2020; Schmid et al., 2015). HoPs are usually synthesized either by the synthase-based pathway or the sucrase-dependent pathway. Meanwhile, HePs are produced by the Wzx/Wzy-dependent pathway and the ABC transporter-dependent pathway (Angelin and Kavitha, 2020).

9.2   STRATEGY FOR ENHANCEMENT OF EPS PRODUCTION BY MICROORGANISMS

To enhance EPS production, several approaches have been performed such as optimization of process condition and medium composition, and modification of genes involved in EPS biosynthesis (Barcelos et al., 2020). The process condition and medium composition directly affect EPS production, including the carbon and nitrogen sources, temperature, pH, aeration, and agitation.

The availability of carbon and nitrogen sources is the most affecting factor in EPS production by microorganisms, especially the ratio between the two sources. The authors mentioned that the greater carbon source improves the biomass yield, however, the greater nitrogen source can result in a higher EPS degrading enzyme, thus yielding a lower amount of accumulated EPS. The most favorable ratio of C/N of 10:1 was reported (de Souza and Sutherland, 1994). Bacterial EPS is normally synthesized and accumulated high level in the late log phase to the stationary phase (Barcelos et al., 2020). The suitable temperature for EPS production has been reported as lower than the optimal temperature for the growth of microorganisms

(Kumar et al., 2007). For the high production of EPS, the pH should be maintained since pH can affect the activity of enzymes involving EPS biosynthesis (Kumar et al., 2007). Most EPS-producing strains are aerobic, therefore high agitation and aeration have been reported that can result in better EPS production. EPS accumulated in the medium can lead to the high viscosity and reduce the mass transfer of oxygen (Kumar et al., 2007). To optimize the production of EPS, several authors successfully applied the statistical tools such as response surface methodology (Barcelos et al., 2020).

Genetic modification can be applied to improve the production of EPS by the microorganism, in which the encoding genes for glycosyltransferases (GTs) are modified and over-expressed to aim for the high production of EPS (Barcelos et al., 2020). However, this approach can affect the mechanism pathway of microbial cells. Furthermore, the shortage of genetic tools (Shin et al., 2016), in combination with GMO regulation, this approach is not popularly reported until now.

9.3 APPLICATIONS OF MICROBIAL EXOPOLYSACCHARIDES

Microbial exopolysaccharides have applications in the food, medical industry, and many other industries. Most popular, microbial EPS is used to change the rheological properties of food products. It can also be used as a viscosity enhancer, emulsifier, stabilizer, gelling agent, or water-remaining agent (Suryawanshi et al., 2022). Some microbial EPS types are water-insoluble and able to retain water, thus EPS can be potentially used for drug delivering systems in the medical industry (Mohd Nadzir et al., 2021). EPS from microbes are normally hydrophilic and charged. Subsequently, they can be used in waste water treatment to absorb heavy metals, aromatic compounds, and dyes (Huang et al., 2022). Common microbial EPSs and their applications in the food industry are discussed as follows.

DEXTRAN

Dextrans, certified GRAS status by FDA (US), are homopolysaccharides composed of repeating units with a main chain of α-1-6-linked D-glucopyranosyl and branched by α-1-2, α-1-3, or α-1-4 bonds (Díaz-Montes, 2021; Mohd Nadzir et al., 2021) (Figure 9.1A). Dextran molecular weight can be varied from 3 to 2000 kDa. Lactic acid bacteria such as *Leuconostoc mesenteroides*, *Streptococcus mutans*, *Lactobacillus sanfrancisco, L. plantarum*, and yeast *Saccharomyces cerevisiae* are now reported as main dextran producers (Barcelos et al., 2020; Díaz-Montes, 2021). The structure of dextran is significantly varied between species and strains of bacteria.

Though dextran can be applied in photography, textile, paper, cosmetics, and pharmaceutical industry, it is the most common EPS commercially employed in the food industry as additive for the improvement of gelling status, viscosity, and texture of products. In food products, dextran having a long chain improve the rheology and physicochemical properties (Mollakhalili and Mohammadifar, 2015; Suryawanshi et al., 2022).

Xanthan

Xanthan is produced by *Xanthomonas campestris* and one of the most used microbial EPSs. It is a heteropolysaccharide and mainly composed of D-glucose via β-1–4 linkage (Figure 9.1B). The side chains which are built from one glucuronic acid and two mannose residues are branched at the C3 position of glucose residue (Kumar et al., 2007; Mollakhalili and Mohammadifar, 2015). Xanthan has been certified GRAS by US FDA. Many authors reported that xanthan is highly water soluble and stable in acidic/alkaline solution or in the presence of salts. It is also proven as resistant to enzymes and temperature fluctuation. Therefore, xanthan gum is the most used in the food processing sector as a thickener, emulsifier, and stabilizer (Chaturvedi et al., 2021; Suryawanshi et al., 2022).

Pullulan

Pullulan is a neutral heteroexopolysaccharide produced by *Aureobasidium pullulans*. It consists of units linked via α-1-6 glycosidic bonds (Muthusamy et al., 2022) (Figure 9.1C). Pullulan is employed in the food industry for different purposes, such as emulsion stabilizer, filler for beverages and sauces, binder, moisture retention, crystallization prevention, etc. (Suryawanshi et al., 2022).

Gellan

Produced by *Pseudomonas* or *Sphingomonas*, gellan a linear anionic heteroexopolysaccharide. Its repeating unit is comprised of two β-D-glucose, one β-D-glucuronic acid and one α-L-rhamnose (Camelia et al., 2015) (Figure 9.1E). In the food industry, gellan gum is employed as a gelling and thickening agent. Gellan gums are also used to suspend protein, minerals, vitamins, etc. in beverages. They are popularly used in confectionery products and dairy products (Camelia et al., 2015; Mollakhalili and Mohammadifar, 2015).

Curdlan

The curdlan is a water-insoluble homoexopolysaccharide which is composed of glucose units linked via β-1-3 glycosidic bond. Curdlan can be produced by *Rhizobium* spp., *Agrobacterium, Alcaligenes, Cellulomonas,* and *Bacillus* (Aquinas et al., 2022). In aqueous solutions, it can form a thermo-reversible gel (low-set gel) or a thermo-irreversible gel (high-set gel) when the temperature of the solution is heated up to 55°C or 80°C with subsequent cooling, respectively (Aquinas et al., 2022). Curdlan has a broad scope of utilization in different fields, particularly in food in which it serves as a stabilizer, gelling agent, and thickener in many food products (Suryawanshi et al., 2022).

SCLEROGLUCAN

Scleroglucan is extracellularly produced by various filamentous fungi, for example, *S. rolfsii*, *S. glucanicum*. It also can be produced by *Schizophyllum commune* and *Epicoccum nigrum* (Schmid et al., 2011).

Scleroglucan is a neutral homopolysaccharide with the main chain consisting of glucospyranoses connected by β-1-3 linkage and the size chain branched by β-1-6 bond on every third subunit. With this structure, scleroglucan shows highly viscoelastic behavior and water-soluble. The viscosity of scleroglucan solutions is consistent with temperature variations. The scleroglucan solution shows a higher thermostability, and its viscosity is not significantly affected by temperature variation even at 120°C for 20 h. Due to its neutral character, it is only affected by higher salt concentrations or high acidic/alkaline conditions (Schmid et al., 2011). In food industries, scleroglucan is employed as a stabilizing agent, gelling agent, or a thickener in several products such as jams, ice cream, soups, water-based gels, solidified foods, dairy items, non-fat items (Suryawanshi et al., 2022).

LEVAN

Levan is an extracellular polysaccharide, naturally found in plants and microorganisms. The main chain of levan is made of fructose with β-(2-6) fructofuranosidic bonds and randomly branched with β-(2-1) linkage (Barcelos et al., 2020) (Figure 9.1D). From the source of microorganisms, levan can be produced by species belonging to *Zymomonas, Bacillus, Acetobacter, Aerobacter, Pseudomonas, Erwinia, Gluconobacter, Streptococcus, Corynebacterium* (Barcelos et al., 2020). Microbial levan generally have molecular weight varied depending on the strain, ranging from 6–9 kDa up to above 5000 kDa (Mohd Nadzir et al., 2021).

In the food industry, levan can be used as sweeteners (e.g., as a fructose source), fat substitutes, fillers (bulking agents), emulsifiers and texture-forming compounds, enclosing agents (encapsulation) and flavor carriers to enhance aroma, food-coating materials (e.g., food bio-based packaging materials), and stabilizers or thickeners (Öner et al., 2016).

MICROBIAL CELLULOSE

Cellulose can be produced and secreted by a few bacterial species. Among them the best-known one is *Gluconacetobacter xylinus*, a gram-negative bacterium which is also known as *Acetobacter xylinum*. The polymerization degree of bacterial cellulose varies from 2,000–6,000, meanwhile that of plant cellulose is 13,000–14,000. In general, bacterial cellulose can be considered as nanocellulose since its diameter and length are 10–50 nm and 100–1000 nm, respectively (Mohd Nadzir et al., 2021).

Microbial cellulose is considered high purity; therefore it is potentially applied in different sectors including food, biomedical, nanomaterial industry. Bacterial nanocellulose which is edible, biodegradable, and mechanically firm can be used in the development of packaging materials in the food industry. For this purpose, microbial

TABLE 9.1

Common microbial EPS and their application in the food industry (Barcelos et al., 2020; Ludwicka et al., 2020; Mohd Nadzir et al., 2021; Mollakhalili and Mohammadifar, 2015; Suryawanshi et al., 2022; Trček et al., 2021)

EPS	Common microbial sources	Monomeric unit	Type	Linkage	Average molecular weight	Solubility	Food industrial applications
Dextran	Lactic acid bacteria: *Lactobacillus* spp., *Leuconostoc* spp., *Weissella* spp.	glucose	Homo	a-1-6, a-1-2, a-1-3, a-1-4	5–500 kDa	Aqueous/nonaqueous soluble	Enhances the softness, volume of loaf, mouthfeel of bakery; Cryoprotectant (ice cream); Improve viscosity, moisture retention, gelling agent, prevent the crystallization (confectionary); Prebiotics; Moisture retention (fat reduced products)
Xanthan	*Xanthomonas campestris*	Glucose, manose, glucoronic acid	Hetero	b-1-4, 1-3, 2-1, 4-1	2000–20000 kDa	Water soluble	Enhance texture (bakery); Emulsifier, improve flavor (dairy products); Stabilizer (juice)
Pullulan	*Aureobasidium pullulans*	glucose	Homo	a-1-6, a-1-4)		Water soluble	Water retention; Low viscosity filler; Binder, stabilizer (paste food); Dietary fiber
Gellan	*Sphingomonas, Pseudomonas*	Glucosse, glucuronic acid, rhamnose	Hetero	b-1-4, a-1-3	About 500 kDa	Water soluble	Gelling agent (desserts, jelly drink); Water retention (bakery); Binding agent (nutritious bar, high protein bar); Stabilizer (cloudy beverages)

(Continued)

TABLE 9.1 (Continued)
Common microbial EPS and their application in the food industry

EPS	Common microbial sources	Monomeric unit	Type	Linkage	Average molecular weight	Solubility	Food industrial applications
Curdlan	*Alcaligensis faecais*	Glucose	Homo	b-1-3	1–240 kDa	Water soluble	Texture modified (noodles, hams) Water retention (sausage, hams) Gelling agent (jellies) Binding agent, moisture retention
Alginate	*Azotobacter, Pseudomonas*	Manuronic acid, guluronic acid	Hetero	b-1-4	33–400 kDa		Stabilizer, thickener in food products
Scleroglucan	*Sclerotium*	Glucose	Homo	b-1-3, b-1-6	500 kDa	Water soluble	Gelling agent stabilizer, thickener (jam, soups, dairy, etc.)
Levan	*Bacillus, Pseudomonas, Zymomonas*	Fructose	Homo	b-2-6, b-2-1	<10^5 kDa	Water soluble	Prebiotics Improve the bread texture and volume
Cellulose	*Gluconacetobacter, Agrobacterium, Pseudomonas*	Glucose	Homo	b-1-4	Polymerization degree of 2000–6000		Water retention Functional ingredients for food packaging material
Acetan	*Komagataeibacte*	Rhamnose, Glucose, glucuronic acid, manose	Hetero	b-1-6 (main chain)	1–1000 kDa	Water soluble	Similar to xanthan
Kefiran	*Lactobacillus kefir-anofaciens*	Galactose, glucose	Hetero	1-4, 1-3, 1-2, 1-6	50–15000 kDa	Water soluble	Stabilizer, gelling agent

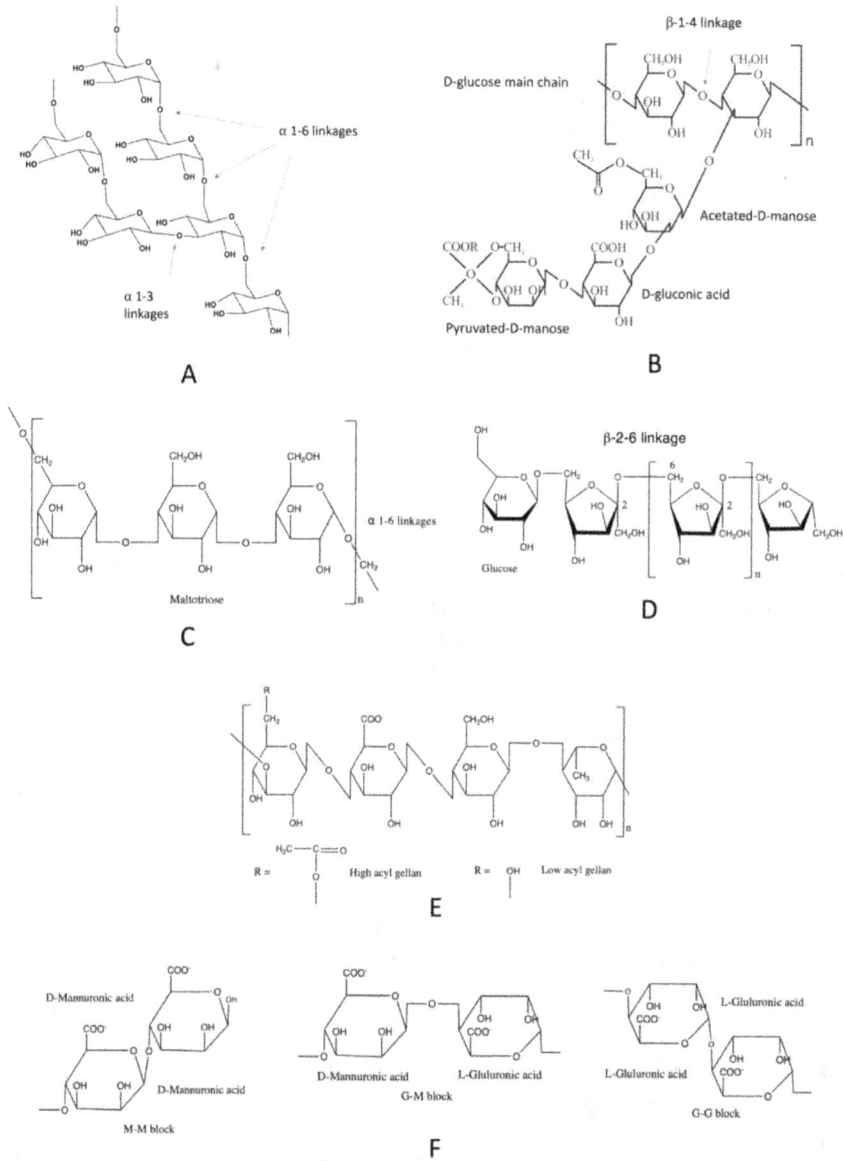

FIGURE 9.1 Common EPS for food application dextran (A), xanthan (B), pullulan (C), levan (D), gellan (E), alginate (F)

cellulose can be integrated with antimicrobial, antioxidant, stabilizers, and other additives to create new functions for packaging (Ludwicka et al., 2020). Bacterial cellulose can be also used to stabilize and thicken the food product texture (Perumal et al., 2022).

ALGINATE

Alginate, well-known EPS, is normally extracted from marine algae. However, it can be obtained from bacterial sources such as *Pseudomonas* and *Azotobacter genera* (Pacheco-Leyva et al., 2016). In bacteria, especially *Azotobacter vinelandii*, alginate plays a role of a protective barrier against toxic and harsh environmental conditions. Bacterial alginates are built up from a random amount of manuronic (M), guluronic (G) residues which are linked by 1-4 glycosidic bonds. These units can link to others to build up G-G, M-M, or G-M blocks (Figure 9.1F) in alginate structure. The O-2 and/or O-3 of the manuronic residues are linked with O-acetyl groups. Similarly to alginate from seaweed, bacterial alginate is applied in the food industry as a gelling agent, stabilizer, and moisture maintainer (Nagarajan et al., 2015).

ACETAN

Acetan is a water-soluble polysaccharide produced by *Komagataeibacter xylinus* (former name *Acetobacter xylinum)* is acetan. Acetan helps bacterial cells to contact oxygen. It is composed of repeating unit of α, L-rhamnose-(1-6)-β, D-glucose-(1-6)-α, D-glucose-(1-4)-β, D-glucuronic acid-(1-2)-α, D-mannose-(1-3)-β, D-glucose-(1-4)-β D-glucose. These repeating units link via β-1-4- in the main chain. The molecular weight of acetan varies from 10^3-10^6 Da, depending on the strain (Trček et al., 2021). The viscoelasticity of acetan solution has been investigated by several researchers. The viscosity of aqueous solutions of acetan decreases with increasing of shear rate (shear-thinning behavior). Acetan is negatively charged, thus salt presence affects the viscosity of acetan solution. The rheological properties of the acetan solution are due to the pentasaccharide side chain and the deacetylation. Currently, the application of acetan is not well researched, except for the purified acetan AC-1 from *K. xylinus* NBI 1005. This acetan was shown useful for preventing allergy response (Trček et al., 2021). However, acetan has been considered as having good potential for commercial use in the food industry similar to the well-known xanthan (Trček et al., 2021).

9.4   DIFFERENT GENERALLY RECOGNIZED AS SAFE (GRAS) MICROBES FOR EPS PRODUCTION

For application in pharmaceutical sectors and especially in the food industry, the generally recognized as safe (GRAS) microbes including *Bacillus*, lactic acid bacteria, and recently medicinal mushrooms have received the great attention from many researchers.

EPSs FROM *BACILLUS* STRAINS

The different types of EPS, such as glucan, γ-polyglutamic acid (γ-PGA), and fructan (inulin and levan) are found in the Bacillus strain, and their functions and properties are summarized in Table 9.2.

TABLE 9.2
EPSs from *Bacillus* strains

EPSs	Strains	Resources	EPS components	Mw*(Da)	Properties	Ref
γ-PGA	*Bacillus* sp. 20.2	Soil, straws (Vietnam)	ND**	ND	Heavy metal absorption	(Thanh et al., 2018)
PGA	*Bacillus* sp. B11, *Bacillus* sp. B33	Fermented soybean (Vietnam)	ND	ND	Biological coagulant in water treatment	(Tuan et al., 2020)
γ-PGA	*B. velezensis* VCN56	Fermented foods (Vietnam)	ND	98×10^3	Antioxidant (DPPH, hydroxyl, and superoxide radicals)	(Quach et al., 2022)
γ-PGA	*B. velezensis* CAU263	Douchi (China)	L-glutamic acid (26.1%) and D-glutamic acid (73.9%)	3.8×10^6	Improve the bread quality	(Liu et al., 2022)
Levan	*B. subtilis* natto D	Vietnam	ND	ND	Strong adhesion, good biocompatibility, anticancer, and film-forming ability	(Dung et al., 2017)
Levan	*B. velezensis* VTX20	Soybean fermented pastes (Vietnam)	Fructose, glucose	ND	Antioxidant activity (DPPH, hydroxyl radicals)	(Vu et al., 2021)
Levan	*Bacillus* sp. SGD-03	Marine sediment (India)	Fructose	1.0×10^4	ND	(Wagh et al., 2022)
Fructan (EPS-B108)	*Bacillus* sp. SCU-E108	Saline Soils (China)	Solely fructose	3.58×10^7	Nontoxic, thermal stability, solubility in water	(Gan et al., 2021)
EPS	*Bacillus* sp. BES19	Sludge (Vietnam)	ND	ND	Flocculation activity	(Van et al., 2014)
EPS	*B. licheniform-is* MS3	Pakistan	Mannose (20.60%), glucose (46.80%), and fructose (32.58%)	ND	Emulsifying properties	(Asgher et al., 2020)
BEPS	*Bacillus* sp. NRC5	Seawater (Egypt)	Glucose: mannose: mannuronic acid = 1.0:1.7:0.8	3.59×10^5	Cytotoxic, anti-inflammatory, antioxidant, antitumor, meta chelation, and inhibition of lipid peroxidation	(Mohamed et al., 2021)

(Continued)

TABLE 9.2 (Continued)
EPSs from Bacillus strains

EPSs	Strains	Resources	EPS components	Mw*(Da)	Properties	Ref
EPS5SH	Bacillus sp. H5	Seawater (China)	Mannose: glucosamine: glucose: galactose = 1.00: 0.02: 0.07: 0.02	89×10^3	Cytotoxic, immunostimulatory activity	(Wei et al., 2021)
EPS	B. xiamenensis RT6	River sediments (Spain)	Glucose (60%), mannose (20%), and galactose (20%)	2.71×10^4	Antioxidant activity, reduce heavy metal concentration and emulsifying natural polysaccharide oils	(Huang-Lin et al., 2022)
EPSR4	B. subtilis AG4	Marine sediment (Egypt)	Glucose: rhamnose: arabinose = 5: 1: 3	1.48×10^4	Cytotoxic, anti-inflammatory, antioxidant, anti-Alzheimer, and membrane stability	(Abdel-Wahab et al., 2022)
EPS	B. subtilis xztubd1	Housefly's body (China)	ND	ND	Antioxidant and antitumor	(Zhang and Yi, 2022)
EPSR3	B. cereus AG3	Red Sea (Saudi Arabia)	Glucose: galacturonic acid: arabinose = 2.0: 0.8: 1.0	ND	Antioxidant, cytotoxic, anti-inflammatory, and antimicrobial	(Selim et al., 2022)
EPS	B. haynesii CamB6	Hot water (Chile)	Mannose (66%), glucose (20%), and galactose (14%)	ND	Antioxidant, thermostability, emulsifier, water and oil holding, and flocculation capacities	(Banerjee et al., 2022)
EPS	B. aerophilus rk1	Soil (India)	ND	ND	Antioxidant (DPPH and hydrogen peroxide scavenging activities)	(Gangalla et al., 2021)

EPS	B. thuringiens-is SMJR	Mango leaves (India)	Mainly glucose	ND	Antibacterial activity, enhancing the shelf life of foods	(Natarajan et al., 2022)
EPS-M1	B. marcoresti-nctum QDR3–1	Seawater (China)	Mannose, glucose, galactose, and fucose	33.8×10^3	Cytotoxicity and repair ability of UVR-mediated	(Li et al., 2022b)
BPS-2	B. thuringiensis IX-01	Bean paste (China)	D-galactosamine: arabinose: glucosamine: glucose: mannose = 5.53: 1.77:4.74:3.24:1	29.36×10^3	Anti-digestive capacity (scavenge of DPPH, hydroxyl, ABTS, and superoxide anions radicals), anti-inflammation and antioxidant	(Gao et al., 2022)
BL-P1	B. subtilis LR-1	Fermented meat (China)	Glucose (82.95%) and mannose (13.58%)	41.11×10^3	Antioxidant and additive in functional food	(Zhao et al., 2022)

MW*: Molecular weight; ND**: not detected

γ-PGA is a natural anionic polymer of D/L-glutamic acids linked together via amide bonds. The degree of monomer polymerization directly related to the molecular weight which is determined by the producers type media used and cultivation conditions (Li et al., 2022a). Its molecular weight is generally in the range of 10^3 Da -10^6 Da (Liu et al., 2022; Quach et al., 2022). γ-PGA has applications in food, medical, and environmental sectors owing to its various biological characteristics (water-solubility, nontoxic and edible nature, non-immunogenic, antioxidant activities, and heavy metal absorption) (Liu et al., 2022; Quach et al., 2022; Thanh et al., 2018; Tuan et al., 2020).

Bacillus species produce levan biopolymer utilizing sucrose as a sole substrate through the activity of levansucrase (E.C. 2.4.1.10) (Al-Qaysi et al., 2021). Its molecular weight of levan produced by *Bacillus* sp. SCU-E108 strain is 3.58×10^7 Da (Gan et al., 2021), while levan from *Bacillus* sp. SGD-03 has a low molecular weight (1.0×10^4 Da) (Wagh et al., 2022). Levan produced by *B. velezensis* VTX20, isolated from Vietnamese traditional soybean fermented pastes, showed fructose and glucose as the main monosaccharides. This EPS was described for its strong antioxidant activity with scavenging activity for 2,2-diphenyl-1-picrylhydrazyl (DPPH) and hydroxyl radical values of 40.1–64.0% and 16.0–40.0%, respectively (Vu et al., 2021).

According to Table 9.2, besides the homopolysaccharides, there are many other types of EPSs produced by *Bacillus* strains, complex EPSs, known as heteropolysaccharides. *Bacillus* sp. BES19, isolated from wastewater sludge in Hanoi Brewery (Vietnam), can synthesize EPS. This strain showed the highest flocculation activity (Van et al., 2014). Marine *Bacillus* sp. strain NRC5 produce BEPS, which is composed of glucose, mannose, and mannuronic acid and have a molecular weight of 3.59×10^5 Da. BEPS shows antioxidant activity against DPPH and anti-inflammatory activity via COX-2 inhibition and also able to inhibit the proliferation of Ehrlich ascites carcinoma cells (EACC). Its significantly reduced tumor volume in mice and effective in the 5-fluorouracil (FU) drug treatment for tumoral cells (Mohamed et al., 2021). Banerjee et al., extracted a low molecular weight heteropolymer composed of mannose (66%), glucose (20%), and galactose (14%) from thermophilic *B. haynesii* strain CamB6 which presented antioxidant, emulsifier, water-holding, oil-holding, and flocculation properties. The EPS was reported to be a (Banerjee et al., 2022) BPS-2, an exopolysaccharide of molecular weight 29.36×10^3 Da synthesized by *B. thuringiensis* strain IX-composed of D-galactosamine: arabinose: glucosamine: glucose: mannose. In vitro upper gastrointestinal simulations results showed that it has a strong anti-digestive capacity, with scavenging of DPPH, hydroxyl, ABTS, and superoxide anions radicals (Gao et al., 2022).

EPS FROM LACTIC ACID BACTERIA

Most lactic acid bacteria (LAB) can produce EPS. Due to the GRAS status of LAB, EPS from this type of bacteria have received great attention from research, especially species from the genus *Fructilactobacillus, Lacticaseibacillus, Lactiplantibacillus, Lactobacillus, Lactococcus, Latilactobacillus, Lentilactobacillus, Leuconostoc, Limosilactobacillus, Pediococcus, Streptococcus,* and *Weissella* (Korcz and Varga, 2021). Common homo-EPS from LAB are dextran, mutan, alternan, reuteran (Table 9.3). Many hetero-EPS also can be produced by LAB with varied structural linkages, monomeric units, molecular weight, and characteristics (Jurášková et al., 2022; Korcz and Varga, 2021).

The production yield of EPS by LAB depends on the genera, strains, culture media, and carbon sources (Table 9.4). The most common culture medium for EPS production investigated are skim milk or MRS medium with carbon sources, including lactose, sucrose, and glucose. Strains from some genera can produce a rather high yield of EPS up to 5–6 g/L such as *Lactobacillus* or *Limosilactobacillus*. Interestingly, *L. pseudomesenteroides* and *L. mesenteroides* can produce very high EPS concentrations ranging from 18.08–61.90 g/L (Abdalrahim et al., 2019).

The use of EPS-producing LAB as a starter culture in food products can be applied instead of the conventional addition of EPS as additives in the products. Apart from lactic acid, EPS generated by LAB introduce benefits for food products (Table 9.5) in which EPS plays the role of gelling agent, viscosity regulator, moisture retention, or thickener (Korcz and Varga, 2021; Prete et al., 2021).

Besides the application in the food industry, EPS from LAB has also attracted great attention from the pharma, nutraceutical sectors due to its biocompatibility, safety, and biodegradability features. EPS from LAB exhibit several health benefits

TABLE 9.3
Homoexopolysaccharides from LAB
(Korcz and Varga, 2021)

EPS	LAB producer	Main chain linkage	Molecular weight
Dextran	*Leuconostoc* species *Lentilactobacillus parabuchneri* *Limosilactobacillus fermentum, Lim. reuteri* *Latilactobacillus sakei, Lat. curvatus* *Lactobacillus hordei, L. nagelli, L. mali, L. satsumensis* *Weissella confusa, W. cibaria* *Streptococcus mutans, S. salivarius*	α-D-Glc (1-6)	1–1000 kDa
Mutan	*Limosilactobacillus reuteri* *Streptococcus downei, S. mutans, S. salivarius*	α-D-Glc (1-3)	>1000 kDa
Alternan	*Leuconostoc mesenteroides, L. citreum* *Streptococcus salivarius*	α-D-Glc (1-6)/α-D-Glc (1-3)	>1000 kDa
Reuteran	*Limosilactobacillus reuteri*	α-D-Glc (1-4)/α-D-Glc (1-6)	10000 kDa
Levans	*Leuconostoc mesenteroides* *Limosilactobacillus reuteri* *Streptococcus mutans*	β-D-Fru (2-6)	10–10^5 kDa
Inulin-type	*Streptococcus mutans* *Limosilactobacillus reuteri* *Leuconostoc citreum* *Lactobacillus johnsonii*	β-D-Fru (2-1)	1–10^4 kDa

TABLE 9.4

The production yield of EPS in different LAB genera

Genera	EPS yield (mg/L)	Common culture media/carbon sources
Streptococcus	55–1950	Skim milk/lactose
Lactobacillus	24– ~6000	Skim milk/lactose
Lactiplantibacillus	75– ~ 400	Modified MRS/sorbitol
Limosilactobacillus	85.5–5000	Modified MRS/sucrose or Skim milk/lactose
Lacticaseibacillus	298–1275	MRS/glucose
Lactococcus	204–354	Skim milk (whey)/lactose or MRS/glucose (sucrose)

TABLE 9.5

Application of EPS producing LAB in food products

(Korcz and Varga, 2021; Prete et al., 2021)

Food products	Strain	Purposes
Yoghurt	*L. bulgaricus, S. thermophilus*	To thicken and stabilize products
Cheese	*Lactobacillus*	Moisture retention; improve the quality and texture of low-fat cheese
Kefir	*Lactobacillus kefiranofaciens*	Viscosity regulator
Plant-based beverage	*Lactiplantbacillus plantarum*	Improve the texture (viscosity), mouthfeel and syneresis
Bakery	*Weissella, Limosilactobacllus reuteri*	Enhance the stability of dough and texture/volume of bread
		Overcome the quality loss of frozen dough products
		Improve the structure/volume of gluten-free bread
Meat products	*L. plantarum, L. curvatus*	Water retention, gelling agent for low fat products

including antioxidant, prebiotic, immunomodulatory, cholesterol-lowering, antimicrobial, and anticancer activity (Table 9.6). Moreover, it should be noted that these health-promoting effects are correlated with structure (monomeric composition, linkage) and molecular weight, which can be varied with conditions of fermentation of EPS by LAB (Prete et al., 2021).

EPS FROM FUNGAL OR MEDICINAL MUSHROOM

In Vietnam, fungal polysaccharides focused on medicinal mushroom strains such as *Cordycep* spp. and *Trameter versicolor. Cordyceps sinensis* is a well-known traditional medicinal mushroom used in Chinese medicine containing bioactive compounds like (adenosine, cordycepin, polysaccharides, proteins) and have antioxidant activity and anti-inflammatory effect (Das et al., 2021; Yan et al., 2014).

TABLE 9.6
Health-promoting effects of EPS from LAB
(Abdalla et al., 2021; Jurášková et al., 2022; Prete et al., 2021)

Health benefits	Mechanism
Prebiotic effects	Provide protection against acidity of gastric environment Promote the growth of *Bifidobacterium*
Immunomodulatory activity	Phosphate (negative charge) on EPS can trigger the immune response
Antioxidant activity	Superoxide anions and hydroxyl radical scavenging activity; reduction of lipid peroxidation; inhibition of metal ion chelating activity
Cholesterol lowering ability	Adsorption of cholesterol in the intestinal system to regulate the cholesterol level in the serum
Antimicrobial activity	Inhibit the formation of biofilms of pathogens by interrupting communication among cells or by weaking cell membrane; carbonyl, hydroxyl, sulfate, phosphate groups on EPS structure exhibit the antimicrobial effect
Anticancer activity	The antitumor activity of LAB EPS has been observed, but the mechanism is not elucidated

Authors used sunflower oils and coconut to enhance the biosynthesis of EPS in *C. sinensis* and able to achieve EPS yields with olive oil (5.94 g/L), sunflower oil (2.56 g/L), and coconut oil (2.43 g/L), respectively (Le et al., 2017b). To improve the EPS production, these authors screened the main factors influencing the EPS production by Plackets-Burman design. The results showed that olive oil, saccharose, and peptone were the three main factors enhancing EPS production (Le et al., 2017a).

Trametes versicolor, known as *Coriolus versicolor* or Yunzhi in Chinese, is one of the best investigated medical mushrooms in recent years (Habtemariam, 2020). It has received a high attention due to high biological activities of its polysaccharides (He et al., 2022; Smith et al., 2002). Polysaccharides is normally extracted from the fruiting body or mycelia. However, the main obstacle on the solid medium has low growing rate. The submerged cultivation of the mushroom, therefore, has received great interest as it seems a promising approach for PSK (polysaccharide-Krestin) or PSP (polysaccharopeptide) synthesis. The submerged cultivation saves time, and exopolysaccharides (EPS) are excreted into the cultures which do not require the extraction of the EPS. The biological activities of these polysaccharides are currently attentively examined and discussed (Dou et al., 2019; Santos Arteiro et al., 2012). The polysaccharide Krestin (PSK) is the first commercial protein-bound polysaccharide isolated from *T. versicolor* and used in chemotherapy, radiotherapy, blood transfusion, and antitumor agent (Cui and Chisti, 2003; Hobbs, 2004; Standish et al., 2008). Studies have shown that PSP significantly relieves pain and boosts immunity in 70–97% of patients with stomach, esophageal, lung, uterus, ovarian, and neck cancers (Dou et al., 2019; Kidd, 2000; Ng, 1998). Side effects caused by tumors and cancer treatment with chemotherapy and radiotherapy are also mitigated with the use of PSP adjuvant therapy (Cui and Chisti, 2003). Moreover, EPS from *T. versicolor* presents the potential for prebiotic and anti-inflammatory activities (Angelova et al., 2022).

EPS, PSK (polysaccharide-Krestin), and PSP (polysaccharopeptide) of *T. versicolor* were successfully produced in a submerged cultivation. Dextrin and peptone were the two main nutrients affecting the PSK and PSP production. The *T. versicolor* crude polysaccharopeptide showed the antioxidant activity and antitumor activity against breast cancer cell line MCF-7 (Tran et al., 2019). The extract process from mycelium was also studied. The optimal PSK extraction was obtained with biomass treated for 30 seconds by handheld homogenizer (IKA), biomass/water ratio = 1/30, and extraction temperature of 121°C for 30 min. The highest PSK recovery rate was 16.8 mg/g of dry biomass. Polysaccharide-Krestin has antioxidant activity with IC_{50} of 0.04 mg/mL and can inhibit 90.6% of breast cancer cell line MCF-7 (Tran et al., 2021). These results promise the possibility of simple PSK extraction from the biomass of Yunzhi submerged fermentation towards application in the production of functional products.

For improvement of PSK production, MPEC (microparticle enhanced cultivation) was used for the first time in submerged fermentation using *T. versicolor* for PSK production. In flask scale, the addition of microparticles such as Talc increased the biomass and the PSK content to 1.07 times and 1.1 times higher than the control sample without MPEC, respectively. In 10 L bioreactor, the submerged fermentation resulted in the highest biomass and PSK productivity of 8.07 g/L and 0.45 g/L, which were 1.58 times and 4.6 times higher than the control group, respectively. Furthermore, with MPEC technique, the time for producing biomass and PSK was almost a half in comparison to the control sample (Nguyen et al., 2022).

Moreover, the effects of LEDs light color, including green, blue, white, red, and purple on PSK production were examined. In the first stage, white and purple LEDs color have a positive effect on increasing biomass and PSK productions, hence should be discussed as a strategy for further improvement (Nguyen et al., 2022).

9.5 CONCLUSIONS

Different types of exopolysaccharides can be obtained from microorganisms. Production yield and characteristics of EPS are not only strain dependent but also cultivation condition dependent. In the food industry, EPS is employed as an emulsifier, gelling agent, stabilizer, and moisture retention agent. Furthermore, EPS from *Bacillus*, lactic acid bacteria, and medical mushrooms is receiving great attention due to their bioactive properties which are beneficial to health such as antimicrobial, immunomodulatory, antioxidant, and anticancer activities. Research is still ongoing to exploit new EPSs, to clarify the biological mechanism of EPS, as well as to optimize microbial EPS biosynthesis towards industrial production and application.

REFERENCES

Abdalla, A.K., Ayyash, M.M., Olaimat, A.N., Osaili, T.M., Al-Nabulsi, A.A., Shah, N.P., Holley, R., 2021. Exopolysaccharides as antimicrobial agents: Mechanism and spectrum of activity. Front. Microbiol. 12, 664395.

Abdalrahim, S., Zohri, A.-N., Khider, M., El-Dean, A., Abulreesh, H., Ahmad, I., Elbanna, K., 2019. Phenotypic and genotypic characterization of exopolysaccharide producing bacteria isolated from fermented fruits, vegetables and dairy products. J. Pure Appl. Microbiol. 13, 1349.

Abdel-Wahab, B.A., Abd El-Kareem, H.F., Alzamami, A., Fahmy, C.A., Elesawy, B.H., Mostafa Mahmoud, M., Ghareeb, A., El Askary, A., Abo Nahas, H.H., Attallah, N.G.M., Altwaijry, N., Saied, E.M., 2022. Novel exopolysaccharide from marine *Bacillus subtilis* with broad potential biological activities: Insights into antioxidant, anti-inflammatory, cytotoxicity, and anti-Alzheimer activity. Metabolites. 12(8), 715.

Al-Qaysi, S.A.S., Al-Haideri, H., Al-Shimmary, S.M., Abdulhameed, J.M., Alajrawy, O.I., Al-Halbosiy, M.M., Moussa, T.A.A., Farahat, M.G., 2021. Bioactive levan-type exopolysaccharide produced by *Pantoea agglomerans* ZMR7: Characterization and optimization for enhanced production. J. Microbiol. Biotechnol. 31(5), 696.

Angelin, J., Kavitha, M., 2020. Exopolysaccharides from probiotic bacteria and their health potential. Int. J. Biol. Macromol. 162, 853.

Angelova, G., Brazkova, M., Mihaylova, D., Slavov, A., Petkova, N., Blazheva, D., Deseva, I., Gotova, I., Dimitrov, Z., Krastanov, A., 2022. Bioactivity of biomass and crude exopolysaccharides obtained by controlled submerged cultivation of medicinal mushroom *Trametes versicolor*. J. Fungi (Basel). 8(7).

Aquinas, N., Bhat, M.R., Selvaraj, S., 2022. A review presenting production, characterization, and applications of biopolymer curdlan in food and pharmaceutical sectors. Polym. Bull. 79(9), 6905.

Asgher, M., Urooj, Y., Qamar, S.A., Khalid, N., 2020. Improved exopolysaccharide production from *Bacillus licheniformis* MS3: Optimization and structural/functional characterization. Int. J. Biol. Macromol. 151, 984.

Banerjee, A., Mohammed Breig, S.J., Gómez, A., Sánchez-Arévalo, I., González-Faune, P., Sarkar, S., Bandopadhyay, R., Vuree, S., Cornejo, J., Tapia, J., Bravo, G., Cabrera-Barjas, G., 2022. Optimization and characterization of a novel exopolysaccharide from *Bacillus haynesii* CamB6 for food applications. Biomol. 12(6), 834.

Barcelos, M.C.S., Vespermann, K.A.C., Pelissari, F.M., Molina, G., 2020. Current status of biotechnological production and applications of microbial exopolysaccharides. Crit. Rev. Food Sci. Nutr. 60(9), 1475.

Camelia, I., Savin, A., Apetrei, C., Martin, P., Popa, M., 2015. Gellan. Food applications. Cellul. Chem. Technol. 50.

Chaturvedi, S., Kulshrestha, S., Bhardwaj, K., Jangir, R., 2021. A review on properties and applications of xanthan gum. In: A. Vaishnav, D.K. Choudhary (Eds.), *Microbial Polymers: Applications and Ecological Perspectives*. Springer, p. 87.

Ciszek-Lenda, M., 2011. Review paper Biological functions of exopolysaccharides from probiotic bacteria. Centr. Eur. J. Immunol. 36(1), 51.

Cui, J., Chisti, Y., 2003. Polysaccharopeptides of *Coriolus versicolor*: Physiological activity, uses, and production. Biotechnol. Adv. 21(2), 109.

Das, G., Shin, H.-S., Leyva-Gómez, G., Prado-Audelo, M.L.D., Cortes, H., Singh, Y.D., Panda, M.K., Mishra, A.P., Nigam, M., Saklani, S., Chaturi, P.K., Martorell, M., Cruz-Martins, N., Sharma, V., Garg, N., Sharma, R., Patra, J.K., 2021. *Cordyceps* spp.: A review on its immune-stimulatory and other biological potentials. Front. Pharma. 11.

de Souza, A.M., Sutherland, I.W., 1994. Exopolysaccharide and storage polymer production in Enterobacter aerogenes type 8 strains. J. Appl. Bacteriol. 76(5), 463.

Díaz-Montes, E., 2021. Dextran: Sources, structures, and properties. Polysaccharides. 2, 554.

Dou, H., Chang, Y., Zhang, L., 2019. *Coriolus versicolor* polysaccharopeptide as an immunotherapeutic in China. Prog. Mol. Biol. Transl. Sci. 163, 361.

Dung, V.K., Thuy, H.T., Nhan, N.T., 2017. Levan production by *Bacillus subtilis* natto D strain. J. Forest. Sci. Technol. 11.

Gan, L., Jiang, G., Li, X., Zhang, S., Tian, Y., Peng, B., 2021. Structural elucidation and physicochemical characteristics of a novel high-molecular-weight fructan from halotolerant Bacillus sp. SCU-E108. Food Chem. 365, 130496.

Gangalla, R., Sampath, G., Beduru, S., Sarika, K., Kaveriyappan Govindarajan, R., Ameen, F., Alwakeel, S., Thampu, R.K., 2021. Optimization and characterization of exopoly-saccharide produced by *Bacillus aerophilus* rk1 and its in vitro antioxidant activities. J. King Saud Univ. Sci. 33(5), 101470.

Gao, Z., Wu, C., Wu, J., Zhu, L., Gao, M., Wang, Z., Li, Z., Zhan, X., 2022. Antioxidant and anti-inflammatory properties of an aminoglycan-rich exopolysaccharide from the submerged fermentation of *Bacillus thuringiensis*. Int. J. Biol. Macromol. 220, 1010.

Habtemariam, S., 2020. *Trametes versicolor* (Synn. *Coriolus versicolor*) polysaccharides in cancer therapy: Targets and efficacy. Biomedicines. 8(5).

He, Z., Lin, J., He, Y., Liu, S., 2022. Polysaccharide-peptide from *Trametes versicolor*: The potential medicine for colorectal cancer treatment. Biomedicines. 10(11), 2841.

Hobbs, C., 2004. Medicinal value of turkey tail fungus *Trametes versicolor* (L.:Fr.) Pilat (Aphyllophoromycetideae). A literature review. Int. J. Med. Mushrooms. 6, 195.

Huang, L., Jin, Y., Zhou, D., Liu, L., Huang, S., Zhao, Y., Chen, Y., 2022. A review of the role of extracellular polymeric substances (EPS) in wastewater treatment systems. Int. J. Environ. Res. Public Health. 19, 12191.

Huang-Lin, E., Sánchez-León, E., Amils, R., Abrusci, C., 2022. Potential applications of an exopolysaccharide produced by *Bacillus xiamenensis* RT6 isolated from an acidic environment. Polymers. 14(18), 3918.

Joulak, I., Finore, I., Nicolaus, B., Leone, L., Moriello, A.S., Attia, H., Poli, A., Azabou, S., 2019. Evaluation of the production of exopolysaccharides by newly isolated *Halomonas* strains from Tunisian hypersaline environments. Int. J. Biol. Macromol. 138, 658.

Jurášková, D., Ribeiro, S.C., Silva, C.C.G., 2022. Exopolysaccharides produced by lactic acid bacteria: From biosynthesis to health-promoting properties. Foods. 11(2).

Kidd, P.M., 2000. The use of mushroom glucans and proteoglycans in cancer treatment. Altern. Med. Rev. 5(1), 4.

Korcz, E., Varga, L., 2021. Exopolysaccharides from lactic acid bacteria: Techno-functional application in the food industry. Trends Food Sci. Technol. 110, 375.

Kumar, A.S., Mody, K., Jha, B., 2007. Bacterial exopolysaccharides: A perception. J. Basic Microbiol. 47(2), 103.

Le, T.T.H., Bach, T.B.P., Nguye, T.T.T., Tran, M.H., Huynh, T., Nguyen, T.T., Dinh, M.H., 2017a. Optimizing culture medium composition with added olive oil for isolation of extracellular polysaccharides in *Ophiocordyceps sinensis* fungus. VNU J. Sci. Natl. Sci. Technol. 1S, 174.

Le, T.T.H., Nguyen, T.T.T., Bach, T.B.P., Tran, M.T., Huynh, T., Dinh, M.H., Nguyen, T.T., 2017b. Effect of plant oils on mycelium biomass production, biosynthesis and antioxi-dants of exopolysaccharide by *Ophiocordyceps sinensis*. J. Sci. Technol. Food. 13(1), 3.

Li, D.H., Lizhen, G., Yaxin, T., Zhiliang Fan, B., Wang, F., Li, S., 2022a. Recent advances in microbial synthesis of poly-gamma-glutamic acid: A review. Foods. 11(5), 739.

Li, F., Hu, X., Qin, L., Li, H., Yang, Y., Zhang, X., Lu, J., Li, Y., Bao, M., 2022b. Characterization and protective effect against ultraviolet radiation of a novel exopolysaccharide from *Bacillus marcorestinctum* QDR3–1. Int. J. Biol. Macromol. 221, 1373.

Liu, H., Yan, Q., Wang, Y., Li, Y., Jiang, Z., 2022. Efficient production of poly-γ-glutamic acid by *Bacillus velezensis* via solid-state fermentation and its application. Food Biosci. 46, 101575.

Ludwicka, K., Kaczmarek, M., Białkowska, A., 2020. Bacterial nanocellulose—A biobased polymer for active and intelligent food packaging applications: Recent advances and developments. Polymers. 12.

Lynch, K.M., Zannini, E., Coffey, A., Arendt, E.K., 2018. Lactic acid bacteria exopolysac-charides in foods and beverages: Isolation, properties, characterization, and health ben-efits. Annu. Rev. Food Sci. Technol. 9, 155.

Mata, J.A., Béjar, V., Llamas, I., Arias, S., Bressollier, P., Tallon, R., Urdaci, M.C., Quesada, E., 2006. Exopolysaccharides produced by the recently described halophilic bacteria *Halomonas ventosae* and *Halomonas anticariensis*. Res. Microbiol. 157(9), 827.

Mohamed, S.S., Ibrahim, A.Y., Asker, M.S., Mahmoud, M.G., El-Newary, S.A., 2021. Production, structural and biochemical characterization relevant to antitumor property of acidic exopolysaccharide produced from *Bacillus* sp. NRC5. Arch. Microbiol. 203(7), 4337.

Mohd Nadzir, M., Nurhayati, R.W., Idris, F.N., Nguyen, M.H., 2021. Biomedical applications of bacterial exopolysaccharides: A review. Polymers (Basel). 13(4).

Mollakhalili, N., Mohammadifar, M.A., 2015. Microbial exopolysaccharides: A review of their function and application in food sciences. J. Food Qual. Hazards Cont. 2, 112.

Muthusamy, S., Anandharaj, S.J., Kumar, P.S., Meganathan, Y., Vo, D.-V.N., Vaidyanathan, V.K., Muthusamy, S., 2022. Microbial pullulan for food, biomedicine, cosmetic, and water treatment: A review. Environ. Chem. Lett. 20(5), 3199.

Nagarajan, A., Anandakumar, S., Zackaria, A., 2015. Mini review on Alginate: Scope and future perspectives. J. Algal. Biom. Util. 2016, 45.

Natarajan, M., Suresh Babu, S.P., Balasubramanian, M., Ramachandran, R., Jesteena, J., 2022. Bioactive exopolysaccharide from endophytic *Bacillus thuringiensis* SMJR inhibits food borne pathogens and enhances the shelf life of foods. Bioact. Carbohydr. Diet. Fibre. 27, 100297.

Ng, T.B., 1998. A review of research on the protein-bound polysaccharide (polysaccharopeptide, PSP) from the mushroom *Coriolus versicolor* (Basidiomycetes: Polyporaceae). Gen. Pharmacol. 30(1), 1.

Nguyen, T.A., Nguyen, T.L.A., To, K.A., Pham, T.A., 2022. Enhancement of polysaccharide Krestin production in Trametes versicolor by using microparticle enhanced cultivation and led light colors. Proc. Vietnam Natl. Conf. Biotechnol. 2022, 1268.

Öner, E.T., Hernández, L., Combie, J., 2016. Review of Levan polysaccharide: From a century of past experiences to future prospects. Biotechnol. Adv. 34(5), 827.

Pacheco-Leyva, I., Guevara Pezoa, F., Díaz-Barrera, A., 2016. Alginate biosynthesis in *Azotobacter vinelandii*: Overview of molecular mechanisms in connection with the oxygen availability. Int. J. Pol. Sci. 2016, 2062360.

Perumal, A.B., Nambiar, R.B., Moses, J.A., Anandharamakrishnan, C., 2022. Nanocellulose: Recent trends and applications in the food industry. Food Hydrocol. 127, 107484.

Prete, R., Alam, M.K., Perpetuini, G., Perla, C., Pittia, P., Corsetti, A., 2021. Lactic acid bacteria exopolysaccharides producers: A sustainable tool for functional foods. Foods. 10(7).

Quach, N.T., Vu, T.H.N., Nguyen, T.T.A., Ha, H., Ho, P.H., Chu Ky, S., Nguyen, L.H., Van Nguyen, H., Thanh, T.T.T., Nguyen, N.A., Chu, H.H., Phi, Q.T., 2022. Structural and genetic insights into a poly-γ-glutamic acid with in vitro antioxidant activity of *Bacillus velezensis* VCN56. W. J. Microbiol. Biotechnol. 38(10), 173.

Rana, S., Upadhyay, L.S.B., 2020. Microbial exopolysaccharides: Synthesis pathways, types and their commercial applications. Int. J. Biol. Macromol. 157, 577.

Santos Arteiro, J.M., Rosário Martins, M., Salvador, C., Fátima Candeias, M., Karmali, A., Teresa Caldeira, A., 2012. Protein—polysaccharides of *Trametes versicolor*: Production and biological activities. Med. Chem. Re. 21(6), 937.

Schmid, J., Meyer, V., Sieber, V., 2011. Scleroglucan: Biosynthesis, production and application of a versatile hydrocolloid. Appl. Microbiol. Biotechnol. 91(4), 937.

Schmid, J., Sieber, V., Rehm, B., 2015. Bacterial exopolysaccharides: Biosynthesis pathways and engineering strategies. Front. Microbiol. 6, 496.

Selim, S., Almuhayawi, M.S., Alharbi, M.T., Nagshabandi, M.K., Alanazi, A., Warrad, M., Hagagy, N., Ghareeb, A., Ali, A.S., 2022. In vitro assessment of antistaphylococci, antitumor, immunological and structural characterization of acidic bioactive exopolysaccharides from marine *Bacillus cereus* isolated from Saudi Arabia. Metabolites. 12(2), 132.

Shin, H.-D., Liu, L., Kim, M.-K., Park, Y.-I., Chen, R., 2016. Metabolic engineering of *Agrobacterium* sp. ATCC31749 for curdlan production from cellobiose. J. Ind. Microbiol. Biotechnol. 43(9), 1323.

Smith, J.E., Rowan, N., Sullivan, R., 2002. *Medicinal Mushrooms: Their Therapeutic Properties and Current Medical Usage with Special Emphasis on Cancer Treatments.* University of Strathclyde, Cancer Research UK, pp. iv and 253.

Standish, L.J., Wenner, C.A., Sweet, E.S., Bridge, C., Nelson, A., Martzen, M., Novack, J., Torkelson, C., 2008. *Trametes versicolor* mushroom immune therapy in breast cancer. J. Soc. Integr. Oncol. 6(3), 122.

Suryawanshi, N., Naik, S., Jujjawarapu, S.E., 2022. Exopolysaccharides and their applications in food processing industries. Food Sci. Appl. Biotechnol. 5(1).

Thanh, N.S.L., Kimura, K., Tuyen, D.T., Anh, L.T.N., 2018. Isolation, characterization of *Bacillus* sp. producing heavy metal absorption γ-PGA. J. Viet. Environ. 9, 49.

Tran, T.H., Le, M.H., To, K.A., Pham, T.A., 2019. Optimization of polysaccharopetides production in *Trametes versicolor* submerged fermentation Industr. Trad. Magaz. 42.

Tran, T.H., Phung, T.T., To, K.A., Pham, T.A., 2021. Recovery of polysaccharide -krestin (PSK) from *Trametes versicolor* BRG04 biomass of submerged fermentation. Industr. Trad. Magaz. 44, 44.

Trček, J., Dogsa, I., Accetto, T., Stopar, D., 2021. Acetan and acetan-like polysaccharides: Genetics, biosynthesis, structure, and viscoelasticity. Polymers. 13.

Tuan, L.A., Oanh, L.T.H., Quang, P.V., Luu, N.D., Quan, V.H., Anh, L.N., 2020. Isolation and selection of *Bacillus* strains with capability in production of polyglutamic acid. TNU J. Sci. Technol. 225(8), 443.

Van, N.H., Anh, D.T.N., Trang, N.T.H., Trung, N.D., Hoang, N.V., Duong, N.M., Ha, D.T.C., 2014. Extracellular polymeric substances production in sludge by bacteria isolated from Hanoi brewery. J. Biol. 36, 351.

Vu, T.H.N., Quach, N.T., Nguyen, N.A., Nguyen, H.T., Ngo, C.C., Nguyen, T.D., Ho, P.H., Hoang, H., Phi, H.H.C.A.Q.T., 2021. Genome mining associated with analysis of structure, antioxidant activity reveals the potential production of Levan-rich exopolysaccharides by food-rerived *Bacillus velezensis* VTX20. Appl. Sci. 11(15), 7055.

Wagh, V.S., Said, M.S., Bennale, J.S., Dastager, S.G., 2022. Isolation and structural characterization of exopolysaccharide from marine *Bacillus* sp. and its optimization by Microbioreactor. Carbohydr. Polymers. 285, 119241.

Wei, M., Geng, L., Wang, Q., Yue, Y., Wang, J., Wu, N., Wang, X., Sun, C., Zhang, Q., 2021. Purification, characterization and immunostimulatory activity of a novel exopolysaccharide from *Bacillus* sp. H5. Int. J. Biol. Macromol. 189, 649.

Yan, J.K., Wang, W.Q., Wu, J.Y., 2014. Recent advances in *Cordyceps sinensis* polysaccharides: Mycelial fermentation, isolation, structure, and bioactivities: A review. J. Funct. Foods. 6, 33.

Yildiz, H., Karatas, N., 2018. Microbial exopolysaccharides: Resources and bioactive properties. Pro. Biochem. 72, 41.

Zhang, L., Yi, H., 2022. Potential antitumor and anti-inflammatory activities of an extracellular polymeric substance (EPS) from *Bacillus subtilis* isolated from a housefly. Sci. Rep. 12(1), 1383.

Zhao, X., Chen, G., Wang, F., Zhao, H., Wei, Y., Liu, L., Zhang, H., 2022. Extraction, characterization, antioxidant activity and rheological behavior of a polysaccharide produced by the extremely salt tolerant *Bacillus subtilis* LR-1. LWT. 162, 113413.

10 Application of exopolysaccharides in cosmetics

Sourav Saha, Meheria Hazari,
and Surabhi Chaudhuri

10.1 INTRODUCTION

The cosmetics industry can be described as a multibillion dollar one (Tejal et al., 2013). People tend to spend a huge amount of money daily on cosmetics to achieve healthier and more radiant skin and hair. The use of cosmetics is an important part of the everyday life of people for improving their skin quality and hair (Milstein et al., 2001). The use and prevalence of cosmetics among men and women are consistently increasing (Tejal et al., 2013). Cosmetics can be defined as the products which are applied to the body for the beautification and improved quality of the skin or the hair. The human skin harbors a beneficial microbial ecosystem in the body, which plays an important role in attaining homeostasis in the body and affects human health. Therefore, the assessment and correct usage of cosmetics are necessary due to the ability of these products to interfere with the microbial microbiota, which can either prove to be helpful or have adverse impacts on the health of the host respectively (Carvalho et al., 2022). Cosmetics are comprised of different ingredients with specific functions that work together to give distinct results in the final products (Freitas et al., 2015). The different ingredients which are present in cosmetics can be broadly classified as emulsifiers, thickeners, preservatives, colors, fragrances, and water. Among all of the ingredients present in cosmetics, most of the contents tend to have harmful effects on the skin and hair, which can be either temporary or affect the epidermis permanently. Therefore, the use of biocompatible and eco-friendly cosmetics has become a recent trend and need among people across the world. Different cosmetics have different benefits for the skin and hair such as better moisturization, improved shine, and reduced effects of harmful pollutants and toxic chemicals.

10.2 THE HUMAN SKIN

The human skin is made up of three basic layers: the epidermis, the dermis, and the hypodermis (Elias, 2001). The epidermis is the outer layer of the skin and contains several different layers of skin which are regenerated and replaced regularly. The cells of the epidermis are regenerated from the basal layer to the stratum corneum.

DOI: 10.1201/9781003342687-10

In this layer, stratum corneum, the protein rich cells, or the corneocytes are found embedded in the lipid rich layer. This lipid rich layer contains many beneficial components such as cholesterol, ceramides, and fatty acids. The different corneocytes are linked with each other with the help of structures called as corneodesmosomes. This helps in improving the structural integrity of the outer basal layer. The basal layer or the stratum corneum is an impermeable layer, which helps in providing extra protection to the inner layers of the skin from harmful environmental factors (Freitas et al., 2015).

The second layer or the dermis supports the epidermis and the skin appendages and is made up of a connective tissue that is strongly innervated and abundantly vascularized (hair follicles, sebaceous glands, and sweat glands). It contains the blood vessels, glands, nerves, and hair follicles found in the skin (Navarro et al., 2017). The third layer of the skin, or the hypodermis, which is also called as the subcutaneous layer of the skin, mostly comprises the nerves, adipose tissue, and blood vessels as the reservoir of skin progenitor cells (Park, 2022). This layer of adipose tissue helps in protecting the skin from mechanical and thermal impacts (Ammala, 2013). Moreover this layer connects dermis to the organs and associated with major paracrine and endocrine signaling pathways (Park, 2022).

The outer layer contains around 13% moisture, whereas the inner layer of the skin is comparatively more hydrophilic and contains a moisture content of around 50% (found in the epidermis) to 70% (in the dermis). The level of moisture found in the basal layer determines the skin's permeability and the skin barrier functions as it is influenced by other external environmental factors and internal factors like age (Ammala, 2013). The internal factor like age is an important factor affecting skin elasticity as the level of hyaluronic acid (HA) occurring naturally in the skin reduces with age, which affects the level of water retention in the skin. A level of moisture found in skin lower than 20% to 35% makes the skin appear dry and scaly (Freitas et al., 2015). It can also lead to the development of other skin defects such as hyperpigmentation. Therefore, the use of cosmetics like moisturizers and sunscreens on a daily basis is necessary, which helps to retain the moisture content of the skin and also improves the quality of the layers of the skin. The use of such cosmetics can help in the formation of an occlusive layer to improve the water retention of the layers of the skin (Freitas et al., 2015). This layer can also help in improved transport of the active ingredients of cosmetics through different pathways like an intracellular pathway or transcellular pathway.

Only substances having molecular weight lower than 500 Da are capable of being transported across the different layers of skin (Ammala, 2013). Examples of such molecules are moderately lipophilic molecules, which are soluble both in the lipophilic basal layer (the stratum corneum) and the hydrophilic inner epidermal layer of the skin (Freitas et al., 2015). The level of permeation found in skin can be further enhanced with the use of strategies such as disruption of the intracellular lipid matrix found in the basal layer. This disruption can be carried out with the help of chemical enhancers such as biodegradable polymers or with the use of physical techniques such as a microneedle or electroporation (Ammala, 2013). Prolonged exposure of the skin to harmful environmental factors such as UV-radiation can increase the peroxidation process of the skin. This is the major reason for the disruption of the protective function of the layer and also leads to the formation of skin defects such as skin cancer, premature aging of the skin, and mutations (Trommer and Neubert, 2005).

10.3 BENEFITS OF USING COSMETICS

10.3.1 ANTIOXIDANT EFFECT

The skin and hair are exposed to a huge number of external harmful factors such as ultraviolet radiation, pollution, toxic chemicals, and pathogenic microorganisms daily (Sies and Jones, 2007). All of such degradative factors lead to the formation of free radicals and reactive species. The internal factors of the body additionally contribute to the development of such free radicals, through the different metabolic pathways of the body, which leads to the formation of reactive oxygen and nitrogen species (Kehrer, 2000). The body has the mechanism and ability to protect the skin through the maintenance of the correct balance between the formation and the neutralization of such reactive species. However, the extended period of exposure to degradative factors such as toxic chemicals and ultraviolet radiation can hamper the protective mechanism. This can lead to an increase in oxidative and nitrosative stress in the skin, which alters cell homeostasis and speeds up the degenerative mechanism of the skin (Rahman et al., 2012). The free radicals and reactive oxygen and nitrogen species damage the stratum corneum and the other living layers of the epidermis (Xu et al., 2018). The skin has a natural ability to fight against free radicals by producing different enzymes and antioxidants having low molecular weight. Better understanding of this mechanism in the body will help the cosmetic industries in developing products containing antioxidants (Heath et al., 2022). There are a number of cosmetics available in the market, which contains antioxidants, that help in improving the balance between the development and the degradation of ROS, therefore stabilizing skin and hair quality.

10.3.2 ANTIAGING EFFECT

Skin aging can be defined as a complex process, which is mostly triggered by both internal and external factors of the body (Freitas et al., 2015 Resende et al., 2022). The internal factors are comprised of the body's metabolism, lifestyle changes, and type of food intake. On the other hand, the extrinsic factors are comprised of pollutants, toxic chemicals in the environment, and extended exposure to the degradative ultraviolet rays of the sun. Skin aging is an essential part of the natural "aging mosaic" in every individual (Ganceviciene et al., 2012). The process of skin aging is greatly catalyzed by the aforementioned factors. This can lead to gradual changes in the different layers of the dermis and epidermis of the skin. Such changes can lead to changes in skin appearance and reduced skin elasticity (Escoffier et al., 1989). Aged skin can make the impacts of external factors more prevalent and lead to greater harmful effects on the skin such as hyperpigmentation and dullness. People have been trying to slow down the process of skin aging for years. The use of cosmetics plays an important role in the antiaging effect due to the presence of various useful components such as antioxidants and cell regulators. The thickness, structure, and strength of collagen fibers within the skin are weakened with age. Because of this, the cosmetics industries have not only emphasized on the ingredients in formulation that support collagen formation or preservation but also incorporated collagen into their products as an ingredient (Heath et al., 2022). Therefore, through regular skin care routine and usage of cosmetics such as moisturizers and sunscreen, cosmetics can prove to help protect the skin against dehydration, allergens, radiation, and harmful microorganisms, which helps in restoring skin elasticity, porosity, and enhance the skin regenerative process (Tabata et al., 2000).

10.3.3 PHOTO PROTECTION

Photo protection can be understood as the reduction of the effects of ultraviolet radiation such as skin aging, skin damage, and skin cancer, caused by photoaging and photo carcinogenesis of the radiation (Matsumura and Ananthaswamy, 2002). Photoaging can make the skin look wrinkly and dull, whereas photo carcinogenesis leads to the damage of the skin's cells and DNA (Young et al., 2017). Therefore, people have been trying to make use of different ways for protecting the skin against the adverse effects of the sun, by using sunglasses, sunscreen lotions, and clothing. Ultraviolet radiation can be classified as ultraviolet A (UVA) and ultraviolet B (UVB). The UVB has more adverse impacts on the skin and can be one of the leading causes of melanoma and non-melanoma skin cancer (Ulrich et al., 2009). Therefore, cosmetics such as sunscreen lotions were made mostly for protecting against UVB. UV ray damage can also accelerate the process of skin aging due to release of free radicals, which can alter the oxidative balance within the skin. The release of the free radical can interact with the important constituents of the skin cells like lipids carbohydrate and nucleic acid, which in turn will lead to the direct disruption of the skin structure and make it appear less radiant and darkened (Heath et al., 2022).

The effect of the sun can be reduced by two mechanisms, primary and secondary. The primary protection against the radiation of the sun includes the use of cosmetics such as sunscreens, which act as primary barriers and help to scatter and reflect the sun's rays. Sunscreen can also be used as a chemical barrier against the sun, by absorbing the incident radiations. Cosmetics can also be used as a secondary barrier against ultraviolet radiation, through the effect of antioxidants and osmolytes, which helps to reduce skin damage, by balancing the photochemical cascade of the skin.

10.3.4 REDUCED HYPERPIGMENTATION

The pigmentary disorder is one of the most common dermatological disorders found in blacks, Hispanics, and Asian populations (Halder and Nootheti, 2003). The effect of such disorder is not only seen just on the appearance of the skin but also has adverse psychosocial impairment in people affected by such disorder. Pigmentary disorders are non-treatable, and therefore, the use of effective skin lightening and regular skin and hair care are the only significant ways of preventing it. Various processes are used for effective depigmentation. Cosmetics ingredients such as tyrosinase inhibitors (hydroquinone, arbutin, azelaic acid, etc.) targets the rate limiting step of melanin production; kojic acid has the ability to chelate copper, an important cofactor tyrosinase thereby inhibiting the synthesis of melanin (Resende et al., 2022). Vitamin C is widely used as an antioxidant in cosmetics, which acts as reducing agent and inhibits different oxidative steps in the process of melanin formation, reducing melanogenesis and accelerating epidermal turnover (Resende et al., 2022; Sarkar et al., 2013).

10.4   INGREDIENTS AND FORMULATIONS OF COSMETICS

Cosmetics formulations are a combination of various substances having different functions in the final product (Freitas et al., 2015). Based on the type of the ingredient and its function, the ingredient can be categorized into three major groups: (1) base substances—includes the natural skin component (such as oil-based substances, moisturizing substances); (2) active agents—imparts the main activity/functionality of the cosmetics (improving the skin condition, antimicrobial activity, etc.); (3) additives—added to increase the shelf life of the product and to be more appealing to the consumer (antioxidants, preservatives, perfumes, etc.) (Freitas et al., 2015).

The use of EPS can help in providing these functions to cosmetics products due to the antioxidative and antimicrobial properties of EPS, which in turn provide extra shelf life to the products, without any significant adverse effects on the human skin and hair. Moreover, some exopolysaccharides are known to be beneficial for the skin's natural texture and health (Freitas et al., 2015; Fiume et al., 2016).

The various substances found in cosmetics products are considered as additives. Except for the ones which are UV filters, the other additives are used for improving the consistency of the products. However, such synthetic chemical additives can cause allergies and irritation to the skin. The active ingredients of the cosmetics

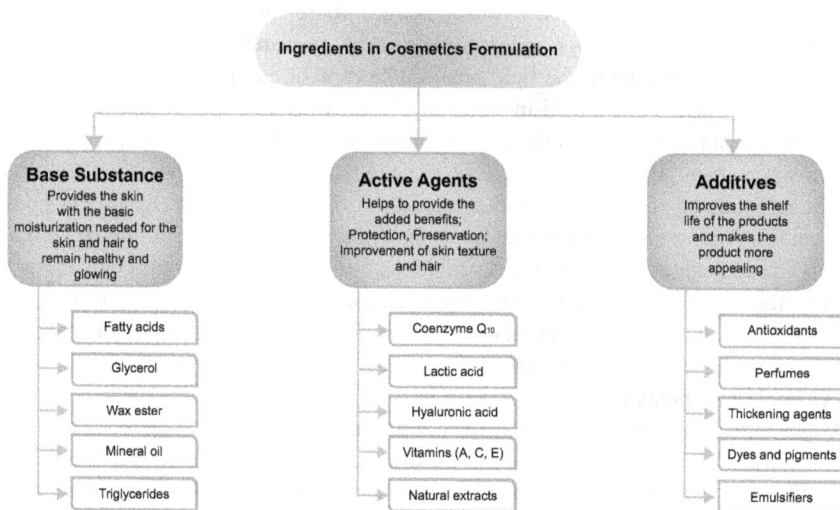

FIGURE 10.1   Ingredients in cosmetics formulation (Freitas et al., 2015)

are added with the purpose of application on the skin for additional benefits such as smoothening of the skin, regeneration of new cells, improvement of skin color, protection from the sun's harmful radiations, and slowing down of the aging process (Freitas et al., 2015). Studies have shown that microbial exopolysaccharides can be used as an alternative to chemical additives due to the benefits described previously.

10.5 COSMETICS VEHICLES

Cosmetics vehicles can be described as the substances used for the optimization of the product's formulation. These cosmetic vehicles act like a matrix into which the other additives are embedded for improved action of the active ingredients. Therefore, the vehicles work as a carrier, which transports the product and its ingredients to the site of action from the site of supplication. These vehicles do not have any significant pharmacological effects but help to serve other purposes such as cleansing, protection, or hydration (Freitas et al., 2015). The significant vehicles, where the different polysaccharides play different roles can include hydrogels, encapsulating agents, emulsions, and suspensions (Buchmann, 2001).

10.5.1 EMULSIONS

Emulsions are the most common type of vehicles used for the transport of active ingredients into the skin. Emulsions are usually oil based, lipids are used most commonly due to their appealing texture on the skin and hair and the ease of applicability. Emulsions are preferred to be flowy such as lotions or in a semi solid form such as creams (Buchmann, 2001). Emulsions are made up of two immiscible liquids, which are most commonly the lipids or lipophilic agents, and water, which is a hydrophilic substance. The lipid is dispersed within water ranging from 0.1 to 100 micrometers (Freitas et al., 2015).

Emulsions can be classified as oil-in-water, where the lipid molecules are dispersed in water or any other aqueous phase, and as water-in-oil, where the water molecules are dispersed in the lipid phase. The type of emulsions used in skin care are water-in-oil-in-water or oil-in-water-in-oil. In water-in-oil-in-water, the internal and the external phases of water are partitioned by the lipid molecules, whereas, in oil-in-water-in-oil, the two oil or lipid phases are partitioned with the use of the water or the aqueous layer (Epstein and Simion, 2001). Thermodynamically this bi-phase system is unstable as the two phases are immiscible. Amphiphilic compounds (emulsifiers) can be added to make it metastable, as the amphiphilic compounds helps to reduce the interfacial tension present between the two immiscible phases. The film formed helps to stabilize the emulsion formulation by inhibiting the coalescence of the droplet phase and reducing phase separation (Epstein and Simion, 2001).

Other components which are present in the emulsions are active substances like UV protective sunscreens and vitamins, emollients that helps to increase the sensory characters of the emulsion, such as silicon oils and isopropyl myristate, antimicrobial reagents, fragrance and coloring agents, moisturizers such as glycerol and viscosity improving agents (Buchmann, 2001). Nano emulsion is also one of the emerging forms of emulsion since it could be designed according to the need, such as oil-in-water or water-in-oil or bicontinuous. The size of the nanoparticle lies between 50 to 200 nanometers, which assures low interfacial tension, a boost in the capacity for solubilization, and the ability to overcome common concerns with macromolecular system, including sedimentation, flocculation, creaming, and coalescence. Nano emulsion allows high skin penetration and decreased water loss across the epidermis. They are excellent cosmetic vehicles because of their pleasing appearance and rich blending texture (Dubey et al., 2022).

Polysaccharides which are obtained from microbial sources also serve as good cosmetic emulsions as it helps to carry the active agents to the site of action, along with providing additional antimicrobial impacts (Nwodo et al., 2012). These exopolysaccharides can help in producing a hydrogel network, which helps in increasing the viscosity between the two phases of the emulsion and slow down the motion of the droplets in the solution phase (Freitas et al., 2015). This can be described as a secondary stabilization stage, which helps to inhibit the occurrence of creaming or the upward movement of the oil or the lipid droplets in the aqueous phase, as a result of the low specific gravity of the oil droplets as compared to the water molecules. The microbial EPS can also help in improving the texture of the cosmetic product, which allows the customers to have better ease of application of the product (Gilbert et al., 2013).

Carbopol® polymers, which are homopolymers and copolymers of acrylic acid, and Veegum, which is an aluminum silicate material, are the most commonly used thickeners. Natural polymers, which are used as thickeners, are alginates, cellulose polymers, etc. (Epstein and Simion, 2001).

10.5.2 HYDROGELS

Gels are the mixtures of two or more components, where the different components produce different continuous phases in the system. The gel's minor components comprise of the polymeric phase, which helps to form a three-dimensional network around the liquid phase. The three-dimensional network inhibits the flowing of the fluid, therefore, preventing the collapsing of the structure (Freitas et al., 2015). The polymeric network along with a liquid phase helps to distinguish gels from solids and fluids. It also helps to provide the gel with its unique structure and texture, along with the viscoelastic properties (Jung and Shinkai, 2005).

Hydrogels are made up of a three-dimensional structure of hydrophilic polymers such as a polysaccharide, which are then surrounded by the molecules of water. The physical and mechanical characteristics of this structure may range from being soft to brittle to elastic. This distinction of mechanical properties is based on the type of the polymer and the interactions between the molecules holding the structure together, such as the covalent bonds or ionic interactions.

Bacterial polysaccharides are useful for making the hydrogels and one of the most commonly used exopolysaccharides is hyaluronic acid (HA). HA is used as a hydrogel for its property to absorb water molecules, which can help to transport the active ingredients of cosmetics. HA is also an excellent agent for deep penetration into the skin and provides better hydration and elasticity to the skin and hair (Berkó et al., 2013). Gellan gum is also widely used in the form of hydrogels due to the ability of flavor release and low viscosity during the production process.

10.5.3 ENCAPSULATING STRUCTURES

Most of the biological compounds are found to be unstable due to variations in pH, temperature, and light. Therefore, these biological compounds have to be protected in order to prevent unwanted degradation of the compounds. In cosmetics, an encapsulating structure is needed in the formulation to improve the storage and applicability of the final product (Freitas et al., 2015). Furthermore, topical and transdermal regulated transport of active cosmetic ingredients is helpful, which requires a safe and nontoxic technique for reaching the target locations without causing irritation (Ammala, 2013). This is possible through the use of an encapsulating structure such as a micro or nanosphere, micro or nanosponges, and liposomes (Patravale and Mandawgade, 2008).

The encapsulation serves many purposes other than providing protection. The other functions served by the encapsulation are the ease of handling of the biological substances, protection of the fragrance of the compounds, and improvement of the dilution process of the substances which can be toxic if used in high quantities (Freitas et al., 2015). Many different active cosmetic ingredients have been encapsulated with a variety of materials. Nylon microparticles containing vitamins, sunscreens, moisturizer, perfumes, retinyl palmitate, D-panthenol, ascorbic acid, tocopheryl acetate, dihydroxyacetone, vitamin E, and dimethicone, are just a few examples (Patravale and Mandawgade, 2008). The use of biodegradable compounds for encapsulating the cosmetic ingredients is in wide usage nowadays. This is because, the use of biodegradable encapsulating structures can be beneficial as they are nonreactive in nature when in contact with the epidermis of the human skin. These biodegradable molecules can also be mobilized easily for the removal of the encapsulating structure from the human body. A few examples of biological encapsulating structures include proteins, polyalkylcyanoacrylates, and poly α-esters (Ammala, 2013). Agar is one of the most commonly commercially manufactured microspheres which can hold vitamins and emollients for giving a better appearance to cosmetic and personal care products (Freitas et al., 2015).

HA is a widely tested bacterial exopolysaccharide, for the commercial usage in nanoemulsion and as a transdermal transporter of the cosmetic's active ingredients. These nanoemulsions have been proven to be good carriers for lipophilic compounds of cosmetics such as vitamin E (Kong et al., 2011b; Kong et al., 2011a). HA is also useful for stabilizing gold particles due to the biocompatibility of HA. These combination of HA with gold nanoparticles serves a great purpose to provide extra glow to skin when used in the personal care and cosmetic products (Hien et al., 2012).

10.5.4 SUSPENSIONS

Suspensions can be understood as the formulation which contains the different particles with functional excipients that are spread across a semisolid or liquid medium. Suspensions are more common in cosmetic products such as sunscreens and nail products which contain distinct appearance and essence. The widely appearing problem in suspensions is the formation of sediments of the particles which have a greater density than the liquid solvent (Freitas et al., 2015). This can hamper the formulation of the product during manufacturing, packaging, and storage. Therefore, the occurrence of sedimentation should be hindered in suspensions through strategies such as improving the viscosity of the liquid. Bacterial polysaccharides such as xanthan gum is useful as a suspension due to its unique properties of thickening and shear-thickening. Gellan gum has also proven to be useful for stabilizing the suspension of cosmetic products such as shampoos and conditioners, due to its essential characteristics such as thickening behavior and gel formation property (Prajapati et al., 2013).

10.6 HARMFUL INGREDIENTS USED IN COSMETICS FOR SHORT-TERM BENEFITS

Cosmetics are products that are used for beautification, cleansing, and the improvement of the texture of the skin. Cosmetics can be of different types like shampoo, mascara, conditioner, creams, perfumes, and nail polish. However, the prolonged use of such cosmetics can be harmful for the epidermis, due to the presence of harmful chemicals and toxins. These chemicals can work miraculously for a short period of time; nevertheless, after the temporary effect fades off, the skin and hair can appear to be dull and unhealthy (Khan and Alam, 2019). The range of chemicals which are used in the cosmetics are as follows:

Hydroquinone (HQ) is a commonly used skin lightening agent which is used in the fairness creams and lotions. However, HQ is one of the most harmful chemicals used in cosmetics. HQ has been reported to be one of the causes for ochronosis and mutagenesis in many people (Khan and Alam, 2019). Ochronosis is mostly caused by HQ through the gradual darkening of the area of application on the skin, if it is applied for years. HQ also interferes with the process of melanin synthesis and can also interfere with the degradation of the melanosomes as it is highly cytotoxic and can be mutagenic to human skin (Smit et al., 2009).

Sunscreens can contain many harmful sensitizers, which can cause irritation and allergy in people. The most commonly used irritant in sunscreens are **benzophenones**, which can be a significant cause of photo-allergic dermatitis of skin (Khan and Alam, 2019). Due to its ability to produce liver hypertrophy in rats at low doses, benzophenone is classified as a toxicant by the European Food Safety Authority. Benzophenone is proven to cause endocrine disruption and cause cancer in human. In rats, it may penetrate the skin and build up in the blood, kidney, and liver (Kale et al., 2022). The different types of fragrances used in many cosmetics such as perfumes, shampoos, and moisturizers can also cause allergic reactions to the epidermis. These fragrances can affect the different body parts such as the lungs, nose, airways, and the brain if the fragrances are ingested in high levels. The chemicals such as coumarins and phethleugenol, found in the fragrances are also reported to be carcinogenic (Khan and Alam, 2019). Among the other chemicals which are used in the products of daily usage, the most abundantly used are BHT (butylated hydroxyl toluene) and BHA (butylated hydroxyl anisole), which are used as preservatives in cosmetics such as lipsticks (Khan and Alam, 2019). These two chemicals are a prominent allergen and carcinogen in humans, as per the International Agency for Research on Cancer (Schrader, 2000). Alcohol, in particular ethanol, is being used by some cosmetics because of its disinfectant properties and ability of acting as a vehicle for dissolving various ingredients. However, the prolonged exposure to ethanol causes skin inflammatory responses, skin cytotoxicity, as well as cutaneous erythema (Al-Halaseh et al., 2022).

Diethanolamine (DEA) is used as a pH adjuster or to make the cosmetics creamier, which reacts with the nitrites content in shampoos and cleansers and cause harmful effects in the skin (Khan and Alam, 2019). The use of DEA is reported to be a cause of liver cancer in humans and cause pre-cancerous damages to the skin and also act as an irritant to the eye and skin (Turkoglu et al., 1999).

Parabens is the most commonly used agent for protecting the cosmetics from microbial contamination and for preservation. However, parabens can permeate the epidermis and cause endocrine disruption by interfering with the functioning of the hormones. Parabens have the potential to act similarly like the female hormone estrogen and affect the reproductive cycle within the body (Khan and Alam, 2019). Parabens, if consumed by accident, can bypass the body's metabolic pathways and reach the different organs through the bloodstream. The prolonged use of parabens has also been reported to lead to cancer and neurotoxicity in women (Smith and Alexander, 2005).

Therefore, natural products are preferred to be used in cosmetics, which will not affect the body even after use for a prolonged period. Exopolysaccharides have the capability to provide the similar effects that the previously mentioned chemical contents do in various cosmetics. EPS has many beneficial properties such as antioxidant, antimicrobial, and antiproliferative property, which allows EPS to be used for a wide spectrum of purposes within cosmetics.

10.7 EXOPOLYSACCHARIDE

Exopolysaccharides (EPSs) are the biopolymers released by various organisms by different metabolic pathways. They are the biopolymers with high molecular weight, which are either tightly attached as bacterial capsules or are released out of the microbial cell and loosely attached as a slimy outer layer. They can be characterized according to the different biological functions like anticancer property, anti-inflammatory property, anti-biofilm property, and immunomodulation (Angelin and Kavitha, 2020). These biochemical properties of EPS make it suitable for various industrial applications such as emulsifier and preservatives in the cosmetics and food industry, bio-flocculant and anti-biofilm agents for removing heavy metals, and can also be used as a drug or an important ingredient in cosmetics due to its antioxidant effect. EPS can be produced by different sources like plants and microbes. Examples of EPS produced from plant sources are guar gum and pectin and from algal sources are carrageenan and alginate (Shankar et al., 2014). However, the yield of EPS from a microbial source such as bacteria, microalgae, and cyanobacteria is reported to be more than that of the EPS retrieved from a plant source. Therefore, the extraction of EPS from a microbial source is proven to be economically beneficial, which makes the use of microbes for EPS extraction more prevalent than a plant source.

10.7.1 BENEFITS OF USING EPS IN COSMETICS

There are various synthetic chemicals that are being currently used as cosmetics ingredients for short term benefit, but they cause various harmful and adverse effect on the user and, in some cases, may cause irreversible damages to the skin. Therefore, the use of natural bio-compatible products as cosmetics ingredients has gained popularity since early 1990s. One of the most common forms of natural and biotechnological substance being used in cosmetics are the exopolysaccharide, such as hyaluronic acid, gellan gum, xanthan gum, and other microbial products (Gomes et al., 2020). Exopolysaccharides or the EPS are the extracellular microbial biopolymers which have various biological properties and benefits which can be useful for replacing harmful chemicals such as parabens and benzophenones (Freitas et al., 2015; Khan and Alam, 2019). The properties which can be beneficial for making use of EPS in cosmetics can be classified as follows:

10.7.1.1 Antioxidant effect

The several cellular metabolisms within the human body gives rise to various reactive oxygen species (ROS), hydrogen peroxide (H_2O_2), and superoxide anions of oxygen (O_2^-) which can lead to oxidative stress in the normal system (Sies and Jones, 2007). The harmful effects of the ROS can be disruption of normal cellular signaling, apoptosis, disrupted gene expression, DNA damage, and base damage (Kehrer, 2000). These harmful effects of ROS can lead to the development of diseases in humans such as cancer, Parkinson's disease, and atherosclerosis. Therefore, the use of an antioxidant in the daily usage cosmetics is prevalent in recent years. The most common chemicals used for reducing oxidative stress are parabens and butylated hydroxytoluene (BHT) (Khan and Alam, 2019). However, such chemicals can be the cause of diseases in humans such as skin and liver cancer (Schrader and Cooke, 2000). Thus, the use of natural antioxidative agents such as EPS can be helpful. EPS helps to reduce the free radicals of oxygen and ROS in the human system, without showing any other adverse health impacts such as cytotoxicity or cancer. The antioxidant effect of EPS also helps to make it a significant part of cosmetics such as sunscreens, due to the reduced effect of ROS produced by the harmful ultraviolet radiations of the sun. The antioxidant effect can also help EPS to be an integral part of the antiaging cosmetics as it helps to restore skin-elasticity and glow (Marcellin et al., 2009).

10.7.1.2 Antimicrobial effect

The use of chemicals such as parabens is prominent in most of the cosmetics, which works as a preservative. However, the use of parabens can be proven to be harmful to both the skin and hair. Parabens can enter the epidermis and lead to diseases such as skin cancer (Khan and Alam, 2019). The ability of parabens to bypass the metabolic system of the body also makes it harmful as it can enter the bloodstream and affect the vital organs (Khan and Alam, 2019). On the other hand, the use of EPS can help to reduce such health risks when used as a preservative due to its film forming ability. The film formed inhibits the growth of any other form of pathogen or unwanted bacteria. This in turn helps to improve shelf life of the products (Jiang et al., 2022).

The antimicrobial impact of EPS inhibits the growth of microbes in the cosmetics and increase shelf life of the product (Abdalla et al., 2021). This property of EPS also helps to make use of it in creams and face washes for reducing pimples and acne-producing bacteria such as *Propionibacterium acnes.*

10.8   MICROBIAL EPS

Microbial exopolysaccharides (mEPS) contain different types of polymers that are secreted by microorganisms such as algae, bacteria, or cyanobacteria (De Philippis et al., 2001). These mEPS can either be tightly attached to the external surface of the microbe in the form of capsules or are loosely attached to the external surface in the form of a slimy outer layer. The microbial EPS are mostly ubiquitous and possess unique properties (Ates, 2015). They can be isolated from the microbes such as bacteria living in different areas and environments such as the fresh water or the marine environment, extreme harsh conditions, or in the soil ecosystem. EPS comprises of similar repetitive units of sugar moieties, which are mostly attached to a lipid carrier molecule. In turn these lipid carriers are associated with proteins, organic and inorganic molecules, DNA, and several metal ions (Bhaskar and Bhosle, 2006). The distinct functions of the mEPS are dependent on the structure of the mEPS molecule and the ecological niches of the corresponding microbe. The bacterial EPS are the most widely used EPS in the commercial field due to the several beneficial physiochemical properties, such as the properties of detoxification, immunomodulation, antioxidant effect, antitumor property, and the ability to stop the growth of other harmful bacteria due to its antimicrobial property. mEPS have paved their way into various industries, such as the food, cosmetics, and the pharmaceutical industry, mostly as a preservative, emulsifier, or a gelling agent (Kanmani et al., 2013). The use of mEPS in cosmetics is prevalent mostly to replace the harmful chemicals found in everyday products, which can have adverse impacts on the skin and hair of people.

Several research works have been performed on the functions and properties of exopolysaccharides, mostly produced by the lactic acid bacteria in the lab conditions since last decade (Holst and Müller-Loennies, 2007). However, the exploitation and understanding of the mEPS in cosmetics is comparatively unexplored, and the interest of people are mostly directed towards the isolation, screening of suitable organisms, characterization, and the possible application of mEPS in various industries, which may include the food and the pharmaceutical industry (Daba et al., 2021).

The EPS from marine isolate *Polaribacter* sp. SM1127 has a huge moisturizing ability and can hold up more water molecule than hyaluronic acid in cosmetics. It can boost fibroblast production, speed up mice's wound healing, and lessen the skin damages due to low temperature. Additionally, this EPS can enhance glutathione (GSH) content, decrease superoxide dismutase (SOD) enzyme activity, decrease lactate dehydrogenase (LDH) and ROS, and preserve cell activity and integrity to fend against UV radiation damages (Ding et al., 2022). Very few works have been done on the application of mEPS in the cosmetics industry. However, the use of this natural microbial product can be an advantageous alternative to the different toxic chemicals found in cosmetics. There is a need for more extensive research to be conducted on the downstream processing and the genetic engineering of mEPS for better biosynthesis (Castillo et al., 2015).

TABLE 10.1

Sources, function, and properties of different bacterial exopolysaccharides used in cosmetics

Polysaccharide	Structure	Monosaccharide unit	Molecular weight	Ionic charge	Solubility	Producing organism	Properties	Function	Reference
Hyaluronic Acid	Two repeating monosaccharide units alternatively linked by β-(1→3) and β-(1→4) glycosidic bonds	D-glucuronic acid N-acetylglucosamine	2×10^6	Anionic	Water soluble	*Streptococcus* spp. *Pasteurella multocida*	Highly hydrophilic Highly viscous hydrocolloid High swelling capacity Viscoelastic Biocompatible Biological activity	Skin hydration Gelling agent Filmogenic Collagen stimulation Face rejuvenation Anti-nasolabial fold Antiaging Skin augmentation Angiogenic	(Liu et al., 2011), (Chong et al., 2005), (Yamada and Kawasaki, 2005), (Freitas et al., 2015)
Cellulose	β-(1→4) glycosidic bonds	D-glucose	1×10^6	Neutral	Water insoluble; Soluble in strong acid/basic condition	*Gluconacetobacter xylinum*, *Acetobacter* spp. *Agrobacterium* spp. *Azotobacter*, *Rhizobium* spp. *Sarcina* sp. *Alcaligenes* sp. *Pseudomonas* sp.	High crystallinity High tensile strength Biological activity	Wound healing Skin moisturizer Tissue-engineered blood vessels Skin regeneration Preparing face mask	(Rehm, 2010), (Lahiri et al., 2021), (Gallegos et al., 2016),

(Continued)

TABLE 10.1 (Continued)

Sources, function, and properties of different bacterial exopolysaccharides used in cosmetics

Polysaccharide	Structure	Monosaccharide unit	Molecular weight	Ionic charge	Solubility	Producing organism	Properties	Function	Reference
Xanthan gum	β-(1→4) linked repeating pentasaccharide units	D-glucose D-mannose D-glucuronic acid	$(2-50) \times 10^6$	Anionic	Cold water soluble	*Xanthomonas* sp.	Highly stable High viscous hydrocolloid	Food additive (emulsifier or thickener) Cosmetic grade thickener	(Moorhouse et al., 1977), (Nwodo et al., 2012), (Rehm, 2010), (Sworn, 2021)
Gellan Gum	β-(1→3) linked repeating tetrasaccharide units	D-glucose L-rhamnose D-glucuronic acid	5×10^5	Anionic	Water soluble	*Sphingomonas paucimobilis*	Gelling agent Highly Stable Hydrocolloid	Filmogenic Encapsulation Emulsion stabilizer Rheology modifiers Alternative to agar in culture media	(Fialho et al., 2008), (Freitas et al., 2011), (Soumiya et al., 2021), (Freitas et al., 2015), (Rehm, 2010),
Levan	Repeating monosaccharide linked with β-(2→6) linked with side chain linked with β-(2→1) linked	D-fructose	3×10^6	Neutral	Water soluble	*Bacillus subtilis* *Halomonas smyrnensis* *Zymomonas mobilis*	Low viscosity High water solubility Biological activity	Skin moisturizer Anti-inflammatory Skin whitening Skin irritation alleviation Filmogenic Anti-inflammatory Antitumor activity	(Freitas et al., 2015), (Feingold and Gehatia, 1957), (Freitas et al., 2011)

10.9 APPLICATION OF MICROBIAL EPS IN COSMETICS

10.9.1 Hyaluronic acid

Hyaluronic acid (HA) is one of the most important components of the intracellular matrix of the mammalian connective tissues as it has the significant functions of hydration, transport of macromolecules, and has an antibacterial activity on the tissues (Freitas et al., 2015). Age-related reductions in the naturally occurring HA levels in human skin have a significant role in the emergence of wrinkles and the aging appearance, decreasing the flexibility and moisture of the tissue (Bogdan Allemann and Baumann, 2008). The sources of the microbial exopolysaccharide, HA are mainly the bacteria *Streptococcus* spp. and *Pasteurella multocida* (Andhare et al., 2014).

The properties of HA are high levels of viscosity, lubricity and immunomodulation, and antigenicity provided by the HA molecules in the human body (Zamboni et al., 2018). The advanced levels of biocompatibility, biodegradability, and the immunogenic property of HA make it suitable for usage in various industries, including the pharmaceutical and cosmetics industry (Berkó et al., 2013).

The activity of hyaluronic acid mostly depends on the molecular size, which has allowed it to become one of the most active ingredients in many products of the pharmaceutical and the cosmetics industry (Freitas et al., 2015). The application of HA and its derivatives in the different industrial sectors have a great market of around 2,000 to 6,000 USD per kg of HA (Pires et al., 2010).

In the cosmetics industry, HA works especially as a moisturizing active ingredient due to its hydrophilic nature. It is usually formulated as an emulsion or in the form of serums, to allow more hydration of the skin and improve skin elasticity and reduce wrinkles and skin aging (Marcellin et al., 2009). Previous studies have been useful to prove that the usage of HA in the cosmetics can be useful for protecting the skin against the harmful UV radiations of the sun (Avadhani et al., 2017). This can help in providing added benefits to the products, such as reducing skin hyperpigmentation and effective moisturization and glow to the skin (Liu et al., 2021). The ability of HA to maintain a more hydrated and firmer skin can help in reducing the injurious impacts of the radiation catalyzed by the antioxidative impacts of HA.

When used in the formulation of cosmetics, HA is mostly used as a viscosity modifier or as a conditioning agent for the skin and hair (Freitas et al., 2015). It has also been found that the lower molecular weight HA has the ability to initiate skin regeneration. On the other hand, the usage of higher molecular weight HA allows the formation of a viscoelastic film on the skin surface, which reduces the loss of moisture from skin due to evaporation and enhance the effect of the active ingredients present in the cosmetics products (Juncan et al., 2021).

HA is mostly used in products such as the antiaging creams, moisturizers, serums, and sunscreens. The use of HA and its derivative salts at a concentration of 1–2% of sodium hyaluronate is found in various products such as lipsticks, lotions, and sprays (Becker et al., 2009). At present, several studies have proven the beneficial impacts of the use of HA as a dermal filler. This has made HA a useful and effective alternative to the various old collagens used in nonsurgical cosmetic procedures (Baser et al., 2021). The addition of HA in dermal filter is gaining more popularity in recent years due to the aesthetic enhancement of HA. This allows better augmentation of the body's soft tissue, and the high hydration property of HA helps in replenishing the moisture and the natural component lost at the cellular level (Al-Halaseh et al., 2022). A few examples of fillers available in the market approved by the FDA are Prevelle Silk® (Mentor Corp., USA), Anika® (Anika Therapeutics, Inc., MA), Restylane® (Medicis, USA), and Juvéderm™ (Allergan, USA) (Ammala, 2013).

The use of HA in hair care products have also been scientifically proven to have the ability to revitalize the hair follicles. The property of HA to combine with several amino acids which in turn help in the anabolic process of the hair papillary cells makes it an important vehicle for the solutions of hair problems like hair fall and help promote better growth of hair (Al-Halaseh et al., 2022).

10.9.2 CELLULOSE

Various cosmetic products, such as face masks, contact lenses, formulas for personal cleansing, and facial scrubs, are formulated with bacterial cellulose. The advantages of employing bacterial cellulose over vegetal cellulose include the fact that bacterial cellulose is more chemically pure, has no lignin or hemicellulose, is more hydrophilic and therefore has a high water holding capacity, has greater tensile strength (Mbituyimana et al., 2021).

Cellulose has the properties of skin cell regeneration and thus can be applied in products for wound healing and skin repair. The exopolysaccharide also has the ability to provide better moisturization to skin, as a result of its high hydration and water retention capability (Ioelovich, 2008). This allows for use of cellulose in moisturizers and creams for adding extra hydration to the skin. Due to its nanofibers, it has a high capacity for absorption. Wet cellulose retains fluids almost ten times more than non-woven masks and 100 times its dry weight. Because of its exceedingly thin thickness and nano-fiber of 20 nm in diameter, bacterial cellulose has good adhesion qualities to uneven skin surfaces, allowing the bacterial cellulose mask's components to penetrate deep into fine creases and wrinkles where ordinary sheet masks fail to reach (Mbituyimana et al., 2021). Another important application of cellulose is as a component of fingernail and the base of artificial nails (Freitas et al., 2015).

Due to the high degree of moldability, cellulose can also be used as an important component of face masks and several types of facial components, which allows to act as a carrier for the additives present in the facial packs (Palaninathan et al., 2014). Bacterial cellulose lacks natural antiaging, whitening, and cleaning properties; its uses in cosmetology are limited. To optimize bacterial cellulose's potential in the cosmetics sector, it must be successfully fabricated with active substances such as plant extracts and essential oils. The active components are held together inside the bacterial cellulose matrix, for example, through hydrogen bonds. This facilitates the impregnation of active ingredients and allows them to function on the skin for a longer period of time than standard cosmetic emulsions. Furthermore, prolonging the release duration allows active compounds to penetrate deeper into the skin. As a result, the bacterial cellulose sheet mask pack provides excellent skin moisturization and conditioning (Mbituyimana et al., 2021). The impact of cellulose-based skin care products was analyzed with the help of human volunteers ranging between the age group of 21 years to 40 years. The study showed that though the cellulose masks did not have any significant effect in the improvement of the skin quality, it helped in increasing the moisturization of skin by 28% (Amnuaikit et al., 2011). Cellulose can also be used as a component of other cosmetics products such as facial scrubs, along with other important ingredients like aloe vera extract, powdered glutinous rice, and ascorbic acid (Freitas et al., 2015). Many companies are now engaged in long-term research and have evaluated the uses of bacterial cellulose membranes in skin treatment and skin care. They created a variety of masks with moisturizing, cleaning, whitening, antiaging, and other properties (Mbituyimana et al., 2021).

10.9.3 XANTHAN GUM

Xanthan gum is one of the most unique types of emulsifying and gelling agent due to its biotechnological origin. This makes the gum suitable for alterations, such as the introduction of new properties, which can make the final product more stable. This makes xanthan gum suitable for being used in making pseudoplastic solution with a better flow rate. It also provides other sensory characteristics like a more pleasant and lighter structure to the end product, which is an attractive characteristic for the cosmetics industry (Furtado et al., 2022). Due to the excellent viscosity-enhancing capabilities at low concentration, xanthan gum is the most often utilized bacterial polysaccharide in the global hydrocolloids market. Moreover, the aqueous solutions formed by xanthan has great low-shear viscosity and shear-thinning property (Freitas et al., 2015).

Xanthan gum is a great alternative to synthetic emulsifiers due to the higher stabilizing property and nontoxic nature of the EPS. These properties of xanthan gum do not change with the variation in climate or seasonal temperature changes. This allows the gum to be used across a wide range of temperature and pH, such that the necessary formulation can be accommodated with both hot and cold formulations (Furtado et al., 2022). Moreover, the production process of xanthan is eco-friendly and does not affect the environment and is therefore a more sustainable emulsifying product for several commercial applications (Furtado et al., 2022). The application of xanthan gum is seen in almost every class of cosmetics product, such as moisturizers, creams, shampoos, deodorants, baby products, and oral hygiene products. The application of xanthan in cosmetics is as an emulsion stabilizer, surfactant/emulsifier, binder, and as a viscosity enhancer. It is used at a concentration of 4% to 6% in the dermal and mucous cells in products like hair colors and nail paints. For other products, a varied range of concentrations are used, such as 0.2% to 0.6% for baby products, 0.6% for deodorants, and 0.05% for other types of sprays (Fiume et al., 2016; McNeely and Kovacs, 1975).

10.9.4 GELLAN GUM

Gellan gum has the ability to form gels when it is cooled down. This allows the products to remain in the gel state after being packaged into bottles. This also allows the manufacturing and packaging procedure of the cosmetics easier as the fragile ingredients which otherwise could get destroyed remains protected at the time of manufacture in a gel form. The ability to impart flavor is also useful for keeping the essence of the different ingredients together in the product without the need to add any external essence to the cosmetics product (Freitas et al., 2015).

Along with the easy manufacturing and packaging, the usage of different blends of low and high acyl gellan can help in designing a range of products such as toothpaste, which contains different characteristics of viscosity and binding (Prajapati et al., 2013). Gellan gum can also be used for the preparation of emulsion-filled gels with distinct structures to differentiate between a gel and an emulsion used in cosmetics products. These types of emulsion-gels are prepared by the dispersion of oil droplets in a continuous semisolid gel, considering a concentration ranging between 2% to 20% (Buchmann, 2001). The elastic nature of the gel helps the oil droplets to remain suspended in the gel without the occurrence of creaming.

Gellan gum is an important ingredient of any cosmetics product mostly due to its properties such as low-cost, safe for the consumption, and external use by humans and as the polymer is environmentally benign. Therefore, the production process and the incorporation of the polymer into cosmetic products is easier (Yamada and Kametani, 2022). Gellan gum is an important ingredient of any cosmetics product, which serves as an emulsion stabilizer and viscosity enhancer. It is used at different concentrations for the type and function of end products. For example, a concentration of 0.3% to 0.5% is used for the leave-on products which come in direct contact with the dermal layer of the skin (Freitas et al., 2015). A concentration of 0.0004% is used for products such as powders and eye products, which come in direct contact with the mucosal layer of the hair and skin (Freitas et al., 2015).

10.9.5 LEVAN

Levan has several properties which sometimes limits the applicability of levan for industrial purposes like in food, cosmetics, or pharmaceuticals. This is mostly because levan does not have the ability to swell when placed in water and has a very narrow intrinsic viscosity value of about 0.14 dl/g (Kasapis et al., 1994). Levan has been approved for usage as a food additive in USA, Japan, Australia, and Europe (Kang et al., 2009). The usage and safety of levan in cosmetic products is also under research and has been proven to have several beneficial impacts when applied on the skin. The advantages of using levan on the skin and hair includes the properties of cell proliferation and skin cell regeneration, along with skin-whitening and irritation-alleviating impacts. Studies with the three-dimensional artificial skin model also reveals the anti-inflammatory and antiaging properties of levan (Freitas et al., 2015). The polymer also has antioxidant and improves skin hydration (Wasilewski et al., 2022). The antioxidant properties can help levan to be used in cosmetic products for reducing the impacts of free oxygen radicals, which can lead to skin problems. Levan is also found to be nontoxic to human cells, which makes it a more beneficial cosmetic ingredient (Wasilewski et al., 2022).

Anti-inflammatory and antiaging properties of levan are due to its ability to reduce the secretion rate of interleukin 1α, which works as a pro-inflammatory mediator in the body. Therefore, the inhibited secretion of this interleukin can prevent the future aging of the skin. The moisturizing impacts of levan was tested in human volunteers, and it was found that levan can reduce the loss of moisture from skin, which helps to keep the skin and hair better hydrated and moisturized (Kang et al., 2009). However, despite the beneficial impacts of levan, the lowered ability of levan to expand in aqueous state limits the applicability of it in cosmetics products. Moreover, levan is partially hydrolyzed into fructose and oligosaccharides with low molecular weight in certain conditions like low pH and high temperatures. This makes the preservation process comparatively difficult (Kang et al., 2009).

10.10   REGULATORY REQUIREMENTS

The level of safety of the cosmetics available in the market mostly depends on the safety level of the ingredients used in the cosmetics. Therefore, the evaluation of the ingredients is important and conducted by the toxicological tests (Freitas et al., 2015). The determination of the toxic potential of any ingredient found in the cosmetics is done through a series of toxicity studies (Freitas et al., 2015). The minimal safety and base set requirements which is needed for the ingredients to be safe is determined by the tests such as acute toxicity test, irritation or corrosiveness caused by the ingredient, dermal absorption rate, skin sensitization, toxicity of repeated dose, and any rate of mutagenicity. Follow-up tests such as that of carcinogenicity, toxicokinetics, and the reproductive toxicity are also needed for understanding the level of further toxicity which can be found in humans through the regular exposure to these ingredients. In certain cases, the data of photo-induced toxicity can also be required when the cosmetics is designed to be used in the presence of sunlight (Chew and Maibach, 2001).

The use of human volunteers in such toxicological tests can be a great issue of ethics; however, the new guideline Scientific Committee on Consumer Safety (SCCS) in-silico toxicity method is to be approached before human testing (SCCS, 2021). When certain tests are not required or are not technically possible, then a scientific justification has to be given for the same.

For the use of any ingredient which is derived from biotechnological procedure, specific information about the ingredient has to be available for the proper justification of the safety of the contents. The data which is needed about the ingredients comprise of EL (observed adverse effect level), NO(A)EL (no adverse effect level), SED (systemic exposure dose), MoS (margin of safety). NO(A)EL is the highest level of dosage, for which no adverse effects are found in humans, which is expressed in the form of mg/kg body weight/day (Freitas et al., 2015).

SED can be described as the amount of the ingredient which can enter the bloodstream of humans in a single day per kg of body weight. On the other hand, the MoS (margin of safety) value is used to extrapolate from test animal group used for testing to that of an average human being and from that of an average human being to sensitive subpopulations (Freitas et al., 2015).

MoS value of any ingredient is important as, for any content of cosmetics to be considered safe for human use, the MoS value of the substance has to be minimum 100 (Freitas et al., 2015). The evaluation of the safety level of any cosmetics is done based on the toxicological profile of the various ingredients present in it, along with the chemical structure of the constituents and the level of exposure to the human skin which can cause any harmful adverse effects, and in certain cases, the evaluation of the finished end product is also needed (SCCS, 2021).

The level of exposure of the cosmetics ingredient depends on several other factors such as the type of the cosmetics product and the type of application of the product, such as whether it is a skin or hair product, the method used for the application of the cosmetics, for example, it has to be sprayed or rubbed on or washed off after usage. It also depends on other categories of factors like the concentration of the particular ingredient in the finished product, the amount of product which will be used in a single application along with the total area of the human body (skin or hair) that will come in direct contact with the product during application, the frequency of usage of the product and the targeted group of consumers, for example, whether it is for adults or children, as it will also affect the type of skin texture and tolerance level of the product (Freitas et al., 2015).

If the final finished product is found to be safe and not causing any specific adverse effects in the human skin under any foreseeable conditions of usage, then the compatibility testing is carried out with the help of a group of human volunteers before the product can be ready to be supplied to the market for commercial usage (Bashir and Maibach, 2001).

10.11 TOXICITY ASSESSMENT OF BACTERIAL POLYSACCHARIDES

The Cosmetic Ingredient Review (CIR) Expert Panels have tested the toxicity levels of hyaluronic acid and reached the conclusion that it is safe for being used in various cosmetics products (Becker et al., 2009). The application of HA had revealed no serious reactions when used for the treatment of tissue augmentation (Becker et al., 2009). Moreover, hyaluronic acid has been successful to act as a better vehicle for the transfer of the other cosmetics ingredients into the skin layers. This has to be considered when using HA in the formulations of cosmetics, which are concerned for the products being absorbed to the epidermal layers.

Cellulose had the properties of being nontoxic to human skin and biocompatible in nature, which further allows it to be used in cosmetics (Moreira et al., 2009). The toxicity analysis of the microbial cellulose done through testing on the human umbilical epithelial cells showed no toxicity, which makes it suitable for human use and formulations of cosmetic products (Moreira et al., 2009).

Levan also shows no forms of adverse impacts on human, however, the lethal dose for levan is found to be close to 7.5 g/kg body weight and has a NOEL value of 1.5 g/kg/day (Kang et al., 2009). Levan was tested for cytotoxicity with the use of cell lines of human fibroblast and keratinocyte, which revealed no forms of cytotoxicity till a level of 100 μg/mL. The cell proliferative impacts of levan were evident in the keratinocyte cell line when applied at a concentration of 1 mg/mL (Kim et al., 2005).

Oral assessments of xanthan gum in rodents and dogs revealed that the exopolysaccharide can be toxic when used at a concentration of 20 g/kg body weight (Eastwood et al., 1987), (McNeely and Kovacs, 1975). The reproductive and developmental toxicity was tested for the application of xanthan gum in albino rats at a concentration of 0–0.5 g/kg body weight. The test showed neither beneficial impacts nor toxic effects on the mouse (Woodard et al., 1973). Gellan gum, when tested on animals, demonstrated its nontoxic behavior (Fiume et al., 2016). It was also found to be non-carcinogenic along with no neoplastic lesions found when tested on dogs and monkeys (Anderson et al., 1988). The tests done on rabbits and humans with gellan gum revealed it is a non-irritant to the eyes, which makes it suitable for application on the skin products of daily use. The ingestion of gellan gum at a concentration of 175–200 mg/kg body weight also revealed no severe dietary problems or allergic impacts on humans (Eastwood et al., 1987; Liu et al., 2010).

10.12   COMMERCIAL APPLICATION OF MICROBIAL EXOPOLYSACCHARIDES

With the increase in the knowledge of the harmful side effect of the chemical ingredients in cosmetics formulation, various cosmetic companies now incorporate the addition of exopolysaccharide as a cosmetic ingredient instead of the harmful chemical components. Environmental Working Group (EWG) is a non-profit organization that specializes in agricultural subsidies, toxic chemicals, drinking water pollutants, and corporate accountability research and advocacy (Environmental Working Group, 2022a). They have created a database of cosmetic products with more than 87,996 products of 3,023 brands (Environmental Working Group, 2022b).

Figure 10.2 reveals that out of the 8,263 products listed in the EWG Skin Deep Cosmetic database, xanthan gum is the highest used exopolysaccharide with 7,050 products followed by hyaluronic acid with 720 products. Levan is the least used cosmetic ingredient of three products only. Although various research is still going on in the cosmetic formulation of better and healthier life.

Figure 10.3 represents the percentage distribution of different cosmetic category like skin care, hair care, personal hygiene, and makeup. Out of 720 products containing hyaluronic acid, 512 products belong to skin care products, which generally include facial moisturizer, serum, face mask, and many others; 141 products belong to makeup products like foundation, concealer, BB cream, lip gloss, and others; 50 products belong to personal hygiene, which includes hand sanitizer, aftershave, bath

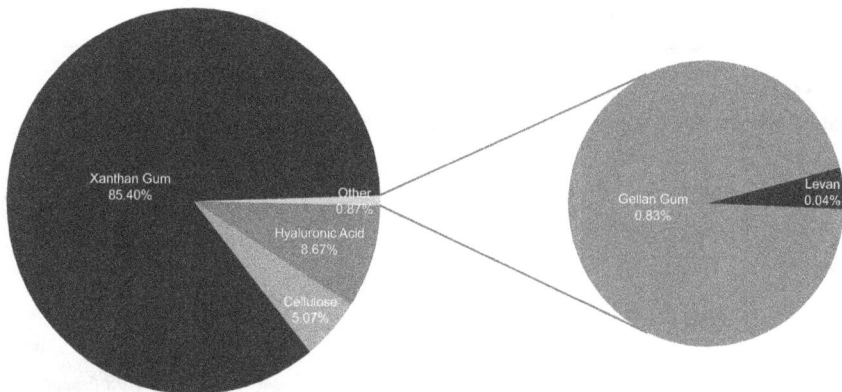

FIGURE 10.2   Distribution of exopolysaccharide as cosmetics ingredient in 8,263 cosmetics products

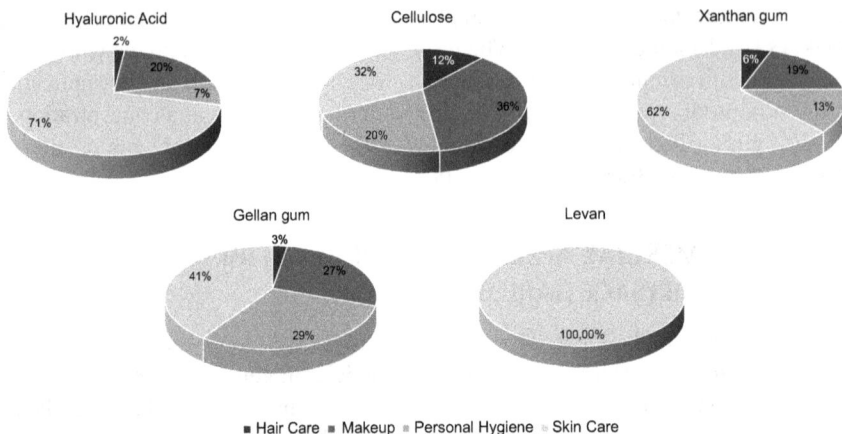

FIGURE 10.3 Percentage distribution of various categories of cosmetics in different exopolysaccharide

oil, bar soap, body spray, and others; and 17 products belongs to hair care products like shampoo, conditioner, detangler, and others. Similarly, out of 421 products containing cellulose, 153 products belong to makeup category, 135 products belong to skin care products, 84 products belong to personal hygiene products, and 49 products belong to hair care category.

Xanthan gum is the highest used exopolysaccharide in cosmetics products. Out of 7,050 products containing xanthan gum, 4,373 products belong to skin care category, 1,319 products belong to makeup category, 951 products belong to personal hygiene category, and 407 products belong to hair care products. There are 69 products that contain gellan gum, out of which 28 products belong to skin care, 20 products belong to personal hygiene, 19 products belong to makeup, and 2 products belong to hair care category products. Only 3 products contain levan, which belong to skin care products. Some of these products are listed in Table 10.2.

TABLE 10.2

List of the commercially available cosmetics products containing EPS

EPS	Product category	Product type	Product name	Company
Hyaluronic Acid	Skin Care	Moisturizer	Hydra Genius Daily Liquid Care	L'Oreal Paris
			Hydro Boost Gel-Cream with Hyaluronic Acid for Extra-Dry Skin	Neutrogena
			Hydralux Extreme Moisture	KPS Essentials
			Deep Hydration Healthy Glow Daily Cream	Cetaphil

TABLE 10.2 (*Continued*)
List of the commercially available cosmetics products containing EPS

EPS	Product category	Product type	Product name	Company
		Serum	Revitalift 1.5% Hyaluronic Acid Serum	L'Oreal Paris
			Peptide Booster Serum	Sally B's
			Cellular Filler Hyaluronic Acid Serum	NIVEA
			Hyaluronic + Peptide 24	Olay
		Cleanser	TONY LAB AC Control Bubble Foam Cleanser	Tony Moly
			Micellar Cleansing Water with Hyaluronic Acid + Aloe	GARNIER
			Cosrx Hydrium Triple Hyaluronic Moisturizing Cleanser	Beauty Barn
			Exfoliating Jelly Cleanser	Hero Cosmetics
		Mask	Moisture Lock Serum Mask	SeSpring
			Light Up Vitamin C & E Flash Brightening Mask	Allies of Skin
			Skin Camp Hydra-Gel Rosy Lip Mask	Skin Gym
			Manuka Honey + Collagen Sheet Face Mask	By Nature
	Makeup	Foundation	Dream Radiant Liquid Hydrating Foundation	Maybelline
			Skin Perfector HD Serum Foundation	VANI-T
			Good Apple Skin-Perfecting Foundation Balm	KVD
		Lip Gloss/ Lipstick	Lifter Gloss	Maybelline
			Superstar Lips	Charlotte Tilbury
			Rouge Tarou Nude	Maréna Beauté
		Eye Liner	Everlasting Waterproof Liquid Liner	Profusion Cosmetics
	Hair Care	Shampoo/ Conditioner	Healthy Scalp Hydro Boost with Hyaluronic Acid	Neutrogena
			Hyaluron Moisture 72H Moisture Filling Shampoo	L'Oreal Paris
			GO SMOOTH	St. botanica
			Soothing Mint Moisturizing Conditioner	Inahsi

(*Continued*)

TABLE 10.2 (*Continued*)
List of the commercially available cosmetics products containing EPS

EPS	Product category	Product type	Product name	Company
		Hair Treatment/ Serum	Natural Moisturizing Factors + HA for Scalp	The Ordinary
			2Chic® ULTRA-REVIVE 3-In-1 HAIR REVIVAL TREATMENT	Giovanni
			Healthy Scalp Hydro Boost Deep Treatment Hair Mask	Neutrogena
			Hyaluronic Acid Hydrating Hair Treatment	The Inkey List
		Hair Styling Aide	2Chic® ULTRA-REVIVE LEAVE-IN CONDITIONING & STYLING ELIXIR	Giovanni
			Rock Your Curls Curl Enhancing Cream	Inahsi
	Personal Hygiene	Sanitizer	Hand Sanitizer Gel	Naples Soap
			Hand Sanitizer	by Humankind
		Body Wash	Coconut Cream Body Wash	Pacifica
			Cleansing & Nourishing Body Wash with Hyaluronic Acid	Olay
			Hydrating Body Wash	CeraVe
Cellulose	Skin Care	Moisturizer	Everyday Radiance + C Moisturizer	Honest
			Kind Hydra Lotion	Madara
			Glow Exfoliating Soft Peel	Aloisia Beauty
		Serum	Blisspro Brightening Serum	Bliss
			Uplift Firming Serum	City Beauty
			Firming Serum	ZO Skin Health
		Cleanser	Enzyme Cleanser	BABOR
			Recharge Regimen	Rodan and Fields
			Clear Complexion Cleanser	Boscia
		Mask	Resurfacing Sleep Mask	Cocokind
			Brightening Bio Cellulose Face Sheet Mask	LuxaDerme
			24K Black Gold Peel Off Mask	masque BAR
	Makeup	Mascara	Think Big All-In-One Mascara	Beautycounter
			Extreme Length Mascara + Lash Primer	Honest
			Expressionist Pro Mascara	W3LL PEOPLE

TABLE 10.2 (*Continued*)

List of the commercially available cosmetics products containing EPS

EPS	Product category	Product type	Product name	Company
		Eyeliner	Liquid Eyeliner	Palladio
			Wings of Love Long Lasting Liquid Liner Pen	Doll Face
		Foundation	Skin Perfector HD Serum Foundation	VANI-T
			Infallible 24H Matte Cover Foundation	L'Oreal Paris
	Hair Care	Shampoo/ Conditioner	Floof Dry Shampoo	Billie
			Cure Liqueur Strengthening Shampoo	Drybar
			The Big Wig Thickening Volume Conditioner	cake
			Strengthening Conditioner	KENRA
		Hair Treatment/ Serum	Dream Big Volume Perfectly Full Thickening Cream	Marc Anthony
			The In So Deep	cake
			2Chic® ULTRA-MOIST HAIR HYDRATION TREATMENT	Giovanni
	Personal Hygiene	Body Wash	Men + Care Avocado Oil + Lime Body Wash	Dove
			Pink Peony & Sea Salt	Softsoap
			Hydration + Sea Kelp & Hyaluronic Acid Scrub & Wash	OGX
Xanthan Gum	Skin Care	Facial Moisturizer/ Treatment	REVITALIFT Micro Hyaluronic Acid + Ceramides Line-Plumping Water Cream	L'Oreal Paris
			Build-A-Tan Gradual Sunless Tan Lotion	Neutrogena
			Raw Nourish Am Treatment	AO Skincare
			Glowstarter Mega Illuminating Moisturizer	Glamglow
		Serum	Regenerist MAXTONE SERUM WITH VITAMIN C	Olay
			Stroke Of Brilliance Brightening Serum	Versed
			Organic Flowers Eye Essence	Whamisa
			Hyper Glow® Serum	DIME

(*Continued*)

TABLE 10.2 (*Continued*)

List of the commercially available cosmetics products containing EPS

EPS	Product category	Product type	Product name	Company
		Mask	Pure Clay Exfoliate & Refining Face Mask	L'Oreal Paris
			Kale-B Balancing Serum Mask 5 Minute Sheet Mask with Niacinamide + Kale	Garnier
			Charcoal Whipped Purifying Detox Mask	BIORÉ
			So Revitalizing Serum Mask	SeSpring
		Cleanser	Blackhead Solutions	Clinique
			Regenerist Whip	Olay
			Ideal All Skin Types Makeup Removing Towelettes	L'Oreal Paris
			Blackhead Eraser® Scrub	Clean & Clear
	Makeup	Foundation	Glowish Multidew Vegan Skin Tint Foundation	HUDA BEAUTY
			Studio Fix Fluid SPF 15	MAC
			Feel Good Skin	Profusion
			True Intentions Hydrating Foundation	Range Beauty
		Mascara	Healthy Volume® Mascara	Neutrogena
			Kind & Free™ Clean, Volumizing & Lengthening Mascara	Rimmel London
			Makeup Revolution 5D Lash Pow Mascara	Revolution
			Lash Sensational® Washable Mascara	Maybelline
		Eye Liner	Ultra Liner® Waterproof Liquid Eyeliner	Maybelline
			Vivid Matte Liquid Liner	NYX Professional
			numero uno bff liquid liner	Colourpop
			Stay Put® Matte Eyeliner	Milani
	Hair Care	Shampoo/ Conditioner	Whole Blends Sulfate Free Remedy Royal Hibiscus & Shea Butter Shampoo	Garnier
			Apple Miracle Restoring Shampoo	Marc Anthony
			Daily Fortifying Conditioner	Baxter of California
		Hair Treatment/ Serum	EVERPURE Sulfate-Free 21-In-1 Color Caring Spray	L'Oreal Paris
			Golden Hour Recovery Cream	Ursa Major

TABLE 10.2 (*Continued*)

List of the commercially available cosmetics products containing EPS

EPS	Product category	Product type	Product name	Company
			Fructis Nourishing Treat 3-in-1 Hair Mask + Coconut Extract Hair Mask	Garnier
			2Chic® ULTRA-MOIST DUAL-ACTION PROTECTIVE LEAVE-IN SPRAY	Giovanni
	Personal Hygiene	Toothpaste	Total Whitening™ Gel Toothpaste	Colgate
			Crest Gum & Enamel Repair Toothpaste Intensive Clean	Oral-B
		Body Lotion	Intensive Care™ Cocoa Radiant® Lotion	VASELINE
			Calm & Centered Cream	Sagely Naturals
			Lift & Firm Body Cream Vinosculpt	Caudalie
		Body Wash/ Cleanser	Age Defying Body Wash	Olay
			Aloe Vera Body Wash	Clean & Clear
Gellan Gum	Skin Care	Facial Moisturizer/ Treatment	Active Hydration Body Replenish	Rodan + Fields
		Serum	Squeezed Lemon Serum	SOO'AE
			Pressed Serum Stick	Olay
			Active Hydration Serum	Rodan + Fields
		Mask	Masks Pore Detox	Olay
			Squeezed Lemon Eye Patch	SOO'AE
			Brightening Enzyme Mask	Malin+Goetz
	Makeup	Highlighter/ Blush	Bio-Radiant Gel-Powder Highlighter	Haus Laboratories
			Blusher	Facetheory
		Eye Shadow	Lid Glaze Hydrating Jelly Eyeshadow	Item Beauty
	Personal Hygiene	Body Wash/ Cleanser	Sudsy	bioClarity
			Body Wash Concentrate	Dove
		Body Firming Lotion	Chill Mama Soothing Jelly	Honest
Levan	Skin Care	Serum	Goopglow 20% Vitamin C + Hyaluronic Acid Glow Serum	Goop
			Superskin™ Face Serum	LIZ EARLE
		Exfoliator	Goopglow Microderm Instant Glow Exfoliator	Goop

10.13   FUTURE PROSPECTS

There are several companies using microbial exopolysaccharides like hyaluronic acid, cellulose, and xanthan gum in their cosmetics formulations. These polysaccharides can help to work as a great alternative to the harmful chemical products like parabens and BHT, which can lead to adverse impacts on the skin and hair of human. These harmful impacts can comprise of skin lesions, hyperpigmentation, and carcinogenic effects on human skin. The use of EPS in the cosmetic products in the form of an emulsion (gellan gum), a hydrogel (hyaluronic acid), and a suspension (xanthan gum) can be helpful in acting as good carriers for the active ingredients in the cosmetics products. The further benefits of using EPS in cosmetics products are antioxidant, antimicrobial, and antiaging activities, which can help in making the skin look better hydrated and plump. Therefore, the incorporation of microbial EPS in the formulation of cosmetic products in skin care products by different brands can help to change the cosmetic industry and move it towards a more biocompatible direction.

10.14   CONCLUSION

The use of cosmetics is an important part of the everyday life of people for improving the quality of the skin and hair. Cosmetics can be applied on the epidermis, such as the skin, lips, eyebrows, nails, and also on the hair. Different cosmetics can have a distinct effect on the skin and hair such as better moisturization, improved shine, and reduced effects of harmful pollutants and toxic chemicals. The use of cosmetics is mostly directed towards getting results such as reduced hyperpigmentation, reduced impacts of the harmful impacts of the UV-radiation of the sun, and better antioxidative impacts, which can help to reverse aging of the skin. However, these cosmetic products can have several harmful ingredients like hydroquinone, BHT (butylated hydroxyl toluene) and BHA (butylated hydroxyl anisole), DEA (diethanolamine), and parabens. These ingredients can lead to adverse impacts like darkening of the skin, faster aging of the skin, and these components can also enter the bloodstream and affect the internal organs of the body. Therefore, natural ingredients like exopolysaccharides are better for daily cosmetic products as these EPS have many beneficial properties such as antioxidant property, antimicrobial property, and antiproliferative properties, which allows EPS to be used for a wide spectrum of purposes within cosmetics. There are a range of microbial EPS, such as hyaluronic acid, cellulose, levan, xanthan gum, and gellan gum. All of these EPS can provide beneficiary effects, such as hyaluronic acid, which helps to keep the skin better hydrated due to its ability to absorb moisture from the surroundings and preventing the loss of hydration from hair and skin, and can be used in moisturizers and creams. Cellulose has regenerative properties and can help to replace the old and dead skin cells with new ones, which makes it a good ingredient for antiaging products and skin lightening products. The use of xanthan gum as an emulsifier makes the formulation of cosmetics easier due to the viscosity improving ability of the EPS. Gellan gum on the other hand has the capability to form gels when cooled down. This allows the production and packaging of the products easier, further increasing the shelf life of the cosmetics. The use of gellan gum in different concentrations helps to make the application of products on hair and skin easier as per the required texture. All of such EPS also have antimicrobial characteristics, which helps to replace the generic preservatives used in cosmetic industries, making the products more biocompatible and biodegradable.

REFERENCES

Abdalla, A.K., Ayyash, M.M., Olaimat, A.N., Osaili, T.M., Al-Nabulsi, A.A., Shah, N.P., Holley, R., 2021. Exopolysaccharides as antimicrobial agents: Mechanism and spectrum of activity. Front. Microbiol. 12. https://doi.org/10.3389/fmicb.2021.664395

Al-Halaseh, L.K., Tarawneh, S.K., Al-Jawabri, N.A., Al-Qdah, W.K., Abu-Hajleh, M.N., Al-Samydai, A.M., Ahmed, M.A., 2022. A review of the cosmetic use and potentially therapeutic importance of hyaluronic acid. J. Appl. Pharm. Sci. 12, 34–41. https://doi.org/10.7324/JAPS.2022.120703

Ammala, A., 2013. Biodegradable polymers as encapsulation materials for cosmetics and personal care markets. Int. J. Cosmet. Sci. 35, 113–124. https://doi.org/10.1111/ics.12017

Amnuaikit, T., Chusuit, T., Raknam, P., Boonme, P., 2011. Effects of a cellulose mask synthesized by a bacterium on facial skin characteristics and user satisfaction. Med. Devices Evid. Res. 4, 77–81. https://doi.org/10.2147/MDER.S20935

Anderson, D.M.W., Brydon, W.G., Eastwood, M.A., 1988. The dietary effects of gellan gum in humans. Food Addit. Contam. 5, 237–249. https://doi.org/10.1080/02652038809373701

Andhare, P., Chauhan, K., Dave, M., Pathak, H., 2014. Microbial exopolysaccharides: Advances in applications and future prospects. Biotechnology. 3, 25. https://doi.org/10.13140/RG.2.1.3518.4484

Angelin, J., Kavitha, M., 2020. Exopolysaccharides from probiotic bacteria and their health potential. Int. J. Biol. Macromol. 162, 853–865. https://doi.org/10.1016/j.ijbiomac.2020.06.190

Ates, O., 2015. Systems biology of microbial exopolysaccharides production. Front. Bioeng. Biotechnol. 3, 200. https://doi.org/10.3389/fbioe.2015.00200

Avadhani, K.S., Manikkath, J., Tiwari, M., Chandrasekhar, M., Godavarthi, A., Vidya, S.M., Hariharapura, R.C., Kalthur, G., Udupa, N., Mutalik, S., 2017. Skin delivery of epigallocatechin-3-gallate (EGCG) and hyaluronic acid loaded nano-transfersomes for antioxidant and anti-aging effects in UV radiation induced skin damage. Drug Deliv. 24, 61–74. https://doi.org/10.1080/10717544.2016.1228718

Baser, B., Singh, P., Shubha, P., Roy, P.K., Chaubey, P., 2021. Non-surgical rhinoplasty and use of hyaluronic acid based dermal filler-user experience in few subjects. Indian J. Otolaryngol. Head Neck Surg. 73, 52–58. https://doi.org/10.1007/s12070-020-02100-8

Bashir, S.J., Maibach, H.I., 2001. In vivo irritation. In: A.O. Barel, H.I. Maibach (Eds.), *Handbook of Cosmetic Science and Technology*. CRC Press, pp. 107–118. https://doi.org/10.1201/9780824741396

Becker, L.C., Bergfeld, W.F., Belsito, D.V., Klaassen, C.D., Marks, J.G., Shank, R.C., Slaga, T.J., Snyder, P.W., Andersen, F.A., 2009. Final report of the safety assessment of hyaluronic acid, potassium hyaluronate, and sodium hyaluronate. Int. J. Toxicol. 28, 5–67. https://doi.org/10.1177/1091581809337738

Berkó, S., Maroda, M., Bodnár, M., Erős, G., Hartmann, P., Szentner, K., Szabó-Révész, P., Kemény, L., Borbély, J., Csányi, E., 2013. Advantages of cross-linked versus linear hyaluronic acid for semisolid skin delivery systems. Eur. Polym. J. 49, 2511–2517. https://doi.org/10.1016/j.eurpolymj.2013.04.001

Bhaskar, P.V., Bhosle, N.B., 2006. Bacterial extracellular polymeric substance (EPS): A carrier of heavy metals in the marine food-chain. Environ. Int. 32, 191–198. https://doi.org/10.1016/j.envint.2005.08.010

Bogdan Allemann, I., Baumann, L., 2008. Hyaluronic acid gel (Juvederm) preparations in the treatment of facial wrinkles and folds. Clin. Interv. Aging. 3, 629–634. https://doi.org/10.2147/CIA.S3118

Buchmann, S., 2001. Main cosmetic vehicles. In: A.O. Barel, H.I. Maibach (Eds.), *Handbook of Cosmetic Science and Technology*. CRC Press, pp. 145–169. https://doi.org/10.1201/9780824741396

Carvalho, M.J., S. Oliveira, A.L., Santos Pedrosa, S., Pintado, M., Pinto-Ribeiro, I., Madureira, A.R., 2022. Skin microbiota and the cosmetic industry. Microb. Ecol. https://doi.org/10.1007/s00248-022-02070-0

Castillo, N.A., Valdez, A.L., Fariña, J.I., 2015. Microbial production of scleroglucan and downstream processing. Front. Microbiol. 6, 1–19. https://doi.org/10.3389/fmicb.2015.01106

Chew, A.-L., Maibach, H.I., 2001. Safety terminology. In: A.O. Barel, H.I. Maibach (Eds.), *Handbook of Cosmetic Science and Technology*. CRC Press, pp. 47–51. https://doi.org/10.1201/9780824741396

Chong, B.F., Blank, L.M., Mclaughlin, R., Nielsen, L.K., 2005. Microbial hyaluronic acid production. Appl. Microbiol. Biotechnol. 66, 341–351. https://doi.org/10.1007/s00253-004-1774-4

Daba, G.M., Elnahas, M.O., Elkhateeb, W.A., 2021. Contributions of exopolysaccharides from lactic acid bacteria as biotechnological tools in food, pharmaceutical, and medical applications. Int. J. Biol. Macromol. 173, 79–89. https://doi.org/10.1016/j.ijbiomac.2021.01.110

De Philippis, R., Sili, C., Paperi, R., Vincenzini, M., 2001. Exopolysaccharide-producing cyanobacteria and their possible exploitation: A review. J. Appl. Phycol. 13, 293–299. https://doi.org/10.1023/A:1017590425924

Ding, J., Wu, B., Chen, L., 2022. Application of marine microbial natural products in cosmetics. Front. Microbiol. 13. https://doi.org/10.3389/fmicb.2022.892505

Dubey, S.K., Dey, A., Singhvi, G., Pandey, M.M., Singh, V., Kesharwani, P., 2022. Emerging trends of nanotechnology in advanced cosmetics. Colloids Surf. B Biointerfaces. 214, 112440. https://doi.org/10.1016/j.colsurfb.2022.112440

Eastwood, M.A., Brydon, W.G., Anderson, D.M.W., 1987. The dietary effects of xanthan gum in man. Food Addit. Contam. 4, 17–26. https://doi.org/10.1080/02652038709373610

Elias, J.J., 2001. The microscopic structure of the epidermis and its derivatives. In: A.O. Barel, H.I. Maibach (Eds.), *Handbook of Cosmetic Science and Technology*. CRC Press, pp. 19–27. https://doi.org/10.1201/9780824741396

Environmental Working Group, 2022a. No Title [WWW Document]. URL www.ewg.org/who-we-are

Environmental Working Group, 2022b. Skin Deep Database [WWW Document]. URL www.ewg.org/skindeep/

Epstein, H., Simion, F.A., 2001. Emulsion-based skincare products: Formulating and measuring their moisturizing benefit. In: A.O. Barel, H.I. Maibach (Eds.), *Handbook of Cosmetic Science and Technology*. CRC Press, pp. 511–529. https://doi.org/10.1201/9780824741396

Escoffier, C., de Rigal, J., Rochefort, A., Vasselet, R., Lévêque, J.L., Agache, P.G., 1989. Age-related mechanical properties of human skin: An in vivo study. J. Invest. Dermatol. 93, 353–357.

Feingold, D.S., Gehatia, M., 1957. The structure and properties of levan, a polymer of D-fructose produced by cultures and cell-free extracts of aerobacter levanicum. J. Polym. Sci. 23, 783–790. https://doi.org/10.1002/pol.1957.1202310421

Fialho, A.M., Moreira, L.M., Granja, A.T., Popescu, A.O., Hoffmann, K., Sá-Correia, I., 2008. Occurrence, production, and applications of gellan: Current state and perspectives. Appl. Microbiol. Biotechnol. 79, 889–900. https://doi.org/10.1007/s00253-008-1496-0

Fiume, M.M., Heldreth, B., Bergfeld, W.F., Belsito, D.V., Hill, R.A., Klaassen, C.D., Liebler, D.C., Marks, J.G., Shank, R.C., Slaga, T.J., Snyder, P.W., Andersen, F.A., 2016. Safety assessment of microbial polysaccharide gums as used in cosmetics. Int. J. Toxicol. 35, 5S–49S. https://doi.org/10.1177/1091581816651606

Freitas, F., Alves, V.D., Reis, M.A.M., 2011. Advances in bacterial exopolysaccharides: From production to biotechnological applications. Trends Biotechnol. 29, 388–398. https://doi.org/10.1016/j.tibtech.2011.03.008

Freitas, F., Alves, V.D., Reis, M.A.M., 2015. Bacterial polysaccharides: Production and applications in cosmetic industry. In: K.G. Ramawat, J.-M. Mérillon (Eds.), *Polysaccharides*. Springer International Publishing, pp. 2017–2043. https://doi.org/10.1007/978-3-319-16298-0_63

Furtado, I.F.S.P.C., Sydney, E.B., Rodrigues, S.A., Sydney, A.C.N., 2022. Xanthan gum: Applications, challenges, and advantages of this asset of biotechnological origin. Biotechnol. Res. Innov. 6, e202204. https://doi.org/10.4322/biori.202205

Gallegos, A.M.A., Herrera Carrera, S., Parra, R., Keshavarz, T., Iqbal, H.M.N., 2016. Bacterial cellulose: A sustainable source to develop value-added products: A review. BioResources. 11, 5641–5655. https://doi.org/10.15376/biores.11.2.Gallegos

Ganceviciene, R., Liakou, A.I., Theodoridis, A., Makrantonaki, E., Zouboulis, C.C., 2012. Skin anti-aging strategies. Dermatoendocrinol. 4, 308–319. https://doi.org/10.4161/derm.22804

Gilbert, L., Savary, G., Grisel, M., Picard, C., 2013. Predicting sensory texture properties of cosmetic emulsions by physical measurements. Chemom. Intell. Lab. Syst. 124, 21–31. https://doi.org/10.1016/j.chemolab.2013.03.002

Gomes, C., Silva, A.C., Marques, A.C., Sousa Lobo, J., Amaral, M.H., 2020. Biotechnology applied to cosmetics and aesthetic medicines. Cosmetics. 7, 33. https://doi.org/10.3390/cosmetics7020033

Halder, R.M., Nootheti, P.K., 2003. Ethnic skin disorders overview. J. Am. Acad. Dermatol. 48, S143–S148. https://doi.org/10.1067/mjd.2003.274

Heath, R.S., Ruscoe, R.E., Turner, N.J., 2022. The beauty of biocatalysis: Sustainable synthesis of ingredients in cosmetics. Nat. Prod. Rep. 39, 335–388. https://doi.org/10.1039/D1NP00027F

Hien, N.Q., Van Phu, D., Duy, N.N., Quoc, L.A., 2012. Radiation synthesis and characterization of hyaluronan capped gold nanoparticles. Carbohydr. Polym. 89, 537–541. https://doi.org/10.1016/j.carbpol.2012.03.041

Holst, O., Müller-Loennies, S., 2007. Microbial polysaccharide structures. In: H. Kamerling (Ed.), *Comprehensive Glycoscience*. Elsevier, pp. 123–179. https://doi.org/10.1016/B978-044451967-2/00004-0

Ioelovich, M., 2008. Cellulose as a nanostructured polymer: A short review. BioResources. 3, 1403–1418. https://doi.org/10.15376/biores.3.4.Ioelovich

Jiang, G., He, J., Gan, L., Li, X., Xu, Z., Yang, L., Li, R., Tian, Y., 2022. Exopolysaccharide produced by pediococcus pentosaceus E8: structure, bio-activities, and its potential application. Front. Microbiol. 13, 1–16. https://doi.org/10.3389/fmicb.2022.923522

Juncan, A.M., Moisă, D.G., Santini, A., Morgovan, C., Rus, L.L., Vonica-țincu, A.L., Loghin, F., 2021. Advantages of hyaluronic acid and its combination with other bioactive ingredients in cosmeceuticals. Molecules. 26, 1–43. https://doi.org/10.3390/molecules26154429

Jung, J.H., Shinkai, S., 2005. Gels as Templates for Nanotubes, in: Topics in Current Chemistry. Springer, Berlin, Heidelberg, pp. 223–260. https://doi.org/10.1007/b99915

Kale, N., Bajpai, N., Wasule, D., 2022. Toxic chemicals in cosmetics. World J. Pharm. Res. 11, 409–427. https://doi.org/10.20959/wjpr202210-24882

Kang, S.A., Jang, K.-H., Seo, J.-W., Kim, K.H., Kim, Y.H., Rairakhwada, D., Seo, M.Y., Lee, J.O., Ha, S. Do, Kim, C.-H., Rhee, S.-K., 2009. Levan: Applications and perspectives. In: B.H.A. Rehm (Ed.), *Microbial Production of Biopolymers and Polymer Precursors: Applications and Perspectives.* Caister Academic Press, pp. 145–161. https://doi.org/10.21775/9781910190395

Kanmani, P., Satish Kumar, R., Yuvaraj, N., Paari, K.A., Pattukumar, V., Arul, V., 2013. Probiotics and its functionally valuable products: A review. Crit. Rev. Food Sci. Nutr. 53, 641–658. https://doi.org/10.1080/10408398.2011.553752

Kasapis, S., Morris, E.R., Gross, M., Rudolph, K., 1994. Solution properties of levan polysaccharide from *Pseudomonas syringae* pv. phaseolicola, and its possible primary role as a blocker of recognition during pathogenesis. Carbohydr. Polym. 23, 55–64. https://doi.org/10.1016/0144-8617(94)90090-6

Kehrer, J.P., 2000. The Haber-Weiss reaction and mechanisms of toxicity. Toxicology. 149, 43–50. https://doi.org/10.1016/s0300-483x(00)00231-6

Khan, A.D., Alam, M.N., 2019. Cosmetics and their associated adverse effects: A review. J. Appl. Pharm. Sci. Res. 2, 1–6. https://doi.org/10.31069/japsr.v2i1.1

Kim, K.H., Chung, C.B., Kim, Y.H., Kim, K.S., Han, C.S., Kim, C.H., 2005. Cosmeceutical properties of levan produced by *Zymomonas mobilis*. J. Cosmet. Sci. 56, 395–406.

Kong, M., Chen, X.G., Kweon, D.K., Park, H.J., 2011b. Investigations on skin permeation of hyaluronic acid based nanoemulsion as transdermal carrier. Carbohydr. Polym. 86, 837–843. https://doi.org/10.1016/j.carbpol.2011.05.027

Kong, M., Chen, X.G., Park, H., 2011a. Design and investigation of nanoemulsified carrier based on amphiphile-modified hyaluronic acid. Carbohydr. Polym. 83, 462–469. https://doi.org/10.1016/j.carbpol.2010.08.001

Lahiri, D., Nag, M., Dutta, B., Dey, A., Sarkar, T., Pati, S., Edinur, H.A., Abdul Kari, Z., Mohd Noor, N.H., Ray, R.R., 2021. Bacterial cellulose: Production, characterization, and application as antimicrobial agent. Int. J. Mol. Sci. 22, 12984. https://doi.org/10.3390/ijms222312984

Liu, L., Liu, Y., Li, J., Du, G., Chen, J., 2011. Microbial production of hyaluronic acid: Current state, challenges, and perspectives. Microb. Cell Fact. 10, 99. https://doi.org/10.1186/1475-2859-10-99

Liu, Y., Han, Y., Zhu, T., Wu, X., Yu, W., Zhu, J., Shang, Y., Lin, X., Zhao, T., 2021. Targeting delivery and minimizing epidermal diffusion of tranexamic acid by hyaluronic acid-coated liposome nanogels for topical hyperpigmentation treatment. Drug Deliv. 28, 2100–2107. https://doi.org/10.1080/10717544.2021.1983081

Liu, Y., Liu, J., Zhang, X., Zhang, R., Huang, Y., Wu, C., 2010. In situ gelling Gelrite/alginate formulations as vehicles for ophthalmic drug delivery. AAPS PharmSciTech. 11, 610–620. https://doi.org/10.1208/s12249-010-9413-0

Marcellin, E., Chen, W., Nielsen, L.K., 2009. *Microbial Hyaluronic Acid Biosynthesis, Microbial Production of Biopolymers and Polymer Precursors: Applications and Perspectives.* Caister Academic Press. https://doi.org/10.21775/9781910190395

Matsumura, Y., Ananthaswamy, H.N., 2002. Short-term and long-term cellular and molecular events following UV irradiation of skin: Implications for molecular medicine. Expert Rev. Mol. Med. 2002, 1–22. https://doi.org/10.1017/s146239940200532x

Mbituyimana, B., Liu, L., Ye, W., Ode, B.O., Zhang, K., Chen, J., Thomas, S., Victor, R., Shi, Z., Yang, G., 2021. Bacterial cellulose-based composites for biomedical and cosmetic applications: Research progress and existing products. Carbohydr. Polym. 273, 118565. https://doi.org/10.1016/j.carbpol.2021.118565

McNeely, W.H., Kovacs, P., 1975. The physiological effects of alginates and xanthan gum. In: A. Jeanes, J. Hodge (Eds.), *Physiological Effects of Food Carbohydrates*. American Chemical Society, pp. 269–281. https://doi.org/10.1021/bk-1975-0015.ch017

Milstein, S., Bailey, J.E., Halper, A.R., 2001. Definition of cosmetics. In: A.O. Barel, H.I. Maibach (Eds.), *Handbook of Cosmetic Science and Technology*. CRC Press, pp. 5–18. https://doi.org/10.1201/9780824741396

Moorhouse, R., Walkinshaw, M.D., Arnott, S., 1977. Xanthan gum—Molecular conformation and interactions. In: *Extracellular Microbial Polysaccharides*, pp. 90–102. https://doi.org/10.1021/bk-1977-0045.ch007

Moreira, S., Silva, N.B., Almeida-Lima, J., Rocha, H.A.O., Medeiros, S.R.B., Alves, C., Gama, F.M., 2009. BC nanofibres: In vitro study of genotoxicity and cell proliferation. Toxicol. Lett. 189, 235–241. https://doi.org/10.1016/j.toxlet.2009.06.849

Navarro, M., Ruberte, J., Carretero, A., 2017. Common integument. In: J. Ruberte, A. Carretero, M. Navarro (Eds.), *Morphological Mouse Phenotyping*. Academic Press, pp. 541–562. https://doi.org/10.1016/B978-0-12-812972-2.50016-7

Nwodo, U.U., Green, E., Okoh, A.I., 2012. Bacterial exopolysaccharides: Functionality and prospects. Int. J. Mol. Sci. 13, 14002–14015. https://doi.org/10.3390/ijms131114002

Palaninathan, V., Chauhan, N., Poulose, A.C., Raveendran, S., Mizuki, T., Hasumura, T., Fukuda, T., Morimoto, H., Yoshida, Y., Maekawa, T., Sakthi Kumar, D., 2014. Acetosulfation of bacterial cellulose: An unexplored promising incipient candidate for highly transparent thin film. Mater. Express. 4, 415–421. https://doi.org/10.1166/mex.2014.1191

Park, S., 2022. Biochemical, structural and physical changes in aging human skin, and their relationship. Biogerontology. 23, 275–288. https://doi.org/10.1007/s10522-022-09959-w

Patravale, V.B., Mandawgade, S.D., 2008. Novel cosmetic delivery systems: An application update. Int. J. Cosmet. Sci. 30, 19–33. https://doi.org/10.1111/j.1468-2494.2008.00416.x

Pires, A.M.B., Macedo, A.C., Eguchi, S.Y., Santana, M.H.A., 2010. Microbial production of hyaluronic acid from agricultural resource derivatives. Bioresour. Technol. 101, 6506–6509. https://doi.org/10.1016/j.biortech.2010.03.074

Prajapati, V.D., Jani, G.K., Zala, B.S., Khutliwala, T.A., 2013. An insight into the emerging exopolysaccharide gellan gum as a novel polymer. Carbohydr. Polym. 93, 670–678. https://doi.org/10.1016/j.carbpol.2013.01.030

Rahman, T., Hosen, I., Islam, M.M.T., Shekhar, H.U., 2012. Oxidative stress and human health. Adv. Biosci. Biotechnol. 03, 997–1019. https://doi.org/10.4236/abb.2012.327123

Rehm, B.H.A., 2010. Bacterial polymers: Biosynthesis, modifications and applications. Nat. Rev. Microbiol. 8, 578–592. https://doi.org/10.1038/nrmicro2354

Resende, D.I.S.P., Ferreira, M.S., Lobo, J.M.S., Sousa, E., Almeida, I.F., 2022. Skin Depigmenting agents in anti-aging cosmetics: A medicinal perspective on emerging ingredients. Appl. Sci. 12, 775. https://doi.org/10.3390/app12020775

Sarkar, R., Arora, P., Garg, K., 2013. Cosmeceuticals for hyperpigmentation: What is available? J. Cutan. Aesthet. Surg. 6, 4. https://doi.org/10.4103/0974-2077.110089

SCCS, T., 2021. The SCCS notes of guidance for the testing of cosmetic ingredients and their safety evaluation, 11th revision, 30–31 March 2021, SCCS/1628/21. Regul. Toxicol. Pharmacol. 127, 105052. https://doi.org/10.1016/j.yrtph.2021.105052

Schrader, T.J., 2000. Examination of selected food additives and organochlorine food contaminants for androgenic activity in Vitro. Toxicol. Sci. 53, 278–288. https://doi.org/10.1093/toxsci/53.2.278

Schrader, T.J., Cooke, G.M., 2000. Examination of selected food additives and organochlorine food contaminants for androgenic activity in vitro. Toxicol. Sci. 53, 278–288. https://doi.org/10.1093/toxsci/53.2.278

Shankar, T., Arts, V., Pandian, V., Sivakumar, T., 2014. Screening of exopolysaccharide producing bacterium frateuria aurentia from elephant dung. Appl. Sci. Rep. 1. https://doi. org/10.15192/PSCP.ASR.2014.1.3.105109

Sies, H., Jones, D., 2007. Oxidative stress. In: *Encyclopedia of Stress*. Elsevier, pp. 45–48. https://doi.org/10.1016/B978-012373947-6.00285-3

Smit, N., Vicanova, J., Pavel, S., 2009. The hunt for natural skin whitening agents. Int. J. Mol. Sci. 10, 5326–5349. https://doi.org/10.3390/ijms10125326

Smith, C.N., Alexander, B.R., 2005. The relative cytotoxicity of personal care preservative systems in Balb/C 3T3 clone A31 embryonic mouse cells and the effect of selected preservative systems upon the toxicity of a standard rinse-off formulation. Toxicol. Vitr. 19, 963–969. https://doi.org/10.1016/j.tiv.2005.06.014

Soumiya, S., Santhiagu, A., Manjusha Chemmattu, M., 2021. Optimization of cultural conditions of gellan gum production from recombinant *Sphingomonas paucimobilis* ATCC 31461 and its characterization. J. Appl. Biol. Biotechnol. 9(1), 58–67. https://doi. org/10.7324/JABB.2021.9108

Sworn, G., 2021. Xanthan gum. In: *Handbook of Hydrocolloids*. Elsevier, 833–853. https:// doi.org/10.1016/B978-0-12-820104-6.00004-8

Tabata, N., O'Goshi, K., Zhen, Y.X., Kligman, A.M., Tagami, H., 2000. Biophysical assessment of persistent effects of moisturizers after their daily applications: Evaluation of corneotherapy. Dermatology. 200, 308–313. https://doi.org/10.1159/000018393

Tejal, P., Nishad, D., Amisha, J., Umesh, G., Desai, K.T., Bansal, R.K., 2013. Cosmetics and health: Usage, perceptions and awareness. Bangladesh J. Med. Sci. 12, 392–397. https:// doi.org/10.3329/bjms.v12i4.13330

Trommer, H., Neubert, R.H.H., 2005. The examination of polysaccharides as potential antioxidative compounds for topical administration using a lipid model system. Int. J. Pharm. 298, 153–163. https://doi.org/10.1016/j.ijpharm.2005.04.024

Turkoglu, M., Pekmezci, E., Sakr, A., 1999. Evaluation of irritation potential of surfactant mixtures. Int. J. Cosmet. Sci. 21, 371–382. https://doi.org/10.1046/j.1467-2494.1999.211920.x

Ulrich, C., Jürgensen, J.S., Degen, A., Hackethal, M., Ulrich, M., Patel, M.J., Eberle, J., Terhorst, D., Sterry, W., Stockfleth, E., 2009. Prevention of non-melanoma skin cancer in organ transplant patients by regular use of a sunscreen: A 24 months, prospective, case-control study. Br. J. Dermatol. 161, 78–84. https://doi.org/10.1111/j.1365-2133.2009.09453.x

Wasilewski, T., Seweryn, A., Pannert, D., Kierul, K., Domżał-Kędzia, M., Hordyjewicz-Baran, Z., Łukaszewicz, M., Lewińska, A., 2022. Application of levan-rich digestate extract in the production of safe-to-use and functional natural body wash cosmetics. Molecules. 27, 2793. https://doi.org/10.3390/molecules27092793

Woodard, G., Woodard, M.W., McNeely, W.H., Kovacs, P., Cronin, M.T.I., 1973. Xanthan gum: Safety evaluation by two-year feeding studies in rats and dogs and a three-generation reproduction study in rats. Toxicol. Appl. Pharmacol. 24, 30–36. https://doi. org/10.1016/0041-008X(73)90178-6

Xu, H., Zheng, Y.-W., Liu, Q., Liu, L.-P., Luo, F.-L., Zhou, H.-C., Isoda, H., Ohkohchi, N., Li, Y.-M., 2018. Reactive oxygen species in skin repair, regeneration, aging, and inflammation. In: *Reactive Oxygen Species (ROS) in Living Cells*. InTech. https://doi. org/10.5772/intechopen.72747

Yamada, M., Kametani, Y., 2022. Preparation of gellan gum-inorganic composite film and its metal ion accumulation property. J. Compos. Sci. 6, 42. https://doi.org/10.3390/ jcs6020042

Yamada, T., Kawasaki, T., 2005. Microbial synthesis of hyaluronan and chitin: New approaches. J. Biosci. Bioeng. 99, 521–528. https://doi.org/10.1263/jbb.99.521

Young, A.R., Claveau, J., Rossi, A.B., 2017. Ultraviolet radiation and the skin: Photobiology and sunscreen photoprotection. J. Am. Acad. Dermatol. 76, S100–S109. https://doi.org/10.1016/j.jaad.2016.09.038

Zamboni, F., Vieira, S., Reis, R.L., Miguel Oliveira, J., Collins, M.N., 2018. The potential of hyaluronic acid in immunoprotection and immunomodulation: Chemistry, processing and function. Prog. Mater. Sci. 97, 97–122. https://doi.org/10.1016/j.pmatsci.2018.04.003

11 Biomedical applications of exopolysaccharides

Thaaranni Bashkeran, Shinji Sakai,
Retno Wahyu Nurhayati, Minh Hong Nguyen,
Wildan Mubarok, Ryota Goto,
Dinda Shezaria Hardy Lubis,
Auzan Luthfi, and Masrina Mohd Nadzir

11.1 INTRODUCTION

Microbial exopolysaccharides (EPSs) are polymers of high molecular weight that are composed of long molecular chains of sugar units. The EPS are secreted by microorganisms such as bacteria, fungi, and archaea into their surrounding environment, and surrounds the cell membrane of a microbial cell (Rana and Upadhyay, 2020). EPSs are produced by microbes in nature for the protection of cells, cell adhesion, and for the formation of biofilm (Donot et al., 2012). Homo- and heteropolysaccharides are the forms in which the EPS are discovered (Chaisuwan et al., 2020). Hyaluronic acid, alginate, and xanthan gum are examples of heteropolysaccharides, while cellulose, dextran, curdlan, pullulan, and levan are homopolysaccharides.

The microbial EPSs' diverse functionalities and structural options that can be modified for a specific application make the EPSs a popular alternative to synthetic polymers in the biomedical industries. Not only the EPSs are degradable via biological processes and regarded as generally safe, they are also inexpensive and could be obtained from many sources (Laubach et al., 2021). EPSs such as alginate, chitosan, curdlan, dextran, gellan gum, hyaluronic acid, pullulan, xanthan, and levan are among the EPSs that are commonly studied and used in drug delivery, tissue engineering, and wound healing applications. Kefiran and polyglucuronic acid are rarely utilized, but their physicochemical properties showed great potential for wide usage in the biomedical industries.

11.2 MICROBIAL BIOSYNTHESIS OF EPS

In bacteria including cyanobacteria, it is typically accepted that the EPS assembly mechanism includes four stages: (i) the nucleotide sugar activation and transport in the cytoplasm, (ii) the repeat units assembly by linking sugars onto a lipid carrier, (iii) the polymerization, (iv) the EPS extracellular export (Schmid, 2018). The EPSs intracellular biosynthesis is a complex process, which is influenced by environmental conditions such as pH, incubation period, temperature, and culture media. The differences can

DOI: 10.1201/9781003342687-11

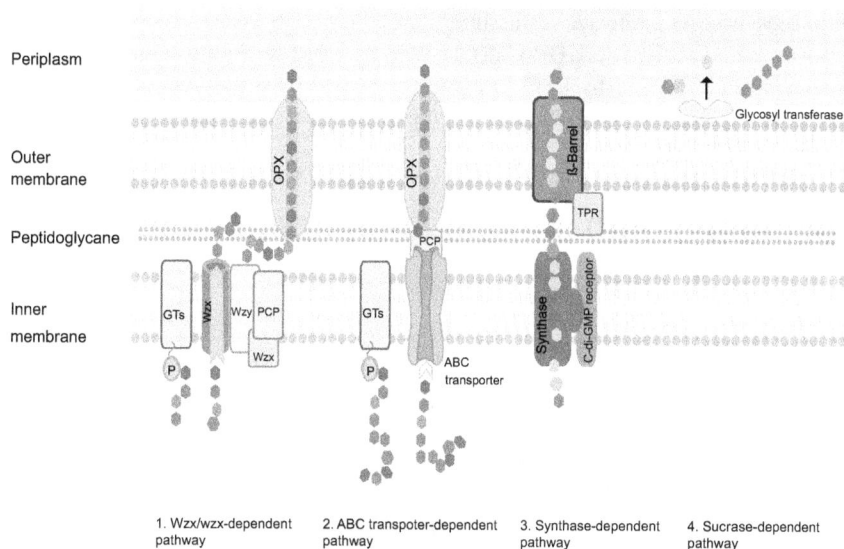

FIGURE 11.1 The EPS synthesis microbial pathways—GTs: glycosyltransferases; PCP: polysaccharide co-polymerase; OPX: outer membrane polysaccharide export; TPR: tetratri-copeptide repeat protein

Figure and caption reused from Nadzir et al. (2021), used under the Creative Commons License (http://creativecommons.org/licenses/by/4.0/)

be attributed to the dissimilarity between the heteropolysaccharides and homopoly-saccharides structures and the biosynthesis enzymes. Four general EPS biosynthesis mechanisms (Figure 11.1) have been known in bacterial cells: (i) Wzx/Wzy-dependent pathway, (ii) ABC transporter-dependent pathway, (iii) Synthase-dependent pathway, and (iv) extracellular biosynthesis via sucrase protein (Angelin and Kavitha, 2020; Ates, 2015; Rana and Upadhyay, 2020; Schmid, 2018; Schmid et al., 2015).

Fungi have great prospects for the synthesis of EPSs for biomedical applications due to their uniform particle size and antimicrobial activities. The typical EPSs obtained from the fungal species are chitin, chitosan, elsinan, pullulan, pleuran, len-tinan, scleroglucan, schizophyllan, and grifolan (Merzendorfer, 2011; Selvasekaran et al., 2021). The mechanism of fungal EPS production is dependent on the specific EPS and fungal species. It is difficult to assume the common pathway for EPS bio-synthesis in fungi. Among common EPS, chitin and chitosan synthesis have been investigated over the past years. Through the data of genome sequencing, CHS genes categorized into five or seven classes are responsible for chitin formation. As a model organism for the study of chitin synthesis, the genome of *Saccharomyces cerevisiae* was performed. *S. cerevisiae* genome contains three CHS genes (ScCHS1, ScCHS2, and ScCHS3). However, the genome of *Aspergillus nidulans* harbours eight CHS genes, AnCHSA, AnCHSB, AnCHSC, AnCHSD, AnCHSF, AnCHSG, AnCSMA, and AnCSMB, encode the class II, III, I, IV, III, VII, V, and VI enzymes, respectively (Horiuchi, 2009). Although the structural studies were insufficient, the chitin biosynthesis pathway was assumed based on CHS genes and related glycosyltransferases. Firstly, the UDP-GlcNAc is transferred to the growing chitin polymer's non-reducing end by CHS enzymes, followed by a polymerization reaction akin to inverting glycosyltransferases from bacteria (Lairson and Withers, 2004; Merzendorfer, 2011).

11.3 EPS TYPES AND CHARACTERISTICS FOR BIOMEDICAL APPLICATION

11.3.1 ALGINATE

Alginate is a water-soluble polysaccharide possessing high viscosity and can be ionically cross-linked with divalent ions forming a hydrogel. The microbes mainly produce D-mannuronic acid, in which L-guluronic acid is later formed through epimerization of the D-mannuronic acid using C-5-epimerase (Laksen et al., 1986; Pindar and Bucke, 1975). Microbial alginate is characterized by the presence of O-acetyl groups in the D-mannuronic acid residues, which is absent in the alginate from brown algae (Davidson et al., 1977; Evans and Linker, 1973). The presence of O-acetyl groups increases the interaction of the alginate with the water molecules which results in higher solubility. More importantly, the O-acetyl groups also affect the ionic cross-linking, viscoelasticity, water capacity, and swelling of the microbial alginate (Castillo et al., 2013; Skjåk-Bræk et al., 1989).

Microbial alginate was first reported from *Pseudomonas* isolated from cystic fibrosis patients (Linker and Jones, 1964, 1966). Following this report, microbial alginate was also reported to be produced by a variety of *Pseudomonas*, such as *P. aeruginosa* (Evans and Linker, 1973), *P. mendocina* (Sengha et al., 1989), *P. fluorescens* (Gimmestad et al., 2003), *P. putida* (Conti et al., 1994), and *P. syringae* (Krishna et al., 2022). Apart from *Pseudomonas* spp., alginate is also reported to be produced by *Azotobacter vinelandii*, a nitrogen-fixing soil microbe (Clementi, 1997).

In recent years, microbial alginate has received great interest. The main advantage of microbial alginate compared to algal alginate is that the environmental condition of the microbial culture can be controlled, which results in less variation in the alginate produced by the microbes (Valentine et al., 2020). More importantly, microbial alginate can be designed to possess a certain M/G ratio (e.g., increasing C-5 epimerase AlgE1 to generate very long guluronic acid residues) (Aarstad et al., 2019), molecular weight (e.g., culturing *A. vinelandii* under microaerophilic conditions) (García et al., 2020), and O-acetylation degree (Chanasit et al., 2020), which allows tailoring of the properties of the alginate.

However, several challenges remain, for example, the low yield of alginate and the release of toxins by the microbes. Efforts have been made to address these challenges, such as deleting the *mucG* gene, a negative regulator of alginate polymerization in *A. vinelandii*, which increases the production of the alginate (Ahumada-Manuel et al., 2017). Generation of mutants devoid of exotoxin A, haemolytic toxin phospholipase C, and pyocyanin also can increase the safety of the alginate produced by the bacteria (Valentine et al., 2020). Upon addressing these issues, microbial alginate can be a promising source of alginate for commercial or biomedical applications.

11.3.2 CELLULOSE

French chemist Anselme Payen took the first systematic effort to identify cellulose in 1837 where a substance that could be broken down into glucose just like starch was discovered while examining the different types of woods. This novel chemical was named "cellulose" as it was extracted from the plant cell wall (Mohanty et al., 2000). Other than plants, cellulose could also be obtained from gram-negative bacteria of the genera *Aerobacter, Pseudomonas, Acetobacter, Azobacter, Achromobacter, Salmonella, Agrobacterium, Alcaligenes, Gluconacetobacter, Rhizobium,* and *Sarcina* (Nadzir et al., 2021).

Using various methods, naturally occurring cellulose is converted to micro- or nanoscale materials to create microcrystalline cellulose, microfibrillar cellulose, and whiskers. The crystallinity, form, and level of polymerization of these nanostructures set them apart from one another (Kondo, 1997). Because of its great crystallinity, native cellulose is resistant to disintegration. Cellulose solvents should compete for available intermolecular H-bond interactions to detach polymer chains from one another, resulting in biopolymer dissolution (Kondo, 1997). In addition to good dissolving power, an ideal cellulose solvent must also meet other criteria, such as low toxicity and viscosity, easy to recycle, low melting temperature, and great thermal stability. The biopolymer's crystallinity and degree of polymerization, as well as mixing time and temperature, all have a role in the dissolving process (Pinkert et al., 2010). Due to all these properties, cellulose and its derivatives are widely used in bioadhesive and mucoadhesive drug delivery systems. Adding to that, these polysaccharides also contribute to pharmaceutical coating processes to defend degradation of sensitive drugs (Arca et al., 2018). Besides, nanocellulose-reinforced biocomposites are created for biomedical applications. According to a study, a fiber made of electrospun poly and nanocellulose was developed to serve as a scaffold for tissue engineering applications (Si et al., 2016).

11.3.3 CURDLAN

Curdlan is a linear glucan that was brought to light in the year 1966 when Harada and colleagues made the initial discovery of *Alcaligenes faecalis* var. myxogenes 10C3 strain (Harada et al., 1966). According to viscometry measurements, curdlan is easily made up of 12,000 glucose units (Futatsuyama et al., 1999). This fact aids to a particular importance as it boosts the solubility in dilute alkalis such as diluted sodium hydroxide, formic acid, and aprotic solvents in relative to other commonly used linear glucans. However, water and most organic solvents cannot dissolve curdlan. Regarding application, its rheological characteristics are particularly intriguing. Upon heating an aqueous suspension of curdlan, two different types of gels are resulted. The first type of gel would form at a comparatively higher temperature (80°C), which is the high-set thermal nonreversible gel, while the second type of gel forms at 55°C, known as the low-set thermal reversible gel (Zhang and Edgar, 2014). The modification and adaptation of curdlan's helical structure at varying temperatures consisting of single helices and loosely entangled triple helices at room temperature and concentrated rodlike triple helices at elevated temperature is highly

related to this gelation property of curdlan. A previous study reported that this curdlan gel undergoes a sharp melting process at a temperature ranging from 140°C to
160°C (Zhang et al., 2000).

Due to curdlan's exceptional rheological capabilities, gel forming ability, edibil-
ity, and thermal properties, curdlan and its derivatives may be used in the field of
regenerative medicine, where it induces proliferation and migration of human keratinocytes that amplifies wound healing (Wang et al., 2004). Other than that, curdlan plays a vital role in the pharmaceutical industry. The backbone, branch chains,
conformation, solubility, and degree of polymerization of curdlan highly influence
its biological activity (Han et al., 2020). In a recent study on curdlan and its derivatives, the expression of T cell proliferation and the cytokine 1L-2 stimulated by
the cell receptors alleviates cancer metastasis and dramatically boost anticancer
immunity (Bao et al., 2021). Additionally, curdlan helps to prevent the proliferation
of human retrovirus without the need to encapsulate drugs with high concentrations
(Su et al., 2021).

11.3.4 DEXTRAN

Dextran is an EPS synthesized by the lactic acid bacteria. The major types of lactic acid bacteria used in the synthesis of dextran are *Leuconostoc*, *Lactobacillus*,
Weissella, and *Streptococcus* (Kothari et al., 2014). This EPS is classified as soluble
polysaccharide because of its tendency to absorb substantial amount of water and
form hydrogel (Campos et al., 2016). Studies have reported that dextran, which presents a completely linear structure, would be perfectly soluble in water. This is attributed by the accessibility of hydroxy groups to interact with water molecules (Gil
et al., 2008). Molecular weight equally donates in the solubility of dextran where
higher molecular weight polysaccharide exhibits lesser solubility degree in relative
to low molecular weight polysaccharides (Whistler, 1973).

Dextran and derivatives are extensively used in specialized drug delivery systems. Dextran and the correct therapeutic moiety can be coupled to create macromolecule conjugates which extends drug efficacy, improved targeting, reduce
immunogenicity, and reduces toxicity (Chen et al., 2020). Furthermore, dextran is
utilized in the medical industry to lessen thrombosis in the arteries. Dextran binds to
erythrocytes, platelets, and vascular endothelium, enhancing their electronegativity,
lowering erythrocyte aggregation and platelet adhesiveness in the process (Bhavani
and Nisha, 2010).

11.3.5 GELLAN

A meticulous screening of more than 30,000 bacterial isolates in quest of a few polysaccharides that could exhibit high solution viscosity led to the discovery of gellan in
year 1977 (Milas et al., 1990). *Sphingomonas paucimobilis* (ATCC 31461) or previously known as *Pseudomonas elodea* is the bacterium responsible to produce gellan.
It is vital to note that the gellan's conformation and structure is highly dependent on
factors such as temperature, polymer content, aqueous environment, and the monovalent or divalent cations' existence in the solution. Gellan develops an organized

helix of double strands at low temperatures, whereas a single-stranded polysaccharide forms at elevated temperatures, greatly reducing its solution viscosity. Below the transition temperature ($30°C$ to $50°C$), a stiff structure is produced, which is roughly at $35°C$, also known as the setting point (Yuguchi et al., 1993). More chains are accumulated at the junction zones as the temperature falls below the setting point, whereas some chains are released as the temperature rises. Additionally, the number of connection zone is significantly increased, and they become more heat-resistant with monovalent or divalent cations, increasing the gelling potential of gellan (Watase and Nishinari, 1993).

Gellan has mostly been employed in oral drug administration as a matrix-forming excipient for prolonged release or as a disintegrating agent in immediate release tablets. While the behavior of swelling is the basis for both applications, the concentration of gellan is essential for the outcome. In this case, delayed release tablets often consist of higher levels of gellan, whereas fast drug release tablets only need a small amount of gellan (Morris et al., 2012; Osmałek et al., 2014).

11.3.6 HYALURONIC ACID

Hyaluronic acid possesses high viscosity even in low diluted concentrations (Chong et al., 2005; Laurent et al., 1996). In addition, this EPS has viscoelastic and hygroscopic properties (Gupta et al., 2019b). The unique property of hyaluronic acid is that it exerts different biological responses based on its molecular weight. High-molecular-weight hyaluronic acid ($>10^6$ Da) is reported to play a role in mucoadhesion and inhibition of cancer aggressiveness (Hansen et al., 2017; Ooki et al., 2019). Meanwhile, low-molecular-weight hyaluronic acid ($10^4 \sim 10^5$ Da) and small-fragment hyaluronic acid (10^3 Da) promote cancer aggressiveness and induce inflammation response (Mascarenhas et al., 2004; Ohkawara et al., 2000; Ooki et al., 2019; Termeer et al., 2002). In addition, as the native extracellular matrix's natural component, cells possess specific receptors for hyaluronic acid: CD44 and receptor for hyaluronic acid-mediated motility (Kouvidi et al., 2011; Misra et al., 2015).

Microbial hyaluronic acid is widely known to be synthesized as extracellular capsule by pathogenic bacteria *Streptococcus* spp., including *S. zooepidemicus*, *S. equi*, *S. pyogenes*, and *S. uberis* (Blank et al., 2008; Field et al., 2003; Singh et al., 2014). In addition, hyaluronic acid is also found in other pathogenic bacteria, such as *Pasteurella multocida* (Guan et al., 2020) and *Cryptococcus neoformans* (Jong et al., 2007). Naturally, the hyaluronic acid in the microbial capsule is used to protect the bacteria from phagocytic killing (Guan et al., 2020; Wessels et al., 1991; Wibawan et al., 1999) and anchor the bacteria to the mucosal tissue via CD44 interaction (Schrager et al., 1998).

Industrially, microbial hyaluronic acid has been produced mainly using the fermentation of *S. zooepidemicus* and *S. equi* (Chong et al., 2005). However, concern remains as these microbes are categorized as Group C Streptococci that could induce hemolysis and diseases such as meningitis and pneumonia (Kłos and Wójkowska-Mach, 2017). In addition, the release of exotoxins and hyaluronidase also poses a risk (Izawa et al., 2009). To solve these issues, several recombinant strains have been developed using *Bacillus subtilis* (Amjad Zanjani et al., 2022; Westbrook et al., 2018), *Corynebacterium glutamicum* (Cheng et al., 2019), *Escherichia coli* (Woo et al., 2019), *Lactococcus lactis* (Sunguroğlu et al., 2018), and *S. thermophilus* (Izawa et al., 2011). In addition to the improved safety, hyaluronic acid production using recombinant strains can be tuned to possess a certain molecular weight. With the high demand of hyaluronic acid, microbial hyaluronic acid is a promising source for biomedical application.

11.3.7 KEFIRAN

Kefiran is an EPS which is majorly produced by *Lactobacillus kefiranofaciens* that consists of multiple hydroxyl groups obtained from microorganisms present in kefir grains. Viscoelastic is an important rheological property of kefiran. However, the viscosity of kefiran is highly dependent on different levels of polymerization of the polysaccharide. At 25°C, the inherent viscosity is evaluated at 6.0 dL/g and 5.95 dL/g, respectively (Micheli et al., 1999). In terms of gelling properties of kefiran, viscoelastic gels are produced at comparatively lower temperature. Consequently, kefiran has tremendous potential in transforming to film forming agent (Ghasemlou et al., 2011). At low concentrations that ranges up to 4 g/l, kefiran exhibits Newtonian behavior, whereas at higher concentrations, it exhibits pseudoplastic behavior. Other than that, at a particular shear rate, lower biopolymer concentration results in a decrease in apparent viscosity (Piermaria et al., 2009).

In the medical industry, kefiran could elevate the mechanical and physical qualities of nanocomposites. This allows the development of kefiran usage in many nanotechnology domains such as kefiran-based bio-nanocomposites synthesized along with starch and zinc oxide that possessed diameter of ranging from 20–30 mm (Babaei-Ghazvini et al., 2018; Niska et al., 2018).

11.3.8 LEVAN

The discovery of levan by Lipmann took place between 1870 to 1881, and Greig-Smith and Steel reported that levan was made by microbes isolated from *Eucalyptus stuarini* secretion in the year 1902 (Zhang et al., 2014). Levan is soluble in water and oil engendered by β-(2,6) linkages presented in its chemical structure. High solubility is achieved in room temperature water. However, levan is insoluble in organic solvents such as acetone, ethanol, toluene, n-propanol, isopropanol, ethyl lactate, ethylene carbonate, d-limonene, propylene carbonate, methylethylketone, and methanol (Tomulescu et al., 2016). Most other polymers swell in water, whereas levan does not. Levan's intertwined branches increase its cohesive strength, and the substance's abundance of hydroxyl groups aids in the formation of adhesion bonds with a range of substrates (Tomulescu et al., 2016). This polysaccharide is strongly adhesive and easily eliminated with water (Kazak Sarilmiser and Toksoy Oner, 2014; Srikanth et al., 2015). Levan has lower intrinsic viscosity in comparison to other large molecular weight compounds. In specific, levans with molecular weight ranging from 16 to 24 million Da results in viscosity ranging from 0.07 to 0.18 dL/g (Stivala and Bahary, 1978), and stable viscosity is achieved at room temperature ranges from pH 4 to 11 (Harada, 1965). Adding to that, levan is very stable against heat as it has a decomposition point of 225°C and acid and alkali media (Kazak Sarilmiser and Toksoy Oner, 2014; Silbir et al., 2014).

Levan is a versatile chemical molecule with diversified roles due to its special mix of characteristics. Films and membranes are among the uses for levan's adhesive qualities that have garnered a lot of interest.

11.3.9 PULLULAN

The early observation on the formation of pullulan by a yeast like fungus, *Aureobasidium pullulans* was made by Bauer (1938). This EPS was first isolated from *A. pullulans* and characterized about two decades later by Bernier (1958). Other than *A. pullulans*, other species belonging to the genus *Aureobasidium*, namely, *A. leucospermi, A. melanogenum, A. nambiae, A. proteae*, and *A. thailandense* are also associated with pullulan production (Cheng et al., 2011; Chi et al., 2009; Hamidi et al., 2019; Prasongsuk et al., 2018; Wei et al., 2021). Other microorganisms that have been identified as having the capacity to produce pullulan includes *Tremella mesenterica* (Fraser and Jennings, 1971), *Cryphonectria parasitica* (Forabosco et al., 2006), and *Teloschistes flavicans* (Reis et al., 2002).

Pullulans is soluble in hot or cold water and flexible, owing to its a-linkages. Furthermore, pullulan is non-hygroscopic, non-carcinogenic, non-immunogenic, non-ionic, and nontoxic; herein contributing to pullulan application as hydrogels, beads, fibers, and nanoparticles in the biomedical industries (Coltelli et al., 2020; Leathers, 2003; Li et al., 2017; Rai et al., 2021; Singh et al., 2016).

11.3.10 XANTHAN

Xanthan was discovered in the 1950s by Allene Rosalind Jeanes (Petri, 2015). This EPS is produced by bacteria of the genus *Xanthomonas*, specifically *X. campestris*. Xanthan is soluble in cold and hot water and has an extremely pseudoplastic behavior. Furthermore, it is stable over a wide range of pH (2–12) and has a greater thermal stability against hydrolysis compared to other water-soluble polysaccharides (Kumar et al., 2018; Riaz et al., 2021; Stokke and Christensen, 1996). This EPS is known for its excellent biocompatibility and nontoxicity (Chellat et al., 2000; Han et al., 2012). It is a well-known natural polymer in the biomedical field, in which its application includes as an excipient in tablets as supporting hydrogels for drug release, and as scaffolds for cells (Petri, 2015).

11.3.11 POLYGLUCURONIC ACID

Polyglucuronic acid is a natural polysaccharide found in the cell walls and extracellular matrix of *Mucor rouxii* species (De Ruiter et al., 1992; Dow et al., 1983), in the cell wall of green algae, and as an EPS of a mutated strain of *Ensifer* (formerly *Sinorhizobium*) *meliloti* M5N1CS (Heyraud et al., 1993).

An interesting property of polyglucuronic acid is the gelling property of its aqueous solution in the presence of monovalent, divalent, or trivalent cations depending on the concentration of polyglucuronic acid and ionic strength of these cations:

thermo-reversible gels can be obtained under high content of sodium ion, and thermally stable gels can be obtained in the presence of divalent cations such as calcium and barium ions (Elboutachfaiti et al., 2011). The gelling property would be useful for biomedical applications as alginate aqueous solution has been widely used due to its similar gelling property (Augst et al., 2006).

11.4 BIOMEDICAL APPLICATIONS

Alginate, dextran, gellan, and xanthan are the most widely mentioned EPSs for biomimetic materials development due to their novel features and functionality, chemical and physical properties, reproducibility, and cost production stabilities (Falconer et al., 2011; Goncalves et al., 2009), including targeted drug delivery, tissue engineering, and wound healing. Numerous EPSs have been reported to exhibit bioactivities such as antitumor (Calazans et al., 2000; Yoo et al., 2004), antioxidant (He et al., 2008), antibacterial (He et al., 2010), antiulcer (Nagaoka et al., 1994), and cholesterol-lowering activity (Rico et al., 2011).

11.4.1 DRUG DELIVERY

Microbial EPSs offer significant benefits over synthetic biopolymers when used as a core component for the development of new drug delivery systems due to their superior characteristics including biocompatibility, biodegradability, nontoxicity, and affordable manufacturing. The EPSs can enhance absorption of the encapsulated medicine by prolonging bioavailability in the body. Because of their considerable advantages and form flexibility (e.g., beads, fibers, films, powder, capsules, particles,

TABLE 11.1
Drug delivery systems from microbial EPSs

Route for drug delivery	Delivery system	Type of drug	Reference
Nasal	Xanthan/guar gum nasal insert	Metoclopramide hydrochloride	(Dehghan and Girase., 2012)
Oral	Dextran porous particle	Rifampicin	(Kadota et al., 2019)
	Gellan microparticle coated with starch/pectin	Insulin	(Meneguin et al., 2018)
	Gellan/glucosamine film	Clioquinol	(Tsai et al., 2018).
In vitro	Alginate fibers	Chlorhexidine gluconate	(Shi et al., 2020)
	Hyaluronic acid-modified carbon dot nanoparticles	Doxorubicin	(Li et al., 2020)
	Dextran hydrogel	Polydopamine	(Zhang et al., 2021a)
	Levan hydrogel	Amphotericin B	(Demirci et al., 2020)

and membranes), EPSs are well-known as carrier materials in drug delivery systems (Table 11.1). The routes of delivery can be diverse such as oral, nasal, and others. Various drugs have been successfully delivered with the use of microbial EPSs, including indomethacin (Shiraishi et al., 1993), papaverin hydrochloride, griseofulvin (Thanoo et al., 1992), and prednisolone (Berthold et al., 1996).

Several review articles have discussed the EPSs' potential as materials for the synthesis of drug delivery systems. The published reports commonly used chitosan/chitin (Kato et al., 2003), alginates (George and Abraham, 2006; Kim et al., 2002), pullulan (Mocanu et al., 2002; Na and Bae, 2002), gellan gum (Kubo et al., 2003; Miyazaki et al., 1999), dextran (Martins et al., 2007; Simonsen et al., 1995), hyaluronic acid, and their derivatives (Lim et al., 2000) as delivery platforms for active agent administration.

Chitosan is one of the microbial EPSs that is widely used as a carrier for drugs, particularly for recombinant-based pharmaceuticals (e.g., peptide and protein medications, vaccines, and genes) delivery. Because of its attractive features (i.e., biodegradability, low toxicity, biocompatibility, and relatively low manufacturing cost from available natural sources), chitosan has piqued the interest of researchers for usage as a polymeric drug carrier material in dosage-form design (Efthimiadou et al., 2014).

Another EPS, dextran, was previously used as a drug delivery system and blood plasma polymer expander (Kamoun et al., 2017; Yildiz and Karatas, 2018). Dextran-based polymers are also being studied for their potential as nanofibers for controlled drug release (Maslakci et al., 2017), anticancer therapies (Rai et al., 2018), and wound healing hydrogels (Alibolandi et al., 2017). As for levan, the low viscosity of this EPS encourages its usage in pharmaceuticals to make capsules or coatings, as well as in a variety of medicinal applications. Peptides and protein medicines might be delivered using levan nanoparticles. Levan is also applied in the environmentally friendly production of silver and gold nanoparticles (González-Aramundiz et al., 2018; Srikanth et al., 2015).

The potential uses of hyaluronic acid in biomedical applications can be shown in osteoarthritis treatment and tissue engineering (Allemann and Baumann, 2008; Benedini and Santana, 2013; Berkó et al., 2013). In addition, hyaluronic acid is applied in a variety of biological applications, including cartilage regeneration (Tan et al., 2009), ophthalmology (Zhang et al., 2020), and cancer treatment (Hu et al., 2020). It may interact with the stratum corneum barrier, allowing drugs to get through. Hyaluronic acid may potentially address CD44 fibroblasts, and cancer treatment using hyaluronic acid as a carrier is being developed (Misra et al., 2011). Ongoing research on hyaluronic acid and its many modifications/blends indicate the creation of materials with better characteristics for drug delivery and tissue engineering technologies (Rezaeeyiazdi et al., 2018; Tiwari et al., 2019).

Various forms of microbial polymers such as powders, solutions, microparticles, nanoparticles, coated beads, film, membrane, and liposomes are currently used or studied to enhance drug delivery. Such systems should enable the regulation of drug delivery rate, as well as extending the therapeutic impact duration and, possibly, the targeting of the medication to specific areas. Novel, more complicated drug delivery technologies, such as composite systems from polymer/liposome and targeted micro- and nanoparticles, have been created to increase the efficiency of EPSs as drug carriers. To extend contact duration and increase medication absorption, composite systems from polymer/liposome integrated polymer mucoadhesive characteristics with a targeted delivery method provided by liposomes. The use of liposomes in conjunction with polymers such as alginate, chitosan, and dextran significantly increased the stability of liposome and regulated the *in vivo* encapsulated bioactive substances release behavior. The last system has several benefits, including its compact size, capacity to deliver a wide variety of bioactive compounds, prolonged release of drug, minimal cytotoxicity, and preservation of encapsulated pharmaceuticals from degradation (Takeuchi et al., 2003). Although it is unclear whether the particulate systems' potential benefits outweigh the more complex preparation procedures, several appealing concepts, such as coated beads and liposomes, targeted micro- and nanoparticles, films, and membranes have been developed over the last decade.

11.4.2 TISSUE ENGINEERING

Tissue engineering is a rapidly evolving technology that combines cells, scaffolds, and/or biologically active molecules such as growth factors for constructing functional bioartificial tissues and organs *in vitro*. This technique also aims to alter the growth and function of cells *in vivo* by implanting cells obtained from donor tissue with biocompatible scaffolds. In this technique, scaffolds prepared from biomaterials play a vital role in the growth, functionalization, and differentiation of cells. In addition, in general, the scaffolds should be biodegradable for applications *in vivo*. A variety of polysaccharides obtained from various bioresources have been used as the materials for scaffolds.

The typical microbial EPSs which have been used for tissue engineering purposes are alginate, hyaluronic acid, dextran, cellulose, and xanthan. Especially, hyaluronic acid including its derivatives is the most widely used EPS in tissue engineering (Zhu et al., 2017). Attractive points of hyaluronic acid are the high histocompatibility and biodegradability as well as the functionality of affecting cell behaviors such as growth, migration, survival, and differentiation attributing to the interaction with cell surface receptors, for example, CD44 (Zhu et al., 2017). It was reported that cardiomyogenic differentiation of rat mesenchymal stem cells was induced on the scaffold containing hyaluronic acid through the interaction with CD44 (Yang et al., 2010). It was also reported that the migration of human umbilical vein endothelial cells was enhanced in the presence of low molecular weight hyaluronic acid derivative immobilized in a hydrogel (Khanmohammadi et al., 2017).

For the use of microbial EPSs as scaffold materials, it is preferred that insoluble forms like hydrogel can be obtained with ease. From this viewpoint, alginate is the most preferable. Alginate as a hydrogel material for tissue engineering lack bioactivity and less cell-adhesiveness; however, the mild gelling property under the mild condition for mammalian cells is useful for fabricating cell-laden constructs such as cell-enclosing microcapsules (Augst et al., 2006). The less cell-adhesiveness of alginate hydrogel is known as useful in nerve tissue engineering and culturing non-adherent cells in its foam-based scaffold (Karimpoor et al., 2018). Currently, alginate from algae is more frequently used in tissue engineering, but microbial

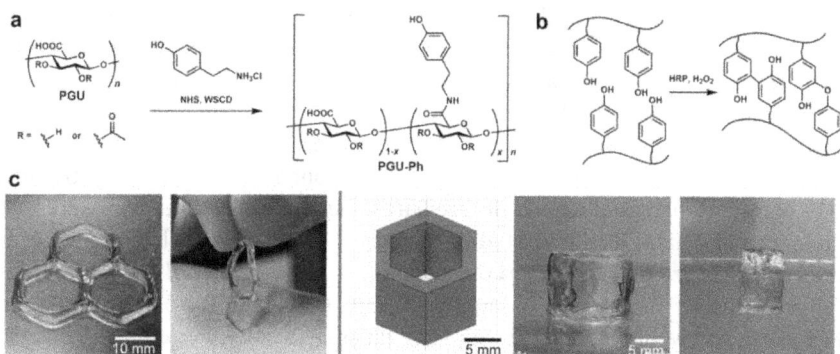

FIGURE 11.2 Schemes of (a) phenolic hydroxyl moieties incorporation into polyglucuronic acid, and (b) cross-linking between phenolic hydroxyl moieties through horseradish peroxidase (HRP)-mediated reaction; (c) 3D hydrogel constructs obtained through 3D printing from polyglucronic acid derivative possessing phenolic hydroxyl moieties through HRP-mediated cross-linking

Adapted from Sakai et al. (2022), used under the Creative Commons License (http://creativecommons.org/licenses/by/4.0/)

alginate produced through fermentation would be used more in the future due to their quick and high-yielding production with a less lot-to-lot variation (Ng et al., 2020). Scaffolds can be also obtained from aqueous solutions of xanthan and gellan, and the effectiveness in engineering various tissues such as bone, cartilage, and the intervertebral disc has been reported (Rahman and Arafat, 2021).

Even from the EPSs, such as dextran and hyaluronic acid, which do not form hydrogels in their native state through cytocompatible routes, hydrogels have been prepared from their derivatives obtained by introducing the moieties cross-linkable under mild reaction for cells. The introduction of methacrylate moieties to hyaluronic acid and dextran resulted in photocurable derivatives (Kim and Chu, 2000; Moller et al., 2007), and the hydrogels obtained from these derivatives have been successfully used for tissue engineering (Moller et al., 2007; Wang et al., 2014). The introduction of phenolic hydroxyl moieties cross-linkable through horseradish peroxidase-catalyzed reaction or visible light irradiation in the presence of sodium persulfate and Tris(2,2'-bipyridyl) ruthenium(II) chloride has also been reported as an effective approach for the purpose (Jin et al., 2010; Kurisawa et al., 2005), and the derivative of microbial hyaluronic acid was successfully used as an ink component of bioprinting (Sakai et al., 2018, 2019). The usefulness of this approach including the application to bioprinting was also reported for microbial polyglucuronic acid (Sakai et al., 2022) (Figure 11.2).

11.4.3 WOUND TREATMENT

Commonly used microbial EPSs in wound healing include hyaluronic acid, levan, xanthan, alginate, and curdlan. These materials have numerous structural and physicochemical properties that can be modified to improve the treatment of wounds.

Hyaluronic acid mediates cellular and matrix events relating to the wound healing process caused by its natural angiogenic property when it is degraded to small fragments. Hyaluronic acid promotes early inflammation that is critical for wound healing initiation, then moderates later stages of the process, reducing long-term inflammation and allowing matrix stabilization. Therefore, hyaluronic acid is effective for the treatment of various chronic and acute wounds such as postoperative incisions, abrasions, first- and second-degree burns, metabolic ulcers, and pressure sores (Saranraj and Naidu, 2013). A study (Alvarez et al., 2021) developed angiogenic hyaluronic acid from cyanobacteria. Wound dressings biomaterial from hyaluronic acid produced by the Nostoc sp. strains PCC7936 and PCC7413 were evaluated. MTT assay and wound healing *in vitro* scratch assay results showed hydrogels biocompatibility and the ability to promote fibroblast migration and proliferation. The Nostoc hyaluronic acid presented promising properties for skin injury treatment.

Xanthan has widely been used in biomedical applications, including for wound treatment. It is a good hydrogel precursor material and can be combined with other components such as chitosan (Chellat et al., 2000), silver nanoparticles (Rao et al., 2016), and chlorohexidine for wound treatment. The composite gels could adhere well to injured tissue and promote the proliferation of fibroblasts (Han et al., 2021). It was reported that the hydrogel's blend of xanthan with konjac glucomannan (Figure 11.3) promoted full cover of the wound site of an *in vitro* wound healing assay after 12 h compared to 24 h for the control group. Furthermore, a higher xanthan to konjac glucomannan ratio (60/40) resulted in improved and quick wound closure (Alves et al., 2020).

Alginate is well-known as a good wound dressing and has been used clinically in wound treatment due to its biocompatibility, mild gelation conditions, and simple modification method for its derivative with new properties (Lee and Mooney, 2012). However, alginate has several limitations such as low availability, high cost, and absence of antimicrobial activity. Pourali and Yahyaei (2018) improved the wound healing capacity by combining alginate with herb extract from *Alhagi maurorum* and examined toxicity to the Wistar rats. Results showed that the alginate aqueous solution-*A. maurorum* extract complex group showed no toxicity effect and improvement in wound healing activity. Other bioactive alginate-based wound dressings have also been reported by Diniz et al. (2020). Their study showed that the addition of silver nanoparticles into alginate dressings improved antimicrobial activity and increased the binding affinity for elastase, matrix metalloproteases-2, and proinflammatory cytokines (e.g., TNF-α, IL-8).

The study by Sturzoiu et al. (2011) investigated the use of levan for wound healing in Wistar rats. The results showed that levan produced by *Z. mobilis* seems to have significant benefits in wounds and burns, applied either individually or in combination with either natural or artificial halotherapy. Although the way of levan contributes to the healing process is still unknown, it was assumed that levan activated tissue metalloproteinases which then promoted tissue regeneration (Sturzoiu et al., 2011).

Another EPSs used for wound treatment is curdlan. Curdlan is extensively used for biomedical applications due to its immunomodulatory effects, good gelling ability, thermal stability, biodegradability, and nontoxicity (Rafigh et al., 2014). Curdlan's chemical structure enables it to be recognized by macrophages and other

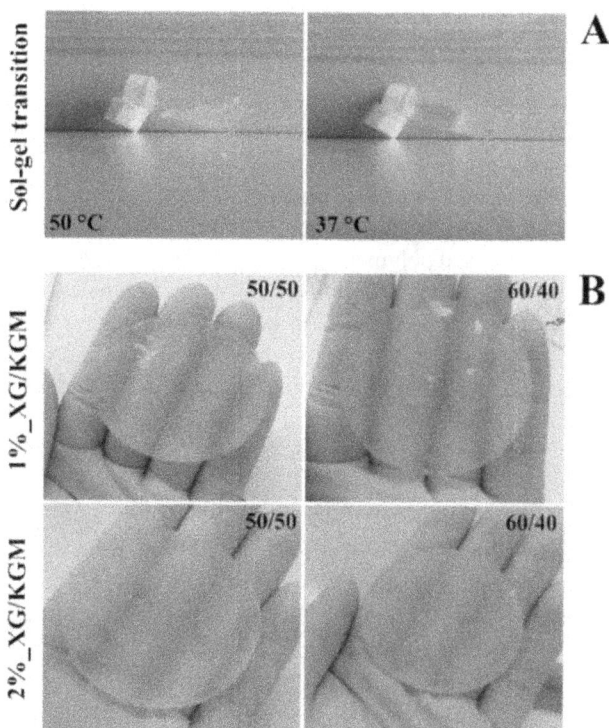

FIGURE 11.3 The solid and the gel states of hydrogels observed at different temperatures (A) and the macroscopic images of the hydrogels produced (B) XG/KGM: xanthan gum—konjac glucomannan

Figure and caption reused from Alves et al. (2020), used under the Creative Commons License (http://creativecommons.org/licenses/by/4.0/)

immune cells, thus promoting the immune system, and exerting immune-related biological activities, which resulted in an improvement in the efficiency of wound healing (Zhang and Edgar, 2014). Curdlan showed abundant potential for enhancing wound healing, but there are still no ideal curdlan-based hydrogels to support the entire wound healing stages.

Zhou et al. (2022) constructed curdlan-tannic acid-based hydrogels (with a ratio of 1:1) through an annealing technique and examined the gel's properties *in vitro*. Tannic acid was used as an additive due to its abundant phenolic hydroxyl groups that can form hydrogels with various polymers (Zhang et al., 2021) as well as improve the hydrogels' antioxidant, anti-inflammatory, and antibacterial properties (Gao et al., 2021). The hybrid hydrogel showed exceptional physicochemical properties (i.e., degradability, porosity, swellability, water retention, and rheology), clinical properties (antibacterial, antioxidant, and rapid hemostasis effects), and did not display toxicity effects. In addition, the curdlan-tannic acid hydrogel was also examined *in vivo* by testing in a full-thickness skin defect animal model, and it is suggested as a candidate for the full-thickness wounds clinical treatment (Zhou et al., 2022).

11.5   CONCLUSION AND PERSPECTIVE

As been highlighted in this chapter, microbial EPSs offer a diverse range of functions and uses. Their versatile physicochemical composition could be tailored to fit any applications. Perhaps the most important features microbial EPSs have are their nontoxicity, biocompatibility, and biodegradability without harmful by-products. Synthetic polymers used in biomedical applications have been related to immunogenicity and toxicity issues. Indeed, for certain polymers such as cellulose and alginate, utilization of microbial-based polymers is still limited compared to the conventional sources. Herein, more research is required on the biomedical potential of levan, kefiran, and polyglucuronic acid. Nevertheless, microbial EPSs have shown promising biomedical use, as drug delivery system, in tissue engineering and wound treatment, and have been recorded as a major contributing factor to the advancement of the present biomedical situations.

ACKNOWLEDGMENTS

The authors thank the Ministry of Higher Education Malaysia for their financial support under the Fundamental Research Grant Scheme (account number: FRGS/1/2021/TK0/USM/02/16). The Short Term Grant (KHAS) from Universiti Sains Malaysia (account number: 304/PJKIMIA/6315703) is also acknowledged.

REFERENCES

Aarstad, O.A., Stanisci, A., Sætrom, G.I., Tøndervik, A., Sletta, H., Aachmann, F.L., Skjåk-Bræk, G., 2019. Biosynthesis and function of long guluronic acid-blocks in alginate produced by *Azotobacter vinelandii*. Biomacromolecules. 20(4), 1613–1622.

Ahumada-Manuel, C.L., Guzmán, J., Peña, C., Quiroz-Rocha, E., Espín, G., Núñez, C., 2017. The signaling protein MucG negatively affects the production and the molecular mass of alginate in *Azotobacter vinelandii*. Appl. Microbiol. Biotechnol. 101(4), 1521–1534.

Alibolandi, M., Mohammadi, M., Taghdisi, S.M., Abnous, K., Ramezani, M., 2017. Synthesis and preparation of biodegradable hybrid dextran hydrogel incorporated with biodegradable curcumin nanomicelles for full thickness wound healing. Int. J. Pharm. 532(1), 466–477.

Allemann, I.B., Baumann, L., 2008. Hyaluronic acid gel (Juvéderm) preparations in the treatment of facial wrinkles and folds. Clin. Interv. Aging. 3(4), 629–634.

Alvarez, X., Alves, A., Ribeiro, M.P., Lazzari, M., Coutinho, P., Otero, A., 2021. Biochemical characterization of *Nostoc* sp. exopolysaccharides and evaluation of potential use in wound healing. Carbohydr. Polym. 254, 117303.

Alves, A., Miguel, S.P., Araujo, A.R.T.S., de Jesús Valle, M.J., Sánchez Navarro, A., Correia, I.J., Ribeiro, M.P., Coutinho, P., 2020. Xanthan gum—konjac glucomannan blend hydrogel for wound healing. Polymers. 12(1), 99.

Amjad Zanjani, F.S., Afrasiabi, S., Norouzian, D., Ahmadian, G., Hosseinzadeh, S.A., Fayazi Barjin, A., Cohan, R.A., Keramati, M., 2022. Hyaluronic acid production and characterization by novel *Bacillus subtilis* harboring truncated hyaluronan synthase. AMB Express. 12(1), 88.

Angelin, J., Kavitha, M., 2020. Exopolysaccharides from probiotic bacteria and their health potential. Int. J. Biol. Macromol. 162, 853–865.

Arca, H.C., Mosquera-Giraldo, L.I., Bi, V., Xu, D., Taylor, L.S., Edgar, K.J., 2018. Pharmaceutical applications of cellulose ethers and cellulose ether esters. Biomacromolecules. 19(7), 2351–2376.

Ates, O., 2015. Systems biology of microbial exopolysaccharides production. Front. Bioeng. Biotechnol. 3, 200.

Augst, A.D., Kong, H.J., Mooney, D.J., 2006. Alginate hydrogels as biomaterials. Macromol. Biosci. 6(8), 623–633.

Babaei-Ghazvini, A., Shahabi-Ghahfarrokhi, I., Goudarzi, V., 2018. Preparation of UV-protective starch/kefiran/ZnO nanocomposite as a packaging film: Characterization. Food Packag. Shelf Life. 16, 103–111.

Bao, M., Ehexige, E., Xu, J., Ganbold, T., Han, S., Baigude, H., 2021. Oxidized curdlan activates dendritic cells and enhances antitumor immunity. Carbohydr. Polym. 264, 117988.

Bauer, R., 1938. Physiology of Dematium pullulans de Bary. Zentralbl. Bacteriol. Parasitenkd. Infektionskr. Hyg. Abt. 98(2), 133–167.

Benedini, L.J., Santana, M.H., 2013. Effects of soy peptone on the inoculum preparation of *Streptococcus zooepidemicus* for production of hyaluronic acid. Bioresour. Technol. 130, 798–800.

Berkó, S., Maroda, M., Bodnár, M., Erős, G., Hartmann, P., Szentner, K., Szabó-Révész, P., Kemény, L., Borbély, J., Csányi, E., 2013. Advantages of cross-linked versus linear hyaluronic acid for semisolid skin delivery systems. Eur. Polym. J. 49(9), 2511–2517.

Bernier, B., 1958. The production of polysaccharides by fungi active in the decomposition of wood and forest litter. Can. J. Microbiol. 4(3), 195–204.

Berthold, A., Cremer, K., Preparation, K.J., 1996. Preparation and characterization of chitosan microspheres as drug carrier for prednisolone sodium phosphate as model for anti-inflammatory drugs. J. Control Release. 39, 17–25.

Bhavani, A.L., Nisha, J., 2010. Dextran—the polysaccharide with versatile uses. Int. J. Pharm. Bio. Sci. 1(4), 569–573.

Blank, L.M., Hugenholtz, P., Nielsen, L.K., 2008. Evolution of the hyaluronic acid synthesis (has) operon in *Streptococcus zooepidemicus* and other pathogenic streptococci. J. Mol. Evol. 67(1), 13–22.

Calazans, G.M., Lima, R.C., de França, F.P., Lopes, C.E., 2000. Molecular weight and antitumour activity of *Zymomonas mobilis* levans. Int. J. Biol. Macromol. 27(4), 245–247.

Campos, F.D.S., Ferrari, L.Z., Cassimiro, D.L., Ribeiro, C.A., De Almeida, A.E., Gremião, M.P.D., 2016. Effect of 70-kDa and 148-kDa dextran hydrogels on praziquantel solubility. J. Therm. Anal. Calorim. 123(3), 2157–2164.

Castillo, T., Galindo, E., Peña, C.F., 2013. The acetylation degree of alginates in *Azotobacter vinelandii* ATCC9046 is determined by dissolved oxygen and specific growth rate: Studies in glucose-limited chemostat cultivations. J. Ind. Microbiol. Biotechnol. 40(7), 715–723.

Chaisuwan, W., Jantanasakulwong, K., Wangtueai, S., Phimolsiripol, Y., Chaiyaso, T., Techapun, C., Phongthai, S., You, S.G., Regenstein, J.M., Seesuriyachan, P., 2020. Microbial exopolysaccharides for immune enhancement: Fermentation, modifications and bioactivities. Food Biosci. 35, 100564.

Chanasit, W., Gonzaga, Z.J.C., Rehm, B.H.A., 2020. Analysis of the alginate O-acetylation machinery in *Pseudomonas aeruginosa*. Appl. Microbiol. Biotechnol. 104(5), 2179–2191.

Chellat, F., Tabrizian, M., Dumitriu, S., Chornet, E., Magny, P., Rivard, C.-H., Yahia, L., 2000. In vitro and in vivo biocompatibility of chitosan-xanthan polyionic complex. J. Biomed. Mater. Res. 51(1), 107–116.

Chen, F., Huang, G., Huang, H., 2020. Preparation and application of dextran and its derivatives as carriers. Int. J. Biol. Macromol. 145, 827–834.

Cheng, F., Yu, H., Stephanopoulos, G., 2019. Engineering *Corynebacterium glutamicum* for high-titer biosynthesis of hyaluronic acid. Metab. Eng. 55, 276–289.

Cheng, K.-C., Demirci, A., Catchmark, J.M., 2011. Pullulan: Biosynthesis, production, and applications. Appl. Microbiol. Biotechnol. 92(1), 29–44.

Chi, Z., Wang, F., Chi, Z., Yue, L., Liu, G., Zhang, T., 2009. Bioproducts from *Aureobasidium pullulans*, a biotechnologically important yeast. Appl. Microbiol. Biotechnol. 82(5), 793–804.

Chong, B.F., Blank, L.M., Mclaughlin, R., Nielsen, L.K., 2005. Microbial hyaluronic acid production. Appl. Microbiol. Biotechnol. 66(4), 341–351.

Clementi, F., 1997. Alginate production by *Azotobacfer Vinelandii*. Crit. Rev. Biotechnol. 17(4), 327–361.

Coltelli, M.B., Danti, S., De Clerck, K., Lazzeri, A., Morganti, P., 2020. Pullulan for advanced sustainable body-and skin-contact applications. J. Funct. Biomater. 11(1), 20.

Conti, E., Flaibani, A., O'Regan, M., Sutherland, I.W., 1994. Alginate from *Pseudomonas fluorescens* and *P. putida*: Production and properties. Microbiology. 140, 1125–1132.

Davidson, I.W., Sutherland, I.W., Lawson, C.J., 1977. Localization of O acetyl groups of bacterial alginate. J. Gen. 98, 603–606.

Dehghan, M.H., Girase, M., 2012. Freeze-dried xanthan/guar gum nasal inserts for the delivery of metoclopramide hydrochloride. Iran. J. Pharm. Res. 11(2), 513.

Demirci, T., Hasköylü, M.E., Eroğlu, M.S., Hemberger, J., Öner, E.T., 2020. Levan-based hydrogels for controlled release of Amphotericin B for dermal local antifungal therapy of Candidiasis. Eur. J. Pharm. Sci. 145, 105255.

De Ruiter, G.A., Josso, S.L., Colquhoun, I.J., Voragen, A.G.J., Rombouts, F.M., 1992. Isolation and characterization of β(1–4)-d-glucuronans from extracellular polysaccharides of moulds belonging to Mucorales. Carbohydr. Polym. 18(1), 1–7.

Diniz, F.R., Maia, R.C.A., de Andrade, L.R.M., Andrade, L.N., Vinicius Chaud, M., da Silva, C.F., Corrêa, C.B., de Albuquerque Junior, R.L.C., Pereira da Costa, L., Shin, S.R., Hassan, S., 2020. Silver nanoparticles-composing alginate/gelatine hydrogel improves wound healing in vivo. Nanomaterials. 10(2), 390.

Donot, F., Fontana, A., Baccou, J.C., Schorr-Galindo, S., 2012. Microbial exopolysaccharides: Main examples of synthesis, excretion, genetics and extraction. Carbohyd. Polym. 87(2), 951–962.

Dow, J.M., Darnall, D.W., Villa, V.D., 1983. Two distinct classes of polyuronide from the cell walls of a dimorphic fungus, *Mucor rouxii*. J. Bacteriol. 155(3), 1088–1093.

Efthimiadou, E.K., Metaxa, A.F., Kordas, G., 2014. Modified polysaccharides for drug delivery. In: Kishan Gopal Ramawat and Jean-Michel Merillon (Eds.), *Polysaccharides, 1805–1835*. Springer.

Elboutachfaiti, R., Delattre, C., Petit, E., Michaud, P., 2011. Polyglucuronic acids: Structures, functions and degrading enzymes. Carbohydr. Polym. 84(1), 1–13.

Evans, L.R., Linker, A., 1973. Production and characterization of the slime polysaccharide of *Pseudomonas aeruginosa*. J. Bacteriol. 116, 915–924.

Falconer, D.J., Mukerjea, R., Robyt, J.F., 2011. Biosynthesis of dextrans with different molecular weights by selecting the concentration of *Leuconostoc mesenteroides* B-512FMC dextransucrase, the sucrose concentration, and the temperature. Carbohydr. Polym. 364, 280–284.

Field, T.R., Ward, P.N., Pedersen, L.H., Leigh, J.A., 2003. The hyaluronic acid capsule of *Streptococcus uberis* is not required for the development of infection and clinical mastitis. Infect. Immun. 71(1), 132–139.

Forabosco, A., Bruno, G., Sparapano, L., Liut, G., Marino, D., Delben, F., 2006. Pullulans produced by strains of *Cryphonectria parasitica*—I. Production and characterisation of the exopolysaccharides. Carbohydr. Polym. 63(4), 535–544.

Fraser, C.G., Jennings, H.J., 1971. A glucan from *Tremella mesenterica* NRRL-Y6158. Can. J. Chem. 49(11), 1804–1807.

Futatsuyama, H., Yui, T., Ogawa, K., 1999. Viscometry of curdlan, a linear (1→3)-β-D-glucan, in DMSO or alkaline solutions. Biosci. Biotechnol. Biochem. 63(8), 1481–1483.

Gao, X., Xu, Z., Liu, G., Wu, J., 2021. Polyphenols as a versatile component in tissue engineering. Acta Biomater. 119, 57–74.

García, A., Castillo, T., Ramos, D., Ahumada-Manuel, C.L., Núñez, C., Galindo, E., Büchs, J., Peña, C., 2020. Molecular weight and viscosifying power of alginates produced by mutant strains of *Azotobacter vinelandii* under microaerophilic conditions. Biotechnol. Rep. 26, e00436.

George, M., Abraham, T.E., 2006. Polyionic hydrocolloids for the intestinal delivery of protein drugs: Alginate and chitosan-a review. J. Control. Release. 114, 1–14.

Ghasemlou, M., Khodaiyan, F., Oromiehie, A., 2011. Physical, mechanical, barrier, and thermal properties of polyol-plasticized biodegradable edible film made from kefiran. Carbohydr. Polym. 84(1), 477–483.

Gil, E.C., Colarte, A.I., El Ghzaoui, A., Durand, D., Delarbre, J.L., Bataille, B., 2008. A sugar cane native dextran as an innovative functional excipient for the development of pharmaceutical tablets. Eur. J. Pharm. Biopharm. 68(2), 319–329.

Gimmestad, M., Sletta, H., Ertesvåg, H., Bakkevig, K., Jain, S., Suh, S. Jin, Skjåk-Bræk, G., Ellingsen, T.E., Ohman, D.E., Valla, S., 2003. The *Pseudomonas fluorescens* AlgG protein, but not its mannuronan C-5-epimerase activity, is needed for alginate polymer formation. J. Bacteriol. 185, 3515–3523.

Goncalves, V.M.F., Reis, A., Domingues, M.R.M., Lopes-da-Silva, J.A., Fialho, A.M., Moreira, L.M., Sá-Correia, I., Coimbra, M.A., 2009. Structural analysis of gellans produced by *Sphingomonas elodea* strains by electrospray tandem mass spectrometry. Carbohydr. Polym. 77, 10–19.

González-Aramundiz, J.V., Peleteiro, M., González-Fernández, Á., Alonso, M.J., Csaba, N.S., 2018. Protamine nanocapsules for the development of thermostable adjuvanted nanovaccines. Mol. Pharm. 15(12), 5653–5664.

Guan, L., Zhang, L., Xue, Y., Yang, J., Zhao, Z., 2020. Molecular pathogenesis of the hyaluronic acid capsule of *Pasteurella multocida*. Microb. Pathog. 149, 104380.

Gupta, P.L., Rajput, M., Oza, T., Trivedi, U., Sanghvi, G., 2019a. Eminence of microbial products in cosmetic industry. Nat. Prod. Bioprospect. 9, 267–278.

Gupta, R.C., Lall, R., Srivastava, A., Sinha, A., 2019b. Hyaluronic acid: Molecular mechanisms and therapeutic trajectory. Front. Vet. Sci. 6, 1–24.

Hamidi, M., Kennedy, J.F., Khodaiyan, F., Mousavi, Z., Hosseini, S.S., 2019. Production optimization, characterization and gene expression of pullulan from a new strain of *Aureobasidium pullulans*. Int. J. Biol. Macromol. 138, 725–735.

Han, B., Baruah, K., Cox, E., Vanrompay, D., Bossier, P., 2020. Structure-functional activity relationship of β-glucans from the perspective of immunomodulation: A mini-review. Front. Immunol. 11, 1–8.

Han, G., Wang, G., Zhu, X., Shao, H., Liu, F., Yang, P., Ying, Y., Wang, F., Ling, P., 2012. Preparation of xanthan gum injection and its protective effect on articular cartilage in the development of osteoarthritis. Carbohydr. Polym. 87(2), 1837–1842.

Han, X., Hua, W., Liu, Y., Ao, Z., Han, D., 2021. In situ self-organizing materials for local stress-responsive reconstruction of skin interstitium. Macromol. Biosci. 21, 8.

Hansen, I.M., Ebbesen, M.F., Kaspersen, L., Thomsen, T., Bienk, K., Cai, Y., Malle, B.M., Howard, K.A., 2017. Hyaluronic acid molecular weight-dependent modulation of mucin nanostructure for potential mucosal therapeutic applications. Mol. Pharm. 14(7), 2359–2367.

Harada, T., 1965. Succinoglucan 10C3: A new acidic polysaccharide of *Alcaligenes faecalis* var. myxogenes. Arch. Biochem. Biophys. 112(1), 65–69.

Harada, T., Masada, M., Fujimori, K., Maeda, I., 1966. Production of a firm, resilient gel-forming polysaccharide by a mutant of *Alcaligenes faecalis* var. myxogenes 10 C3. Agric. Biol. Chem. 30(2), 196–198.

He, F., Yang, Y., Yang, G., Yu, L., 2008. Components and antioxidant activity of the polysaccharide from *Streptomyces virginia* H03. Z. Naturforsch. C J. Biosci. 63(3–4), 181–188.

He, F., Yang, Y., Yang, G., Yu, L., 2010. Studies on antibacterial activity and antibacterial mechanism of a novel polysaccharide from *Streptomyces virginia* H03. Food Control. 21, 1257–1262.

Heyraud, A., Courtois, J., Dantas, L., Colin-Morel, P., Courtois, B., 1993. Structural characterization and rheological properties of an extracellular glucuronan produced by a *Rhizobium meliloti* M5N1 mutant strain. Carbohydr. Res. 240, 71–78.

Horiuchi, H., 2009. Functional diversity of chitin synthases of *Aspergillus nidulans* in hyphal growth, conidiophore development and septum formation. Med. Mycol. 47(sup 1), S47–S52.

Hu, Z., Wang, S., Dai, Z., Zhang, H., Zheng, X., 2020. A novel theranostic nano-platform (PB@ FePt—HA-g-PEG) for tumor chemodynamic—photothermal co-therapy and triple-modal imaging (MR/CT/PI) diagnosis. J. Mater. Chem. B. 8(24), 5351–5360.

Izawa, N., Hanamizu, T., Iizuka, R., Sone, T., Mizukoshi, H., Kimura, K., Chiba, K., 2009. *Streptococcus thermophilus* produces exopolysaccharides including hyaluronic acid. J. Biosci. Bioeng. 107(2), 119–123.

Izawa, N., Serata, M., Sone, T., Omasa, T., Ohtake, H., 2011. Hyaluronic acid production by recombinant *Streptococcus thermophilus*. J. Biosci. Bioeng. 111(6), 665–670.

Jin, R., Teixeira, L.S., Dijkstra, P.J., van Blitterswijk, C.A., Karperien, M., Feijen, J., 2010. Enzymatically-crosslinked injectable hydrogels based on biomimetic dextran-hyaluronic acid conjugates for cartilage tissue engineering. Biomaterials. 31, 3103–3113.

Jong, A., Wu, C.H., Chen, H.M., Luo, F., Kwon-Chung, K.J., Chang, Y.C., LaMunyon, C.W., Plaas, A., Huang, S.H., 2007. Identification and characterization of CPS1 as a hyaluronic acid synthase contributing to the pathogenesis of *Cryptococcus neoformans* infection. Eukaryot. Cell. 6(8), 1486–1496.

Kadota, K., Yanagawa, Y., Tachikawa, T., Deki, Y., Uchiyama, H., Shirakawa, Y., Tozuka, Y., 2019. Development of porous particles using dextran as an excipient for enhanced deep lung delivery of rifampicin. Int. J. Pharm. 555, 280–290.

Kamoun, E.A., Kenawy, E.S., Chen, X., 2017. A review on polymeric hydrogel membranes for wound dressing applications: PVA-based hydrogel dressings. J. Adv. Res. 8, 217–233.

Karimpoor, M., Yebra-Fernandez, E., Parhizkar, M., Orlu, M., Craig, D., Khorashad, J.S., Edirisinghe, M., 2018. Alginate foam-based three-dimensional culture to investigate drug sensitivity in primary leukaemia cells. J. R. Soc. Interface. 15(141), 20170928.

Kato, Y., Onishi, H., Machida, Y., 2003. Application of chitin and chitosan derivatives in the pharmaceutical field. Curr. Pharm. Biotechnol. 4, 303–309.

Kazak Sarilmiser, H., Toksoy Oner, E., 2014. Investigation of anti-cancer activity of linear and aldehyde-activated levan from *Halomonas smyrnensis* AAD6T. Biochem. Eng. J. 92, 28–34.

Khanmohammadi, M., Sakai, S., Taya, M., 2017. Impact of immobilizing of low molecular weight hyaluronic acid within gelatin-based hydrogel through enzymatic reaction on behavior of enclosed endothelial cells. Int. J. Biol. 97, 308–316.

Kim, B., Bowersock, T., Griebel, P., Kidane, A., Babiuk, L.A., Sanchez, M., Attah-Poku, S., Kaushik, R.S., Mutwiri, G.K., 2002. Mucosal immune responses following oral immunization with rotavirus antigens encapsulated in alginate microspheres. J. Control. Release. 85(1–3), 191–202.

Kim, S.H., Chu, C.C., 2000. Synthesis and characterization of dextran—methacrylate hydrogels and structural study by SEM. J. Biomed. Mater. Res. 49(4), 517–527.

Kłos, M., Wójkowska-Mach, J., 2017. Pathogenicity of virulent species of Group C Streptococci in human. Can. J. Infect. Dis. Med. Microbiol. 2017.

Kondo, T., 1997. The relationship between intramolecular hydrogen bonds and certain physical properties of regioselectively substituted cellulose derivatives. J. Polym. Sci. B Polym. Phys. 35(4), 717–723.

Kothari, D., Das, D., Patel, S., Goyal, A., 2014. Dextran and food application. In: K. Ramawat, J.M. Mérillon (Eds.), *Polysaccharides*. Springer, Cham.

Kouvidi, K., Berdiaki, A., Nikitovic, D., Katonis, P., Afratis, N., Hascall, V.C., Karamanos, N.K., Tzanakakis, G.N., 2011. Role of receptor for hyaluronic acid-mediated motility (RHAMM) in low molecular weight hyaluronan (LMWHA)-mediated fibrosarcoma cell adhesion. J. Biol. Chem. 286(44), 38509–38520.

Krishna, P.S., Woodcock, S.D., Pfeilmeier, S., Bornemann, S., Zipfel, C., Malone, J.G., 2022. *Pseudomonas syringae* addresses distinct environmental challenges during plant infection through the coordinated deployment of polysaccharides. J. Exp. Bot. 73(7), 2206–2221.

Kubo, W., Miyazaki, D., Attwood, D., 2003. Oral sustained delivery of paracetamol from in situ-gelling gellan and sodium alginate formulations. Int. J. Pharm. 258, 55–64.

Kumar, A., Rao, K.M., Han, S.S., 2018. Application of xanthan gum as polysaccharide in tissue engineering: A review. Carbohydr. Polym. 180, 128–144.

Kurisawa, M., Chung, J.E., Yang, Y.Y., Gao, S.J., Uyama, H., 2005. Injectable biodegradable hydrogels composed of hyaluronic acid—tyramine conjugates for drug delivery and tissue engineering. Chem. Commun. 34, 4312–4314.

Lairson, L.L., Withers, S.G., 2004. Mechanistic analogies amongst carbohydrate modifying enzymes. Chem. Commun. 20, 2243–2248.

Laksen, B., Skjåk-Bræk, G., Painter, T., 1986. Action pattern of mannuronan C-5-epimerase: Generation of block-copolymeric structures in alginates by a multiple-attack mechanism. Carbohydr. Res. 146, 342–345.

Laubach, J., Joseph, M., Brenza, T., Gadhamshetty, V., Sani, R.K., 2021. Exopolysaccharide and biopolymer-derived films as tools for transdermal drug delivery. J. Control. Release. 329, 971–987.

Laurent, T.C., Laurent, U.B.G., Fraser, J.R.E., 1996. The structure and function of hyaluronan: An overview. Immunol. Cell Biol. 74(2), a1–a7.

Leathers, T.D., 2003. Biotechnological production and applications of pullulan. Appl. Microbiol. Biotechnol. 62(5–6), 468–473.

Lee, K.Y., Mooney, D.J., 2012. Alginate: Properties and biomedical applications. Prog. Polym. Sci. 37(1), 106–126.

Li, J., Li, M., Tian, L., Qiu, Y., Yu, Q., Wang, X., Guo, R., He, Q., 2020. Facile strategy by hyaluronic acid functional carbon dot-doxorubicin nanoparticles for CD44 targeted drug delivery and enhanced breast cancer therapy. Int. J. Pharm. 578, 119122.

Li, R., Tomasula, P., de Sousa, A., Liu, S.-C., Tunick, M., Liu, K., Liu, L., 2017. Electrospinning pullulan fibers from salt solutions. Polymers. 9(12), 32.

Lim, S.T., Martin, G.P., Berry, D.J., Brown, M.B., 2000. Preparation and evaluation of the in vitro drug release properties and mucoadhesion of novel microspheres of hyaluronic acid and chitosan. J. Control. Release. 66(2–3), 281–292.

Linker, A., Jones, R.S., 1964. A polysaccharide resembling alginic acid from a *Pseudomonas* micro-organism. Nature. 204(4954), 187–188.

Linker, A., Jones, R.S., 1966. A new polysaccharide resembling alginic acid isolated from pseudomonads. J. Biol. 241, 3845–3851.

Martins, S., Sarmento, B., Souto, E.B., Ferreira, D.C., 2007. Insulin-loaded alginate microspheres for oral delivery – Effect of polysaccharide reinforcement on physicochemical properties and release profile. Carbohydr. Polym. 69, 725–731.

Mascarenhas, M.M., Day, R.M., Ochoa, C.D., Choi, W. Il, Yu, L., Ouyang, B., Garg, H.G., Hales, C.A., Quinn, D.A., 2004. Low molecular weight hyaluronan from stretched lung enhances interleukin-8 expression. Am. J. Respir. Cell Mol. Biol. 30(1), 51–60.

Maslakci, N.N., Ulusoy, S., Uygun, E., Çevikbaş, H., Oksuz, L., Can, H.K., Uygun Oksuz, A., 2017. Ibuprofen and acetylsalicylic acid loaded electrospun PVP-dextran nanofiber mats for biomedical applications. Polym. Bull. 74, 3283–3299.

Meneguin, A.B., Beyssac, E., Garrait, G., Hsein, H., Cury, B.S., 2018. Retrograded starch/pectin coated gellan gum-microparticles for oral administration of insulin: A technological platform for protection against enzymatic degradation and improvement of intestinal permeability. Eur. J. Pharm. Biopharm. 123, 84–94.

Merzendorfer, H., 2011. The cellular basis of chitin synthesis in fungi and insects: Common principles and differences. Eur. J. Cell Biol. 90, 759–769.

Micheli, L., Uccelletti, D., Palleschi, C., Crescenzi, V., 1999. Isolation and characterisation of a ropy *Lactobacillus* strain producing the exopolysaccharide kefiran. Appl. Microbiol. Biotechnol. 53(1), 69–74.

Milas, M., Shi, X., Rinaudo, M., 1990. On the physicochemical properties of gellan gum. Biopolymers. 30(3–4), 451–464.

Misra, S., Hascall, V.C., Markwald, R.R., Ghatak, S., 2015. Interactions between hyaluronan and its receptors (CD44, RHAMM) regulate the activities of inflammation and cancer. Front. Immunol. 6, 201

Misra, S., Heldin, P., Hascall, V.C., Karamanos, N.K., Skandalis, S.S., Markwald, R.R., Ghatak, S., 2011. Hyaluronan-CD44 interactions as potential targets for cancer therapy. Febs J. 278, 1429–1443.

Miyazaki, S., Aoyama, H., Kawasaki, N., Kubo, W., Attwood, D., 1999. In situ gelling gellan formulations as vehicles for oral drug delivery. J. Control. Release. 60, 287–295.

Mocanu, G., Mihai, D., Picton, L., LeCerf, D., Muller, G., 2002. Associative pullulan gels and their interaction with biological active substances. J. Control. Release. 83, 41–51.

Mohanty, A.K., Misra, M., Hinrichsen, G., 2000. Biofibres, biodegradable polymers and biocomposites: An overview. Macromol. Mater. Eng. 276–277, 1–24.

Moller, S., Weisser, J., Bischoff, S., Schnabelrauch, M., 2007. Dextran and hyaluronan methacrylate based hydrogels as matrices for soft tissue reconstruction. Biomol. Eng. 24(5), 496–504.

Morris, E.R., Nishinari, K., Rinaudo, M., 2012. Gelation of gellan—A review. Food Hydrocoll. 28(2), 373–411.

Na, K., Bae, Y.H., 2002. Self-assembled hydrogel nanoparticles responsive to tumor extracellular pH from pullulan derivative/sulphonamide conjugate: Characterization, aggregation, and adriamycin release in vitro. Pharmaceut. Res. 19, 681–688.

Nadzir, M.M., Nurhayati, R.W., Idris, F.N., Nguyen, M.H., 2021. Biomedical applications of bacterial exopolysaccharides: A review. Polymers. 13, 1–25.

Nagaoka, M., Hashimoto, S., Watanabe, T., Yokokura, T., Mori, Y., 1994. Anti-ulcer effects of lactic acid bacteria and their cell wall polysaccharides. Biol. Pharm. Bull. 17(8), 1012–1017.

Ng, J.Y., Obuobi, S., Chua, M.L., Zhang, C., Hong, S., Kumar, Y., Gokhale, R., Ee, P.L.R., 2020. Biomimicry of microbial polysaccharide hydrogels for tissue engineering and regenerative medicine: A review. Carbohydr. Polym. 241(11634), 5.

Niska, K., Zielinska, E., Radomski, M.W., Inkielewicz-Stepniak, I., 2018. Metal nanoparticles in dermatology and cosmetology: Interactions with human skin cells. Chem. Biol. Interact. 295, 38–51.

Ohkawara, Y., Tamura, G., Iwasaki, T., Tanaka, A., Kikuchi, T., Shirato, K., 2000. Activation and transforming growth factor-β production in eosinophils by hyaluronan. Am. J. Respir. Cell Mol. Biol. 23(4), 444–451.

Ooki, T., Murata-Kamiya, N., Takahashi-Kanemitsu, A., Wu, W., Hatakeyama, M., 2019. High-molecular-weight hyaluronan is a hippo pathway ligand directing cell density-dependent growth inhibition via PAR1b. Dev. Cell. 49(4), 590–604.

Osmałek, T., Froelich, A., Tasarek, S., 2014. Application of gellan gum in pharmacy and medicine. Int. J. Pharm. 466(1–2), 328–340.

Petri, D.F.S., 2015. Xanthan gum: A versatile biopolymer for biomedical and technological applications. J. Appl. Polym. Sci. 132(23).

Piermaria, J.A., Pinotti, A., Garcia, M.A., Abraham, A.G., 2009. Films based on kefiran, an exopolysaccharide obtained from kefir grain: Development and characterization. Food Hydrocoll. 23(3), 684–690.

Pindar, D.F., Bucke, C., 1975. The biosynthesis of alginic acid by *Azotobacter vinelandii*. Biochem. J. 152, 617–622.

Pinkert, A., Marsh, K.N., Pang, S., 2010. Reflections on the solubility of cellulose. Ind. Eng. Chem. Res. 49(22), 11121–11130.

Pourali, P., Yahyaei, B., 2018. Wound healing property of a gel prepared by the combination of *Pseudomonas aeruginosa* alginate and *Alhagi maurorum* aqueous extract in rats. Dermatol. Ther. 32, 1.

Prasongsuk, S., Lotrakul, P., Ali, I., Bankeeree, W., Punnapayak, H., 2018. The current status of *Aureobasidium pullulans* in biotechnology. Folia Microbiol. 63(2), 129–140.

Rafigh, S.M., Yazdi, A.V., Vossoughi, M., Safekordi, A.A., Ardjmand, M., 2014. Optimization of culture medium and modeling of curdlan production from *Paenibacillus polymyxa* by RSM and ANN. Int. J. Biol. Macromol. 70, 463–473.

Rahman, M.W., Arafat, M.T., 2021. Gellan and xanthan-based nanocomposites for tissue engineering. In: *Polysaccharide-Based Nanocomposites for Gene Delivery and Tissue Engineering*. Woodhead Publishing, pp. 155–190.

Rai, M., Wypij, M., Ingle, A.P., Trzcińska-Wencel, J., Golińska, P., 2021. Emerging trends in pullulan-based antimicrobial systems for various applications. Int. J. Mol. Sci. 22(24), 13596.

Rai, S., Kureel, A.K., Dutta, P.K., Mehrotra, G.K., 2018. Phenolic compounds based conjugates from dextran aldehyde and BSA: Preparation, characterization and evaluation of their anti-cancer efficacy for therapeutic applications. Int. J. Biol. Macromol. 110, 425–436.

Rana, S., Upadhyay, L.S.B., 2020. Microbial exopolysaccharides: Synthesis pathways, types and their commercial applications. Int. J. Biol. Macromol. 157, 577–583.

Rao, K.M., Kumar, A., Haider, A., Han, S.S., 2016. Polysaccharides based antibacterial polyelectrolyte hydrogels with silver nanoparticles. Mater. Lett. 184, 189–192.

Reis, R.A., Tischer, C.A., Gorin, P.A., Iacomini, M., 2002. A new pullulan and a branched (1→3)-,(1→6)-linked β-glucan from the lichenised ascomycete *Teloschistes flavicans*. FEMS Microbiol. Lett. 210(1), 1–5.

Rezaeeyiazdi, M., Colombani, T., Memic, A., Bencherif, S.A., 2018. Injectable hyaluronic acid-co-gelatin cryogels for tissue-engineering applications. Materials. 11, 8.

Riaz, T., Iqbal, M.W., Jiang, B., Chen, J., 2021. A review of the enzymatic, physical, and chemical modification techniques of xanthan gum. Int. J. Biol. Macromol. 186, 472–489.

Rico, C.W., Shin, J.H., Um, I.C., Kang, M.Y., 2011. Cholesterol-lowering action and antioxidative effects of microbial gum in C57BL/6N mice fed a high fat diet. Biotechnol. Bioprocess Eng. 16, 167–172.

Sakai, S., Kotani, T., Harada, R., Goto, R., Morita, T., Bouissil, S., Dubessay, P., Pierre, G., Michaud, P., El Boutachfaiti, R., Nakahata, M., Kojima, M., Petit, E., Delattre, C., 2022. Development of phenol-grafted polyglucuronic acid and its application to extrusion-based bioprinting inks. Carbohydr. Polym. 277, 118820.

Sakai, S., Mochizuki, K., Qu, Y., Mail, M., Nakahata, M., Taya, M., 2018. Peroxidase-catalyzed microextrusion bioprinting of cell-laden hydrogel constructs in vaporized ppm-level hydrogen peroxide. Biofabrication. 10(4), 45007.

Sakai, S., Ohi, H., Taya, M., 2019. Gelatin/hyaluronic acid content in hydrogels obtained through blue light-induced gelation affects hydrogel properties and adipose stem cell behaviors. Biomolecules. 9(8), 342.

Saranraj, P., Naidu, M.A., 2013. Hyaluronic acid production and its applications: A review. Int. J. Pharm. Biol. Arch. 4(5), 853–859.

Schmid, J., 2018. Recent insights in microbial exopolysaccharide biosynthesis and engineering strategies. Curr. Opin. Biotechnol. 53, 130–136.

Schmid, J., Sieber, V., Rehm, B., 2015. Bacterial exopolysaccharides: Biosynthesis pathways and engineering strategies. Front. Microbiol. 6, 496.

Schrager, H.M., Albertí, S., Cywes, C., Dougherty, G.J., Wessels, M.R., 1998. Hyaluronic acid capsule modulates M protein-mediated adherence and acts as a ligand for attachment of Group A *Streptococcus* to CD44 on human keratinocytes. J. Clin. Invest. 101(8), 1708–1716.

Selvasekaran, P., Mahalakshmi, Roshini, F., Angalene, L.A., Chandini, Sunil, T., Chidambaram, R., 2021. Fungal exopolysaccharides: Production and biotechnological industrial applications in food and allied sectors. In: A.N. Yadav (Ed.), *Recent Trends in Mycological Research. Fungal Biology*. Springer, Cham.

Sengha, S.S., Anderson, A.J., Hacking, A.J., Dawes, E.A., 1989. The production of alginate by *Pseudomonas mendocina* in batch and continuous culture. Microbiology. 135, 795–804.

Shi, S.S., Lei, S.J., Fu, C.X., 2020. Studies of the properties of CHG-Loaded alginate fibers for medical application. Polym. Test. 83, 106141.

Shiraishi, S., Imai, T., Otagiri, M., 1993. Controlled release of indomethacin by chitosan-polyelectrolyte complex: Optimization and in vivo/in vitro evaluation. J. Control. Release. 25, 217–225.

Si, J., Cui, Z., Wang, Q., Liu, Q., Liu, C., 2016. Biomimetic composite scaffolds based on mineralization of hydroxyapatite on electrospun poly(ε-caprolactone)/nanocellulose fibers. Carbohydr. Polym. 143, 270–278.

Silbir, S., Dagbagli, S., Yegin, S., Baysal, T., Goksungur, Y., 2014. Levan production by *Zymomonas mobilis* in batch and continuous fermentation systems. Carbohydr. Polym. 99, 454–461.

Simonsen, L., Hovgaard, L., Mortensen, P.B., Brøndsted, H., 1995. Dextran hydrogels for colon-specific drug delivery. V. degradation in human intestinal incubation models. Eur. J. Pharm. Sci. 3(6), 329–337.

Singh, R.S., Kaur, N., Rana, V., Kennedy, J.F., 2016. Recent insights on applications of pullulan in tissue engineering. Carbohydr. Polym. 153, 455–462.

Singh, S.K., Malhotra, S., Akhtar, M.S., 2014. Characterization of hyaluronic acid specific hyaluronate lyase (HylP) from *Streptococcus pyogenes*. Biochimie. 102(1), 203–210.

Skjåk-Bræk, G., Zanetti, F., Paoletti, S., 1989. Effect of acetylation on some solution and gelling properties of alginates. Carbohydr. Res. 185, 131–138.

Srikanth, R., Reddy, C.H.S.S.S., Siddartha, G., Ramaiah, M.J., Uppuluri, K.B., 2015. Review on production, characterization and applications of microbial levan. Carbohydr. Polym. 120, 102–114.

Stivala, S.S., Bahary, W.S., 1978. Some dilute-solution parameters of the levan of streptococcus salivarius in various solvents. Carbohydr. Res. 67(1), 17–21.

Stokke, B.T., Christensen, B.E., 1996. Release of disordered xanthan oligomers upon partial acid hydrolysis of double-stranded xanthan. Food Hydrocoll. 10(1), 83–89.

Sturzoiu, C., Petrescu, M., Galateanu, B., 2011. *Zymomonas mobilis* levan is involved in metalloproteinases activation in healing of wounded and burned tissues. Sci. Pap. Anim. Sci. Biotechnol. 44, 453–458.

Su, Y., Chen, L., Yang, F., Cheung, P.C.K., 2021. Beta-D-glucan-based drug delivery system and its potential application in targeting tumor associated macrophages. Carbohydr. Polym. 253, 117258.

Sunguroğlu, C., Sezgin, D.E., Aytar Çelik, P., Çabuk, A., 2018. Higher titer hyaluronic acid production in recombinant *Lactococcus lactis*. Prep. Biochem. Biotechnol. 48(8), 734–742.

Takeuchi, H., Matsui, Y., Yamamoto, H., Kawashima, Y., 2003. Mucoadhesive properties of carbopol and chitosan-coated liposomes and their effectiveness in the oral administration of calcitonin to rats. J. Control. Release. 86, 235–242.

Tan, H., Chu, C.R., Payne, K.A., Marra, K.G., 2009. Injectable in situ forming biodegradable chitosan-hyaluronic acid based hydrogels for cartilage tissue engineering. Biomaterials. 30, 2499–2506.

Termeer, C., Benedix, F., Sleeman, J., Fieber, C., Voith, U., Ahrens, T., Miyake, K., Freudenberg, M., Galanos, C., Simon, J.C., 2002. Oligosaccharides of hyaluronan activate dendritic cells via Toll-like receptor 4. J. Exp. Med. 195(1), 99–111.

Thanoo, B.C., Sunny, M.C., Jayakrishnan, A., 1992. Cross-linked chitosan microspheres: Preparation and evaluation as a matrix for the controlled release of pharmaceuticals. J. Pharm. Pharmacol. 44(4), 283–286.

Tiwari, S., Patil, R., Bahadur, P., 2019. Polysaccharide based scaffolds for soft tissue engineering applications. Polymers. 11, 1.

Tomulescu, C., Stoica, R., Sevcenco, C., Căşărică, A., Moscovici, M., Vamanu, A., 2016. Levan-a mini review. Sci. Bull. Ser. F. Biotechnol. 20, 309–317.

Tsai, W., Tsai, H., Wong, Y., Hong, J., Chang, S., Lee, M., 2018. Preparation and characterization of gellan gum/glucosamine/clioquinol film as oral cancer treatment patch. Mater. Sci. Eng. C. 82, 317–322.

Valentine, M.E., Kirby, B.D., Withers, T.R., Johnson, S.L., Long, T.E., Hao, Y., Lam, J.S., Niles, R.M., Yu, H.D., 2020. Generation of a highly attenuated strain of *Pseudomonas aeruginosa* for commercial production of alginate. Microb. Biotechnol. 13, 162–175.

Wang, H., Zhou, L., Liao, J., Tan, Y., Ouyang, K., Ning, C., Ni, G., Tan, G., 2014. Cell-laden photocrosslinked GelMA-DexMA copolymer hydrogels with tunable mechanical properties for tissue engineering. J. Mater. Sci. Mater. Med. 25, 2173–2183.

Wang, Y., Fei, D., Vanderlaan, M., Song, A., 2004. Biological activity of bevacizumab, a humanized anti-VEGF antibody in vitro. Angiogenesis. 7(4), 335–345.

Watase, M., Nishinari, K., 1993. Effect of potassium ions on the rheological and thermal properties of gellan gum gels. Top. Catal. 7(5), 449–456.

Wei, X., Liu, G.-L., Jia, S.-L., Chi, Z., Hu, Z., Chi, Z.-M., 2021. Pullulan biosynthesis and its regulation in *Aureobasidium* spp. Carbohydr. Polym. 251, 117076.

Wessels, M.R., Moses, A.E., Goldberg, J.B., Dicesare, T.J., 1991. Hyaluronic acid capsule is a virulence factor for mucoid Group A *Streptococci*. Proc. Natl. Acad. Sci. U S A. 88(19), 8317–8321.

Westbrook, A.W., Ren, X., Oh, J., Moo-Young, M., Chou, C.P., 2018. Metabolic engineering to enhance heterologous production of hyaluronic acid in *Bacillus subtilis*. Metab. Eng. 47, 401–413.

Whistler, R.L., 1973. Solubility of polysaccharides and their behavior in solution. *ACS*. 242–255.

Wibawan, I.W.T., Pasaribu, F.H., Utama, I.H., Abdulmawjood, A., Lämmler, C., 1999. The role of hyaluronic acid capsular material of Streptococcus equi subsp. zooepidemicus in mediating adherence to HeLa cells and in resisting phagocytosis. Res. Vet. Sci. 67(2), 131–135.

Woo, J.E., Seong, H.J., Lee, S.Y., Jang, Y.S., 2019. Metabolic engineering of *Escherichia coli* for the production of hyaluronic acid from glucose and galactose. Front. Bioeng. Biotechnol. 7, 1–9.

Yang, M.C., Chi, N.H., Chou, N.K., Huang, Y.Y., Chung, T.W., Chang, Y.L., Liu, H.C., Shieh, M.J., Wang, S.S., 2010. The influence of rat mesenchymal stem cell CD44 surface markers on cell growth, fibronectin expression, and cardiomyogenic differentiation on silk fibroin—Hyaluronic acid cardiac patches. Biomaterials. 31(5), 854–862.

Yildiz, H., Karatas, N., 2018. Microbial exopolysaccharides: Resources and bioactive properties. Process Biochem. 72, 41–46.

Yoo, S.H., Yoon, E.J., Cha, J., Lee, H.G., 2004. Antitumor activity of levan polysaccharides from selected microorganisms. Int. J. Biol. Macromol. 34, 37–41.

Yuguchi, Y., Mimura, M., Kitamura, S., Urakawa, H., Kajiwara, K., 1993. Structural characteristics of gellan in aqueous solution. Top. Catal. 7(5), 373–385.

Zhang, H., Huang, L., Nishinari, K., Watase, M., Konno, A., 2000. Thermal measurements of curdlan in aqueous suspension during gelation. Food Hydrocoll. 14(2), 121–124.

Zhang, M., Huang, Y., Pan, W., Tong, X., Zeng, Q., Su, T., Qi, X., Shen, J., 2021a. Polydopamine-incorporated dextran hydrogel drug carrier with tailorable structure for wound healing. Carbohydr. Polym. 253, 117213.

Zhang, R., Edgar, K.J., 2014. Properties, chemistry, and applications of the bioactive polysaccharide curdlan. Biomacromolecules. 15(4), 1079–1096.

Zhang, T., Li, R., Qian, H., Mu, W., Miao, M., Jiang, B., 2014. Biosynthesis of levan by levansucrase from *Bacillus methylotrophicus* SK 21.002. Carbohydr. Polym. 101(1), 975–981.

Zhang, X., Li, Z., Yang, P., Duan, G., Liu, X., Gu, Z., Li, Y., 2021b. Polyphenol scaffolds in tissue engineering. Mater. Horiz. 8(1), 145–167.

Zhang, Z., Suner, S.S., Blake, D.A., Ayyala, R.S., Sahiner, N., 2020. Antimicrobial activity and biocompatibility of slow-release hyaluronic acid-antibiotic conjugated particles. Int. J. Pharm. 576, 119024.

Zhou, Z., Xiao, J., Guan, S., Geng, Z., Zhao, R., Gao, B., 2022. A hydrogen-bonded antibacterial curdlan-tannic acid hydrogel with an antioxidant and hemostatic function for wound healing. Carbohydr. Polym. 285, 119235.

Zhu, Z., Wang, Y.-M., Yang, J., Luo, X.-S., 2017. Hyaluronic acid: A versatile biomaterial in tissue engineering. Plast. 4, 12.

12 Exopolysaccharides for bioremediation

*Ramesh Sharma, Pinku Chandra Nath,
Shubhankar Debnath, Amiya Ojha, Biswajit Sarkar,
Tarun Kanti Bandyopadhyay, and Biswanath Bhunia*

12.1 INTRODUCTION

The excessive contamination in the present scenario occurred due to industrialization, urbanization, and different agricultural practices. The intentions for the achievement of comfort, food security, and growth with industrialization, urbanization, and vicious agrarian progression ultimately converted to disasters and unrest conditions. Considering these issues, the major problem arrived with their contamination in water and soil. The WHO has mentioned the list of heavy metals including As, Cd, F, Ld, and Mg as the top chemicals associated with awareness of public health (Jiang et al., 2023; Lin et al., 2023). Other heavy metals are also included, which are hazardous to human health. Various organizations quoted regarding the application of bioremediation for lowering the metal availability to below permissible limits (He et al., 2023; Sun et al., 2023). The researchers are trying to develop protocols for the identification of bacterial species to solve the problems associated with heavy metals (Kielak et al., 2017). Microbes can fight the metal stress by releasing the EPS. EPS is mainly obtained from unpleasant stressful environmental conditions with different nutrient supplementation (Mohite et al., 2017). The EPS are produced with lower cost of operation also produced biofilm with maintaining sustainability. Considering the merits point of EPS, it is believed to be an effective tool to remove impurities than comparing to conventional methods. The conventional methods are ion exchange, reverse osmosis, ultrafiltration, nanofiltration, ultrafiltration, nanofiltration, coagulation, chemical precipitation, flotation, and electrodialysis for dealing with the detection of these heavy metals (Feng et al., 2023). Hence, for the fulfilment of current challenges, the EPS must be used as a potential biosorbent for scavenging heavy metals. Metal scavenging capacity is altered for both capsular polysaccharides and EPS. The metal removing capacity depends on the size of the colonies. For the case of binding the metal ions, uronic acid performed the major role. The acidic nature of EPS allowed it to tolerate heavy metal stress and is hence considered an effective tool for removing heavy metals (Bhunia et al., 2018). Microbial EPS play a greater role due to its unique features with comparing to EPS extracted from conventional sources. Therefore, for its commercialization through the industrialization process, there is a need to improve the production rate and quality by reducing the overall cost. The contamination with heavy metals is unmitigated problem associated with the environment. For this consideration, EPS is considered as the best option to remove heavy metals (Kuyucak and Volesky, 1990). The biomass cell walls play a crucial role in the biosorption of heavy metals. The substances retaining the structure of the fiber is considered an established medium to attach the heavy metals with active absorption (Ozdemir et al., 2005b). This chapter emphasize overapplication of EPS in bioremediation of heavy metals.

12.2   GENERAL PROPERTIES OF EPS

Chemically, the structure EPS obtained from bacteria may be homo- or hetero-polysaccharides (Sun and Zhang, 2021). The various bacterial groups released the polysaccharides with molecular weight ranging from 0.2–2 9 104 kDa, which are potentially used due to their higher solubility and ease of removing from solutions (Pereira et al., 2009). The EPS obtained from bacteria have distinct features which make it distinct from other derived polysaccharides. The anionic feature is because of attachment of uronic acid and sugar with the sulfate group (Pereira et al., 2009) that functions by imparting the formation of hydrated gels (Kehr and Dittmann, 2015). Secondly, the EPS originating from bacteria are complex hetero-polysaccharides with 75% constituents of six or various kinds of monosaccharides units (Pereira et al., 2009). The obtained polysaccharides from bacteria may be intracellular or extracellular in nature (Pereira et al., 2009). A type of polysaccharide known as released polysaccharides is used by some bacterial groups to release extracellular components into the surrounding media (De Philippis and Vincenzini, 1998). The aggregate's formation under the dense condition with its propagation for blooming is associated with the formation of loose and tightly attached polymeric substances in extracellular region (Xu et al., 2013). The bacterial EPS also retains the polypeptides with the excess presence of valine, leucine, glycine, alanine, phenylalanine, and isoleucine (Flaibani et al., 1989). The few EPS isolated from bacteria retained with higher content of glutamic and aspartic acids (Flaibani et al., 1989; Kawaguchi and Decho, 2002). Therefore, EPS originating from bacteria has greater potential for application in biotechnology (De Philippis et al., 2011). The availability of negative charges in EPS is considered as potential chelating agent for metals in solutions (De Philippis et al., 2011; Tran et al., 2016). The use of bacterial EPS has been redesigned in the field of drug delivery, coatings, or metal remediation (Pereira et al., 2019).

12.3   HEAVY METALS BIOREMEDIATION USING EPS

12.3.1   BIOREMEDIATION OF LEAD (PB)

Automotive industries, metal recycling industries, and processing facilities all released lead metals. Lead is a non-bioelementary, dangerous, and widespread metal that has an impact on all living things in the environment (Ojuederie and Babalola, 2017). A number of studies have focused on the lead adsorption by EPS. The *Enterobacter* A47 derived polysaccharide FucoPol has excellent potential as a biodegradable and secure biosorbent for the control of Pb^{2+} contaminated water (Concórdio-Reis et al., 2020). The EPS's binding ability to Pb was 303.03 mg/g, indicating the positive interaction results (Nicolaus et al., 2010).

12.3.2 BIOREMEDIATION OF CADMIUM (CD)

The environmental contaminant, Cd is a powerful pollutant that has the potential to negatively impact human health. In the crust of the planet, Cd occurs naturally. The majority of the time, it is present in minerals as a mixture with other substances like oxygen, chlorine, or sulfur. A portion of Cd is present in most soils, rocks, coal, and mineral fertilizers (Bhatla et al., 2018). The majority of today's Cd production comes from recovered nickel-cadmium batteries and zinc by-products (Morrow, 2000). High Cd exposure affects people's lungs and pulmonary disease. Renal disease and brittle bones are two conditions that are made more likely by long-term exposure to low levels of Cd in the environment, such as through cigarette smoke. The Cd is also responsible for causing cancer diseases (Mezynska et al., 2018; Zheng et al., 2021). The various heavy metals associated with the diseases to human health is shown in Figure 12.1. This heavy metal is harmful to people even at very low quantities of roughly 0.001–0.1 mg/L (Trivedi, 2020). The synthesis of EPS by microorganisms is one of their most effective strategies for fending off the toxicity. Arthrobacter viscous bacterial EPS can considerably boost the biosorption of Cd (Singh and Kumar, 2020). An EPS-producing halophilic bacteria called *Halomonas* sp. may be utilized to bioremediate Cd in contaminated salty soils by absorbing more than 50% of it (Amoozegar et al., 2012). Many bacterial species, such as *Pseudomonas, Bacillus,* and *Streptomyces* have a substantial capacity to produce EPS which break down the Cd metals and then destroy them (Al-Dhabi et al., 2019).

12.3.3 BIOREMEDIATION OF ARSENIC (AS)

As per the WHO, the As concentration in drinking water is 10 μg/L (Yu and Fein, 2015). As is poisonous metalloid emitted into the atmosphere by anthropogenic and natural processes. Mining, weathering, fossil fuel combustion, and volcanic events

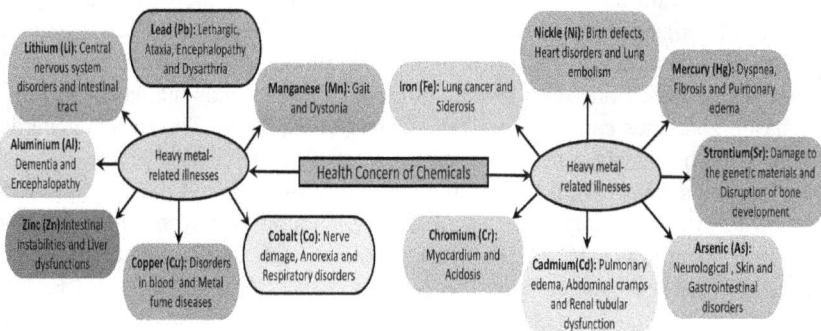

FIGURE 12.1 Heavy metals associated with various diseases to human health

are some of the main activities. Numerous deadly diseases in both humans and animals are brought on by arsenic (Butt and Rehman, 2011; Wang and Zhao, 2009). Long-term As exposure may potentially cause bladder and lung cancer. As also causes neurological and gastrological disorder (Chikkanna et al., 2019). The biofilm made from polyanionic EPS is thought to be an excellent biosorbance to bioremediate arsenic metals (Saranya et al., 2017). The As bioremediation using microbial species community involves their oxidation, reduction, intracellular bioaccumulation, and methylation. *Pseudomonas, Acinetobacter, Klebsiella*, and *Comamonas* sp. produced the EPS and also oxidized arsenic (Rong et al., 2022).

12.3.4 BIOREMEDIATION OF MERCURY (HG)

Hg is widely dispersed across the environment and is present in the soil, air, and water. Remediation of this metal is absolutely necessary because it is highly toxic to people, animals, and plants at low concentration. Minamata disease is caused by methylmercury toxicity in humans. Neurological and behavioral disorders may be observed after Hg inhalation (Dos Santos et al., 2018). The primary sources of Hg contamination include metal mining, the burning of fossil fuels, the acetaldehyde and paint industries, amalgamation plants, and chloralkali plants (Mani and Kumar, 2014). Specifically, the speciation of Hg determines the poisonous nature of Hg (Abd AL-MAnhEL, 2017). Hg is mainly exposed to the environment in the form of Hg^{2+} from industrial sources (Salehizadeh and Shojaosadati, 2003). EPS plays a significant part in the heavy metal bioremediation method's detoxification of Hg. At four different concentrations (100, 150, 200, and 250 μg/ml) of Hg, the *Vibrio fluvialis* was examined for evaluating Hg removing capacity, and optimum results were obtained for 250 μg/ml (with removing 60% of Hg) (Shuhong et al., 2014). In order to bioremediate Hg pollution, *Bacillus thuringiensis* strain RGN1.2 was utilized. At doses of 10, 25, and 50 ppm, it removed 96.72%, 90.67%, and 90.10% of the mercury, respectively (Rasulov et al., 2013).

12.3.5 BIOREMEDIATION OF CHROMIUM (CR)

Since it has high toxicity and carcinogenicity, Cr is hazardous even in very small amounts. Cr contamination in the environment is caused by industrial development (Iyer et al., 2006). Supplemental of Cr may result in hypoglycemia, dermatitis, anemia, thrombocytopenia, liver dysfunction, renal failure, myocardium, rhabdomyolysis, and weight loss. Long-term consumption of Cr compounds increases the risk of lung, nasal, and sinus cancer (Wu, 2020; Yatera et al., 2018). The EPS is generated from bacterium *Enterobacter sp.* in presence of Cr (VI). The concentration used for the previous experiments was in the range of 25–100 ppm, and it accumulated about 60 to 70% Cr (Kalpana et al., 2018). A potential mutagen and carcinogen, hexavalent chromium [Cr(VI)] is released into the atmosphere by human activity (Das et al., 2019). Cr(VI) content was reduced by 31.7% at 10 mg/l by EPS from *Enterobacter cloacae* (Iyer et al., 2004). In addition to bacterial polysaccharides, fungal polysaccharides are the subject of research in a number of biotechnological disciplines. Moreover, *Aspergillus niger, A. terreus, Fusarium solani, Lentinula edodes*, and *Fusarium oxyporum* species produce some fungal EPS (Castro et al., 2019; Hu, 1996; Subudhi et al., 2016).

12.3.6 BIOREMEDIATION OF ZN

The Zn is obtained through environment from soil, water, and air, and it also obtained from commercial Zn containing products, burning of coal, burning of wastage, car tire, Zn mining and metallurgical operations. It is crucial for several enzymes that power numerous metabolic processes. High concentration of Zn can be toxic to both plants and animals. Although Zn poisoning in humans is curable and non-hazardous, a number of symptoms, such as fever, breathing problems, nausea, chest pain, and cough, can interfere with daily activities and lifestyle. Many gram-positive and negative species have been identified as potential Zn bioremediation agents, including *Staphylococcus* sp., *Bacillus* sp., *Pseudomonas* sp., *Streptococcus* sp., *Klebsiella* sp., *E. coli*, and *Enterobacter* species.

12.3.7 BIOREMEDIATION OF CO

The major source of Co contamination comes from its mining, however, it is not found in pure form; it is typically found as $CoAsS$, $CoCO_3$, Co_3S_4, Co $(Ni)As_3$, Co_3S_4, $CuCo_2S_4$, $CoAs_2$, $CoAs_3$, and $Co_3(AsO_4)_2$. Coal-fired power stations, incinerators, and vehicle exhaust all have the potential to discharge trace amounts of Co into the atmosphere. Crop productivity decreases and reproductive and respiratory health problems occur from Co contamination. The cobalt concentration in soil varies considerably, typically falling between 1 and 40 ppm, with an average level of 7 ppm. As people and animals consume these plants, cobalt enters their bodies and builds up. The following negative health impacts are brought on when Co concentration exceeds its greatest limit. For six hours, workers who breathed air 0.038 mg cobalt/m^3 experienced problems with breathing. Individuals who were exposed to 0.007 mg cobalt/m^3 at work also experienced asthma attacks and skin rashes due to Co allergy. Those that were exposed to 0.5 mg cobalt/kg for a few weeks experienced effects on the thyroid. The EPS is one of its vital components and is released as a self-defense mechanism against extreme famine situations.

12.3.8 BIOREMEDIATION OF NI

The main source of nickel in the environment is the chemical industry, ship cruise effluents, weathering of rocks and soils, forest fires, and commercial waste. As the concentration of Ni in the environment increases with time, its results adversely impact the quality of the environment for flora and fauna. Excess concentration of Ni in the soil causes chlorosis and necrosis as with other heavy metals and causes disruption of Fe uptake and metabolism. Excess consumption of Ni by animals causes inflammation of lung with damaging nasal cavity to animal after exposing to Ni retaining compound. So it is necessary to recover or remediate the contamination of Ni. Intact microbial cells and cellular EPS is extensively used to remediate heavy metals in industrial and environmental like Ni in wastewater sources.

Among microorganisms—algae, bacteria, and fungi all are considered as EPS-producing organisms. The basic mechanism behind the bioremediation of Ni by EPS is absorption.

12.4 RELATIONSHIP BETWEEN STRUCTURE ACTIVITY OF EPS TOWARDS THE ABSORPTION OF HEAVY METALS

The EPS components include carbohydrates, nucleic acid, proteins. The EPS may be available in free or capsular form. It is reported that EPS exhibits a higher affinity to the wastewater retaining the heavy metals (Kantar et al., 2011; Liu and Fang, 2002; Sheng et al., 2013). The mechanism of heavy metal interaction with EPS is shown by Figure 12.2.

The methods including complexation, ion exchange, surface micro-precipitation, etc. are applied for attachment of the heavy metals to EPS. The capacity of binding is determined by EPS compositions, binding sites available on the surface of EPS (Aquino and Stuckey, 2004). Moreover, the release of EPS is occurred due to excretion, which took place as a result of toxic metal availability in the environment and added protection to microorganisms through inhibition from heavy metals. The organo-metal complex is formed between metal and protein, polysaccharides and phospholipids retaining carboxyl, hydroxyl, and phosphoric amine groups which

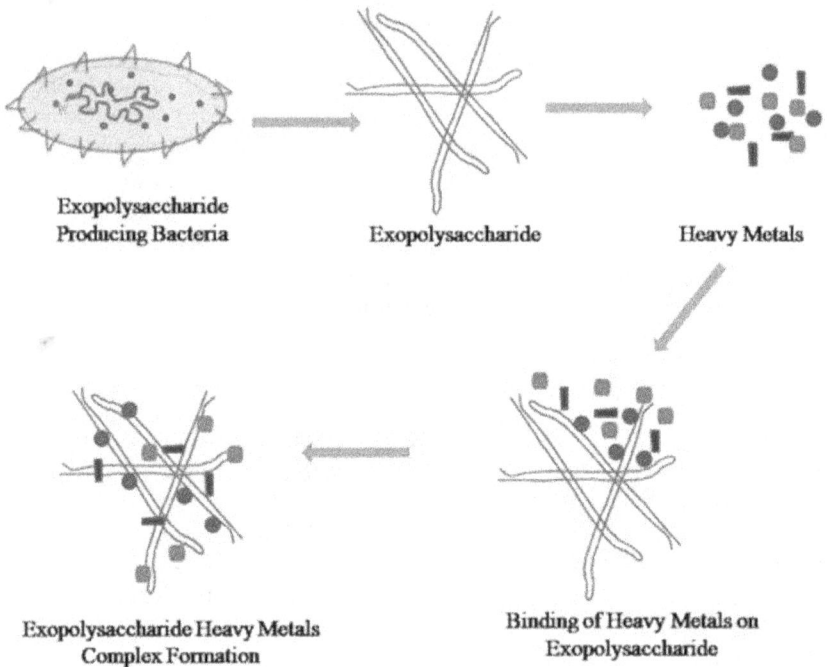

Exopolysaccharide
Producing Bacteria   Exopolysaccharide   Heavy Metals

Exopolysaccharide Heavy Metals   Binding of Heavy Metals on
Complex Formation   Exopolysaccharide

FIGURE 12.2   Mechanism of exopolysaccharide interaction with heavy metals

are contained in EPS (Pagnanelli et al., 2009; Song et al., 2014). The complexity occurred by interaction through hydrophilicity between phosphoric groups in EPS and heavy metals (Kantar et al., 2011; Sheng et al., 2013). The interaction with hydrophobicity resulted to stronger attachment of heavy metals on the EPS surface (Hou et al., 2013; Späth et al., 1998). Additionally, the pH added a crucial role in forming the organo-metal complex as it facilitates overcharge distribution on the EPS surface (Zhang et al., 2014). The capacity of absorption of EPS with heavy metals is dependent on the types of EPS used (Wei et al., 2017). The adsorption capacity improves with decreasing of heavy metal's hydrated radius because the lower value of hydration radii resulted to accelerate absorption with a higher quantity of absorption to EPS (Bhatti et al., 2009; Erdem et al., 2004; Zhang et al., 2014). Additionally, the higher rate of absorption is dependent on EPS hydrophilic characteristics as EPS retain the higher protein content which ultimately exaggerate with forming complex bond through attraction with electrostatic force between heavy metals and EPS (Maurice et al., 2002). Additionally, the EPS amphiphilic properties added the vital role to improve the capacity of heavy metal absorption (Liu et al., 2015). The polyvalent metal ions are efficient for binding with carbohydrates and proteins (Liu et al., 2015), and the ratio of available proteins and carbohydrates is regarded as index of adsorption capacity for EPS. It is mentioned that divalent metal adsorption capacity is dependent on the ratio of protein to carbohydrates present in EPS, and divalent metal included Pb(II), Cd(II), Cu(II), and Zn(II) etc. (Liu et al., 2015) and concluded that protein is a key component for performing absorption in EPS. It is revealed that Zn(II) suitably attached to carbohydrates rather than to proteins (Wei et al., 2017). The comprehensive changes that occurred in compositions of EPS after binding the heavy metal to EPS can be characterized with analysis by excitation-emission matrix (Li and Yu, 2014). After the transfer of Zn^{2+}, the spectra for activated sludge and anaerobic granular sludge was changed (Pan et al., 2010; Zhang et al., 2010). It was noticed that the EPS fluorophore intensity value was decreased randomly as a result of entrapping the metal ions to EPS (Wei et al., 2017). The experiment was performed with investigating the absorption quality by anaerobic flocculent sludge, activated sludge, and anaerobic granular sludge, with their fraction to absorb the Zn^{2+} and Cu^{2+} ions (Wei et al., 2017). At the initial conditions, the extraction of EPS was done from three sludges. The dilution with fractionation was performed for EPS using XAD-4 resins. The collected five fractions retained the molecular weight higher by 3.5KDa. EEM was used for evaluating the compositions changes that occurred in EPS after heavy metal absorption. The fluorescence intensity was decreased under reduced manner for HPI fraction with absorption of Zn^{2+} and Cu^{2+}. Among the fraction, HPI relatively performed the better absorption of heavy metals, considering other fractions indicated that non-fluorophores group, including hydrocarbon, simple carbon chain, and fatty chains, added the crucial role for absorption of heavy metal. Although, the capacity of absorption for heavy metals considering other fraction were lower than HPI, whereas the value of fluorescent intensity was decreased equally in all fractions, which indicated the presence of a higher quantity of substances featuring aromatic protein, SMP-related fluorophore, and fulvic acid-like substances. A remarkable decrease in fluorescence intensity was seen for all fractions, which revealed that heavy metals were easily bound by EPS. The EPS with its surface charge is changed as a result of the interaction between EPS and heavy metal (Tedetti et al., 2013). The fluctuation over zeta potential was estimated before and after binding of heavy metals (Wei et al., 2017). It is revealed over structural characteristics of EPS to study the bioremediation properties shown by the strain called *Enterobacter ludwigii* (Pau-Roblot et al., 2013). The produced EPS from *E. ludwigii* had higher molecular weight, which was heteropolysaccharide. The EPS is constituted with repeating units of d-glucuronic acid d-glucose, d-galactose, and l-fucose, at the ratio of 2:1:2:1 and is substituted with acyl and pyruvyl groups. The presence of pyruvate galactose residue in the heteropolysaccharide mainly responded to binding with heavy metals (Pau-Roblot et al., 2013). The absorption capacity of heavy metal with EPS using sulfate reducing bacteria was discussed (Yue et al., 2015). The Fourier transformed infrared spectroscopy (FTIR) analyses revealed that functional group expression on the EPS surface is not dependent on the heavy metal used. However, group density on the EPS surface is dependent on the heavy metal dose. Additionally, pH added the significant role to absorbed heavy metals in EPS because several buffering zones are available for each EPS. Each buffering zone is associated with the EPS retaining specific functional group. The various zone is determined by potentiometric titration, and peaks are obtained and represented the highest pH variation with respect to equivalence points. Similarly, the local minima signify the lower pH variation with equivalence points which represented the buffering zone (Braissant et al., 2007).

12.5 TOOLS OF UNDERSTANDING THE EPS-METAL INTERACTION

Different approaches were applied including biophysical techniques which perform with interaction of functional group with heavy metals. The merits and demerits of heavy metals absorption by EPS is shown in Table 12.1. The most commonly used biophysical techniques included X-ray spectroscopy and FTIR and microscopic approach such as AFM and SEM which delivered the light to the functional group involved with interaction to heavy metals.

The understanding of functional group attachment to heavy metals is a crucial aspect of designing the system for bioremediation. X-ray spectroscopy help to determine the complexity of EPS metal interaction including involvement of the functional group, metal adsorbed speciation and its oxidation number, metal localization in cell surface, and isotopic content. The application of spectroscopy based on X-ray performed with estimating the correlations between element energy values at the particular atomic number and the edge of X-ray absorption spectra provided the strategy to identify the elements with their states and oxidation degree (Minkina et al., 2016). The use of X-ray spectroscopy in the field of microenvironments has enabled research with improved chemical sensitivity and specificity with respect to elements with the ability to investigate over heterogeneous elements (Potnis et al., 2021). These techniques have been successfully used in cyanobacterial mats/biofilm. In the previous studies, the cyanobacterial biofilm was detected with X-ray fluorescence and estimated the sequestered amount of Zn and Mn from the cyanobacterial mats, whereas the speciation pattern was not cleared (Bender et al., 1994). The X-ray

TABLE 12.1

Major benefits and drawbacks of pure and cellular EPS associated for adsorption of heavy metals

Sl. No.	Benefits	Drawbacks
1	Environmentally friendly and economical adsorbent	Real industrial samples have poor biosorption efficiency.
2	Requires less harsh operating conditions than traditional physiochemical techniques	Weak desorption chemicals must be used for reusability, which alternatively leads to decreased desorption efficiency.
3	No production of harmful by-products	Metal ion selectivity or specificity is absent.
4	It is recycled with following the desorption of deposited metal ions, and the recovered metal ions is used later	Sorption—desorption with 5 to 10 cycles are the most consecutive reusability permitted.
5	Dead cells are free of the risks that living cells present, hence both live and dead cell and live cell bound EPS is used to absorb the heavy metals	Adsorption is difficult to apply to real samples because it is highly affected by the operating parameters including pH, temperature, ionic strength, as well as availability of biological ligands.
6	Sensitive and able to sequester even at low metal ion concentrations	It is difficult to keep living cells viable at increased metal or toxicant concentrations.

spectroscopy in combination with electron microscopy was successfully used for the identification of Fe, Al-silicate complex with trace amount of K in epilithic biofilms consisting in the bacterial cell and constituents of cyanobacteria after collection from sites of Brahmani River (Konhauser et al., 1998). Using Scanning Transmission X-ray Microscopy, the species of Fe, including Fe, that is, Fe(II) and Fe(III), Mn and Ni were found in water-constituted algae and bacteria, and these organisms were attached with polysaccharides (Dynes et al., 2006). In a research carried out using cyanobacterial biofilm (*Lyngbya* sp. and *Leptolyngbya* sp.) using X-ray spectroscopy, it mentioned minor components diatom *Achnanthes exigua* and concluded with the fact that Cu was preferentially bound with O and N as well as a small fraction with S (Coutaud et al., 2018). Another fact revealed that the fractionation of Cu isotopes and the fractionation favor for heavier isotopes during the process of adsorption phase, and carboxylate ligands are involved with forming fractionation of lighter isotopes after incorporating Cu with biofilm (Coutaud et al., 2018). The heavier isotope adsorption was influenced by the formation of a covalent bond with surface moieties with biofilm (Coutaud et al., 2014). The application of X-ray Photoelectron Spectroscopy in the biofilm of *Nostoc muscorum* with exposure to Cd detected the availability of Cd in second state of oxidation at the time of interacting with biofilm. The SEM with incorporation to energy dispersive spectroscopy is used for the analysis of surface materials and used for identification of surface metals and is used for biological purpose for analyzing the cell surface. The interacting effect of SEM-EDS is used to estimate the metal distribution concept in biofilm and is mentioned for Cd(II) and other various metals. The analysis with SEM-EDX revealed with a binding capacity

of metal with RPS, which was isolated from *Cyanothece*, and other metals including Ca, Mg, Na, and Cl were also detected (Mota et al., 2016). With contact to metal
ions, the C concentration was increased, whereas the concentration of O, Mg, S, and
Ca were lowered on the surface of RPS (Mota et al., 2016). FTIR is an important
tool for the identification of functional groups with their absorption peak, which are
identical for specific functional groups. Similarly, the FTIR revealed in algal mats
with the mechanism of Cr(IV) uptake with identifying the changes that occurred in
functional groups mainly the occurrence of complexation involved with carboxylate
group with obtaining reduced Cr, involved with formation of C-O group in the polysaccharides during the Cr biosorption and the chromium (VI)-O bond is appeared
giving the specific signal which was shown in surface of algal consortium mat after
attachment of Cr through analysis using FTIR (Shukla et al., 2012). Additionally, the
biofilm derived from *N. Muscorum* bind with Cd with forming the linkages including Cd-O and C-N linkages, which was revealed by FTIR with combining to XPS
(Raghavan et al., 2020). The RPS derived from *Cyanothece* treated with deionized
H_2O, alkaline, or acid by exposing to mono metals [Cu(II), Cd(II), Pb(II)] revealed
the distinct features showing shift under three conditions, with the optimum shift
at the absorbance peak of O-H stretching in RPS, which are treated with base than
comparing to water and acid (Mota et al., 2016). Therefore, the information provided
with the facts that the changes over functional groups is noticed with finding the
differences for the mono metals remediation under basic conditions (Mota et al.,
2016). The exopolymeric substances were used for identifying the oxygen of carboxyl groups in the protein components and considered as the key sites for complex
formation of EPS-Cd (Xie et al., 2020). The changes that occurred on topography of
cell surface due to challenges associated with metals occurred at microscopic level
and up to manometer which can be measured using AFM. The application of AFM
approaches for understanding the interaction between metal-biofilm-EPS played the
important role. The various parameters evaluated using AFM are (1) size of cells
(Raghavan et al., 2020; Tekaya et al., 2013), (2) morphology of cell or lysis of cell in
Arthospira platensis biofilm with exposure to heavy metals (Tekaya et al., 2013), and
(3) cell roughness measurement in biofilm found in *N. muscorum* after exposing to
Cd (Raghavan et al., 2020) and Pb.

12.6   EPS FUNCTION OVER SOIL CONTAMINATION

12.6.1   EPS FUNCTION OVER CONTAMINATED SOIL OBTAINED
THROUGH INDUSTRIAL WASTE

The polluted soil retains the Cr(VI), which is a pollutant metal and is considered as
carcinogenic element. It can be eliminated through reductive absorption with the
following microbes by changing it from Cr(VI) to Cr(III). The organism including *Klebsiella pneumoniae*, *Bacillus firm*, and *Mycobacterium* sp. efficiently
absorb the Cr(VI), whereas in the fungal groups which included the *Aspergillus*
sp., *Aspergillus flavus*, and *A. niger* which are competent to change it from Cr(VI)
to Cr(III). The microorganism including *Klebsiella pneumoniae* can survive at the

Cr(VI) concentration of 550 mg/L. Similarly, the *Mycobacterium* sp. had a higher tolerance level of 750 mg/L, and fungal groups can withstand at the Cr(VI) concentration of 600 mg/L at 35 to 37°C, and pH 7. The reduction of Cr(VI) was obtained (>0.05%) (Bennett et al., 2013). The strain called *Providencia vermicola*, which retains with bmtA known as metallothionein gene, functions by facilitating the lead sequestration from intracellular regions. Additionally, the analysis from TEM concluded with presence of periplasmic space, which is utilized to sequestrate the lead form called $PbSO_4$ by using *Providencia vermicola* strain S (Sharma et al., 2017). Bacterial species were separated from rhizosphere by selecting plant groups. These selected species were used to check with the capacity of bioaccumulation towards Cr, Ni, and Cd. The capacity of bio-absorption towards heavy metals were dependent on pH with value of 5 for Cr, 7 for Cd, and under the incubation period of 24 h and 5 min. At the Cr(VI) concentration of 250 mg/L the following, including KA24, KA18, and KA25, could accumulate at the concentration of 118.2, 121.87, and 90 mg metal ion per biomass (g) respectively. Additionally, 135, 127.5, and 146.25 mg of Ni (II) was accumulated by KA24, KA18, and KA25 respectively after the experiment performed at the same initial concentration of Ni. Additionally, the selective three strains were efficient to accumulate at the concentration of 120, 129.37, 135 mg Cd(II) with considering gram of biomass respectively (Akhter et al., 2017). The bacterial strains including Phaseolus *vulgaris* AR6 and *Phaseolus vulgaris* AR8 after isolating from *Phascolus vulgaris*, function as bio-adsorption of Cr. The accumulation of Cr occurred in two ways. In one way, bacterial un-inoculated Cr is used to treat root and shoots. In another case, the bacterial inoculated Cr is used to treat the root and shoot. The accumulation of Cr under bacterial inoculation was found with 7.92 μg/g and 4.28 and 4.28 μg/g for the root and shoot respectively. The higher Cr aggregation achieved using *Phaseolus vulgaris* bacteria with accumulating 71% by root and 66% by shoots (Karthik et al., 2016). The properties bio-flocculant in the sludges is altered with presence of heavy metals. Therefore, removing this metal pollutants is the primary task. The bacterial group called *Achromobacter* sp. TL-3 strain could deposit the Pb(II) at pH ranging from 5.2 to 7 at temperature of 37°C. The efficiency of bio-flocculation found to increase with decreasing the pH from 7 to 5.2 (Batta et al., 2013). Strontium (Sr) is considered as a nonradioactive heavy metal which can be stabilized using *Cupriavidus metallidurans* CH34. The growth of *Cupriavidus metallidurans* CH34 was observed in 60 mm $SrCl_2$, whereas the precipitation was also visualized. During the process of precipitation, the metal pollutants changed from soluble nature to lesser soluble form by reducing the bioavailability. Hence, the removing of heavy metals is required from the contaminate areas. The removing of Sr mainly occurred at pH region of 7–5 (Ben Salem et al., 2013). *Rhizobium radiobacter* strain was isolated from species called vignaradiata. The polluted soil as bioproducts of chemical and fertilizer industries were considered as the sample. The strain was isolated which was resistant to Arwhich, which has higher inhibitory concentration of 10 Mm compared to other metals. The sequestration of arsenate using EPS therefore applied *Rhizobium radiobacter* strain with greater efficiencies for uptake of arsenate. The evidence was found with the fact with uptake of 0.068 mg of arsenate considering gram of biomass (Deepika et al., 2016).

12.6.2 EPS FUNCTION OVER CONTAMINATED SOIL PRODUCED FROM MINES INDUSTRY

Azotobacter bd39 produced the phytohormone, and fixation of nitrogen is performed with non-symbiotic nature. *Azotobacter* is associated with plant growth and also functions by removing the Hg. The organism can overcome from contamination of Hg due to synthesis of EPS (Hindersah et al., 2017). The presence of *Bacillus* sp. was reported in lead ore and reported towards lead tolerance and has capacity up to 74.5 mg/gm and 30°C at pH 4. In the process, 1% of inoculation size was considered with lead initial concentration of 500 ppm after duration of 24 h and the lead removal efficiency achieved was 98% with considering industrial waste (Cephidian et al., 2016). *Anabaena doliolum* was considered the effective cyanobacterium for removing the Cd at 25°C with considering at pH 7 and the removal percent achieved was 92 with duration of one week for contact period (Goswami et al., 2015).

12.6.3 EPS FUNCTION OVER CONTAMINATED SOIL IN AGRICULTURAL FARMING

The biofilm produced by *Achyranthus aspera* strains has been used to purify soil contaminated with Ni(II) (Karthik et al., 2016). *Pseudomonas Psd* strain is considered a growth-enhancing strain in plants. The EPS produced can be used for bio-absorption of zinc. The important component involved in formation of EPS is the alge8 gene, an alginate polymerase subunit. The mentioned research revealed the fact that the unavailability of alge8 gene resulted with reducing the Zn adsorption capacity (Upadhyay et al., 2017a). *Pseudomonas stutzeri* AS22 resulted with heavy metals adsorption after one day of contact period at incubation temperature of 30°C at pH 8. It has the capacity for bio-absorption with 460 mg/g of lead after using the produced crude EPS (Maalej et al., 2016). The promotors of plant growth called *rhizobacteria* PGPR resulted with improved accumulation of Cd in the leaf and root of *Helianthus annus*, which function with elongation of root in sunflower. For achieving the highest rate of accumulation, the plant needs to be treated with micrococcus and EDTA at 28°C, 150 rpm. The efficiency for removing the Cd was at 25 mg/L ion concentration with 97% results (Prapagdee et al., 2013). The studies have mentioned overapplication of light-emitting diodes for the promotion of growth in algae (Kwan et al., 2015). Cr(VI) has been considered as carcinogenic and mutagenic agent. The availability of Cr(VI) metal pollutants in Hindan River lowers the survival of mustard plant. The irrigation of *Pseudomonas aeruginosa* in the *rhizospheric* soil growing the mustard crop containing the Cr(VI) at 1800 mg/ml of concentrations. The strain was capable of reducing Cr(VI) with 100% results at pH value of 6 to 8 at operating temperatures of 30 to 40°C. The results of reduction for Cr(VI) were increased at 35°C with decreasing the value at 40°C. At temperature ranging from 30°C–35°C, the reduction over Cr(VI) was around 20%, and reduction reached to 25% at 40°C. The Ni(II) concentration decreases the productivity of EPS, which is reserved by bacterium *Cupriavidus* and has good tolerate capacity towards Ni(II) metal at the pH value of 7.5–7.7 at 25°C. The pH with optimum results obtained for Ni(II) was

6.5 at 25°C. In various global regions, the soil is contaminated with Se, which is chronic for human health and environment. The strains *Bacillus* RS1, *Bacillus* RS2, *Bacillus* RS3 were reported with bio-sequestration capacity in divalent Pb (Gupta et al., 2014). During the cultivation of rice, the sulfuric acid is available for crops in field retaining the acidic soil. The group of Hans bacteria, including genus *Desulfovibrio*, which are mainly the sulfide or sulfate-reducing bacteria, cannot be utilized in acidic conditions. The concentration of Cd in the roots of garden peas was lowered with treatment by mycorrhizal fungi (Koga and Masuda, 2013). The excluders are considered the plants which can withstand the concentration of heavy metals, which are mainly lethal to other groups of plants. The concentration of soil metal gets evaded into the shoots through the process called translocation (Rascio and Navari-Izzo, 2011), and similarly, *Anthyllis vulneraria* can tolerate the concentration of metals with fixing the nitrogen. The symbiosis of Legume-rhizobium interaction affected overaccumulation of zinc in roots shoots and nodules. Four subspecies of *Anthyllis vulneraria* associated for their growth with restriction of Zinc constraints which symbiotically function with two mezorhizobiue strains (Soussou et al., 2013).

12.7 EPS FUNCTION ON CONTAMINATION OF WATER

12.7.1 EPS FUNCTION OVER CONTAMINATED WATER PRODUCED BY INDUSTRIAL WASTE

Two bacterial sets are used for testing the heavy metals, including Cd, Ni, Hg, Cu, and Pb (Table 12.2). The isolated MF1 and MF2 are considered as type 1, whereas the group, including MF3, MF4, and MF5 belonged to type 2. The type 2 isolates are considered as gram-negative groups and type 1 as gram-positive groups of bacteria (Chauhan et al., 2017). The alginate, which are the polymer of guluronic acid, and mannuronic acid, which are obtained after harvesting from *Azotobacter* and *Pseudomonas* bacteria are efficient in removing the Cu ions from wastewater obtained under synthetic manner. Synthetically produced alginate using cheaper carbon sources including molasses can improve its production under an industrial scale. The formed or obtained alginate is used to combine with natural zeolite and clinoptilolite. The combination Pb with improved absorption of Co ions up to concentration of 131.6 mg/g at pH of 4.5. The six strains including *Enterococci* MC1, MC2, BT1, BT2 and *Bacillus acidiproducens* SM1, SM2 were selected for studying it resistivity towards lead, cadmium, mercury, and chromium. With the consideration of all strains, *Enterococci* MC1 performed with cadmium removal of 46.19% and lead discharge with 43%, and similarly the *Bacillus acidiproducens* SM2 function with chromium removal at the 43% (Huët and Puchooa, 2017). Despite of its application to heal animal and human body, the biomass of *Spirulina platensis* are used for efficient removal of heavy metals. The optimal conditions achieved for *Spirulina platensis* for efficient bio-absorption was noticed at pH5 at the initial concentration of 9.4 mg/L at 15 min period of contact time (Zinicovscaia et al., 2016). Similarly, *Anthrospira platensis* was used for reducing the chromium ion with considering tannery effluents. Chelation was considered the primary process involved with reduction potential by 73.5% for chromium ions at pH 9.5 and room temperature (Balaji et al., 2015). Similarly, the recovery of chromium using *Spirulina platensis* improved the results threefold with treatment under acidic conditions. With considering the various acids, including HCL, HNO$_3$, and H$_2$SO$_4$, the HCL showed highest efficiency at pH 3 with contact period of 15 h. With the addition of acids, the EPS retaining the negative charges is protonated with decreases of negative charge surface. The involved two-stage process which included the surface absorption initially and followed by transportation of metal ions towards cytoplasm (Hegde et al., 2016).

12.7.2 EPS FUNCTION OVER CONTAMINATED WATER PRODUCED FROM AGRICULTURAL OPERATIONS

The *Pseudomonas aeruginosa* strain efficiently catabolizes the Hexadecane substrate for production of bio-flocculant for removing various heavy metals, including Ni(II), Zn(II), Cd(II), Pb(II), and Cu(II). Its efficiency was higher with Ni(II) giving 79.30% bio-flocculant activity results (Pathak et al., 2017). The EPS-producing strain called *Rhodococcus opacus* and *Rhodococcus rhodochrous*, which are gram-positive bacterium, showed efficient absorption capacity to Cd, Cr, Ni, Pb, and Co. But the operating conditions, including pH, time, and temperature added the difference in the results. The conditions of sorption equilibrium for the strain called *Rhodococcus opacus* and *Rhodococcus rhodochrous* after 30 and 1 min interval of time. For EPS derived from specific bacteria retained with highest adsorption capacity at pH ranging between 2 to 7.5 pH with considering various heavy metals. But these groups of bacterial strain retained with poor adsorption capacity for following heavy metals including Cd, Cr, Ni, Pb, and Co ions at 25°C, while Ni ions have optimum results at 35°C (Dobrowolski et al., 2017). *Plesiomonas shigelloides* (H-5) is a gram-negative bacterial group which have cadmium tolerance capacity of 150 mg/L. Similarly, it has the cadmium adsorption capacity of 106.6 mg per gram with optimum absorption rate of 42.72%. At a temperature of 35°C and a pH of 7, this specialized strain functioned most effectively. *P. shigelloides* was considered as diseases-producing bacterial groups to human and animal body, but it efficiently absorbed the metals, hence used for reducing the contamination of heavy metals (Xue et al., 2016). The Cr(VI) is actively present in wastewater with the form called Cr$_2$O$_7$, which is a cancer-causing agent. The produced exopolysaccharide Hz-7 from *Klebsiella* sp. is efficient for absorption of hexavalent Cr(VI) with efficiency of 99.2% and favorable conditions for absorption occurred at pH ranging from 7–12 (Qiang et al., 2013). Heavy metals are widely viewed as environmental hazards due to their toxicity. The heavy metal containing waste materials were released by the industrial sector, which includes electroplating, mining, and metal mechanical processing. Among the various heavy metals, Cr and Cd are the groups of heavy metals associated with discharge wastewater in the environment. The strain with metal tolerance called *Pseudomonas aeruginosa* is used to remove heavy metals during biological waste treatment, and sludge volume is the major problem associated on it. Research was carried out in which Fe° was important element to enhance the removal of Cd and Cr. The bioremediation is primarily carried out at pH 7, 37°C for 30 min (Singh et al., 2013).

12.7.3 EPS FUNCTION OVER CONTAMINATED WATER IN RIVER AND LAKES

The strain called *Bacillus mucilagino* was effective to remove remediate the Fe(III) heavy metals in the waste under aqueous phase conditions (Yang et al., 2017). The mercury are mainly present in the river at the range of 0.5–3 mgL^{-1}. Additionally, it ranges higher towards coastal sea water with ranges from 2 to 15 mgL^{-1}. The *B. Lichenifonnis* strain with EPS producing capacity absorbed the Hg at maximum rate by removing 70% of 25 mg at 7 pH (Upadhyay et al., 2017b). The EPS secreted by *Raphidocelis subcapitata* towards the external environment performed with bioremediation to Cu and Zn by reducing its excess concentration (Expósito et al., 2017). Athelia rolfsii is a well-known fungi, and EPS was extracted from it and used to bioremediate the metals including Cd, Zn, and Cu with functioning at 5–7 pH at 25°C considering the time of reaction for 1 h. The uptake capacity for various metals including Zn, Cu, and Cd were 153.84, 103.08, and 116.27 mg per gram, respectively. With the involved process, it is mentioned that the nitrogen ions and aldehyde group present in EPS get reacted with cadmium and hydroxyl group and nitrogen get reacted with zinc and copper ions (Li et al., 2016). Similarly, the *B. vallismortis* of bacteria was used for removing the uranium. It works with MIC of 15 mgL^{-1} and 18 mgL^{-1} considering the medium like solid and liquid state. The production of Beta-amylase at the concentration of 1, 2.5, and 5 mgL^{-1} function with improving the uranium removing percentages from 86.28% to 100% and 82.47% to 100% (Ozdemir et al., 2017). The biosorption capacity of the metals Cd(II), Cu(II), and Ni(II) was measured using EPS at 37°C, pH 7, and 200 rpm for four days at pH-5 and 40°C. Among the biosorption capacity of Ni, Pb, and Cu the capacity with following order were obtained including Ni > Pb > Cu. The tainted river water was selected from a polluted river in Argentina called Matanzas Riachuelo and was treated with different algal species (*Chlorella ellipsoidea, Monoraphidium contortum, Ankistrodesmus fusiformis*, and *Scenedesmus acuminatus*) to determine the noxious metals including Cu(II), Cd(II), and Zn(II). Among which Cu and Zn are associated with growth as essential micronutrients for different enzymes. Phytohormones and polyamine were involved with algal adaption ability for pollutant heavy metals in the marine system and function by inhibiting the bio-absorption of heavy metals and simulation of antioxidant activity in enzymes system. The selected four strains function differently by showing their activity after exposure to Cu(II), Cd(II), and Zn(II) during 96 h under the experimental conditions (Magdaleno et al., 2014).

12.7.4 EPS FUNCTION OVER CONTAMINATED WATER IN MARINE ECOSYSTEM

The EPS producing strain called *Chlorella vulgaris, Skeletonema* sp., *Phaeodactylum tricornutum, Nitzschia* sp. have been employed in marine ecosystem for removing heavy metals. Red algae called *Porphyridium cruentum* (S.F. Gray) Nägeliis has been known to remove the heavy metals pollutants. It functions with antibacterial property and has the capacity for accumulation of Cr, Pb, Cd, and Cu that also dependent on the metals concentration. The bioremediation of metal using *Porphyridium cruentum* showed with optimum results at pH ranging between 7 and 8. The time for accumulation of heavy metals is shorter in this organism. *Porphyridium cruentum* organism is also used to estimate the seawater salinity at pH7 to 8, 28 to 38°C and light 400 lux of light intensity. The salinity values found was 28 ppts (Soeprobowati and Hariyati, 2013). In addition to other pollutants, salt water is contaminated by Cr(III) heavy metal. *Phanerochaete chrysosporium* are efficient in removing heavy metals at 35°C, at pH-5, and for 26 h, optimum conditions of growth was achieved, and Cr(III) was efficiently removed by obtaining 98% results. *P. chrysoporium* has the capacity of tolerance with limit to 600 mgL^{-1} of Cr(III) (Sepehr et al., 2013).

12.8   THE BACTERIA AND FUNGI FOR EPS PRODUCTION

Fungi and bacteria are involved in the production of EPS. The advantages in the formation of EPS using a fungus is better than with bacteria, which is due to random availability. The morphological features and its characteristics have revealed its surviving capacity under higher temperatures, pH, and higher concentrations of metal. Fungal strains are useful with acclimatized interactions which added the suitability for the formation of consortiums from various strains. For these reasons, the sensitive strains are also capable of surviving against the higher concentration of metals, which improved with the better and efficient process in bioremediations. The fungi with constant growth rate against the extreme environmental conditions improved the stability for larger scale production. It provides the best remedy to avoid multiple metal contamination in the sites. The formed hydrocarbon called Benzo(a)Pyrene (BaP) is the formed organic components after fossil fuel combustion under incomplete manner. The white rot fungus called *Pleurotus ostreatus* was associated with the degradation of BaP. The two kinds of mediators are used to remediate the BaP which included artificial and natural methods. Including a synthetic mediator termed 2,2′-azinobis-(3-ethylbenzothiazoline-6-sulfonate) with *P. ostreatus* yielded 79% degradation at 1 mM, and vanillin, a natural mediator also used with *P. ostreatus*, demonstrated 84% degradation at 5 mM concentration. *A. terreus* (AML-02), *P. Fumosovoseus*-4099, *B. Bassiana*-4580, *A. terreus* (PD-17), and *A. fumigatus* (PD-18) were applied for accumulation of metals including Cd, Cr, Cu, Ni, Pb, and Zn. The index measuring the metal tolerance using *A. fumigates* (PD-18) showed the best results among four strains with 500 mg/L obtained at the range of 0.72–1.02. The capacity of multi-metal removal form the 2-strain called *B. bassiana*-4580 and *A. fumigatus* (PD-1) ranging from 89–94% and 91–95% respectively. These strains were considered best with their highest rate of metal accumulation and their highest growth rate to the environment with extreme multi-metal concentration thereby considered as the best strain for accumulation of multi-metals (Dey et al., 2016). *B. Thuringiensis* (UEAB3), *B. Cereus* (UEAB6), *Geomyces pannorum* (UAEFr), *Geomyces* sp. (UEAFg) considering these fungi and bacteria. The fungus called *Geomyces* sp. (UEAFg) performed with 100% removing ability for copper at 5 mg/L for the period of one week, and similarly *Geomyces pannorum* (UAEFr) removed the metals with 93% results with mentioned conditions. The bacteria species including *B. cereus* and *B. thuringiensis* performed with absorption capacity of 36.2 and 7.8% which clearly demonstrated higher performance of fungus (Maddela et al., 2015). The fungus *Acremonium* sp. was highly tolerant towards Cr(VI) at 50 μg per mL (Herath et al., 2014). The interesting features for incorporation of fungus biofilm

TABLE 12.2

Microbial approach for remediation of heavy metals, with efficiencies using EPS

Microbial sources with their specification	Strains	Heavy metals adsorbed	Remediation efficiency	Reference
Cyanobacteria (Homogenous Consortial EPS)	*Calothrix marchica*	Pb	65 mg Pb^{2+}/g (CPS)	(Ruangsomboon et al., 2007)
	Gloeocapsa gelatinosa	Pb	82.22 ±4.82 mg Pb^{2+}/g (CPS)	(Ruangsomboon et al., 2007)
	Cyanospira capsulate (Cells + EPS)	Cu	At 12.3 ppm initial metal load, 115 mg Cu^{2+}/g (EPS)	(De Philippis et al., 2007)
	Anabaena spiroides	Mn	8.52 mg Mn^{2+}/g (EPS)	(Freire-Nordi et al., 2005).
	Gloeocapsa calcarea	Cr	At 20 ppm initial metal load 36 mg Cr^{6+}/g (EPS)	(Sharma et al., 2008)
	Nostoc punctiforme	Cr	90.05 mg Cr^{6+}/g (EPS)	(Gupta and Diwan, 2017)
	Nostoc PCC7936	Cu	At 12.3 ppm initial metal load, 85.0 ±3.2 mg Cu^{2+}/g (EPS)	(Gupta and Diwan, 2017)
	Lyngbya putealis	Cr	At 30 ppm initial load, 157 mg/g Cr^{6+}(EPS)	(Kiran and Kaushik, 2008)
	Rhizobium. tropici	Cd	At 10 ppm initial load, 80% Cd^{2+}	
	Pseudomonas putida	Cd	At 10 ppm initial load, 80% Cd^{2+}	(Kenney, 2011)
Marine microbes are isolated via activated sludge (Homogenous Consortial EPS)	*Shewenella oneidensis*	Cd	80% Cd^{2+}	(Ha et al., 2010)
	Herminiimonas arsenicoxydans	As	Uptake of metal ions of up to 5 mmol/L	(Ha et al., 2010)
	Enterobacter cloaceae	Cd and Cu	Reduction of 75% Cr^{6+}, 20% Cu^{2+}, and 65% Cd^{2+} from the initial metal load of 100 ppm	(Gutierrez et al., 2012)
	Pseudomonas sp.	Cu	N.A	(Lau et al., 2005)
	Methylobacterium organophilum	Cu and Pb	Removal of 21% Cu^{2+} and 18% Pb^{2+} from a metal load of 0.04 ppm	(Lau et al., 2005)

(Continued)

TABLE 12.2 (*Continued*)
Microbial approach for remediation of heavy metals, with efficiencies using EPS

Microbial sources with their specification	Strains	Heavy metals adsorbed	Remediation efficiency	Reference
Generally Recognized as Safe status (Homogenous Consortial EPS)	*Paenibacillus polymyxa*	Cu	1602 mg Cu^{2+}/g (EPS)	(Prado Acosta et al., 2005)
	Lactobacillus plantarum	Pb	At 1000 ppm initial metal load, 276.44 mg Pb^{2+}/g (EPS)	(Feng et al., 2012)
	Paenibacillus jamilae	Pb and Cd	200–300 mg Pb^{2+}/g (EPS), 21 mg Cd^{2+}/g of (EPS)	(Morillo et al., 2006)
Soil isolates (Homogenous Consortial EPS)	*Azotobacter chroococcum*	Pb and Hg	33.5 mg Pb^{2+}/g (EPS), removed 40.48% 38.9 mg of Hg^{2+}/g (EPS), removed 47.87%	(Rasulov et al., 2013)
	Ensifer meliloti	Pb, Ni, and Zn	89% Pb^{2+}, 85% Ni^{2+}, Reduction of 66% Zn^{2+}, at 50 ppm initial load	(Lakzian et al., 2008)
	Bacillus firmus	Pb, Zn, and Cu	1103 mg Pb^{2+}/g (EPS) removed 98.3% 860 mg Cu^{2+}/g (EPS) removed 74.9% 722 mg Zn^{2+}/g (EPS) removed 61.8%	(Salehizadeh and Shojaosadati, 2003)
Activated sludge isolate (Dead Biomass EPS)	*Bacillus pumilus*	Cr	Reduction of 89.23% from the starting metal load of 50 ppm	(Gupta and Diwan, 2017)
	Pantoea agglomerans	Cr	Reduction of 85.5% from the initial metal load of 50 ppm	(Gupta and Diwan, 2017)
	Bacillus cereus	Cr	Reduction of 89.87% from the starting metal load of 50 ppm	(er et al., 2012)

Phosphorylated bacterial EPS (Modified EPS)	*Ochrobactrum anthropi*	Cr, Cd, and Cu	At initial metal load of 280 ppm, 57.8 mg Cr^{6+}/g (EPS) At initial metal load of 91.6 ppm, 26 mg Cu^{2+}/g (EPS) At 100.6 ppm initial metal load, 29.5 mg Cd^{2+}/g (EPS)	(Ozdemir et al., 2003)
Activated sludge mixed consortia (Heterogenus Consortial EPS)	*Acetobacter*	Pb, Cu, Mn, Zn, and Co	90% reduction from the initial metal load of 0.1 mmol/dm³ (Fe^{3+}> Cu^{2+}> Mn^{2+} > Zn^{2+}; Co^{2+})	(Oshima et al., 2008)
		Pb and Cd	N.A	(Guibaud et al., 2008)
		Zn, Cu, Cr, and Cd	Reduction of 85 to 95% from the initial metal load of 10 to 100 ppm	(Liu et al., 2001)
Hydrocarbon contaminated water microbial consortium (Heterogenus Consortial EPS)		Zn, Cu, and Cd	Reduction of 87.12% Cd^{2+}, 19.82% Zn^{2+}, and 37.64% Cu^{2+} from a metal load of 1 ppm	(Martins et al., 2008)
Gram-negative microbial consortia (Heterogenus Consortial EPS)		Zn, Pb, Cr, Ni and Cu	Reduction of 75 to 78% n metal load	(Gawali Ashruta et al., 2014)
Agar beads immobilized (Immobilized EPS)	*Paenibacillus polymyxa*	Pb	111.11 mg Pb^{2+}/g (EPS)	(Mokaddem et al., 2014)
Alginate bead immobilized (Immobilized EPS)	*Chryseomonas luteola*	Cd, Co, Cu, and Ni	64.10 mg Cd^{2+}/g (EPS) 55.25 mg Co^{2+}/g of (EPS) 1.989 mmol Cu^{2+}/g (EPS) 1.224 mmol Ni^{2+}/g (EPS)	(Ozdemir et al., 2005a) (Ozdemir et al., 2005b)

with bacterial biofilm with forming new biofilm called fungal-bacterial biofilm was investigated. The analysis was done from day 1 to day 10. For the day 1 case, 43% of Cr(VI) removal occurred by FBBs for which around 10 to 16% of Cr(VI) removal took place using mono cultures. Similarly for day 10 case, around 90% of Cr(VI) removal took place while mono culture could remove with 60% results. The application of two-phase phenomenon for removal of Cr (VI) with FBB showed more efficient results than bacterial isolates. This result was due to the higher production of EPS by fungal *mycelium* when compared to bacteria and proved the efficiency over EPS production by fungi (Herath et al., 2014).

12.9 CONCLUSION

The rise in concentration of heavy metals in soil and water lead to elevate the contamination to human health and environment. The application of traditional methods for remediation has not been up to the mark situation with adding risk or danger consequences. Therefore, application of EPS is the best option to overcome the problems associated with traditional methods. The efficiencies of EPS are dependent on the associated organic and inorganic components including lipids, polysaccharides, uronic acid, and amino acids. The contamination with heavy metals is revealed due to association of human activities; therefore, the EPS production could be resourceful practices to heavy metal bioremediation. The contaminated soil builds up due to industrial waste, mining, and disposed chemicals in agricultural sectors can be bioremediate with application of EPS under competent manner. Similarly, the EPS that originated from different microorganisms can remove the heavy metal from contaminated water. Comprehensive overview towards application of EPS for removing the heavy metals from water and soil is presented in this chapter. Considering all these factors, the EPS is regarded as a reliable material to overcome the heavy metals contamination with conservation of environment and human health.

REFERENCES

Abd Al-Manhel, A.J., 2017. Production of exopolysaccharide from local fungal isolate. Curr. Res. Nutr. Food Sci. J. 5(3), 338–346.

Akhter, K., Ghous, T., Andleeb, S., Ejaz, S., Khan, B.A., Ahmed, M.N., 2017. Bioaccumulation of heavy metals by metal-resistant bacteria isolated from Tagetes minuta rhizosphere, growing in soil adjoining automobile workshops. Pak. J. Zool. 49(5).

Al-Dhabi, N.A., Esmail, G.A., Mohammed Ghilan, A.-K., Valan Arasu, M.J.I.J.O.E.R., Health, P., 2019. Optimizing the management of cadmium bioremediation capacity of metal-resistant *Pseudomonas* sp. strain Al-Dhabi-126 isolated from the industrial city of Saudi Arabian environment. Int. J. Environ. Res. Public Health. 16(23), 4788.

Amoozegar, M.A., Ghazanfari, N., Didari, M., 2012. Lead and cadmium bioremoval by Halomonas sp., an exopolysaccharide-producing halophilic bacterium. Prog. Biol. Sci. 2(1), 1–11.

Aquino, S.F., Stuckey, D.C., 2004. Soluble microbial products formation in anaerobic chemostats in the presence of toxic compounds. Water Res. 38(2), 255–266.

Balaji, S., Kalaivani, T., Rajasekaran, C., Siva, R., Shalini, M., Das, R., Madnokar, V., Dhamorikar, P., 2015. Bioremediation potential of *Arthrospira platensis* (Spirulina) against chromium (VI). CLEAN—Soil Air Water. 43(7), 1018–1024.

Batta, N., Subudhi, S., Lal, B., Devi, A., 2013. Isolation of a lead tolerant novel bacterial species, *Achromobacter* sp. TL-3: assessment of bioflocculant activity. Indian J. Exp. Biol. 51(11), 1004–1011.

Bender, J., Gould, J., Vatcharapijarn, Y., Young, J., Phillips, P., 1994. Removal of zinc and manganese from contaminated water with cyanobacteria mats. Water Environ. Res. 66(5), 679–683.

Bennett, R.M., Cordero, P.R.F., Bautista, G.S., Dedeles, G.R., 2013. Reduction of hexavalent chromium using fungi and bacteria isolated from contaminated soil and water samples. Chem. Ecol. 29(4), 320–328.

Ben Salem, I., Sghaier, H., Monsieurs, P., Moors, H., Van Houdt, R., Fattouch, S., Saidi, M., Landolsi, A., Leys, N., 2013. Strontium-induced genomic responses of Cupriavidus metallidurans and strontium bioprecipitation as strontium carbonate. Ann. Microbiol. 63(3), 833–844.

Bhatla, S.C., A. Lal, M., Kathpalia, R., Bhatla, S.C., 2018. Plant mineral nutrition. In: Satish C Bhatla and Manju A. Lal (Eds.), *Plant Physiology, Development and Metabolism*. Springer, pp. 37–81.

Bhatti, H.N., Khalid, R., Hanif, M.A., 2009. Dynamic biosorption of Zn (II) and Cu (II) using pretreated Rosa gruss an teplitz (red rose) distillation sludge. Chem. Eng. J. 148(2–3), 434–443.

Bhunia, B., Uday, U.S.P., Oinam, G., Mondal, A., Bandyopadhyay, T.K., Tiwari, O.N., 2018. Characterization, genetic regulation and production of cyanobacterial exopolysaccharides and its applicability for heavy metal removal. Carbohydr. Polym. 179, 228–243.

Braissant, O., Decho, A.W., Dupraz, C., Glunk, C., Przekop, K.M., Visscher, P.T., 2007. Exopolymeric substances of sulfate-reducing bacteria: Interactions with calcium at alkaline pH and implication for formation of carbonate minerals. Geobiology. 5(4), 401–411.

Butt, A.S., Rehman, A., 2011. Isolation of arsenite-oxidizing bacteria from industrial effluents and their potential use in wastewater treatment. World J. Microbiol. Biotechnol. 27(10), 2435–2441.

Castro, A.M., Nogueira, V., Lopes, I., Rocha-Santos, T., Pereira, R., 2019. Evaluation of the potential toxicity of effluents from the textile industry before and after treatment. Appl. Sci. 9(18), 3804.

Cephidian, A., Makhdoumi, A., Mashreghi, M., Mahmudy Gharaie, M.H., 2016. Removal of anthropogenic lead pollutions by a potent Bacillus species AS2 isolated from geogenic contaminated site. Int. J. Environ. Sci. Technol. 13(9), 2135–2142.

Chauhan, M., Solanki, M., Nehra, K., 2017. Putative mechanism of cadmium bioremediation employed by resistant bacteria. Jordan J. Biol. Sci. 10(2).

Chikkanna, A., Mehan, L., Sarath, P., Ghosh, D., 2019. Arsenic exposures, poisoning, and threat to human health: Arsenic affecting human health. In: *Environmental Exposures and Human Health Challenges*. IGI Global, pp. 86–105.

Concórdio-Reis, P., Reis, M.A., Freitas, F., 2020. Biosorption of heavy metals by the bacterial exopolysaccharide FucoPol. Appl. Sci. 10(19), 6708.

Coutaud, A., Meheut, M., Viers, J., Rols, J.-L., Pokrovsky, O.S., 2014. Zn isotope fractionation during interaction with phototrophic biofilm. Chem. Geol. 390, 46–60.

Coutaud, M., Méheut, M., Glatzel, P., Pokrovski, G.S., Viers, J., Rols, J.-L., Pokrovsky, O.S., 2018. Small changes in Cu redox state and speciation generate large isotope fractionation during adsorption and incorporation of Cu by a phototrophic biofilm. Geochim. Cosmochim. Acta. 220, 1–18.

Das, S., Sen, I.K., Kati, A., Some, S., Mandal, A.K., Islam, S.S., Bhattacharyya, R., Mukhopadhyay, A., 2019. Flocculating, emulsification and metal sorption properties of a partial characterized novel exopolysaccharide produced by *Rhizobium tropici* SRA1 isolated from Psophocarpus tetragonolobus (L) DC. Int. Microbiol. 22(1), 91–101.

Deepika, K., Raghuram, M., Kariali, E., Bramhachari, P., 2016. Biological responses of symbiotic *Rhizobium radiobacter* strain VBCK1062 to the arsenic contaminated rhizosphere soils of mung bean. Ecotoxicol. Environ. Saf. 134, 1–10.

De Philippis, R., Colica, G., Micheletti, E., 2011. Exopolysaccharide-producing cyanobacteria in heavy metal removal from water: Molecular basis and practical applicability of the biosorption process. Appl. Microbiol. Biotechnol. 92(4), 697–708.

De Philippis, R., Paperi, R., Sili, C., 2007. Heavy metal sorption by released polysaccharides and whole cultures of two exopolysaccharide-producing cyanobacteria. Biodegradation. 18(2), 181–187.

De Philippis, R., Vincenzini, M., 1998. Exocellular polysaccharides from cyanobacteria and their possible applications. FEMS Microbiol. Rev. 22(3), 151–175.

Dey, P., Gola, D., Mishra, A., Malik, A., Kumar, P., Singh, D.K., Patel, N., von Bergen, M., Jehmlich, N., 2016. Comparative performance evaluation of multi-metal resistant fungal strains for simultaneous removal of multiple hazardous metals. J. Hazard. Mater. 318, 679–685.

Dobrowolski, R., Szcześ, A., Czemierska, M., Jarosz-Wikołazka, A., 2017. Studies of cadmium (II), lead (II), nickel (II), cobalt (II) and chromium (VI) sorption on extracellular polymeric substances produced by *Rhodococcus opacus* and *Rhodococcus rhodochrous*. Bioresour. Technol. 225, 113–120.

Dos Santos, A.A., Chang, L.W., Guo, G.L., Aschner, M., 2018. Fetal Minamata disease: A human episode of congenital methylmercury poisoning. In: *Handbook of Developmental Neurotoxicology*. Elsevier, pp. 399–406.

Dynes, J.J., Tyliszczak, T., Araki, T., Lawrence, J.R., Swerhone, G.D., Leppard, G.G., Hitchcock, A.P., 2006. Speciation and quantitative mapping of metal species in microbial biofilms using scanning transmission X-ray microscopy. Environ. Sci. Technol. 40(5), 1556–1565.

Erdem, E., Karapinar, N., Donat, R., 2004. The removal of heavy metal cations by natural zeolites. J. Colloid Interface Sci. 280(2), 309–314.

Expósito, N., Kumar, V., Sierra, J., Schuhmacher, M., Papiol, G.G., 2017. Performance of raphidocelis subcapitata exposed to heavy metal mixtures. Sci. Total Environ. 601, 865–873.

Feng, M., Chen, X., Li, C., Nurgul, R., Dong, M., 2012. Isolation and identification of an exopolysaccharide-producing lactic acid bacterium strain from Chinese Paocai and biosorption of Pb (II) by its exopolysaccharide. J. Food Sci. 77(6), T111–T117.

Feng, W., Xiao, X., Li, J., Xiao, Q., Ma, L., Gao, Q., Wan, Y., Huang, Y., Liu, T., Luo, X., 2023. Bioleaching and immobilizing of copper and zinc using endophytes coupled with biochar-hydroxyapatite: Bipolar remediation for heavy metals contaminated mining soils. Chemosphere. 137730.

Flaibani, A., Olsen, Y., Painter, T.J., 1989. Polysaccharides in desert reclamation: Compositions of exocellular proteoglycan complexes produced by filamentous blue-green and unicellular green edaphic algae. Carbohydr. Res. 190(2), 235–248.

Freire-Nordi, C.S., Vieira, A.A.H., Nascimento, O.R., 2005. The metal binding capacity of Anabaena spiroides extracellular polysaccharide: An EPR study. Process Biochem. 40(6), 2215–2224.

Gawali Ashruta, A., Nanoty, V., Bhalekar, U., 2014. Biosorption of heavy metals from aqueous solution using bacterial EPS. Int. J. Life Sci. 2, 373–377.

Goswami, S., Syiem, M.B., Pakshirajan, K., 2015. Cadmium removal by Anabaena doliolum Ind1 isolated from a coal mining area in Meghalaya, India: Associated structural and physiological alterations. Environ. Eng. Res. 20(1), 41–50.

Guibaud, G., Bordas, F., Saaid, A., D'abzac, P., Van Hullebusch, E., 2008. Effect of pH on cadmium and lead binding by extracellular polymeric substances (EPS) extracted from environmental bacterial strains. Colloids Surf. B Biointerfaces. 63(1), 48–54.

Gupta, P., Diwan, B., 2017. Bacterial exopolysaccharide mediated heavy metal removal: A review on biosynthesis, mechanism and remediation strategies. Biotechnol. Rep. 13, 58–71.

Gupta, S., Goyal, R., Prakash, N.T., 2014. Biosequestration of lead using Bacillus strains isolated from seleniferous soils and sediments of Punjab. Environ. Sci. Pollut. Res. 21(17), 10186–10193.

Gutierrez, T., Biller, D.V., Shimmield, T., Green, D.H., 2012. Metal binding properties of the EPS produced by Halomonas sp. TG39 and its potential in enhancing trace element bioavailability to eukaryotic phytoplankton. Biometals. 25(6), 1185–1194.

Ha, J., Gélabert, A., Spormann, A.M., Brown Jr, G.E., 2010. Role of extracellular polymeric substances in metal ion complexation on Shewanella oneidensis: Batch uptake, thermodynamic modeling, ATR-FTIR, and EXAFS study. Geochim. Cosmochim. Acta. 74(1), 1–15.

He, T., Li, Q., Lin, T., Li, J., Bai, S., An, S., Kong, X., Song, Y.-F., 2023. Recent progress on highly efficient removal of heavy metals by layered double hydroxides. Chem. Eng. J. 142041.

Hegde, S.M., Babu, R., Vijayalakshmi, E., Patil, R.H., Naveen Kumar, M., Kiran Kumar, K., Nagesh, R., Kavya, K., Sharma, S.C., 2016. Biosorption of hexavalent chromium from aqueous solution using chemically modified Spirulina platensis algal biomass: An eco-friendly approach. Desalination Water Treat. 57(18), 8504–8513.

Herath, H., Rajapaksha, A.U., Vithanage, M., Seneviratne, G., 2014. Developed fungal—bacterial biofilms as a novel tool for bioremoval of hexavelant chromium from wastewater. Chem. Ecol. 30(5), 418–427.

Hindersah, R., Mulyani, O., Osok, R., 2017. Proliferation and exopolysaccharide production of Azotobacter in the presence of mercury. Biodivers. J. 8(1), 21–26.

Hou, W., Ma, Z., Sun, L., Han, M., Lu, J., Li, Z., Mohamad, O.A., Wei, G., 2013. Extracellular polymeric substances from copper-tolerance Sinorhizobium meliloti immobilize Cu2+. J. Hazard. Mater. 261, 614–620.

Hu, T.-L., 1996. Removal of reactive dyes from aqueous solution by different bacterial genera. Water Sci. Technol. 34(10), 89–95.

Huët, M.A.L., Puchooa, D., 2017. Bioremediation of heavy metals from aquatic environment through microbial processes: A potential role for probiotics? J. Appl. Biol. Biotechnol. 5(6), 1–3.

Iyer, A., Mody, K., Jha, B., 2004. Accumulation of hexavalent chromium by an exopolysaccharide producing marine Enterobacter cloaceae. Mar. Pollut. Bull. 49(11–12), 974–977.

Iyer, A., Mody, K., Jha, B., 2006. Emulsifying properties of a marine bacterial exopolysaccharide. Enzyme Microb. Technol. 38(1–2), 220–222.

Jiang, H., Wu, S., Zhou, J., 2023. Preparation and modification of nanocellulose and its application to heavy metal adsorption: A review. Int. J. Biol. Macromol. 123916.

Kalpana, R., Angelaalincy, M.J., Kamatchirajan, B.V., Vasantha, V.S., Ashokkumar, B., Ganesh, V., Varalakshmi, P., 2018. Exopolysaccharide from Bacillus cereus VK1: Enhancement, characterization and its potential application in heavy metal removal. Colloids Surf. B Biointerfaces. 171, 327–334.

Kantar, C., Demiray, H., Dogan, N.M., Dodge, C.J., 2011. Role of microbial exopolymeric substances (EPS) on chromium sorption and transport in heterogeneous subsurface soils: I. Cr (III) complexation with EPS in aqueous solution. Chemosphere. 82(10), 1489–1495.

Karthik, C., Oves, M., Thangabalu, R., Sharma, R., Santhosh, S., Arulselvi, P.I., 2016. Cellulosimicrobium funkei-like enhances the growth of Phaseolus vulgaris by modulating oxidative damage under Chromium (VI) toxicity. J. Adv. Res. 7(6), 839–850.

Kawaguchi, T., Decho, A.W., 2002. A laboratory investigation of cyanobacterial extracellular polymeric secretions (EPS) in influencing CaCO3 polymorphism. J. Cryst. Growth. 240(1–2), 230–235.

Kehr, J.-C., Dittmann, E., 2015. Biosynthesis and function of extracellular glycans in cyanobacteria. Life. 5(1), 164–180.

Kenney, J.P., 2011. *Metal Adsorption to Bacterial Cells and Their Products.* University of Notre Dame.

Kielak, A.M., Castellane, T.C., Campanharo, J.C., Colnago, L.A., Costa, O.Y., Corradi da Silva, M.L., Van Veen, J.A., Lemos, E.G., Kuramae, E.E., 2017. Characterization of novel Acidobacteria exopolysaccharides with potential industrial and ecological applications. Sci. Rep. 7(1), 1–11.

Kiran, B., Kaushik, A., 2008. Chromium binding capacity of *Lyngbya putealis* exopolysaccharides. Biochem. Eng. J. 38(1), 47–54.

Koga, K., Masuda, S., 2013. *Bacteria That Reduce Content of Heavy Metals in Plant.* Google Patents.

Konhauser, K., Fisher, Q., Fyfe, W., Longstaffe, F., Powell, M., 1998. Authigenic mineralization and detrital clay binding by freshwater biofilms: The Brahmani River, India. Geomicrobiol. J. 15(3), 209–222.

Kuyucak, N., Volesky, B., 1990. Biosorption by algal biomass. Biosorpt. Heavy Metals. 2(4), 173–198.

Lakzian, A., Berenji, A.R., Karimi, E., Razavi, S., 2008. Adsorption capability of lead, nickel and zinc by exopolysaccharide and dried cell of Ensifer meliloti. Asian J. Chem. 20(8), 6075.

Lau, T., Wu, X., Chua, H., Qian, P., Wong, P.K., 2005. Effect of exopolysaccharides on the adsorption of metal ions by Pseudomonas sp. CU-1. Water Sci. Technol. 52(7), 63–68.

Li, H., Wei, M., Min, W., Gao, Y., Liu, X., Liu, J., 2016. Removal of heavy metal Ions in aqueous solution by Exopolysaccharides from Athelia rolfsii. Biocatal. Agric. Biotechnol. 6, 28–32.

Li, W.-W., Yu, H.-Q., 2014. Insight into the roles of microbial extracellular polymer substances in metal biosorption. Bioresour. Technol. 160, 15–23.

Lin, G., Zeng, B., Li, J., Wang, Z., Wang, S., Hu, T., Zhang, L., 2023. A systematic review of metal organic frameworks materials for heavy metal removal: Synthesis, applications and mechanism. Chem. Eng. J. 141710.

Liu, H., Fang, H.H., 2002. Characterization of electrostatic binding sites of extracellular polymers by linear programming analysis of titration data. Biotechnol. Bioengineer. 80(7), 806–811.

Liu, W., Zhang, J., Jin, Y., Zhao, X., Cai, Z., 2015. Adsorption of Pb (II), Cd (II) and Zn (II) by extracellular polymeric substances extracted from aerobic granular sludge: Efficiency of protein. J. Environ. Chem. Eng. 3(2), 1223–1232.

Liu, Y., Lam, M., Fang, H., 2001. Adsorption of heavy metals by EPS of activated sludge. Water Sci. Technol. 43(6), 59–66.

Maalej, H., Hmidet, N., Boisset, C., Bayma, E., Heyraud, A., Nasri, M., 2016. Rheological and emulsifying properties of a gel-like exopolysaccharide produced by *Pseudomonas stutzeri* AS22. Food Hydrocoll. 52, 634–647.

Maddela, N.R., Reyes, J., Viafara, D., Gooty, J., 2015. Biosorption of copper (II) by the microorganisms isolated from the crude-oil-contaminated soil. Soil Sediment Contam. Int. J. 24(8), 898–908.

Magdaleno, A., Vélez, C.G., Wenzel, M.T., Tell, G., 2014. Effects of cadmium, copper and zinc on growth of four isolated algae from a highly polluted Argentina river. Bull. Environ. Contam. Toxicol. 92(2), 202–207.

Mani, D., Kumar, C., 2014. Biotechnological advances in bioremediation of heavy metals contaminated ecosystems: An overview with special reference to phytoremediation. Int. J. Environ. Sci. Technol. 11(3), 843–872.

Martins, P.S.D.O., Almeida, N.F.D., Leite, S.G.F., 2008. Application of a bacterial extracellular polymeric substance in heavy metal adsorption in a co-contaminated aqueous system. Braz. J. Microbiol. 39, 780–786.

Maurice, P.A., Pullin, M.J., Cabaniss, S.E., Zhou, Q., Namjesnik-Dejanovic, K., Aiken, G.R., 2002. A comparison of surface water natural organic matter in raw filtered water samples, XAD, and reverse osmosis isolates. Water Res. 36(9), 2357–2371.

Mezynska, M., Brzóska, M.M.J.E.S., Research, P., 2018. Environmental exposure to cadmium—A risk for health of the general population in industrialized countries and preventive strategies. Environ. Sci. Pollut. Res. Int. 25, 3211–3232.

Minkina, T., Soldatov, A., Nevidomskaya, D., Motuzova, G., Podkovyrina, Y.S., Mandzhieva, S., 2016. New approaches to studying heavy metals in soils by X-ray absorption spectroscopy (XANES)) and extractive fractionation. Geochem. Int. 54, 197–204.

Mohite, B.V., Koli, S.H., Narkhede, C.P., Patil, S.N., Patil, S.V., 2017. Prospective of microbial exopolysaccharide for heavy metal exclusion. Appl. Biochem. Biotechnol. 183(2), 582–600.

Mokaddem, H., Azouaou, N., Kaci, Y., Sadaoui, Z., 2014. Study of lead adsorption from aqueous solutions on agar beads with EPS produced from *Paenibacillus polymyxa*. Chem. Eng. Tran. 38, 31–36.

Morillo, J.A., Aguilera, M., Ramos-Cormenzana, A., Monteoliva-Sánchez, M., 2006. Production of a metal-binding exopolysaccharide by *Paenibacillus jamilae* using two-phase olive-mill waste as fermentation substrate. Curr. Microbiol. 53(3), 189–193.

Morrow, H.J.K.O.E.O.C.T., 2000. Cadmium and cadmium alloys. In: *Enciclopedia De La Tecnologia Quimica*. Wiley, pp. 1–36.

Mota, R., Rossi, F., Andrenelli, L., Pereira, S.B., De Philippis, R., Tamagnini, P., 2016. Released polysaccharides (RPS) from Cyanothece sp. CCY 0110 as biosorbent for heavy metals bioremediation: Interactions between metals and RPS binding sites. Appl. Microbiol. Biotechnol. 100(17), 7765–7775.

Nicolaus, B., Kambourova, M., Oner, E.T., 2010. Exopolysaccharides from extremophiles: From fundamentals to biotechnology. Environ. Technol. 31(10), 1145–1158.

Ojuederie, O.B., Babalola, O.O., 2017. Microbial and plant-assisted bioremediation of heavy metal polluted environments: A review. Int. J. Environ. Res. Public Health. 14(12), 1504.

Oshima, T., Kondo, K., Ohto, K., Inoue, K., Baba, Y., 2008. Preparation of phosphorylated bacterial cellulose as an adsorbent for metal ions. React. Funct. Polym. 68(1), 376–383.

Ozdemir, G., Ceyhan, N., Manav, E., 2005a. Utilization in alginate beads for Cu (II) and Ni (II) adsorption of an exopolysaccharide produced by *Chryseomonas luteola* TEM05. World J. Microbiol. Biotechnol. 21(2), 163–167.

Ozdemir, G., Ceyhan, N., Manav, E., 2005b. Utilization of an exopolysaccharide produced by Chryseomonas luteola TEM05 in alginate beads for adsorption of cadmium and cobalt ions. Bioresour. Technol. 96(15), 1677–1682.

Ozdemir, G., Ozturk, T., Ceyhan, N., Isler, R., Cosar, T., 2003. Heavy metal biosorption by biomass of Ochrobactrum anthropi producing exopolysaccharide in activated sludge. Bioresour. Technol. 90(1), 71–74.

Ozdemir, S., Oduncu, M.K., Kilinc, E., Soylak, M., 2017. Resistance, bioaccumulation and solid phase extraction of uranium (VI) by Bacillus vallismortis and its UV—vis spectrophotometric determination. J. Environ. Radioact. 171, 217–225.

Pagnanelli, F., Mainelli, S., Bornoroni, L., Dionisi, D., Toro, L., 2009. Mechanisms of heavy-metal removal by activated sludge. Chemosphere. 75(8), 1028–1034.

Pan, X., Liu, J., Zhang, D., 2010. Binding of phenanthrene to extracellular polymeric substances (EPS) from aerobic activated sludge: A fluorescence study. Colloids Surf. B Biointerfaces. 80(1), 103–106.

Pathak, M., Sarma, H.K., Bhattacharyya, K.G., Subudhi, S., Bisht, V., Lal, B., Devi, A., 2017. Characterization of a novel polymeric bioflocculant produced from bacterial utilization of n-hexadecane and its application in removal of heavy metals. Front. Microbiol. 8, 170.

Pau-Roblot, C., Lequart-Pillon, M., Apanga, L., Pilard, S., Courtois, J., Pawlicki-Jullian, N., 2013. Structural features and bioremediation activity of an exopolysaccharide produced by a strain of *Enterobacter ludwigii* isolated in the Chernobyl exclusion zone. Carbohydr. Polym. 93(1), 154–162.

Pereira, S.B., Sousa, A., Santos, M., Araújo, M., Serôdio, F., Granja, P., Tamagnini, P., 2019. Strategies to obtain designer polymers based on cyanobacterial extracellular polymeric substances (EPS). Int. J. Mol. Sci. 20(22), 5693.

Pereira, S.B., Zille, A., Micheletti, E., Moradas-Ferreira, P., De Philippis, R., Tamagnini, P., 2009. Complexity of cyanobacterial exopolysaccharides: Composition, structures, inducing factors and putative genes involved in their biosynthesis and assembly. FEMS Microbiol. Rev. 33(5), 917–941.

Potnis, A.A., Raghavan, P.S., Rajaram, H., 2021. Overview on cyanobacterial exopolysaccharides and biofilms: Role in bioremediation. Rev. Environ. Sci. Bio/Technol. 20(3), 781–794.

Prado Acosta, M., Valdman, E., Leite, S., Battaglini, F., Ruzal, S.M., 2005. Biosorption of copper by Paenibacillus polymyxa cells and their exopolysaccharide. World J. Microbiol. Biotechnol. 21(6), 1157–1163.

Prapagdee, B., Chanprasert, M., Mongkolsuk, S., 2013. Bioaugmentation with cadmium-resistant plant growth-promoting rhizobacteria to assist cadmium phytoextraction by Helianthus annuus. Chemosphere. 92(6), 659–666.

Qiang, L., Yumei, L., Sheng, H., Yingzi, L., Dongxue, S., Dake, H., Jiajia, W., Yanhong, Q., Yuxia, Z., 2013. Optimization of fermentation conditions and properties of an exopolysaccharide from *Klebsiella* sp. H-207 and application in adsorption of hexavalent chromium. PLoS ONE. 8(1), e53542.

Raghavan, P.S., Potnis, A.A., Bhattacharyya, K., Salaskar, D.A., Rajaram, H., 2020. Axenic cyanobacterial (*Nostoc muscorum*) biofilm as a platform for Cd (II) sequestration from aqueous solutions. Algal Res. 46, 101778.

Rascio, N., Navari-Izzo, F., 2011. Heavy metal hyperaccumulating plants: How and why do they do it? And what makes them so interesting? Plant Sci. 180(2), 169–181.

Rasulov, B.A., Yili, A., Aisa, H.A. 2013. Biosorption of metal ions by exopolysaccharide produced by *Azotobacter chroococcum* XU1. J. Environ. Prot. 4(9).

Rong, Q., Ling, C., Lu, D., Zhang, C., Zhao, H., Zhong, K., Nong, X., Qin, X.J.C., 2022. Sb (III) resistance mechanism and oxidation characteristics of *Klebsiella aerogenes* X. Chemosphere. 293, 133453.

Ruangsomboon, S., Chidthaisong, A., Bunnag, B., Inthorn, D., Harvey, N.W., 2007. Lead (Pb2+) adsorption characteristics and sugar composition of capsular polysaccharides of cyanobacterium *Calothrix marchica*. Songklanakarin J. Sci. Technol. 29(2), 529–541.

Salehizadeh, H., Shojaosadati, S., 2003. Removal of metal ions from aqueous solution by polysaccharide produced from Bacillus firmus. Water Res. 37(17), 4231–4235.

Saranya, K., Sundaramanickam, A., Shekhar, S., Swaminathan, S., Balasubramanian, T. 2017. Bioremediation of mercury by *Vibrio fluvialis* screened from industrial effluents. BioMed Res. Int. 2017.

Sepehr, M.N., Zarrabi, M., Amrane, A., Samarghandi, M.R., 2013. Removal of Cr (III) from model solutions and a real effluent by *Phanerochaete chrysosporium* isolated living microorganism: Equilibrium and kinetics. Desalination Water Treat. 51(28–30), 5627–5637.

Sharma, J., Shamim, K., Dubey, S.K., Meena, R.M., 2017. Metallothionein assisted periplasmic lead sequestration as lead sulfite by Providencia vermicola strain SJ2A. Sci. Total Environ. 579, 359–365.

Sharma, M., Kaushik, A., Bala, K., Kamra, A., 2008. Sequestration of chromium by exopolysaccharides of Nostoc and Gloeocapsa from dilute aqueous solutions. J. Hazard. Mater. 157(2–3), 315–318.

Sheng, G.-P., Xu, J., Li, W.-H., Yu, H.-Q., 2013. Quantification of the interactions between Ca2+, Hg2+ and extracellular polymeric substances (EPS) of sludge. Chemosphere. 93(7), 1436–1441.

Shuhong, Y., Meiping, Z., Hong, Y., Han, W., Shan, X., Yan, L., Jihui, W., 2014. Biosorption of Cu2+, Pb2+ and Cr6+ by a novel exopolysaccharide from *Arthrobacter* ps-5. Carbohydr. Polym. 101, 50–56.

Shukla, D., Vankar, P.S., Srivastava, S.K., 2012. Bioremediation of hexavalent chromium by a cyanobacterial mat. Appl. Water Sci. 2(4), 245–251.

Singh, R., Bishnoi, N.R., Kirrolia, A., 2013. Evaluation of Pseudomonas aeruginosa an innovative bioremediation tool in multi metals ions from simulated system using multi response methodology. Bioresour. Technol. 138, 222–234.

Singh, S., Kumar, V., 2020. Mercury detoxification by absorption, mercuric ion reductase, and exopolysaccharides: A comprehensive study. Environ. Sci. Poll. Res. 27(22), 27181–27201.

Soeprobowati, T.R., Hariyati, R., 2013. Bioaccumulation of Pb, Cd, Cu, and Cr by *Porphyridium cruentum* (SF Gray) Nägeli. Int. J. Mar. Sci. 3.

Song, W., Pan, X., Mu, S., Zhang, D., Yang, X., Lee, D.-J., 2014. Biosorption of Hg (II) onto goethite with extracellular polymeric substances. Bioresour. Technol. 160, 119–122.

Soussou, S., Mahieu, S., Brunel, B., Escarré, J., Lebrun, M., Banni, M., Boussetta, H., Cleyet-Marel, J.-C., 2013. Zinc accumulation patterns in four *Anthyllis vulneraria* subspecies supplemented with mineral nitrogen or grown in the presence of their symbiotic bacteria. Plant Soil. 371(1), 423–434.

Späth, R., Flemming, H.-C., Wuertz, S., 1998. Sorption properties of biofilms. Water Sci. Technol. 37(4–5), 207–210.

Subudhi, S., Bisht, V., Batta, N., Pathak, M., Devi, A., Lal, B., 2016. Purification and characterization of exopolysaccharide bioflocculant produced by heavy metal resistant *Achromobacter xylosoxidans*. Carbohydr. Polym. 137, 441–451.

Sultan, S., Khansa, M., Mohammad, F., 2012. Uptake of toxic Cr (VI) by biomass of exopolysaccharides producing bacterial strains. Afr. J. Microbiol. Res. 6(13), 3329–3336.

Sun, Q., Zhang, Y., Ming, C., Wang, J., Zhang, Y., 2023. Amended compost alleviated the stress of heavy metals to pakchoi plants and affected the distribution of heavy metals in soil-plant system. J. Environ. Manage. 336, 117674.

Sun, X., Zhang, J., 2021. Bacterial exopolysaccharides: Chemical structures, gene clusters and genetic engineering. Int. J. Biol. Macromol. 173, 481–490.

Tedetti, M., Joffre, P., Goutx, M., 2013. Development of a field-portable fluorometer based on deep ultraviolet LEDs for the detection of phenanthrene-and tryptophan-like compounds in natural waters. Sens. Actuators B Chem. 182, 416–423.

Tekaya, N., Gammoudi, I., Braiek, M., Tarbague, H., Morote, F., Raimbault, V., Sakly, N., Rebière, D., Ouada, H.B., Lagarde, F., 2013. Acoustic, electrochemical and microscopic characterization of interaction of *Arthrospira platensis* biofilm and heavy metal ions. J. Environ. Chem. Eng. 1(3), 609–619.

Tran, H.T., Vu, N.D., Matsukawa, M., Okajima, M., Kaneko, T., Ohki, K., Yoshikawa, S., 2016. Heavy metal biosorption from aqueous solutions by algae inhabiting rice paddies in Vietnam. J. Environ. Chem. Eng. 4(2), 2529–2535.

Trivedi, R., 2020. Exopolysaccharides: Production and application in industrial wastewater treatment. In: *Combined Application of Physico-Chemical & Microbiological Processes for Industrial Effluent Treatment Plant*. Springer, pp. 15–27.

Upadhyay, A., Kochar, M., Rajam, M.V., Srivastava, S., 2017a. Players over the surface: Unraveling the role of exopolysaccharides in zinc biosorption by fluorescent *Pseudomonas* strain psd. Front. Microbiol. 8, 284.

Upadhyay, K.H., Vaishnav, A.M., Tipre, D.R., Patel, B.C., Dave, S.R., 2017b. Kinetics and mechanisms of mercury biosorption by an exopolysaccharide producing marine isolate *Bacillus licheniformis*. 3 Biotech, 7(5), 1–10.

Wang, S., Zhao, X., 2009. On the potential of biological treatment for arsenic contaminated soils and groundwater. J. Environ. Manage. 90(8), 2367–2376.

Wei, L., Li, Y., Noguera, D.R., Zhao, N., Song, Y., Ding, J., Zhao, Q., Cui, F., 2017. Adsorption of Cu2+ and Zn2+ by extracellular polymeric substances (EPS) in different sludges: Effect of EPS fractional polarity on binding mechanism. J. Hazard. Mater. 321, 473–483.

Wu, G., 2020. Management of metabolic disorders (including metabolic diseases) in ruminant and nonruminant animals. In: *Animal Agriculture*. Elsevier, pp. 471–491.

Xie, Q., Liu, N., Lin, D., Qu, R., Zhou, Q., Ge, F., 2020. The complexation with proteins in extracellular polymeric substances alleviates the toxicity of Cd (II) to Chlorella vulgaris. Environ. Poll. 263, 114102.

Xu, H., Yu, G., Jiang, H., 2013. Investigation on extracellular polymeric substances from mucilaginous cyanobacterial blooms in eutrophic freshwater lakes. Chemosphere. 93(1), 75–81.

Xue, C., Qi, P., Li, M., Liu, Y., 2016. Characterization and sorptivity of the Plesiomonas shigelloides strain and its potential use to remove Cd2+ from wastewater. Water. 8(6), 241.

Yang, M., Liang, Y., Dou, Y., Lan, M., Gao, X., 2017. Characterisation of an extracellular polysaccharide produced by *Bacillus mucilaginosus* MY6–2 and its application in metal biosorption. Chem. Ecol. 33(7), 625–636.

Yatera, K., Morimoto, Y., Ueno, S., Noguchi, S., Kawaguchi, T., Tanaka, F., Suzuki, H., Higashi, T.J.J.O.U., 2018. Cancer risks of hexavalent chromium in the respiratory tract. J. UOEH. 40(2), 157–172.

Yu, Q., Fein, J.B., 2015. The effect of metal loading on Cd adsorption onto Shewanella oneidensis bacterial cell envelopes: The role of sulfhydryl sites. Geochim. Cosmochim. Acta. 167, 1–10.

Yue, Z.-B., Li, Q., Li, C.-C., Chen, T.-H., Wang, J., 2015. Component analysis and heavy metal adsorption ability of extracellular polymeric substances (EPS) from sulfate reducing bacteria. Bioresour. Technol. 194, 399–402.

Zhang, D., Pan, X., Mostofa, K.M., Chen, X., Mu, G., Wu, F., Liu, J., Song, W., Yang, J., Liu, Y., 2010. Complexation between Hg (II) and biofilm extracellular polymeric substances: An application of fluorescence spectroscopy. J. Hazard. Mater. 175(1–3), 359–365.

Zhang, Z., Wang, P., Zhang, J., Xia, S., 2014. Removal and mechanism of Cu (II) and Cd (II) from aqueous single-metal solutions by a novel biosorbent from waste-activated sludge. Environ. Sci. Poll. Res. 21(18), 10823–10829.

Zheng, L., Jiang, Y.-L., Fei, J., Cao, P., Zhang, C., Xie, G.-F., Wang, L.-X., Cao, W., Fu, L., Zhao, H.J.E., Safety, E., 2021. Circulatory cadmium positively correlates with epithelial-mesenchymal transition in patients with chronic obstructive pulmonary disease. Ecotoxicol. Environ. Saf. 215, 112164.

Zinicovscaia, I., Rudi, L., Valuta, A., Cepoi, L., Vergel, K., Frontasyeva, M.V., Safonov, A., Wells, M., Grozdov, D., 2016. Biochemical changes in Nostoc linckia associated with selenium nanoparticles biosynthesis. Ecol. Chem. Eng. 23(4), 559.

13 Exopolysaccharides-based composite materials for industrial applications

*Vishal Ahuja, Sanjeet Mehariya,
Ranjit Gurav, and Shashi Kant Bhatia*

13.1 INTRODUCTION

EPS have gained wide attention for their cost-effective production and diverse applications as pharma-healthcare products, cosmeceutical, nutraceutical, functional food, and biocontrol agents in agriculture (Nwodo et al., 2012), oil recovery in petroleum industries (Ibrahim et al., 2022), heavy metal removal (Gupta and Diwan, 2017) and in drug delivery, and tissue regeneration and repair (Mohd Nadzir et al., 2021). However, in some cases, it was also observed EPS molecules have some structural or functional drawbacks for a specific purpose. Curdlan is well-known EPS produced by microbes, that is, *Alcaligenes* spp., *Agrobacterium* spp., *Aureobasidium pullulan, Candida* spp., *Paenibacillus* spp., *Poria cocos, Rhizobium* spp., and *Saccharomyces cerevisiae*. It is comprised of glucose monomers interconnected with β-(1,3)-glycosidic bonds. Curdlan has been widely used as a prebiotic, thickening agent, immunoregulator, and antitumor compound (Chaudhari et al., 2021). The major challenge with curdlan is its low water solubility and thus poor bioactivity (Hussain et al., 2017). Similarly, xanthan is a heteropolysaccharide composed of D-glucose backbone with mannose and glucuronic acid. It is soluble in water and biocompatible but has poor stability to heat and low electrical conductivity. Hyaluronic acid (HA) has a higher water retention capacity but has poor mechanical strength (Hussain et al., 2017).

Such structural and functional defects restricted the applications and commercial viability of product/s. However, EPS composite/blend with other EPS or other natural or synthetic molecules or metals have shown improved performance and efficiency in comparison to native molecules. HA and gelatin and collagen have higher mechanical strength and flexibility which can be used in tissue engineering. HA blend with tyrosine and humanlike collagen have higher mechanical strength and good biological stability. The matrix used for the preparation of multi-functionalized hydrogel scaffolds cross-linking the components via 1,4-butanedioldiglycidyl ether can be used as soft-tissue fillers (Liu and Fan, 2019). HA blend based hydrogel with

DOI: 10.1201/9781003342687-13

silk fibroin and gelatin has high mechanical strength and biocompatibility and addition of boric acid further improved the wound-healing behavior of wound dressings (Özen et al., 2022). Mixture of gellan–xanthan gum showed commendable gelling properties in presence of alkaline and alkaline earth metal salts. The gelation followed the order: $BaCl_2 > CaCl_2 \approx MgCl_2 > KCl > NaCl$. The hydrodynamic behavior and young modulus suggested its possible use in drilling fluids as well as oil recovery (Nurakhmetova et al., 2018).

Comparative study for rheological performance of welan gum, xanthan gum, and a mixture revealed that in the presence of inorganic salts, the apparent viscosity and viscoelasticity of welan decreased, while the presence of K^+ and Al^{3+} increased both for xanthan and gellan properties that induced gelation. For welan/xanthan mixture, the viscosity retention rate was higher than both components alone even in presence of Na^+, K^+, and Ca^{2+}, while welan/gellan blend has higher viscosity retention rate in presence of Ca^{2+} (Xu et al., 2015). Xanthan blend was prepared with a non-ionic green surfactant "alkyl polyglucoside" and two different solvents, that is, acetone and butanone were compared for oil recovery from the core. The blend with acetone has 32% tertiary oil production from sandstone, while the other one with butanone has 25%. The acetone formulation also has ability to recover residual oil from the reservoir core (Haq, 2021).

The chapter emphasized on EPS-composite and their blends with synthetic and natural EPS and non-EPS materials with special consideration to their application in food, packaging, and environmental conservation.

13.2 EPS-BASED MATERIALS

EPSs have been blended with natural polymers including other EPS, EPS derivatives, proteins, and/or synthetic polymers like poly acrylamide, polyethylene glycol, poly aniline, etc., as per the target application and required improvement in structural and functional characteristics like surface charge, functional groups, hydrophilicity, mechanical strength, etc. (Table 13.1). This section summarizes the information regarding EPS blends with natural and synthetic polymers.

13.2.1 BLEND WITH NATURAL POLYMERS

EPS blends/composites have been prepared with natural polymers like proteins, enzymes, etc., to add additional properties. The most important part of the work is that composite will also be biocompatible and biodegradable. Protein encapsulation in dextran might improve protein's stability. In an experiment five proteins, that is, bovine serum albumin (P1), granulocyte-macrophage colony-stimulating factor (P2), granulocyte colony-stimulating factor (P3), β-galactosidase (P4), and myoglobin (P5) were selected and added to an aqueous mixture of dextran and polyethylene glycol followed by aqueous-aqueous freezing-induced phase separation that induced the protein encapsulation in dextran. The encapsulation efficiency of proteins was >98%, and recovery of extracted proteins from nanoparticles was 65–72%. BSA loaded dextran nanoparticles were spherical shaped with smooth surfaces, having diameters from 200–500 nm (Wu et al., 2013). In another study novel nanogel made up of modified soy protein and dextran was prepared with an aim to use as a carrier for riboflavin with its efficient delivery. Soy protein was denatured by heating at 60°C for 30 min or hydrolyzed with alcalase for 40 min. Dextran was added and mixed to this denatured/modified soy protein followed by ultrasonication for 70 min, which induced self-assembly and formation of nanogels. The addition of riboflavin did not affect the particle size of nanogels. Nanogel exhibited 65.9% encapsulation efficiency for 250 μg/mL of riboflavin. The nanogel withstands the environmental conditions and has higher release rate in simulated intestine fluid than in simulated gastric fluid (Jin et al., 2016).

Xanthan gum–konjac glucomannan (KGM) blend hydrogel was prepared from different concentration combinations of XG and KGM powder solution in distilled water and homogenized by autoclaving for 30 min at 121°C. The temperature of the sample was lowered to 37°C to form hydrogel and stored at 3°C. Hydrogel prepared from XG and KGM with 1:1 ratio has optimum firmness of 3816 g and cohesiveness of 16474 g.s. Adhesiveness was also at its best with −4253 g.s (Alves et al., 2020). An edible packaging material from pullulan and carboxymethyl cellulose by casting method has also been reported. The components solution was prepared by mixing them in different rations, that is, 1:0, 3:1, 1:1, 1:3, and 0:1. The solutions of pullulan and cellulose were prepared separately in distilled water with continuous mixing on a hotplate magnetic stirrer. After the homogenization of individual solutions, both were combined with continuous homogenization with the addition of 30% gelatin for gelatinization. The solution was mixed continuously at 2°C to remove the air bubble and casted to prepare the film. The casted films were dried for up to 24 h at 25°C and 65% relative humidity. Among all the compositions, a blend with 3:1 combination has the lowest percentage of elongation, that is, 5.55%, and the highest tensile strength of 17.30 MPa. The water contact angle for the 3:1 blend film was 63.43°. The properties suggested 3:1 blend film suitable for packaging material (Thangavelu and Kulandhaivelu, 2022). An antibacterial film was prepared from pullulan for the protection of food materials during storage as well as transportation. For the preparation of the film, pullulan was derivatized by sodium periodate oxidation in the dark at 25°C for 5 h followed by dialysis against distilled water for up to 48 h. The recovered material was freeze-dried to recover OxPullulan. OxPullulan solution was prepared by mixing pullulan derivative (5.0 mg/mL in water) followed by the addition of 5% acetone. Similarly, 5 mg/mL sodium alginate aqueous solution was also prepared. Both the solutions were mixed in various ratios, that is, 1:0, 4:1, 3:2, 2:3, 1:4, and 0:1 followed by homogenization for 30 min at room temperature. The solution was dried at 50°C into a Teflon-coated dish to eliminate solvent. The recovered film act as an efficient barrier for water vapor and light with only 11% light transmittance. The addition of sodium alginate composite film resulted in further increase in mechanical strength to 34 MPa (Li et al., 2020).

13.2.2 BLEND WITH SYNTHETIC POLYMERS

EPS composites materials can also be prepared with synthetic polymers, which mainly improved the biocompatibility of polymers composite along with other characteristics of native polymers. The main aim of using synthetic materials is to exploit the mechanical strength, high molding capacity to diverse forms, and low cost of conventional polymers like cotton and polyaniline, polypyrrole, polyethylmethaacrylate, etc., while EPS provide additional advantages like water retention, swelling, antimicrobial potential, and controllable hydrophilicity and solubility in different solvents. Blended material of carboxymethyl derivative of chitosan and pullulan have been prepared via dispersion-assisted polymerization. In process poly (N-isopropylacrylamide) was used as temperature-sensitive polymer, and silane was used as a cross-linker. Chitosan was derivatized O-carboxymethyl chitosan (O-CMC) by treatment with 0.2 M isopropanol, 42% NaOH, and 42% monochloroacetic acid (MCA) step wise and then mixing with (3-Glycidyloxypropyl) trimethoxysilane at 50°C for 3 h. Similarly, pullulan was derivatized with $NaBH_4$, NaOH, and MCA and then mixed with N-isopropylacrylamide and N,N'-methylenebisacrylamide. Both the solutions were mixed together followed by addition of Tetramethylethylenediamine (TEMED), ZnO nanoparticles (9nm), and 2.5% ammonium persulfate (APS) solution and left for 48 h. Polymerization of gel was due to intense interactions between matrices including hydrogen bonds, electrostatic and covalent bonding. In composite nanoparticle size varies with temperature and reduced when temperature increased, that is, 50 nm particle size at 25°C and 7nm at 40°C. On the contrary, the particle size increased with pH and became 118 nm at pH 10. Treatment of cotton fabric with prepared composite by pad-dry-cure technique has confirmed the fine layering of composite on cotton surface and localization of NPs on it. Cotton-composite also exhibited antimicrobial activity against *E. coli* and *Staphylococcus aureus* (Mohamed and Hassabo, 2018).

In another study, a blend of pullulan and poly(vinyl alcohol) was prepared in the presence of sodium trimethaphosphate (STMP). Pullulan was added to aqueous solution of 10% (w/v) PVA47 or 5% (w/v) PVA125 at 90°C followed by cooling and addition of 10 M NaOH and 10% (w/v) STMP aqueous solutions in 1:5 ratio. The final solution had undergone freezing-thawing for three cycles of 20 h at 20°C each followed by 6 h at ~23 ± 1°C. Then the composite disc was collected and lyophilized at 57°C and 5.5×10^{-4} mbar. Composite have interconnected porous structures that were under the influence of preparation conditions as composite prepared at room temperature (C-R) have pores of 30–55 µm, while composite prepared at freeze drying (C-FD) have a pore size of 15–30 µm. Similarly, CR swelled up slowly and took 6 h while C-FD swelled rapidly, and the degradation pattern also suggested that the degree of degradation for C-R was higher than C-FD (Samoila et al., 2019).

A foamlike aerogel was prepared from xanthan gum and sodium montmorillonite clay by freeze-drying process, while 2.5% xanthan, 2.5 agar, and 2.5 wt % clay were mixed with 50 mL deionized water and dispersed. Then 20% glycerol (plasticizer) was also added to polymers suspension for 30 min. Gel mixture was kept in molds and frozen in a solid carbon dioxide/ethanol bath (−80°C) followed by freeze-drying. The presence of agar improved the mechanical strength of aerogels, while clay contributed to heat and mass transfer and improved thermal stability. Aerogel has shown low flammability in comparison to other foam (Wang et al., 2014). Polypyrrole (PPy) colloids and bioamine-crosslinked gellan gum (GG) composite networks were prepared for possible application in drug delivery. The composite hydrogels have interconnected porous structures with tunable viscoelasticity and swelling characteristics

TABLE 13.1

Physiological role of exopolysaccharides

EPS composites and derivatives	Properties	Applications	References
Alginate-arboxyl methyl chitosan	Improved tenacity and elongation. Water retention increased to 130–398% in comparison to 91% in alginate. Carboxymethyl chitosan lowered wet tensile strength and breaking elongation but increased water retention.	Antibacterial fiber for dressing.	(Fan et al., 2006)
Gelatin-pullulan composite nanofibers	Gelatin-pullulan composite prepared by mixing solutions on a magnetic stirrer at 30°C for at least 24 h. Surface tension of all compositions was below 70 mm/m. Viscosity, conductivity, and surface tension increase with gelatin content, but fiber diameter decreased. Gelatin content also lowered the elongation at break.	Tissue engineering.	(Y. Wang et al., 2019)
Chitosan-pullulan-pomegranate peel extract	2% chitosan in 0.5% aqueous citric acid + 2% pullulan in water (1:1), mixed for 2 min at 9000 rpm 23°C. Add 1% glycerol as a plasticizer and 5% of pomegranate peel extract as a natural antioxidant agent. Tomato sensory score was 7.15 ± 0.1 over untreated control 6.41 ± 0.14 at 4°C after 18 days.	Packaging material for tomato.	(Kumar et al., 2021)
Gellan gum-2 hydroxy ethyl cellulose-lignin	Lignin improved the thermal resistance and mechanical strength. 100% UV protection against UVB wavelength (280–320 nm) and 90% against UVA (320–400 nm). Excellent radical scavenging activities. Transmittance was reduced, and the water contact angle increased with an increase in lignin content as 10% lignin has 53.6% transmittance and 70° water contact angle.	Food packaging.	(Rukmanikrishnan et al., 2020)

Material	Findings	Application	Reference
Gellan gum-egg shell membrane	The compressive moduli of all combinations were 244 and 399 kPa. The addition of 4% egg shale membrane reduced the cross-linking and viscosity by 40%. Composite swelling also reduced by 30%. Composite has no cytotoxicity against retinal pigment epithelium.	Regeneration of retinal pigment epithelium.	(Choi et al., 2020)
Hydroxyapatite-embedded levan	0.1–5% hydroxyapatite-levan composite hydrogel improved vivo collagen production and anti-wrinkle efficacy. Increasing in hydroxyapatite concentration reduced in vitro stability and elasticity. Composite has improved cell proliferation for human dermal fibroblast cells in comparison to levan alone. The composite was stable over an 8-week period. Composite with 1% HAp has excellent anti-wrinkle efficacy during 8 weeks by improving the collagen production.	Dermal filler improved collagen production and anti-wrinkle activity.	(Hwang et al., 2021)
Alginate-gelatin	Composite has maximum tenacity of 1.29 cn/dtex. Maximum elongation of 4.41% (composite with 16.7% gelatin). Water absorption and retention increased by 19% (overall 335% and 311% respectively (composite with 16.7% gelatin).	Biomedical applications in wound dressing.	(Q.-Q. Wang et al., 2019)
Hydroxyapatite-chitosan-based hydrogels biomaterials loaded with metronidazole.	Hydroxyapatite reduced the swelling ratio and prevent burst release with respect to pH variation. Drug release was 38–73% as compared to the control. High encapsulation efficiency (96%).	Controlled drug delivery.	(Heragh et al., 2022)
Chitosan-gelatin scaffold loaded with aceclofenac	Porosity—100% allows scaffold degradation and sustainable drug release. Best realized at pH 6.8 (slightly acidic pH is good for drug release in osteopathic conditions). Become liquid in external pressure and reform gel when pressure removed. No floccule in scaffold and no aggregation behaviors.	Controlled drug delivery.	(Rajput et al., 2023)

(Continued)

TABLE 13.1 (*Continued*)

Physiological role of exopolysaccharides

EPS composites and derivatives	Properties	Applications	References
Xanthan gum–lysozyme	Temperature and pH dependent solubility. Heat stability. Better emulsion and foaming behavior. Dose-dependent inhibition for *Escherichia coli* and *Staphylococcus aureus* growth.	Functional food ingredients.	(Hashemi et al., 2014)
L-cysteine–xanthan gum buccal patch	Consistent swelling in simulation saliva (pH 6.75). Non-cytotoxicity to Carey 24 cell lines. Lower matrix erosion via thiolation. Lessened the saliva flow by 33% in sialorrhea.	Sialorrhea treatment.	(Laffleur and Michalek, 2017)
Xanthan gum–adipic acid dihydrazide	pH-dependent swelling with the least swelling at pH 3.0 and pH 13.0. Maximum swelling at neutral pH. Absorbed ~98% of methylene blue in 24 h. Released the absorbed dye in salt solution.	Controlled drug release.	(Bejenariu et al., 2008)
Curdlan-phosphorylated curdlan-ionic hydrogel-Metronidazole	Swelling ratio increased: 9 g/g to 16 g/g. Drug release increased from 50 to 90%.	Controlled drug release.	(Suflet et al., 2021)
Hyaluronic acid–gelatin (0.5% HA-Ph+5% gelatin-Ph)	Diffusion coefficient: 1.2–0.8 (10^{-10} m^2/s). Young modulus: 1.0–1.1 kPa. Viscosity 80–90 mPas.	Adipose stem cells cultivation.	(Sakai et al., 2019)
Dextran-Thyme Magnesium-Doped Hydroxyapatite	As coating material. Prevent formation of biofilm by gram-negative and positive bacteria and even *Candida*.	Antimicrobial coating.	(Iconaru et al., 2020)

Material	Findings	Application	Reference
Methacrylated gelatin–hyaluronic acid	Hybrid hydrogel formed with modified gelatin and hyaluronic acid has higher elastic modulus of 30 kPa. Porosity was around 91% in comparison to the control.	Tissue engineering	(Velasco-Rodriguez et al., 2021)
"Gelatin-hydroxy-phenyl propionic acid"–"hyaluronic acid tyramine"	Gel point 151–167 s. Elastic modulus 3101–4316 Pa. Shear 835–1072 Pa.	Retinal ganglion cells replacement therapy.	(Dromel et al., 2021)
Soy protein–dextran	pH influence the particle size, i.e., smaller particle size at lower pH range. Nanogels with ultrasonication have a lower particle size (32±10 nm). 65.9% efficiency for riboflavin encapsulation. Ultrasonication and alcalase assisted nanogel have tolerance to pH 2–10.	Riboflavin delivery.	(Jin et al., 2016)
Acetone-alkyl polyglucoside-xanthan gum	Viscosity was almost same on increase in temperature. Water flood and SP flood recovery was 45 and 31.6%. Overall recovery 76.26%.	Tertiary oil production and oil recovery.	(Haq, 2021)
Butanone-alkyl polyglucoside-xanthan gum	Temperature has a negligible effect on viscosity. Water flood and SP flood recovery was 50.98 and 25.24%. Overall recovery 76.22%.	Tertiary oil production.	(Haq, 2021)
Curdlan-nanocellulose	The tensile strength: 38.6 Mpa. Elongation at break: 40%. DSC curve has heat absorption peak at 240°C. Blend film has good heat/temperature stability.	Chilled meat preservation.	(Qian et al., 2021)

that can be tuned with filler content. The composite was used as a carrier for ibupro-fen (IBP), and in vitro release was evaluated in the presence and absence of different electrical stimulations. Composite exhibited stimuli assisted release and have 63% release in presence of 5 V stimuli in comparison to 10% in the passive form (Salazar Salas et al., 2022).

13.3 COMMERCIAL APPLICATIONS EXOPOLYSACCHARIDES DERIVE MATERIALS

Exopolysaccharides, natural biopolymers produced by microorganisms are com-prised of various molecules, including polysaccharides as major components. Different molecules contribute to the presence of diverse functional groups account-able for physiological and chemical properties. Its market demand in different sec-tors is increasing day by day (Figure 13.1).

In the upcoming section, the applications of EPS with special consideration to health care, food, and cosmetics have been summarized.

13.3.1 HEALTHCARE

EPS is comprised of biomolecules and hence can act as nontoxic and supportive car-riers and hence preferably used as composites. EPS has healthcare-related applica-tions like modulatory effect on immune system and inhibitory effect against tumor, cancer, and viral infection. Amphiphilic dextran blend material was prepared from carboxymethyl dextran L-cysteine (21.29%) and octadecylamine (19.35%). It is used

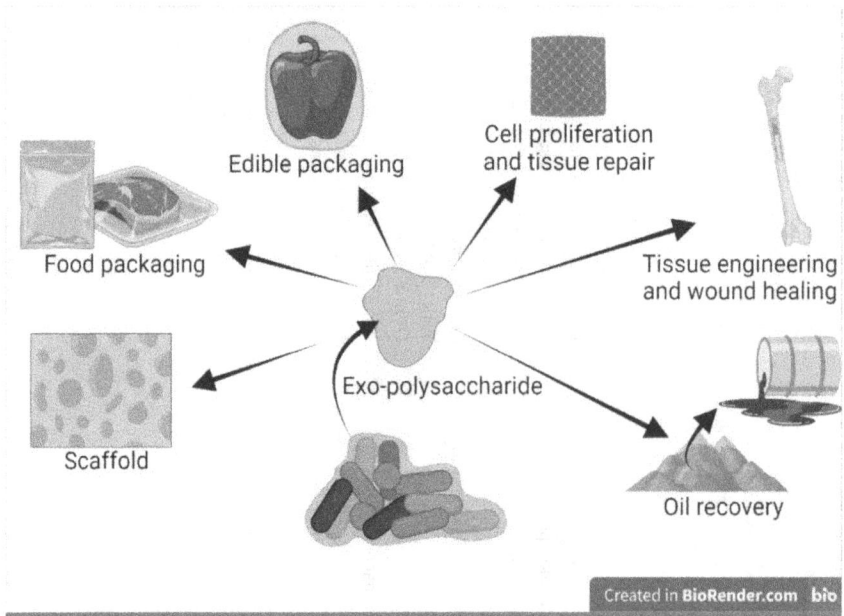

FIGURE 13.1   Multidimensional applications of exopolysaccharides derived materials

as carrier as well as a support material for the preparation of nanomicelles with Quercetin (QCT) via ultrasonic self-assembly method. Cells have shown a higher uptake rate for nanocomposite-based Quercetin in comparison to Quercetin. Under simulated gastric and intestinal fluids environment, the composite exhibited good stability with a release rate of 37.54% and 52.13% within 180 min. The nanocomposite also has scavenging activity against PTIO, OH, and O_2^- even higher than Quercetin itself. As a nanomaterial composite, Quercetin promoted the production of anti-inflammatory cytokines and lowered the production of pro-inflammatory cytokines (He et al., 2022).

Dosage size and controlled release of bioactive components are the major concern for healthcare formulations. *Pueraria lobatae* (Willd) Ohwi is a traditional Chinese medicine known for alcohol intoxication, cerebrovascular ailments, cardiovascular diseases, and diabetes, but the efficiency is suffered due to poor pharmacokinetic and physiological characteristics including low bioavailability, rapid systemic clearance, and poor aqueous solubility. Chitosan/xanthan gum–based hydrogels showed improved drug delivery due to porous structure and regulated release of *Pueraria lobata*-solid dispersion. Composite have maximum drug release rate of 63% at pH 1.2 after 48 h which was higher in comparison to 49% at pH 7.4 (Guan et al., 2022). Similar kind of work also showed the application of hydrogels for tissue engineering. Dialdehyde derivative of alginate and xanthan gum–based composite was used for the bioscaffolds fabrication and tissue regeneration. Hydrogel based scaffolds was thermostable having higher mechanical strength and protein adsorption along with biodegradability. Loading of curcumin on composite added the antibacterial activity against *E. coli*. The scaffolds showed commendable cytocompatibility for 3T3 fibroblast cells without any growth arrest or apoptosis (Jena et al., 2022).

13.3.2 Food packaging

Plastic and synthetically processed polymers were in trend due to low cost and long life, but prolonged application has led to accumulation in the environment and successive pollution. Thus, natural and biologically safe products have gained tremendous popularity and recorded exceptional growth in the last few decades. Due to higher acceptance and future possibilities, the market is moving towards functional packaging materials with multidimensional applications like paraprobiotics, prebiotics, probiotics, and postbiotics along with antimicrobial and antioxidant potential (Moradi et al., 2021).

Microbial exopolysaccharides are one of the possible biodegradable and biologically active materials that can also be used for the manufacturing of biologically active packaging materials from EPS-based composites. Constituting members and available functional groups offer wide opportunities as well as play a crucial role in the formation of composites with other biomolecules and metals ions.

In an effort to reduce plastic used in food packaging industries, cellulose base composite hydrogen film was developed with bacterial cellulose-guar gum based polyvinyl pyrrolidone with carboxymethyl cellulose. In composite, guar gum improved the mechanical strength, barrier properties, as well as hydrophobic properties. In addition, the composite biofilm is biodegradable and degraded by 80%

in 28 days in vermicompost (Bandyopadhyay et al., 2019). Bioactive composites of dextran and nanocellulose along with silver have antimicrobial properties. Dextran-coated silver nanoparticles of 12.0±1.9 nm were loaded on cellulose nanofibrils by solvent casting method. In composites, dextran acts as dispersing media for AgNPs that cumulatively add reduced oxygen permeability as well as microbial growth. The composites reduced the oxygen transmission rates by 32% (2.07 to 1.40–0.78 cm³/m²d) along with hydrophilicity from 35–37° to 62–74° for 3% acetic acid with 0.9% NaCl solutions. In addition, the composites reduced the *Escherichia coli* growth by 99.9% even after five repeated cycles. The film have shown a controlled release of Ag⁺ ions (<0.5 mg/L) (Lazić et al., 2020).

In another study dextran produced by lactic acid bacterium *Leuconostoc mesenteroides* T3 and plasticized with polyglycerol. Among different concentrations, composites developed with 10 wt% of polyglycerol offered maximum tensile strength of 4.6 MPa, while flexibility was maximum when dextran was plasticized with 30 wt% polyglycerol. The composite material film has water vapor permeability in the range of 3.45–8.81 × 10⁻¹² g/m.s.Pa, which suggested that material prevents fruit from drying out during transportation and storage. Application of composite film to coat blueberries followed storage at 8°C for 21 days extended the shelf life of blueberries with reduced weight loss and total sugar solids values along with delay in titratable acidity (Davidović et al., 2021).

Alginate loaded eugenol-tar microparticles where tar microparticles were incorporated into sodium alginate films and cross-linked with eugenol using CaCl₂. Composite has exhibited higher mechanical and humidity-resistance characteristics with antimicrobial activity, especially against *Staphylococcus aureus*. Overall physical, chemical, and biological characteristics suggested its possible use as active packaging application (Taverna et al., 2022). EPS composite with chitosan-dextran-like polymer was developed from lactic acid and plasticized with 1,3-propanediol (1,3-PDO). The plasticized film has 43.33MPa tensile strength over chitosan-exopolysaccharide films (20.08) and 20.73% elongation percentage that was also higher than unplasticized control films. The plasticized films have 51% higher water solubility and 193% moisture absorption. Besides plasticized films also have higher antioxidant properties (Vivek et al., 2021).

13.3.3 COSMETICS

Cosmetics become a major part of the health care and wellness industry. Now it is not only related to appearance but also related to well-being and personal fitness. A porous material for dressing was made from butyric-acetic derivative polyester of chitin (90% of butyryl+10% of acetyl groups) via salt leaching method and categorized into two groups, that is, Medisorb R and Medisorb Ag. Mediosorb Ag was loaded with microsilver while Mediosorb R was control (without silver). In vitro analysis confirmed degradation of Medisorb R within two weeks period when kept in physiological fluids at 37°C only in FTIR tests but no mass loss was detected. Medisorb Ag exhibited antibacterial activity due to the presence of microsilver on biomaterial. In vivo analysis confirmed wound healing without skin irritation or inflammation (Sujka et al., 2019). Alginate-based aerogels was prepared using different concentrations of alginate solutions (5%–15% w/w), which was extruded and coagulated with 5% (w/w) $CaCl_2$ or 1.5% (w/v) $CuSO_4$ followed by supercritical drying (200 bar at 45°C for 4 h). Ca- and Cu-Alg-hydrogels have a porous structure and preserved its nanostructure (Baldino et al., 2019).

Curdlan-based hydrogels are used for wound healing and skin dressing. The curdlan-based biomaterials comprised of 11wt% curdlan having mesoporous structure (pore diameter 14–48 nm) and have commendable wound fluid absorption (swelling ratio 974–1229%) and water vapor transmission rate of >2000 $g/m^2/day$. The biomaterial is Ca^+ enriched and released Ca^+ in the aqueous environment and enhanced fibroblast viability and proliferation and subsequently improves wound healing (Nurzynska et al., 2021). Hyaluronic acid is a glycosaminoglycan commonly studied for wound dressing, skin fillers, tissue engineering, and repair as it promotes movement and differentiation of mesenchymal and epithelial cells to active sites. Its nature of extensive cross-linking and availability of a wide range of functional groups make it resistant to enzymatic degradation and highly adaptable for the fabrication of 3D porous materials, and hydrogels. Besides, it also rejuvenates and moisturizes the skin and acts as a collagen stimulator when used in the form of hydrogel (Sionkowska et al., 2020).

Ultraviolet (UV) radiation increased the metabolism of hyaluronan (HA) and damage the skin cells. Human keratinocytes (HaCaT) cells were immobilized and exposed to UVB radiation followed by treatment with HA. HA treatment was done in two groups, that is, 6 h and 24 h. Exposure to radiation lowered down the expression of CD44 gene and toll-like receptor. Besides, it also reduced the expression of HA synthase-2 and hyaluronidase-2. Cells treatment with HA countered the ill effect of UVB by release of pro-inflammatory cytokines like interleukin-6 and interleukin-8, lowering the production of transforming growth factor β1 and release of soluble CD44. Overall, HA reduced the UVB-induced cell mortality and protect HaCaT cells from UVB irradiation (Hašová et al., 2011). Bisphosphonate-functionalized HA hydrogel have strong tendency to form HA hydrogel–CaP gels after associating with CaP. The composite exhibited adhesion with enamel and hydroxyapatite as well as and self-healing properties (Nejadnik et al., 2014). Glycidyl-methacrylate-conjugated HA hydrogel contained carboxyl group that facilitate synthesis CaP nanoparticles by precipitation mechanism. In comparison to raw HA or hydrogel, HA hydrogel–CaP gels composite have higher biocompatibility and improved cellular responses (Jeong et al., 2016). The composite material made up of decellularized cartilage and demineralized bone matrix with hyaluronic acid and hydroxyapatite nanoparticles in bone defect repair. Rheologically, HA-ECM suspensions as well as HA-HAP-ECM colloidal gels have better yield stresses of 100–1,000 Pa, storage moduli of 100–10,000 Pa and viscoelastic recoveries of ≥ 87% in comparison to controls formulation. The composite mixtures have three-fold higher storage modulus and four-fold increase in yield stress. In vitro assessment revealed that any of the combination of composite did not have any cytotoxicity or adverse effect on cell viability. This might have promising applications for cartilage tissue and bone repair and engineering (Dennis et al., 2017). Biphasic calcium phosphate nanoparticles-hyaluronic acid homogenous composite hydrogels were prepared by loading of biphasic calcium phosphate nanoparticles on self-cross-linked thiolated hyaluronic acid. Composite hydrogel exhibited biocompatibility with osteoblast cells (MC3T3-E1). The composite can be considered for an injectable formulation and for bone tissue related problems (Hong et al., 2021).

13.3.4 ENVIRONMENTAL APPLICATIONS

Microbial exopolysaccharides have shown commendable application in environmental conservation by removing pollutants and heavy metals from the environment. It has been found that some of the *Sphingomonas* species produce and secretes some high-molecular-weight extracellular polymers which are acidic and heteropolysaccharides. These polymers are known as sphingans. This group "sphingans" include a wide range of natural polymer molecules like welan gum, gellan gum, and diutan gum. This group is known for shear thinning properties like other EPS but also resistance to temperature and salt (Huang et al., 2022). The native properties of sphingans have shown a promising future for environment conservation. Sphingan composite with an anionic polyacrylamide and Brij 35 (nonionic surfactant) were prepared separately. In composite, components interacted via hydrogen bonds and electrostatic interactions. The composite with poly-acrylamide (WL-APAM) were tolerant to high temperature and salinity, while other composite "WL-Brij 35" can withstand high temperature. Both composite WL–APAM and WL–Brij 35 improved the displacement of oil with 21.31% and 22.17% efficiencies respectively (Ji et al., 2022).

Carbon quantum dots (CQDs) prepared via hydrothermal method from EPS produced by *Exiguobacterium* sp. VK2 were evaluated for catalytic conversion of waste cooking oil to biodiesel and compared the activity with two conventional catalysts. EPS-CQDs have offered around 96% of recovery of total fatty acid methyl ester (FAME). This suggested the possible use of EPSCQDs as heterogeneous catalyst for biodiesel production as an alternative to conventional chemical catalysts which not only reduced the cost but also chemical pollution (Kalpana et al., 2023).

Xanthan gum nano-polymer suspension composite with anatase TiO_2 nanoparticles have shown improved oil recovery in carbonates. A rise in temperature and presence of salts lowered the shear viscosity of xanthan gum suspension, which was recovered after the addition of NPs. Polymers nanofluids enriched the dispersion and wettability of NPs. Composite exhibited 25% additional oil recovery over 12% and 19% with TiO_2 nanofluid and polymer solution respectively (Keykhosravi et al., 2021). Said et al. have prepared xanthan acrylate with xanthan gum and acrylic acid. For tertiary oil recovery was evaluated with six-inch sandstone core plugs by core flooding experiments with Arabian light crude oil. For extraction two formulations were prepared, that is, xanthan gum + 3% NaCl and modified xanthan gum (xanthan acrylate) + 3% NaCl. First formulation offered 14% residual oil recovery, while the second formulation offered 19% recovery (Said et al., 2021).

13.4 CONCLUSION

Exopolysaccharides (EPS) are among microbial metabolites with exponential market demand and diverse applications. Its wide acceptability is due to nontoxic and biodegradability along with adaptability and use friendly. It has been proved that

EPS have wide acceptability and adaptability for other synthetic as well as metal moieties, which support the composite formation and metal/oil recovery. However, most of the research have been focused in bacteria only, which may restrict the possibilities of diverse molecules from other organisms coming into light. There is need to explore organisms from different habitats for exopolysaccharides having unique compositions and properties that can be used in different applications.

ACKNOWLEDGMENTS

The authors acknowledge the KU Research Professor Program of Konkuk University, Seoul, South Korea. This research was supported by the C1 Gas Refinery Program through the National Research Foundation of Korea (NRF), funded by the Ministry of Science and ICT (2015M3D3A1A01064882), and by the National Research Foundation of Korea (NRF) [NRF-NRF-2022M3I3A1082545, NRF-2022M3J4A1053702, NRF-2022R1A2C2003138 and NRF-2021R1F1A1050325].

REFERENCES

Alves, A., Miguel, S.P., Araujo, A.R.T.S., de Jesús Valle, M.J., Sánchez Navarro, A., Correia, I.J., Ribeiro, M.P., Coutinho, P., 2020. Xanthan gum—Konjac glucomannan blend hydrogel for wound healing. Polymers. 12, 99. https://doi.org/10.3390/polym12010099

Baldino, L., Cardea, S., Scognamiglio, M., Reverchon, E., 2019. A new tool to produce alginate-based aerogels for medical applications, by supercritical gel drying. J. Supercrit. Fluids. 146, 152–158. https://doi.org/10.1016/j.supflu.2019.01.016

Bandyopadhyay, S., Saha, N., Brodnjak, U.V., Sáha, P., 2019. Bacterial cellulose and guar gum based modified PVP-CMC hydrogel films: Characterized for packaging fresh berries. Food Packag. Shelf Life. 22, 100402. https://doi.org/10.1016/j.fpsl.2019.100402

Bejenariu, A., Popa, M., Le Cerf, D., Picton, L., 2008. Stiffness xanthan hydrogels: Synthesis, swelling characteristics and controlled release properties. Polym. Bull. 61, 631–641. https://doi.org/10.1007/s00289-008-0987-6

Chaudhari, V., Buttar, H.S., Bagwe-Parab, S., Tuli, H.S., Vora, A., Kaur, G., 2021. Therapeutic and industrial applications of curdlan with overview on its recent patents. Front. Nutr. 8.

Choi, J., Lee, J., Shin, M.E., Been, S., Lee, D.H., Khang, G., 2020. Eggshell membrane/ Gellan gum composite hydrogels with increased degradability, biocompatibility, and anti-swelling properties for effective regeneration of retinal pigment epithelium. Polymers. 12, 2941. https://doi.org/10.3390/polym12122941

Davidović, S., Miljković, M., Gordic, M., Cabrera-Barjas, G., Nesic, A., Dimitrijević-Branković, S., 2021. Dextran-based edible coatings to prolong the shelf life of blueberries. Polymers. 13, 4252. https://doi.org/10.3390/polym13234252

Dennis, S.C., Whitlow, J., Detamore, M.S., Kieweg, S.L., Berkland, C.J., 2017. Hyaluronic-acid—hydroxyapatite colloidal gels combined with micronized native ECM as potential bone defect fillers. Langmuir. 33, 206–218. https://doi.org/10.1021/acs.langmuir.6b03529

Dromel, P.C., Singh, D., Andres, E., Likes, M., Kurisawa, M., Alexander-Katz, A., Spector, M., Young, M., 2021. A bioinspired gelatin-hyaluronic acid-based hybrid interpenetrating network for the enhancement of retinal ganglion cells replacement therapy. NPJ Regen. Med. 6, 1–12. https://doi.org/10.1038/s41536-021-00195-3

Fan, L., Du, Y., Zhang, B., Yang, J., Zhou, J., Kennedy, J.F., 2006. Preparation and properties of alginate/carboxymethyl chitosan blend fibers. Carbohydr. Polym. 65, 447–452.
https://doi.org/10.1016/j.carbpol.2006.01.031

Guan, Y., Yu, C., Zang, Z., Wan, X., Naeem, A., Zhang, R., Zhu, W., 2022. Chitosan/xan-
than gum-based (Hydroxypropyl methylcellulose-co-2-Acrylamido-2-methylpropane
sulfonic acid) interpenetrating hydrogels for controlled release of amorphous solid dispersion of bioactive constituents of Pueraria lobatae. Int. J. Biol. Macromol. https://doi.
org/10.1016/j.ijbiomac.2022.10.131

Gupta, P., Diwan, B., 2017. Bacterial Exopolysaccharide mediated heavy metal removal: A
review on biosynthesis, mechanism and remediation strategies. Biotechnol. Rep. 13,
58–71. https://doi.org/10.1016/j.btre.2016.12.006

Haq, B., 2021. The role of microbial products in green enhanced oil recovery: Acetone and
butanone. Polymers. 13, 1946. https://doi.org/10.3390/polym13121946

Hashemi, M.M., Aminlari, M., Moosavinasab, M., 2014. Preparation of and studies on the
functional properties and bactericidal activity of the lysozyme—xanthan gum conjugate. LWT—Food Sci. Technol. 57, 594–602. https://doi.org/10.1016/j.lwt.2014.01.040

Hašová, M., Crhák, T., Šafránková, B., Dvořáková, J., Muthný, T., Velebný, V., Kubala, L.,
2011. Hyaluronan minimizes effects of UV irradiation on human keratinocytes. Arch.
Dermatol. Res. 303, 277–284. https://doi.org/10.1007/s00403-011-1146-8

He, Z., Liu, Y., Wang, H., Li, P., Chen, Y., Wang, C., Zhou, C., Song, S., Chen, S., Huang,
G., Yang, Z., 2022. Dual-grafted dextran based nanomicelles: Higher antioxidant,
anti-inflammatory and cellular uptake efficiency for quercetin. Int. J. Biol. Macromol.
https://doi.org/10.1016/j.ijbiomac.2022.10.222

Heragh, B.K., Javanshir, S., Mahdavinia, G.R., Naimi-Jamal, M.R., 2022. Development of
pH-sensitive biomaterial-based nanocomposite for highly controlled drug release.
Results Mater. 16, 100324. https://doi.org/10.1016/j.rinma.2022.100324

Hong, S.Y., Tran, T.V.T., Kang, H.J., Tripathi, G., Lee, B.T., Bae, S.H., 2021. Synthesis and
characterization of biphasic calcium phosphate laden thiolated hyaluronic acid hydrogel based scaffold: Physical and in-vitro biocompatibility evaluations. J. Biomater. Sci.
Polym. Ed. 32, 337–354. https://doi.org/10.1080/09205063.2020.1833816

Huang, H., Lin, J., Wang, W., Li, S., 2022. Biopolymers produced by Sphingomonas strains
and their potential applications in petroleum production. Polymers. 14, 1920. https://
doi.org/10.3390/polym14091920

Hussain, A., Zia, K.M., Tabasum, S., Noreen, A., Ali, M., Iqbal, R., Zuber, M., 2017. Blends
and composites of exopolysaccharides; properties and applications: A review. Int. J.
Biol. Macromol. 94, 10–27. https://doi.org/10.1016/j.ijbiomac.2016.09.104

Hwang, Y., Lee, J.S., An, H., Oh, H., Sung, D., Tae, G., Choi, W.I., 2021. Hydroxyapatite-
embedded levan composite hydrogel as an injectable dermal filler for considerable
enhancement of biological efficacy. J. Ind. Eng. Chem. 104, 491–499. https://doi.
org/10.1016/j.jiec.2021.08.040

Ibrahim, H.A.H., Abou Elhassayeb, H.E., El-Sayed, W.M.M., 2022. Potential functions and
applications of diverse microbial exopolysaccharides in marine environments. J. Genet.
Eng. Biotechnol. 20, 151. https://doi.org/10.1186/s43141-022-00432-2

Iconaru, S.L., Predoi, M.V., Motelica-Heino, M., Predoi, D., Buton, N., Megier, C., Stan, G.E.,
2020. Dextran-thyme magnesium-doped hydroxyapatite composite antimicrobial coatings. Coatings. 10, 57. https://doi.org/10.3390/coatings10010057

Jena, S.R., Dalei, G., Das, S., Nayak, J., Pradhan, M., Samanta, L., 2022. Harnessing the
potential of dialdehyde alginate-xanthan gum hydrogels as niche bioscaffolds for
tissue engineering. Int. J. Biol. Macromol. 207, 493–506. https://doi.org/10.1016/j.
ijbiomac.2022.03.024

Jeong, S.-H., Koh, Y.-H., Kim, S.-W., Park, J.-U., Kim, H.-E., Song, J., 2016. Strong and Biostable hyaluronic acid—Calcium phosphate nanocomposite hydrogel via in situ precipitation process. Biomacromolecules. 17, 841–851. https://doi.org/10.1021/acs.biomac.5b01557

Ji, S., Wei, F., Li, B., Li, P., Li, H., Li, S., Wang, J., Zhu, H., Xu, H., 2022. Synergistic effects of microbial polysaccharide mixing with polymer and nonionic surfactant on rheological behavior and enhanced oil recovery. J. Pet. Sci. Eng. 208, 109746. https://doi.org/10.1016/j.petrol.2021.109746

Jin, B., Zhou, X., Li, X., Lin, W., Chen, G., Qiu, R., 2016. Self-assembled modified soy protein/dextran nanogel induced by ultrasonication as a delivery vehicle for riboflavin. *Molecules.* 21, 282. https://doi.org/10.3390/molecules21030282

Kalpana, R., Sakthi Vignesh, N., Vinothini, K., Rajan, M., Ashokkumar, B., Brindhadevi, K., Thuy Lan Chi, N., Pugazhendhi, A., Varalakshmi, P., 2023. Carbon quantum dots (CQD) fabricated from *Exiguobacterium* sp. VK2 exopolysaccharide (EPS) using hydrothermal reaction and its biodiesel applications. Fuel. 333, 126426. https://doi.org/10.1016/j.fuel.2022.126426

Keykhosravi, A., Vanani, M.B., Aghayari, C., 2021. TiO2 nanoparticle-induced Xanthan Gum Polymer for EOR: Assessing the underlying mechanisms in oil-wet carbonates. J. Pet. Sci. Eng. 204, 108756. https://doi.org/10.1016/j.petrol.2021.108756

Kumar, N., Neeraj, Pratibha, Trajkovska Petkoska, A., 2021. Improved shelf life and quality of tomato (*Solanum lycopersicum L.*) by using chitosan-pullulan composite edible coating enriched with pomegranate peel extract. ACS Food Sci. Technol. 1, 500–510. https://doi.org/10.1021/acsfoodscitech.0c00076

Laffleur, F., Michalek, M., 2017. Modified xanthan gum for buccal delivery—A promising approach in treating sialorrhea. Int. J. Biol. Macromol. 102, 1250–1256. https://doi.org/10.1016/j.ijbiomac.2017.04.123

Lazić, V., Vivod, V., Peršin, Z., Stoiljković, M., Ratnayake, I.S., Ahrenkiel, P.S., Nedeljković, J.M., Kokol, V., 2020. Dextran-coated silver nanoparticles for improved barrier and controlled antimicrobial properties of nanocellulose films used in food packaging. Food Packag. Shelf Life. 26, 100575. https://doi.org/10.1016/j.fpsl.2020.100575

Li, S., Yi, J., Yu, X., Wang, Z., Wang, L., 2020. Preparation and characterization of pullulan derivative antibacterial composite films. Mater. Sci. Eng. C. 110, 110721. https://doi.org/10.1016/j.msec.2020.110721

Liu, Y., Fan, D., 2019. Novel hyaluronic acid-tyrosine/collagen-based injectable hydrogels as soft filler for tissue engineering. Int. J. Biol. Macromol. 141, 700–712. https://doi.org/10.1016/j.ijbiomac.2019.08.233

Mohamed, A.L., Hassabo, A.G., 2018. Composite material based on pullulan/silane/ZnO-NPs as pH, thermo-sensitive and antibacterial agent for cellulosic fabrics. Adv. Nat. Sci. Nanosci. Nanotechnol. 9, 045005. https://doi.org/10.1088/2043-6254/aaeee0

Mohd Nadzir, M., Nurhayati, R.W., Idris, F.N., Nguyen, M.H., 2021. Biomedical applications of bacterial exopolysaccharides: A review. Polymers. 13, 530. https://doi.org/10.3390/polym13040530

Moradi, M., Guimarães, J.T., Sahin, S., 2021. Current applications of exopolysaccharides from lactic acid bacteria in the development of food active edible packaging. Curr. Opin. Food Sci. 40, 33–39. https://doi.org/10.1016/j.cofs.2020.06.001

Nejadnik, M.R., Yang, X., Bongio, M., Alghamdi, H.S., van den Beucken, J.J.J.P., Huysmans, M.C., Jansen, J.A., Hilborn, J., Ossipov, D., Leeuwenburgh, S.C.G., 2014. Self-healing hybrid nanocomposites consisting of bisphosphonated hyaluronan and calcium phosphate nanoparticles. Biomaterials. 35, 6918–6929. https://doi.org/10.1016/j.biomaterials.2014.05.003

Nurakhmetova, Z., Gussenov, I., Aseyev, V., Sigitov, V., Kudaibergenov, S., 2018. Application of sol-gel transition of gellan and xanthan for enhanced oil recovery and as drilling fluids. J. Chem. Technol. Metall. 53, 68–78.

Nurzynska, A., Klimek, K., Palka, K., Szajnecki, Ł., Ginalska, G., 2021. Curdlan-based hydrogels for potential application as dressings for promotion of skin wound healing— preliminary in vitro studies. Materials. 14, 2344. https://doi.org/10.3390/ma14092344

Nwodo, U.U., Green, E., Okoh, A.I., 2012. Bacterial exopolysaccharides: Functionality and prospects. Int. J. Mol. Sci. 13, 14002–14015. https://doi.org/10.3390/ijms131114002

Özen, N., Özbaş, Z., İzbudak, B., Emik, S., Özkahraman, B., Bal-Öztürk, A., 2022. Boric acid-impregnated silk fibroin/gelatin/hyaluronic acid-based films for improving the wound healing process. J. Appl. Polym. Sci. 139, 51715. https://doi.org/10.1002/app.51715

Qian, Y., Bian, L., Wang, K., Chia, W.Y., Khoo, K.S., Zhang, C., Chew, K.W., 2021. Preparation and characterization of curdlan/nanocellulose blended film and its application to chilled meat preservation. Chemosphere. 266, 128948. https://doi.org/10.1016/j.chemosphere.2020.128948

Rajput, I.B., Tareen, F.K., Khan, A.U., Ahmed, N., Khan, M.F.A., Shah, K.U., Rahdar, A., Díez-Pascual, A.M., 2023. Fabrication and in vitro evaluation of chitosan-gelatin based aceclofenac loaded scaffold. Int. J. Biol. Macromol. 224, 223–232. https://doi.org/10.1016/j.ijbiomac.2022.10.118

Rukmanikrishnan, B., Ramalingam, S., Rajasekharan, S.K., Lee, J., Lee, J., 2020. Binary and ternary sustainable composites of gellan gum, hydroxyethyl cellulose and lignin for food packaging applications: Biocompatibility, antioxidant activity, UV and water barrier properties. Int. J. Biol. Macromol. 153, 55–62. https://doi.org/10.1016/j.ijbiomac.2020.03.016

Said, M., Haq, B., Al Shehri, D., Rahman, M.M., Muhammed, N.S., Mahmoud, M., 2021. Modification of xanthan gum for a high-temperature and high-salinity reservoir. Polymers. 13, 4212. https://doi.org/10.3390/polym13234212

Sakai, S., Ohi, H., Taya, M., 2019. Gelatin/hyaluronic acid content in hydrogels obtained through blue light-induced gelation affects hydrogel properties and adipose stem cell behaviors. Biomolecules. 9, 342. https://doi.org/10.3390/biom9080342

Salazar Salas, B.M., Grijalva Bustamante, G.A., Fernández Quiroz, D., Castillo Ortega, M.M., Encinas, J.C., Herrera Franco, P.J., del Castillo Castro, T., 2022. Nanocomposite hydrogels of gellan gum and polypyrrole for electro-stimulated ibuprofen release application. React. Funct. Polym. 176, 105296. https://doi.org/10.1016/j.reactfunctpolym.2022.105296

Samoila, I., Dinescu, S., Pircalabioru, G.G., Marutescu, L., Fundueanu, G., Aflori, M., Constantin, M., 2019. Pullulan/Poly(Vinyl Alcohol) composite hydrogels for adipose tissue engineering. Materials. 12, 3220. https://doi.org/10.3390/ma12193220

Sionkowska, A., Gadomska, M., Musiał, K., Piątek, J., 2020. Hyaluronic acid as a component of natural polymer blends for biomedical applications: A review. Molecules. 25, 4035. https://doi.org/10.3390/molecules25184035

Suflet, D.M., Popescu, I., Prisacaru, A.I., Pelin, I.M., 2021. Synthesis and characterization of curdlan—phosphorylated curdlan based hydrogels for drug release. Int. J. Polym. Mater. Polym. Biomater. 70, 870–879. https://doi.org/10.1080/00914037.2020.1765360

Sujka, W., Draczynski, Z., Kolesinska, B., Latanska, I., Jastrzebski, Z., Rybak, Z., Zywicka, B., 2019. Influence of porous dressings based on butyric-acetic chitin co-polymer on biological processes in vitro and in vivo. Materials. 12, 970. https://doi.org/10.3390/ma12060970

Taverna, M.E., Busatto, C.A., Saires, P.J., Bertero, M.P., Sedran, U.A., Estenoz, D.A., 2022. Bio-composite films based on alginate and rice husk tar microparticles loaded with eugenol for active packaging. Waste Biomass Valorization. 13, 3061–3070. https://doi.org/10.1007/s12649-022-01679-z

Thangavelu, M., Kulandhaivelu, S.V., 2022. Development and characterization of pullulan-carboxymethyl cellulose blend film for packaging applications. Int. J. Polym. Sci. 2022, e9649726. https://doi.org/10.1155/2022/9649726

Velasco-Rodriguez, B., Diaz-Vidal, T., Rosales-Rivera, L.C., García-González, C.A., Alvarez-Lorenzo, C., Al-Modlej, A., Domínguez-Arca, V., Prieto, G., Barbosa, S., Soltero Martínez, J.F.A., Taboada, P., 2021. Hybrid methacrylated gelatin and hyaluronic acid hydrogel scaffolds. preparation and systematic characterization for prospective tissue engineering applications. Int. J. Mol. Sci. 22, 6758. https://doi.org/10.3390/ijms22136758

Vivek, N., Gopalan, N., Das, S., Sasikumar, K., Sindhu, R., Nampoothiri, K.M., Pandey, A., Binod, P., 2021. Synthesis and characterization of transparent biodegradable chitosan: Exopolysaccharide composite films plasticized by bio-derived 1,3-propanediol. Sustain. Chem. 2, 49–62. https://doi.org/10.3390/suschem2010004

Wang, L., Schiraldi, D.A., Sánchez-Soto, M., 2014. Foamlike xanthan gum/clay aerogel composites and tailoring properties by blending with agar. Ind. Eng. Chem. Res. 53, 7680–7687. https://doi.org/10.1021/ie500490n

Wang, Q.-Q., Liu, Y., Zhang, C.-J., Zhang, C., Zhu, P., 2019. Alginate/gelatin blended hydrogel fibers cross-linked by Ca2+ and oxidized starch: Preparation and properties. Mater. Sci. Eng. C. 99, 1469–1476. https://doi.org/10.1016/j.msec.2019.02.091

Wang, Y., Guo, Z., Qian, Y., Zhang, Z., Lyu, L., Wang, Ying, Ye, F., 2019. Study on the electrospinning of gelatin/pullulan composite nanofibers. Polymers. 11, 1424. https://doi.org/10.3390/polym11091424

Wu, F., Zhou, Z., Su, J., Wei, L., Yuan, W., Jin, T., 2013. Development of dextran nanoparticles for stabilizing delicate proteins. Nanoscale Res. Lett. 8, 197. https://doi.org/10.1186/1556-276X-8-197

Xu, L., Dong, M., Gong, H., Sun, M., Li, Y., 2015. Effects of inorganic cations on the rheology of aqueous welan, xanthan, gellan solutions and their mixtures. Carbohydr. Polym. 121, 147–154. https://doi.org/10.1016/j.carbpol.2014.12.030

14 Advances in microalgae exopolysaccharides production and applications

Jorge Alberto Vieira Costa,
Thaisa Duarte Santos, Juliana Botelho Moreira,
Matheus Pereira de Carvalho,
Danielle Rubim Lopes, and
Michele Greque de Morais

14.1 INTRODUCTION

Exopolysaccharides (EPS) are the main groups of biopolymers with significant ecological, industrial, and biotechnological relevance (Cruz et al., 2020; Parwani et al., 2021). EPS are macromolecular polysaccharide compounds that present complex structures. This fact is related to the several structural components in different amounts and configurations. EPS are mostly heteropolysaccharides containing small amounts of peptides, DNA, fatty acids, and uremic acids (Parwani et al., 2021).

Microalgae are photosynthetic microorganisms with the potential to become an industrially sustainable source of EPS. The production and release of EPS from microalgae depend on specific cultivation conditions (culture medium, temperature, pH, carbon dioxide (CO_2) supply, and luminosity) (Moreira et al., 2022a). Besides, the culture medium is related to the monosaccharide composition and the configuration of the glycosidic linkage of microalgae EPS. Therefore, it is crucial to understand how these cultivation parameters affect EPS structures so that it is possible to establish large-scale EPS microalgae production (Colusse et al., 2022).

The scarcity of information regarding their complex structural properties and EPS biosynthesis end up compromising the intensification of the commercialization of microalgal EPS (Parwani et al., 2021). On the other hand, EPS from microalgae has bioactive activities and physical-chemical properties that allow its valorization in several markets, such as pharmaceutical, medical, and agro-industrial (Laroche, 2022; Parwani et al., 2021). Thus, antioxidant, anti-inflammatory, antitumor, and antimicrobial characteristics make EPS from microalgae a welcome addition to the existing market (Moreira et al., 2022b).

DOI: 10.1201/9781003342687-14

In this way, microalgae cultivation systems for obtaining EPS are prepared to improve and expand the applications of the biotechnology sector in the coming decades economically. Although the potential of microalgae to obtain several compounds, including EPS, is established, technological advances in microalgae cultivation and EPS extraction deserve greater attention. In addition, investigating market prices and cultivation bottlenecks is crucial to establish a sustainable commercial production of microalgal EPS. Thus, market challenges in various sectors, such as medical/pharmaceutical, food and agriculture, can be overcome (Colusse et al., 2022).

In this context, this chapter approaches the advances in microalgal EPS production, including their potential for the market. Moreover, the work reports strategies to stimulate EPS biosynthesis and presents their biotechnological applications in several areas.

14.2 MICROALGAL BIOTECHNOLOGY

Microalgae cultivation is carried out for several biotechnological purposes, emphasizing effluent treatment, production of compounds of interest, and CO_2 capture, among others (Kumar et al., 2022; Ma et al., 2022; Oliveira et al., 2022; Villar-Navarro et al., 2022). The use of microalgae in effluent treatment stems from the idea that these waters are rich in nutrients necessary for microalgal cultivation. In addition, there is a need to treat this effluent for its proper disposal in bodies of water (Costa et al., 2021a). In this sense, the microalgal culture demonstrates a sustainable character in line with the sustainable development goals (SDGs) advocated by the United Nations (UN) (Oliveira et al., 2022).

Microalgae are responsible for more than half of global CO_2 fixation. The efficiency in capturing CO_2 is related to the metabolism of the microalgae, which uses this pollutant as a nutrient for the formation of organic matter used for the production of biocompounds, animal and human food (Raja et al., 2018; Ali et al., 2021; Ighalo et al., 2022; Ma et al., 2022).

By using CO_2, the microalgae contribute to mitigating the emissions of greenhouse gas, increasing the yield of biomass and EPS. Furthermore, biomass can be rich in carbohydrates and lipids for producing biofuels (Babiak and Krzemińska, 2021; Ighalo et al., 2022). Thus, during microalgal cultivation, compounds of interest are produced that may be constituted in their biomass or released in the culture medium.

The most investigated biocompounds constituted in microalgal biomass are proteins, lipids, carbohydrates, vitamins, minerals, and pigments (Bernaerts et al., 2019). Microalgae biomass is used as an ingredient in producing foods and feeds due to its high protein content and essential fatty acids (Raja et al., 2018; Gohara-Beirigo et al., 2022). Besides, microalgal biomass can also be applied in the production of biofuels. The carbohydrate is used as a substrate for producing alcohol and biogas, while the lipid fraction can be refined for the production of biodiesel and biokerosene (Kim et al., 2021). At the same time, due presence of minerals (magnesium and iron, for example), microalgae biomass can be used for biofertilizer production (Ali et al., 2021). Among extracellular compounds, polysaccharides are the ones that have attracted the most attention due to their bioactive characteristics and technological potential as thickening agents (Laroche, 2022).

14.3 STRATEGIES TO INCREASE EPS PRODUCTION
BY MICROALGAE

14.3.1 CULTURE MEDIUM

Studies on optimizing the production of EPS by microalgae are still limited. Stress conditions can stimulate the synthesis of EPS since these molecules are related to protecting cells against biotic and abiotic factors (De Philippis and Vincenzini, 1998). The commonly used strategies are changes in nutrient supply or environmental conditions. However, some of these modifications are specific to each species of microalgae (Laroche, 2022).

The nutrient limitation is an inducing factor in producing EPS. Nitrogen limitation induced higher permanence in the stationary phase and higher EPS production (2.5 g L^{-1}) by *Porphyridium marinum* without changes in growth rate and generation time compared to the nitrogen-replete medium (Soanen et al., 2016). Mahesh et al. (2019) also reported maximum EPS production (236 mg L^{-1}) by the microalgae *Scenedesmus abundans* cultivated with nitrogen limitation. According to Esqueda et al. (2022), the reduction in nitrogen concentration induces the beginning of the stationary phase of growth, in which excess carbon is redirected to the formation of energy reserves, such as starch and lipids, or excreted in the form of EPS. On the other hand, Villay et al. (2013) supplemented the culture medium of *Rhodella violacea* with nitrogen and phosphorus. The authors found an increase in the concentration of cells (5.98 × 10^6 cells mL^{-1}) and EPS (594.4 mg L^{-1}).

Although nitrogen is the most studied, phosphorus and sulfur also affect EPS synthesis by microalgae. Phosphorus limitation was related to increased EPS production by cyanobacteria *Spirulina* and *Cyanothece*, green algae *Chlorella* and *Porphyridium*, and diatoms *Cylindrotheca* and *Thalassiosira* (Laroche, 2022). Cultivation of the diatom *Phaeodactylum tricornutum* in a phosphorus-limited medium (12.1 mM of PO_4^{3-}) increased EPS production by 2.5 times compared to cultivation in a phosphorus-replete medium (Abdullahi et al., 2006). Limiting sulfur can reduce the degree of sulfation of synthesized EPS, as reported for the microalgae *Rhodella reticulata* and *Porphyridium cruentum* (Laroche, 2022). This reduction in the degree of sulfation in the EPS chain can influence the biological activity of these molecules (Raposo et al., 2015).

EPS synthesis is also related to carbon supply. CO_2 supplementation during cultivation promotes the growth and production of various metabolites, including EPS (Laroche, 2022). Yin et al. (2022) evaluated the effect of different concentrations of CO_2 on the production of EPS by the microalga *P. cruentum*. The 5% CO_2 concentration resulted in a higher accumulation of EPS by the microalga (218.5 mg L^{-1}).

Changes in other culture medium components stimulate the production of EPS by microalgae. Bafana (2013) evaluated different salts in different concentrations to increase EPS production by *Chlamydomonas reinhardtii*. Enhancing the content of $MgSO_4$ and $NaNO_3$ increased EPS production. However, adding $FeCl_3$ and $NaMoO_4$ and increasing the $CaCl_2$ and K_2HPO_4 concentrations harmed the production of polysaccharides.

High NaCl concentration can improve EPS synthesis by halophilic or halotolerant microalgae (Laroche, 2022). The salinity of the medium impacted the sulfation pattern and yield of EPS produced by *Porphyridium purpureum*, with the highest concentration of EPS (90 mg L^{-1}) obtained in the assay with 32 g L^{-1} of NaCl. Besides, the EPS yield was higher (144 mg L^{-1}) in the pilot scale (800 L). Increased salinity altered the sulfation pattern in xylose and glucose residues (Ferreira et al., 2021). The concentration of NaCl also influenced the production of EPS by the microalga *Dunaliella salina*. The concentration of these molecules increased from 56 mg L^{-1} to 944 mg L^{-1} with increasing NaCl concentration in the medium from 0.5 to 5 M (Mishra and Jha, 2009).

One strategy is using effluents as a culture medium for growing microalgae to reduce nutrient and water costs, making the process sustainable. *Phaeodactylum tricornutum* cultivation with palm oil production effluent showed promising results concerning biomass and EPS production. The highest concentration of EPS (140 mg L^{-1}) was obtained using a medium with 30% effluent, the addition of urea (100 mg L^{-1}), and salinity of 2.6% at 25°C (Nur et al., 2019).

14.3.2 LIGHT AND TEMPERATURE

Changes in temperature and light supply affect EPS production by microalgae. EPS release can be enhanced by stress conditions such as continuous light supply and high irradiance. Furthermore, the light spectrum influences photosynthetic activity and EPS production by microalgae (Trabelsi et al., 2009; Yu et al., 2010; Huang et al., 2021).

The increase in light intensity from 40 to 80 μE m^{-2} s^{-1} in *Nostoc* sp. resulted in increased EPS yield from 155.5 to 206.2 mg g^{-1}. The variation in light intensity did not cause a significant change in the monosaccharide composition of the produced EPS structure (Ge et al., 2014). The highest light intensity evaluated (60 μmol photon m^{-2} s^{-1}) also stimulated EPS production by *Nostoc flagelliform* (106 mg L^{-1}) (Yu et al., 2010). On the other hand, increasing the light intensity (above 400 μmol m^{-2} s^{-1}) in the culture of *Arthrospira platensis* reduced the uronic acid content. Moreover, this physical parameter modified the composition of neutral monosaccharides, increasing the glucose content and reducing the levels of rhamnose and fucose. These alterations may harm the biological activities of the ESP produced under these cultivation conditions (Phélippé et al., 2019).

The green LED wavelength showed significantly increased EPS synthesis (255.6 mg L^{-1}) by the microalga *Porphyridium runtime*, compared to the white LED wavelengths (118.6 mg L^{-1}), red (125.6 mg L^{-1}) and blue (103.2 mg L^{-1}) (Huang et al., 2021). On the other hand, Han et al. (2014) reported that red and blue light favored the synthesis of EPS by *Nostoc flagelliforme*. Under these conditions, the production of EPS by the microalgae was 1.65 (with red light) and 1.59 (with blue light) times higher than in cultivation under white light.

Regarding temperature, it was found that the lowest evaluated value (20°C) resulted in lower EPS production (56 mg L^{-1}) by *Phaeodactylum tricornutum*. The response surface showed that the best conditions to increase EPS production (140 mg L^{-1}) were 25°C, 2.6% salinity, and 100 mg L^{-1} of urea (Nur et al., 2019). Using 25°C, it was verified EPS maximum production (73 mg L^{-1}) when the microalga *Nostoc flagelliforme* was cultivated at 25°C (Yu et al., 2010). Trabelsi et al. (2009) evaluated the effects of light intensity and temperature on EPS production by *Arthrospira platensis*. The authors reported that the best conditions for EPS production were temperature between 33 and 35°C and light intensity of 180 µmol photons $m^{-2} s^{-1}$, which was the highest intensity tested. Under these conditions, the EPS production was approximately 290 mg L^{-1}.

14.3.3 OTHER

In addition to the strategies presented, others have investigated alternatives to induce EPS synthesis by microalgae. Microwave treatment (100 W power for 2 min) increased EPS production by the microalgae *Scenedesmus* sp. by 2.3 times compared to the test without the application of this treatment. According to the authors, the increase is due to the stress induced by the microwave application (Sivaramakrishnan et al., 2020).

The application of a static magnetic field in the cultivation of *Arthrospira platensis* SAG 21.99 and *Spirulina* sp. LEB 18 modified the composition of the EPS produced. There was an increase in the monosaccharide content, such as fucose, rhamnose, and glucuronic acid. These monosaccharides are related to the biological activities of EPS. However, the application of this treatment did not change the concentration of EPS produced (Deamici et al., 2021). Increasing the intensity of the static magnetic field from 200 to 800 Gs reduced the EPS content released into the medium by the microalgae *Chlorella vulgaris* from 2.02 to 1.49 pg $cell^{-1}$ (Luo et al., 2020).

The two-stage system has shown promising results for EPS production by microalgae. In the first stage, optimal cultivation conditions are used to obtain a high microalgal biomass concentration. In the second stage, the microalgae are subjected to conditions that induce the accumulation of the product of interest. These conditions can be included: change in the composition of the medium, temperature, and luminosity (Ho et al., 2014; Nayak et al., 2019; Medina-Cabrera et al., 2020b). This strategy increased EPS production by 3.4 times for *Porphyridium sordidum* and 1.2 times for *P. purpureum*, concerning the levels obtained in the first stage. Furthermore, the sulfate content was higher in the EPS produced by *P. purpureum* in two-stage. The conditions used in the second stage were a temperature of 20°C and a time of exposure to light of 24 h (Medina-Cabrera et al., 2020b). Two-stage cultivation also promoted EPS production by *Arthrospira* sp. subjected to combined stress factors in the second stage. The high concentration of NaCl (40 g L^{-1}) and low light intensity (10 µmol photons $m^{-2} s^{-1}$) used enhanced EPS production by 1.67 times compared to cultivation under optimal conditions (first stage) (Chentir et al., 2017).

Studies show that stress factors that stimulate the production of biomolecules (carbohydrates and lipids) can also influence EPS production by microalgae. However, the mechanisms by which these factors promote the synthesis of these molecules are still unclear. Furthermore, it has been reported that the effect depends on the genus or species investigated (Mahesh et al., 2019; Medina-Cabrera et al., 2020b; Laroche, 2022).

14.4   PROPERTIES AND COMPOSITION OF MICROALGAL EPS

The properties of microalgal EPS are related to its composition. The profile of monosaccharides and functional groups (carboxyls, sulfates, acids) and the high molecular weight of EPS give viscous character and non-Newtonian fluids (Alvarez et al., 2021; de Jesus et al., 2019). The composition of the EPS is directly related to the microalgae species and cultivation conditions (Costa et al., 2021b). Thus, these compounds have drawn the attention of researchers due to their eco-friendly, nontoxic, technological, and bioactive character (Zhang et al., 2019a).

EPS protects and acclimatizes the cell against biotic and abiotic stresses, creating a viscous medium as a physical barrier to prevent the action of these stress factors on the cell (Naveed et al., 2019). The EPS produced by microalgae are mostly heteropolymers with high molecular weight ($>10^6$ Da), formed by combinations up to 12 different monosaccharides glycosidically linked. This heterogeneous composition of EPS is related to biological activities (antioxidant, antiviral, immunoregulatory) and their potential as hydrocolloids for the food industry. In its composition, neutral sugars, acids, and aminos are found along with functional groups (methyl, acetyl, and sulfate groups) (Delattre et al., 2016; Pierre et al., 2019).

The main monosaccharides reported in EPS are L-rhamnose, fucose, arabinose, D-glucose, D-mannose, xylose, and galactose (Naveed et al., 2019; Babiak and Krzemińska, 2021). These monosaccharides are linked by α-(1–3), β (1–3), β (1–4), and β (2–1) glycosidic bonds forming complex linear or branched configurations resulting in spirals, sheets, and single/triple helix structures (Delattre et al., 2016; Babiak and Krzemińska, 2021; Laroche, 2022). Table 14.1 shows the sugar profiles of microalgal EPS and their properties.

Among the monosaccharides present in EPS are galacturonic and glucuronic acids, which correspond to up to 40% of the polymeric composition of EPS produced by the microalgae *Closterium* sp. (Domozych et al., 1993), *Amphora* sp. (Jin et al., 2018), *Anabaena spiroides* (Gouvêa et al., 2005), *Aphanothece halophytica* (Li et al., 2001), *Arthrospira platensis* (Filali Mouhim et al., 1993), *Cyanospira capsulata* (Vincenzini et al., 1993), and *Fischerella maior* (Bellezza et al., 2006). The presence of these acidic sugars and sulfated groups in the composition of EPS is related to antiviral, antibacterial, antioxidant, and immunoregulatory properties capable of helping skin regeneration and respiratory problems (Barboríková et al., 2019; Alvarez et al., 2021; Costa et al., 2021b). Besides, these compounds have negative charges that give an anionic profile to the polymer that interacts with cationic groups in the medium and increases the viscosity (Laroche, 2022).

The polarity of these acidic sugars, sulfate, and carboxyl groups grant the hydrophilic characteristic to these polymers, while acetyl and methyl ester groups linked to peptides confer hydrophobic character. Therefore, these polysaccharides have an amphipathic profile making them potential biosurfactants, flocculants, thickeners, and emulsifiers (Rahman et al., 2019; Laroche, 2022).

Most studies related to rheological properties were performed on EPS from microalgae of the genus *Rhodella* and *Porphyridium* (Laroche, 2022). These EPS mostly showed characteristics of non-Newtonian fluids (Sun et al., 2009), high

TABLE 14.1

Production, composition, and properties of microalgal EPS

Genus	Species	EPS production (g L⁻¹)	Main composition	Properties	References
Rhodella	*violacea*	0.223–0.595	Rhamnose, galactose, arabinose, **xylose**, glucuronic acid, and galacturonic acid	Anti-microsporidian activity	(Villay et al., 2013; Roussel et al., 2015)
	reticulata	3.6	**Galactose**, glucose, glucuronic acid, and xylose	Antioxidant, antiviral, hypoglycemic, and antitumor activities, viscosity increase	(Arad et al., 1992; Eteshola et al., 1998; Xiao and Zheng, 2016)
Porphyridium	sp.	–	Galactose, glucose, mannose and **xylose**	Thickening agent	(Bernaerts et al., 2019)
	sordidum	0.16–1.91	Galactose, glucose, glucuronic acid and **xylose**	Antioxidant, antiviral, hypocholesterolemic, inmunoregulador, and antitumoral activities	(Medina-Cabrera et al., 2020a; Borjas Esqueda et al., 2022)
	marinum	4.52	Galactose, glucose, glucuronic acid, fucose, and **xylose**	Anti-microsporidian activity	(Soanen et al., 2016)
	purpureum	0.24	**Galactose**, glucose, glucuronic acid, fucose, and xylose	Anti-microsporidian and antiviral activities	(Medina-Cabrera et al., 2020a)
	cruentum	0.18–0.95	Rhamnose, **galactose**, arabinose, glucose, mannose, fucose, and xylose	Antioxidant, antiviral, hypoglycemic, and antitumor activities	(Wang et al., 2021)
Nostoc	sp.	0.2–1.58	Rhamnose, galactose, **glucose**, glucuronic acid, galacturonic acid, arabinose, mannose, and xylose	Cough treatments, bronchodilator, immunoregulator	(Tiwari et al., 2015; Uhliariková et al., 2020)
	linckia	0.04–0.17	Galactose, **glucose**, glucuronic acid, mannose, and xylose	Chelating agent	(Mona and Kaushik, 2015)
Dunaliella	*salina*	0.94	**Fructose**, galactose, glucose, glucuronic acid, galacturonic acid, and xylose	Agricultural biostimulant, vitamins, antioxidant, betacarotene, and immunomodulator	(Mishra and Jha, 2009; El Arroussi et al., 2018; Goyal et al., 2019)

Genus	species		Monosaccharide composition	Property/activity	References
	tertiolecta	—	Rhamnose, galactose, **glucose**, and xylose	Antioxidant property	(Brown, 1991)
Chlorella	*vulgaris*	0.36	Rhamnose, **galactose**, arabinose, glucose, galacturonic, mannose, glucosamine, galactosamine, fucose, and xylose	Antioxidant, antitumor, bioflocculant, and anti-asthmatic activities	(Alam et al., 2014; Mohamed, 2008; Zhang et al., 2019a; Barboříková et al., 2019)
	pyrenoidosa	0.05	**Galactose**, arabinose, glucosamine, rhamnose, glucose mannose, fucose, galactosamine, ribose, and galacturonic acid	Antitumor property	(Liu et al., 2022; Zhang et al., 2019b)
	zofingiensis	0.21	Rhamnose, galactose, arabinose, glucose, glucuronic acid, galacturonic acid, **mannose**, and fucose	Antioxidant and antitumor activities	(Zhang et al., 2019a)
	sp.	0.03–0.129	**Galactose**, glicose, ácido galacturónico e glicurônico, manose, galactosamina, arabinose, ramnose e xilose	Antioxidant and anticoagulant properties	(Jakhu et al., 2021; Mousavian et al., 2022)
Neorhodella	*cyanea*	2.32	Rhamnose, galactose, glucose, glucuronic acid, and **xylose**		(Borjas Esqueda et al., 2022)
Corynoplastis	*japonica*	1.59	Rhamnose, galactose, glucose, glucuronic acid, and **xylose**		(Borjas Esqueda et al., 2022)

Monosaccharide in higher concentration highlighted in bold. -: not available.

viscosity (Bernaerts et al., 2018), and viscoelastic behavior (Medina-Cabrera et al., 2020a). EPS from *Porphyridium* sp. has an anionic character due to glucuronic acid and sulfate ester groups that grant gelling behavior similar to xanthan gum. It also presents high intrinsic viscosity due to its high molecular weight (10^{5}-10^{6} Da), electrostatic density, and its low rate of branching of its chains. All these characteristics support the potential as a thickening agent of these biopolymers (Bernaerts et al., 2018; Medina-Cabrera et al., 2020a). EPS from *Porphyrium* sp. was also stable at temperatures of 25–70°C and pH from 2 to 9, increasing the viscosity and organization of its chains as the temperature was raised (Arad et al., 2006). Alvarez et al. (2021) characterized EPS from *Nostoc* sp. as an anionic pseudoplastic fluid capable of promoting gelation with the addition of $FeCl_3$ for the production of hydrogels.

14.5 RECENT BIOTECHNOLOGICAL APPLICATIONS OF MICROALGAL EPS

14.5.1 MEDICAL AND PHARMACEUTICAL FIELDS

EPS extracted from microalgae have a high potential for application in the pharmaceutical and cosmetic industry due to their physicochemical and functional characteristics (Morais et al., 2022). Studies carried out in vitro and in vivo associate these biocompounds with biological activities, such as antioxidant (Sun et al., 2014; Yingying et al., 2014), anti-inflammatory (Uhliariková et al., 2020), immunomodulatory (Ferreira et al., 2021), antimicrobial (Raposo et al., 2014; Hlima et al., 2021), and antitumor (Zhang et al., 2019b; Zhong et al., 2021).

In this context, Ferreira et al. (2021) evaluated the immunostimulatory activity of EPS extracted from the red microalgae *P. purpureum* grown at different salinities (18, 32, and 50 g L^{-1} NaCl). The stimulation assay was carried out in vitro, using purified lymphocytes from rats and applied in their native and desulfated form. As verified by the authors, the percentage of stimulated cells was significantly higher than in untreated controls. Sulfated EPS also induced greater early activation of tested cells compared to the control assay (77 and 96%, respectively). The EPS produced showed an immunostimulatory effect on lymphocytes in vitro, showing potential for application in increasing immune activity.

EPS from microalgae has the potential for application as an immune modulator. EPS produced by *Nostoc* sp. demonstrated antitussive and bronchodilator effect similar to the antitussive drug codeine and the anti-asthmatic salbutamol. EPSs stimulated macrophages to produce pro-inflammatory cytokines and other immunological factors, such as interleukins, prostaglandins, and nitric oxide, without significant cytotoxicity to the evaluated cells (Uhliariková et al., 2020).

EPS extracted from the microalgae cultivation showed antitumor activity against human colon cancer cells. The results showed that EPS of microalgae *Chlorella pyrenoidosa*, *Chlorococcum* sp., and *Scenedesmus* sp. suppressed HCT116 colony formation by 21.0, 42.8, and 31.3% at a concentration of 1.5 mg mL^{-1}, respectively. Furthermore, at the same concentration of EPS, inhibition of HCT8 colony formation was observed in 11.5, 29.3, and 35.2%, respectively. The EPS obtained also showed significant antitumor effects in both cell lines considering the inhibitory effects on cell viability (between 17.2 and 19.2% in HCT116, and between 22.9, and 38.6% in HCT8) at a concentration of 0.6 mg mL^{-1}. Therefore, the EPS obtained has the potential for antitumor agent use.

Zhang et al. (2019a) evaluated the anti-colorectal cancer activity of EPS extracted from the cultivation of the microalgae *Chlorella zofingiensis* 30412 (EPS-CZ) and *Chlorella vulgaris* UTEX 395 (EPS-CV). The effects of EPS-CZ and EPS-CV were tested on the viability of the HCT8 colon cancer cell line. EPS-CZ and EPS-CV 0.6 mg mL^{-1} showed an inhibitory effect of 28.3 and 18.0% on cell viability, respectively.

Extracted EPS from the microalga *Chlorella* sp. presents antitumor activity against human cervical cancer belonging to the HeLa cells and induction of cell apoptosis. HeLa cells treated with *Chlorella* sp. EPS show morphological changes such as atrophy, morphological abnormalities, and reduced intercellular connections (Zhong et al., 2021). Moreover, EPS from microalgae has been reported for its antimicrobial and antiviral properties. EPS was obtained from the microalga *P. cruentum* grown in three sulfate concentrations in the growing medium. EPS showed antiviral activity against HSV virus types 1 and 2, *Vaccinia* virus, and *Vesicular stomatitis* virus. EPS obtained in medium culture containing 21 mM MgSO$_4$ also showed antibacterial activity against *Salmonella enteritidis* (Raposo et al., 2014). Polysaccharides produced by microalgae are emerging compounds with several biomedical and pharmacological properties. EPS-derived hydrogels have the potential for biomedical applications. New wound dressing was produced from the EPS of the microalgae *Nostoc* sp. The wound dressing obtained by gelling the EPS was evaluated in a wound healing test from scratches. The authors verified that the hydrogels showed biocompatibility and the ability to promote the migration and proliferation of fibroblasts, promoting wound healing (Alvarez et al., 2021). Moreover, the discovery of potential bioactive directed the application of EPS in cosmetics (Caetano et al., 2022). There is a growing interest in developing sustainable and ecologically correct cosmetics. Incorporating functional algae-based compounds into cosmetics has grown considerably, especially from the *Chlorella* and *Arthrospira* genera (Tounsi et al., 2022).

In cosmetic formulations, EPS can be used as a moisturizing agent due to the presence of glucuronic acid. The compound that protects skin against dehydration in dry environments and regulate water content (Yarkent et al., 2020). EPS can also act as surfactants, affecting the rheological properties of cosmetic formulations. The application decreases the surface tension between immiscible components and improves the stability of the gel at different temperatures, preventing the precipitation of compounds (Gansbiller et al., 2020). According to Caetano et al. (2022), using EPS as a bioactive ingredient also needs to be better explored in research for better application. The polysaccharide's bioactivities are directly related to their complex structural characteristics and mechanisms of action.

14.5.2 FOOD AND AGRICULTURE

There is a trend in the search for food supplements containing bioactive compounds to prevent and treat chronic diseases. Microalgae EPS have the potential for use in the development of innovative food with anti-inflammatory, immunomodulatory, antioxidant, antilipidemic, and hypoglycemic properties (Zaitseva et al., 2022).

The presence of non-sugar substituents in microalgae EPS—such as sulfate, pyruvate, methyl, and acetyl groups—confers biological activities (antioxidant and antimicrobial) for food preservation (Morais et al., 2022). Sulfated EPS obtained from the *P. cruentum* strain presents a high potential to be used in ground beef to improve microbiological quality and primary and secondary lipid oxidation. This potential is related to antimicrobial and antioxidant activities, extending foods' shelf life (Hlima et al., 2021).

The chemical and rheological characteristics of EPS allow their use in the food industry as a functional ingredient. EPS helps regulate texture, prevent syneresis, stabilize high-fat foods, and improve shelf life (Yuan et al., 2018). EPS from *P. cruentum* showed functional, foaming, and emulsion characteristics, demonstrating a vast potential for application in the food industry (Hlima et al., 2021).

EPS also has potential application in agriculture as a biostimulant. Applying EPS for this purpose is a friendly alternative to replace the use of petrochemical products that present toxicity and non-biodegradable properties. EPS produced by *Gloeothece verrucosa* can be applied as foliar spraying on tomato (*Arabidopsis thalianae*), increasing the activity of components involved in the plant's defense mechanisms, such as phenylalanine ammonia-lyase, free proline, and anthocyanin content (Camp et al., 2022). Sulfated EPS extracted from the microalgae *Porphyridium sordidum* can stimulate the natural defense system of date palms in vitro plants, accumulating H_2O_2—whose release occurs in response to cellular defense—and expression of genes associated with the plant defense system (PR1, SOD, PAL, and WRKY genes) (Drira et al., 2022). The foliar application of EPS can increase the viability of several plants, allowing greater production efficiency and acting as an alternative to conventional chemical products.

14.5.3 OTHER

In addition to pharmaceutical and food industries, EPS may have other industrial applications, such as wastewater treatment (Cheirsilp and Maneechote, 2022; Morais et al., 2022; Priya et al., 2022), stabilization of nanoemulsions, and obtaining nanostructures (Morais et al., 2022). Furthermore, EPS from microalgae is promising in alternative fuel synthesis. EPS from *Synechocystis* sp. (PCC 6803) extracted with surfactants improves subsequent hydrolysis and fermentation processes, allowing higher yields of fermentable sugar for bioethanol production (Velmurugan et al., 2022).

Some of the EPS secreted by microalgae have groups that allow the adsorption of pollutants, helping to remediate heavy metals from wastewater (Cheirsilp and Maneechote, 2022). The uronic groups present in these biopolymers can bind to heavy metals, forming covalent bonds or adhesion of heavy metal cations to negatively charged uronic acids (Cheirsilp and Maneechote, 2022; Priya et al., 2022).

Nanotechnology enables the production of innovative products with better thermal stability, solubility, and bioavailability (Sahani and Sharma, 2020). *Chlorella vulgaris* EPS immobilized on nanoparticles demonstrated a high ability for application in wastewater treatment (Govarthanan et al., 2020). Microalgal EPS also can be applied to develop packaging materials, food additives, and nanosensors

(Kumar et al., 2020). Gallón et al. (2019) used EPS extracted from the microalgae *Botryococcus braunii* and *Chlorella pyrenoidosa* as a polymeric solution for nanotechnological applications. The authors produced silver nanoparticles with antimicrobial properties against gram-positive (*Staphylococcus aureus*), gram-negative (*Escherichia coli*), and antibiotic-resistant (methicillin-resistant *Staphylococcus aureus*). The nanoparticles showed dose-dependent antibacterial activity against all strains tested. Through cytotoxicity assays, the authors did not verify damage to fibroblasts in the same tested concentration range (0.5–10.0 µg mL^{-1}). The study demonstrates the potential nanotechnological use of EPS.

14.6   POTENTIALITY OF MICROALGAL EPS ON THE MARKET

The biological activities of microalgae EPS offer a wide range of market potential in diverse sectors such as food, pharmaceutical, medical, agro-industrial, and biotechnology (de Jesus et al., 2019). The antioxidant, antitumor, immunomodulatory, and bio-stimulating properties make these microalgal biocompounds a promising approach to the existing market (Colusse et al., 2022) (Figure 14.1).

The market for products developed from microalgae has expanded over the years, and the global market for algal products is estimated to be around US\$44.6 billion by 2023 (Colusse et al., 2022). The industrial exploration of EPS has been carried out with a focus on chemical and rheological characterizations. The increase in international market demand for natural hydrocolloids and biological agents contributes to expanding the production of EPS on a large scale (Pierre et al., 2019). Among the commercially available microalgal EPS are Spirulan from *Arthrospira platensis*, Immulan from *Aphanotece halophytica*, Nostoflan from *Nostoc flagelliform*, and Emulcyan from *Phormidium* (Cruz et al., 2020).

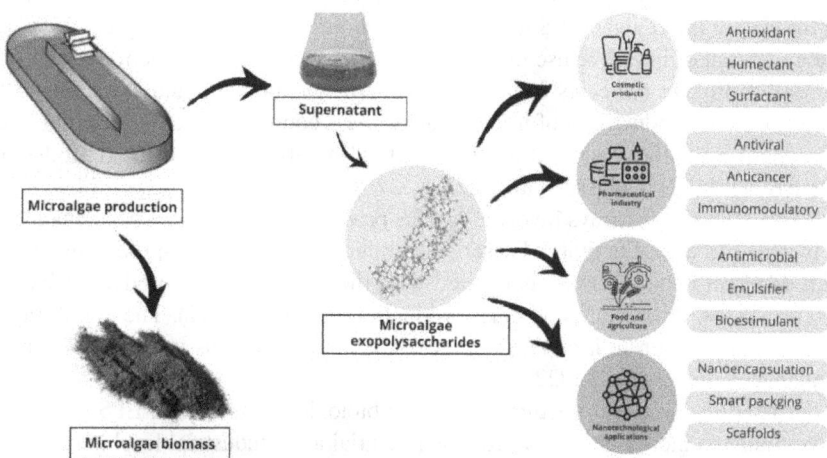

FIGURE 14.1   Illustrative representation of EPS obtained from microalgae cultivation and their potential applications

The industrial economic viability of EPS production is related to the selection of strains, the configuration of photobioreactors, and the technologies used in the downstream process. The *Arthrospira* genus is recognized worldwide for nutritional purposes. Among the biocompounds constituted in the biomass of *Arthrospira* species is the pigment phycocyanin, widely used as a natural dye. The extraction of phycocyanin leads to the production of co-products such as polysaccharides. In this way, the industrial production of Spirulan as a co-product of phycocyanin becomes possible and profitable (Pierre et al., 2019).

P. purpureum is a red microalga well-known for excreting EPS. Silidine® is a mixture of trace elements and small EPS commercially available and comes from the red microalgae *P. purpureum*. Silidine® is an ingredient cosmetic sold to combat redness and heavy legs syndrome. Moreover, the photobioreactor is crucial for innovating and drastically reducing the culture costs by applying oxidative stress to microalgae by tightly closing the air in a 1 m^3 culture batch. The authors also mention Epsiline®, another EPS from *P. purpureum*. This commercial EPS enhances the tanning effect. Concerning economic viability, with high-producing strains and optimized conditions of nutrients, luminosity, and carbon dioxide injection is possible a reduction of 1/3 of the time for obtaining this EPS. The selling price of these two active cosmetic ingredients is around €100/kg, and the cultivation step appears to represent ¼ of the EPS price (Pierre et al., 2019). *Porphyridium* EPS are exploited in the cosmetics segment, including Frutarom (www.frutarom.com), Algosource Technology (www.algosource.com), and Microperi Blue Growth (http://www.micoperibg.com), Greensea (http://greensea. fr/en) (Prybylski et al., 2020).

14.7 CHALLENGES

The main challenges of EPS industrial production by microalgae are the operational costs for obtaining biomass and extracting the biomolecule. The limited information on the cost of microalgae EPS is related to the bottlenecks of large-scale cultivation. The reduction of intensive use of energy for biomass cultivation, recovery, and drying will minimize the costs of producing microalgae biomass and extracting EPS. Moreover, low productivity of microalgal biomass compared to other microbial fermentations is another contributing factor to the limitation of commercial production of microalgal EPS (Colusse et al., 2022).

The metabolic pathways involved in EPS production have not yet been elucidated to increase EPS productivity. There is a limited number of EPS-producing strains with different behavior that, associated with the still expensive production/extraction processes, limit economic mass production (Laroche, 2022). The possibility of cross-contamination in large-scale microalgae culture systems is also considered a challenge (Colusse et al., 2022).

There is also a need to better explore the biological activities of EPS through in vivo studies to increase their production potential at an industrial level (Chen et al., 2019). Furthermore, most EPS have been described only by their monosaccharide composition. The complexity of studies related to the structural aspects of EPS limits their elucidation (Laroche, 2022). The genes that encode the proteins involved in EPS biosynthesis are not well elucidated, nor are the aspects that interfere with structural stability. These factors limit the commercialization of microalgal EPS in several areas (Parwani et al., 2021).

14.8 CONCLUSIONS

The biological and rheological properties of EPS from microalgae clarify the high potential of these compounds for the existing market in several sectors, including cosmetic, medical, pharmaceutical, food, and agriculture. Although some EPS is available in the cosmetics market, using EPS from microalgae in other industrial sectors still deserves further investigation to elucidate the mechanisms of action. EPS production by microalgae highly depends on the culture medium and parameters such as light and temperature. Therefore, conducting more research to optimize microalgae cultivation conditions and extraction procedures will promote greater productivity and yield of EPS.

Microalgae can use industrial effluents, agro-food wastes, and several alternative sources of nutrients for their growth and EPS production. Thus, by minimizing nutrient costs and reducing costs with techniques applied for the extraction and recovery of EPS, the production of microalgal EPS will contribute to expanding the biotechnology sector sustainably and economically. In addition, inserting integrated systems such as those applied in biorefineries is another alternative to reduce costs and increase the competitiveness and economic viability of EPS production by microalgae.

ACKNOWLEDGMENTS

This research was developed within the scope of the Capes-PrInt Program (Process # 88887.310848/2018–00). The authors also are grateful to the Coordenação de Aperfeiçoamento de Pessoal de Nível Superior—Brasil (CAPES)—Finance Code 001.

REFERENCES

Abdullahi, A.S., Underwood, G.J., Gretz, M.R., 2006. Extracellular matrix assembly in diatoms (bacillariophyceae). v. environmental effects on polysaccharide synthesis in the model diatom, *Phaeodactylum tricornutum*. J. Phycol. 42, 363–378.

Alam, M.A., Wan, C., Guo, S.L., Zhao, X.Q., Huang, Z.Y., Yang, Y.L., Chang, J.S., Bai, F.W., 2014. Characterization of the flocculating agent from the spontaneously flocculating microalga *Chlorella vulgaris* JSC-7. J. Biosci. Bioeng. 118, 29–33.

Ali, S., Paul Peter, A., Chew, K.W., Munawaroh, H.S.H., Show, P.L., 2021. Resource recovery from industrial effluents through the cultivation of microalgae: A review. Bioresour. Technol. 337, 125461.

Alvarez, X., Alves, A., Ribeiro, M.P., Lazzari, M., Coutinho, P., Otero, A., 2021. Biochemical characterization of *Nostoc* sp. exopolysaccharides and evaluation of potential use in wound healing. Carbohydr. Polym. 254, 117303.

Arad, S. (Malis), Lerental, Y. (Brown), Dubinsky, O., 1992. Effect of nitrate and sulfate starvation on polysaccharide formation in *Rhodella reticulata*. Bioresour. Technol. 42, 141–148.

Arad, S. (Malis), Rapoport, L., Moshkovich, A., van Moppes, D., Karpasas, M., Golan, R., Golan, Y., 2006. Superior biolubricant from a species of red microalga. Langmuir. 22, 7313–7317.

Babiak, W., Krzemińska, I., 2021. Extracellular polymeric substances (EPS) as microalgal bioproducts: A review of factors affecting EPS synthesis and application in flocculation processes. Energies. 14, 4007.

Bafana, A., 2013. Characterization and optimization of production of exopolysaccharide from *Chlamydomonas reinhardtii*. Carbohydr. Polym. 95, 746–752.

Barboríková, J., Šutovská, M., Kazimierová, I., Jošková, M., Fraňová, S., Kopecký, J., Capek, P., 2019. Extracellular polysaccharide produced by *Chlorella vulgaris*—Chemical characterization and anti-asthmatic profile. Int. J. Biol. Macromol. 135, 1–11.

Bellezza, S., Albertano, P., de Philippis, R., Paradossi, G., 2006. Exopolysaccharides of two cyanobacterial strains from Roman hypogea. Geomicrobiol. J. Taylor & Francis Group, 301–310.

Bernaerts, T.M.M., Gheysen, L., Foubert, I., Hendrickx, M.E., Van Loey, A.M., 2019. The potential of microalgae and their biopolymers as structuring ingredients in food: A review. Biotechnol. Adv. 37, 107419.

Bernaerts, T.M.M., Kyomugasho, C., Van Looveren, N., Gheysen, L., Foubert, I., Hendrickx, M.E., Van Loey, A.M., 2018. Molecular and rheological characterization of different cell wall fractions of *Porphyridium cruentum*. Carbohydr. Polym. 195, 542–550.

Borjas Esqueda, A., Gardarin, C., Laroche, C., 2022. Exploring the diversity of red microalgae for exopolysaccharide production. Mar. Drugs. 20, 246.

Brown, M.R., 1991. The amino-acid and sugar composition of 16 species of microalgae used in mariculture. J. Exp. Mar. Bio. Ecol. 145, 79–99.

Caetano, P.A., Tatiele, C., Fernandes, S., Nass, P.P., Vieira, K.R., Mar, M.R., Jacob-lopes, E., Zepka, L.Q., 2022. Microalgae-based polysaccharides: Insights on production, applications, analysis, and future challenges. Biocat. Agric. Biotech. 45, 102491.

Camp, C. Van, Fraikin, C., Claverie, E., Onderwater, R., Wattiez, R., 2022. Capsular polysaccharides and exopolysaccharides from *Gloeothece verrucosa* under various nitrogen regimes and their potential plant defence stimulation activity. Algal Res. 64, 102680.

Cheirsilp, B., Maneechote, W., 2022. Insight on zero waste approach for sustainable microalgae biorefinery: Sequential fractionation, conversion and applications for high-to-low value-added products. Bioresour. Technol. Rep. 18, 101003.

Chen, Y., Sui, L., Fang, H., Ding, C., Li, Z., Jiang, S., Hou, H., 2019. Superior mechanical enhancement of epoxy composites reinforced by polyimide nanofibers via a vacuum-assisted hot-pressing. Compos. Sci. Technol. 174, 20–26.

Chentir, I., Hamdi, M., Doumandji, A., HadjSadok, A., Ouada, H. Ben, Nasri, M., Jridi, M., 2017. Enhancement of extracellular polymeric substances (EPS) production in *Spirulina* (*Arthrospira* sp.) by two-step cultivation process and partial characterization of their polysaccharidic moiety. Int. J. Biol. Macromol. 105, 1412–1420.

Colusse, G.A., Carneiro, J., Duarte, M.E.R., Carvalho, J.C., Noseda, M.D., 2022. Advances in microalgal cell wall polysaccharides: A review focused on structure, production, and biological application. Crit. Rev. Biotechnol. 42(4), 562–577.

Costa, J.A.V., Cruz, C.G., Rosa, A.P.C. da, 2021a. Insights into the technology utilized to cultivate microalgae in dairy effluents. Biocatal. Agric. Biotechnol. 35, 102106.

Costa, J.A.V., Lucas, B.F., Alvarenga, A.G.P., Moreira, J.B., de Morais, M.G., 2021b. Microalgae polysaccharides: An overview of production, characterization, and potential applications. Polysaccharides. 2, 759–772.

Cruz, D., Vasconcelos, V., Pierre, G., Michaud, P., Delattre, C., 2020. Exopolysaccharides from cyanobacteria: Strategies for bioprocess development. Appl. Sci. 10, 3763.

Deamici, K.M., Morais, M.G., Santos, L.O., Muylaert, K., Gardarin, C., Costa, J.A.V., Laroche, C., 2021. Static magnetic fields effects on polysaccharides production by different microalgae strains. Appl. Sci. 11, 5299.

de Jesus, C.S., de Jesus Assis, D., Rodriguez, M.B., Menezes Filho, J.A., Costa, J.A.V., de Souza Ferreira, E., Druzian, J.I., 2019. Pilot-scale isolation and characterization of extracellular polymeric substances (EPS) from cell-free medium of *Spirulina* sp. LEB-18 cultures under outdoor conditions. Int. J. Biol. Macromol. 124, 1106–1114.

Delattre, C., Pierre, G., Laroche, C., Michaud, P., 2016. Production, extraction and characterization of microalgal and cyanobacterial exopolysaccharides. Biotechnol. Adv. 34, 1159–1179.

De Philippis, R., Vincenzini, M., 1998. Exocellular polysaccharides from cyanobacteria and their possible applications. FEMS Microbiol. Rev. 22, 151–175.

Domozych, C.R., Plante, K., Blais, P., Paliulis, L., Domozych, D.S., 1993. Mucilage processing and secretion in the green alga *Closterium*. I. Cytology and biochemistry1. J. Phycol. 29, 650–659.

Drira, M., Elleuch, J., Hadjkacem, F., Hentati, F., Drira, R., Pierre, G., Gardarin, C., Delattre, C., El Alaoui-Talibi, Z., El Modafar, C., Michaud, P., Abdelkafi, S., Fendri, I., 2022. Influence of the sulfate content of the exopolysaccharides from *Porphyridium sordidum* on their elicitor activities on date palm vitroplants. Plant Physiol. Biochem. 186, 99–106.

El Arroussi, H., Benhima, R., Elbaouchi, A., Sijilmassi, B., El Mernissi, N., Aafsar, A., Meftah-Kadmiri, I., Bendaou, N., Smouni, A., 2018. *Dunaliella salina* exopolysaccharides: A promising biostimulant for salt stress tolerance in tomato (*Solanum lycopersicum*). J. Appl. Phycol. 30, 2929–2941.

Esqueda, A.B., Gardarin, C., Laroche, C., 2022. Exploring the diversity of red microalgae for exopolysaccharide production. Mar. Drugs. 20, 246.

Eteshola, E., Karpasas, M., Arad, S., Gottlieb, M., 1998. Red microalga exopolysaccharides: 2. Study of the rheology, morphology and thermal gelation of aqueous preparations. Acta Polym. 49, 549–556.

Ferreira, A.S., Mendonça, I., Póvoa, I., Carvalho, H., Correia, A., Vilanova, M., Silva, T.H., Coimbra, M.A., Nunes, C., 2021. Impact of growth medium salinity on galactoxylan exopolysaccharides of *Porphyridium purpureum*. Algal Res. 59, 102439.

Filali Mouhim, R., Cornet, J.F., Fontane, T., Fournet, B., Dubertret, G., 1993. Production, isolation and preliminary characterization of the exopolysaccharide of the cyanobacterium *Spirulina platensis*. Biotechnol. Lett. 156(15), 567–572.

Gallón, S.M.N., Alpaslan, E., Wang, M., Larese-Casanova, P., Londoño, M.E., Atehortúa, L., Pavón, J.J., Webster, T.J., 2019. Characterization and study of the antibacterial mechanisms of silver nanoparticles prepared with microalgal exopolysaccharides. Mater. Sci. Eng. C. 99, 685–695.

Gansbiller, M., Schmid, J., Sieber, V., 2020. Rheology of sphingans in EPS—surfactant systems. Carbohydr. Polym. 248, 116778.

Ge, H., Xia, L., Zhou, X., Zhang, D., Hu, C., 2014. Effects of light intensity on components and topographical structures of extracellular polysaccharides from the cyanobacteria *Nostoc* sp. J. Microbiol. 52, 179–183.

Gohara-Beirigo, A.K., Matsudo, M.C., Cezare-Gomes, E.A., Carvalho, J.C.M. de, Danesi, E.D.G., 2022. Microalgae trends toward functional staple food incorporation: Sustainable alternative for human health improvement. Trends Food Sci. Technol. 125, 185–199.

Gouvêa, S.P., Vieira, A.A.H., Lombardi, A.T., 2005. Copper and cadmium complexation by high molecular weight materials of dominant microalgae and of water from a eutrophic reservoir. Chemosphere. 60, 1332–1339.

Govarthanan, M., Jeon, C.H., Jeon, Y.H., Kwon, J.H., Bae, H., Kim, W., 2020. Non-toxic nano approach for wastewater treatment using *Chlorella vulgaris* exopolysaccharides immobilized in iron-magnetic nanoparticles. Int. J. Biol. Macromol. 162, 1241–1249.

Goyal, M., Baranwal, M., Pandey, S.K., Reddy, M.S., 2019. Hetero-polysaccharides secreted from dunaliella salina exhibit immunomodulatory activity against peripheral blood mononuclear cells and RAW 264.7 macrophages. Indian J. Microbiol. 59, 428–435.

Han, P.P., Sun, Y., Jia, S.R., Zhong, C., Tan, Z.L., 2014. Effects of light wavelengths on extracellular and capsular polysaccharide production by *Nostoc flagelliforme*. Carbohydr. Polym. 105, 145–151.

Hlima, H.B., Smaoui, S., Barkallah, M., Elhadef, K., Tounsi, L., Michaud, P., Fendri, I., Abdelkafi, S., 2021. Sulfated exopolysaccharides from *Porphyridium cruentum*: A useful strategy to extend the shelf life of minced beef meat. Int. J. Biol. Macromol. 193, 1215–1225.

Ho, S.H., Ye, X., Hasunuma, T., Chang, J.S., Kondo, A., 2014. Perspectives on engineering strategies for improving biofuel production from microalgae—a critical review. Biotechnol. Adv. 32, 1448–1459.

Huang, Z., Zhong, C., Dai, J., Li, S., Zheng, M., He, Y., Wang, M., Chen, B., 2021. Simultaneous enhancement on renewable bioactive compounds from *Porphyridium cruentum* via a novel two-stage cultivation. Algal Res. 55, 102270.

Ighalo, J.O., Dulta, K., Kurniawan, S.B., Omoarukhe, F.O., Ewuzie, U., Eshiemogie, S.O., Ojo, A.U., Abdullah, S.R.S., 2022. Progress in microalgae application for CO_2 sequestration. Clean. Chem. Eng. 3, 100044.

Jakhu, S., Sharma, Y., Sharma, K., Vaid, K., Dhar, H., Kumar, V., Singh, R.P., Shekh, A., Kumar, G., 2021. Production and characterization of microalgal exopolysaccharide as a reducing and stabilizing agent for green synthesis of gold-nanoparticle: A case study with a *Chlorella* sp. from Himalayan high-altitude psychrophilic habitat. J. Appl. Phycol. 33, 3899–3914.

Jin, C., Yu, Z., Peng, S., Feng, K., Zhang, L., Zhou, X., 2018. The characterization and comparison of exopolysaccharides from two benthic diatoms with different biofilm formation abilities. An. Acad. Bras. Cienc. 90, 1503–1519.

Kim, T.H., Lee, K., Oh, B.R., Lee, M.E., Seo, M., Li, S., Kim, J.K., Choi, M., Chang, Y.K., 2021. A novel process for the coproduction of biojet fuel and high-value polyunsaturated fatty acid esters from heterotrophic microalgae *Schizochytrium* sp. ABC101. Renew. Energy. 165, 481–490.

Kumar, P., Mahajan, P., Kaur, R., Gautam, S., 2020. Nanotechnology and its challenges in the food sector: A review. Mater. Today Chem. 17, 100332.

Kumar, Y., Kaur, S., Kheto, A., Munshi, M., Sarkar, A., Om Pandey, H., Tarafdar, A., Sindhu, R., Sirohi, R., 2022. Cultivation of microalgae on food waste: Recent advances and way forward. Bioresour. Technol. 363, 127834.

Laroche, C., 2022. Exopolysaccharides from microalgae and cyanobacteria: Diversity of strains, production strategies, and applications. Mar. Drugs. 20, 336.

Li, P., Liu, Z., Xu, R., 2001. Chemical characterisation of the released polysaccharide from the cyanobacterium *Aphanothece halophytica* GR02. J. Appl. Phycol. 131(13), 71–77.

Liu, J., Han, X., Xing, H., Nan, Y., Lin, J., He, J., Chen, S., Wei, Y., Guo, P., 2022. Effects of suspended particles on exopolysaccharide secretion of two microalgae in jinjiang estuary (Fujian, China). J. Mar. Sci. Eng. 10.

Luo, X., Zhang, H., Li, Q., Zhang, J., 2020. Effects of static magnetic field on *Chlorella vulgaris*: Growth and extracellular polysaccharide (EPS) production. J. Appl. Phycol. 32, 2819–2828.

Ma, Z., Cheah, W.Y., Ng, I.S., Chang, J.S., Zhao, M., Show, P.L., 2022. Microalgae-based biotechnological sequestration of carbon dioxide for net zero emissions. Trends Biotechnol. 40, 1439–1453.

Mahesh, R., Naira, V.R., Maiti, S.K., 2019. Concomitant production of fatty acid methyl ester (biodiesel) and exopolysaccharides using efficient harvesting technology in flat panel photobioreactor with special sparging system via *Scenedesmus abundans*. Bioresour. Technol. 278, 231–241.

Medina-Cabrera, E.V., Rühmann, B., Schmid, J., Sieber, V., 2020a. Characterization and comparison of Porphyridium sordidum and *Porphyridium purpureum* concerning growth characteristics and polysaccharide production. Algal Res. 49, 101931.

Medina-Cabrera, E.V., Rühmann, B., Schmid, J., Sieber, V., 2020b. Optimization of growth and EPS production in two *Porphyridum* strains. Bioresour. Technol. Rep. 11, 100486.

Mishra, A., Jha, B., 2009. Isolation and characterization of extracellular polymeric substances from microalgae *Dunaliella salina* under salt stress. Bioresour. Technol. 100, 3382–3386.

Mohamed, Z.A., 2008. Polysaccharides as a protective response against microcystin- induced oxidative stress in *Chlorella vulgaris* and *Scenedesmus quadricauda* and their possible significance in the aquatic ecosystem. Ecotoxicology. 17(6), 504–516.

Mona, S., Kaushik, A., 2015. Chromium and cobalt sequestration using exopolysaccharides produced by freshwater cyanobacterium Nostoc linckia. Ecol. Eng. 82, 121–125.

Morais, M.G., Santos, T.D., Moraes, L., Vaz, B.S., Morais, E.G., Costa, J.A.V., 2022. Exopolysaccharides from microalgae: Production in a biorefinery framework and potential applications. Bioresour. Technol. Rep. 18, 101006.

Moreira, J.B., Kuntzler, S.G., Bezerra, P.Q.M., Cassuriaga, A.P.A., Zaparoli, M., Silva, J.L.V., Costa, J.A.V., Morais, M.G., 2022a. Recent advances of microalgae exopolysaccharides for application as bioflocculants. Polysaccharides. 3, 264–276.

Moreira, J.B., Vaz, B.S., Cardias, B.B., Cruz, C.G., Almeida, A.C.A., Costa, J.A.V., Morais, M.G., 2022b. Microalgae polysaccharides: An alternative source for food production and sustainable agriculture. Polysaccharides. 3, 441–457.

Mousavian, Z., Safavi, M., Azizmohseni, F., Hadizadeh, M., Mirdamadi, S., 2022. Characterization, antioxidant and anticoagulant properties of exopolysaccharide from marine microalgae. AMB Express. 12, 1–16.

Naveed, S., Li, C., Lu, X., Chen, S., Yin, B., Zhang, C., Ge, Y., 2019. Microalgal extracellular polymeric substances and their interactions with metal(loid)s: A review. Crit. Rev. Environ. Sci. Technol. 49, 1769–1802.

Nayak, M., Suh, W.I., Chang, Y.K., Lee, B., 2019. Exploration of two-stage cultivation strategies using nitrogen starvation to maximize the lipid productivity in *Chlorella* sp. HS2. Bioresour. Technol. 276, 110–118.

Nur, M.M.A., Swaminathan, M.K., Boelen, P., Buma, A.G.J., 2019. Sulfated exopolysaccharide production and nutrient removal by the marine diatom *Phaeodactylum tricornutum* growing on palm oil mill effluent. J. Appl. Phycol. 31, 2335–2348.

Oliveira, C.Y.B., Jacob, A., Nader, C., Oliveira, C.D.L., Matos, Â.P., Araújo, E.S., Shabnam, N., Ashok, B., Gálvez, A.O., 2022. An overview on microalgae as renewable resources for meeting sustainable development goals. J. Environ. Manage. 320.

Parwani, L., Bhatt, M., Singh, J., 2021. Potential Biotechnological applications of cyanobacterial exopolysaccharides. Braz. Arch. Biol. Technol. 64, e21200401.

Phélippé, M., Gonçalves, O., Thouand, G., Cogne, G., Laroche, C., 2019. Characterization of the polysaccharides chemical diversity of the cyanobacteria *Arthrospira platensis*. Algal Res. 38, 101426.

Pierre, G., Delattre, C., Dubessay, P., Jubeau, S., Vialleix, C., Cadoret, J.-P., Probert, I., Michaud, P., 2019. What is in store for eps microalgae in the next decade? Molecules. 24, 4296.

Priya, A.K., Jalil, A.A., Vadivel, S., Dutta, K., Rajendran, S., Fujii, M., Soto-Moscoso, M., 2022. Heavy metal remediation from wastewater using microalgae: Recent advances and future trends. Chemosphere. 305, 135375.

Prybylski, N., Toucheteau, C., Alaoui, H., Bridiau, N., Maugard, T., Abdelkafi, S., Fendri, I., Delattre, C., Dubessay, P., Pierre, G., Michaud, P., 2020. Bioactive polysaccharides from microalgae. In: *Handbook of Microalgae-Based Processes and Products: Fundamentals and Advances in Energy, Food, Feed, Fertilizer, and Bioactive Compounds.* Academic Press, pp. 533–571.

Rahman, P.K.S.M., Mayat, A., Harvey, J.G.H., Randhawa, K.S., Relph, L.E., Armstrong, M.C., 2019. Biosurfactants and Bioemulsifiers from Marine Algae. Role Microalgae Wastewater Treat. 169–188.

Raja, R., Coelho, A., Hemaiswarya, S., Kumar, P., Carvalho, I.S., Alagarsamy, A., 2018. Applications of microalgal paste and powder as food and feed: An update using text mining tool. Beni-Suef Univ. J. Basic Appl. Sci. 7, 740–747.

Raposo, M.F.J., De Morais, A.M.B., De Morais, R.M.S.C., 2014. Influence of sulphate on the composition and antibacterial and antiviral properties of the exopolysaccharide from *Porphyridium cruentum.* Life Sci. 101, 56–63.

Raposo, M.F.J., De Morais, A.M.B., De Morais, R.M.S.C., 2015. Marine polysaccharides from algae with potential biomedical applications. Mar. Drugs. 13, 2967–3028.

Roussel, M., Villay, A., Delbac, F., Michaud, P., Laroche, C., Roriz, D., El Alaoui, H., Diogon, M., 2015. Antimicrosporidian activity of sulphated polysaccharides from algae and their potential to control honeybee nosemosis. Carbohydr. Polym. 133, 213–220.

Sahani, S., Sharma, Y.C., 2020. Advancements in applications of nanotechnology in global food industry. Food Chem. 342, 128318.

Sivaramakrishnan, R., Suresh, S., Pugazhendhi, A., Pauline, J.M.N., Incharoensakdi, A., 2020. Response of *Scenedesmus* sp. to microwave treatment: Enhancement of lipid, exopolysaccharide and biomass production. Bioresour. Technol. 312, 123562.

Soanen, N., Silva, E., Gardarin, C., Michaud, P., Laroche, C., 2016. Improvement of exopoly-saccharide production by *Porphyridium marinum.* Bioresour. Technol. 213, 231–238.

Sun, L., Wang, C., Shi, Q., Ma, C., 2009. Preparation of different molecular weight poly-saccharides from *Porphyridium cruentum* and their antioxidant activities. Int. J. Biol. Macromol. 45, 42–47.

Sun, L., Wang, L., Li, J., Liu, H., 2014. Characterization and antioxidant activities of degraded polysaccharides from two marine *Chrysophyta.* Food Chem. 160, 1–7.

Tiwari, O.N., Khangembam, R., Shamjetshabam, M., Sharma, A.S., Oinam, G., Brand, J.J., 2015. Characterization and optimization of bioflocculant exopolysaccharide production by *Cyanobacteria nostoc* sp. BTA97 and *Anabaena* sp. BTA990 in culture conditions. Appl. Biochem. Biotechnol. 176, 1950–1963.

Tounsi, L., Hentati, F., Ben, H., Barkallah, M., Smaoui, S., 2022. Microalgae as feedstock for bioactive polysaccharides. Int. J. Biol. Macromol. 221, 1238–1250.

Trabelsi, L., Ben Ouada, H., Bacha, H., Ghoul, M., 2009. Combined effect of temperature and light intensity on growth and extracellular polymeric substance production by the cyanobacterium *Arthrospira platensis.* J. Appl. Phycol. 21, 405–412.

Uhliariková, I., Šutovská, M., Barboríková, J., Molitorisová, M., Kim, H.J., Park, Y. Il, Matulová, M., Lukavský, J., Hromadková, Z., Capek, P., 2020. Structural character-istics and biological effects of exopolysaccharide produced by cyanobacterium *Nostoc* sp. Int. J. Biol. Macromol. 160, 364–371.

Velmurugan, R., Kanwal, S., Incharoensakdi, A., 2022. Non-ionic surfactant integrated extraction of exopolysaccharides from engineered *Synechocystis* sp. PCC 6803 under fed-batch mode facilitates the sugar-rich syrup production for ethanol fermentation. Algal Res. 66, 102772.

Villar-Navarro, E., Ruiz, J., Garrido-Pérez, C., Perales, J.A., 2022. Microalgae biotechnology for simultaneous water treatment and feed ingredient production in aquaculture. J. Water Process Eng. 49, 103115.

Villay, A., Laroche, C., Roriz, D., El Alaoui, H., Delbac, F., Michaud, P., 2013. Optimisation of culture parameters for exopolysaccharides production by the microalga *Rhodella violacea*. Bioresour. Technol. 146, 732–735.

Vincenzini, M., De Philippis, R., Sili, C., Materassi, R., 1993. Stability of molecular and rheological properties of the exopolysaccharide produced by *Cyanospira capsulata* cultivated under different growth conditions. J. Appl. Phycol. 55(5), 539–541.

Wang, W.N., Li, Y., Zhang, Y., Xiang, W.Z., Li, A.F., Li, T., 2021. Comparison on characterization and antioxidant activity of exopolysaccharides from two *Porphyridium* strains. J. Appl. Phycol. 33, 2983–2994.

Xiao, R., Zheng, Y., 2016. Overview of microalgal extracellular polymeric substances (EPS) and their applications. Biotechnol. Adv. 34, 1225–1244.

Yarkent, Ç., Gürlek, C., Oncel, S.S., 2020. Potential of microalgal compounds in trending natural cosmetics: A review. Sustain. Chem. Pharm. 17, 100304.

Yin, H.C., Sui, J.K., Han, T.L., Liu, T.Z., Wang, H., 2022. Integration bioprocess of B-phycoerythrin and exopolysaccharides production from photosynthetic microalga *Porphyridium cruentum*. Front. Mar. Sci. 8, 836370.

Yingying, S., Hui, W., Ganlin, G., Yinfang, P., Binlun, Y., 2014. The isolation and antioxidant activity of polysaccharides from the marine microalgae *Isochrysis galbana*. Carbohydr. Polym. 113, 22–31.

Yu, H., Jia, S., Dai, Y., 2010. Accumulation of exopolysaccharides in liquid suspension culture of *Nostoc flagelliforme* cells. Appl. Biochem. Biotechnol. 160, 552–560.

Yuan, Y., Xu, X., Jing, C., Zou, P., Zhang, C., Li, Y., 2018. Microwave assisted hydrothermal extraction of polysaccharides from *Ulva prolifera*: Functional properties and bioactivities. Carbohydr. Polym. 181, 902–910.

Zaitseva, O.O., Sergushkina, M.I., Khudyakov, A.N., Polezhaeva, T.V., Solomina, O.N., 2022. Seaweed sulfated polysaccharides and their medicinal properties. Algal Res. 68, 102885.

Zhang, J., Liu, L., Chen, F., 2019a. Production and characterization of exopolysaccharides from *Chlorella zofingiensis* and *Chlorella vulgaris* with anti-colorectal cancer activity. Int. J. Biol. Macromol. 134, 976–983.

Zhang, J., Liu, L., Ren, Y., Chen, F., 2019b. Characterization of exopolysaccharides produced by microalgae with antitumor activity on human colon cancer cells. Int. J. Biol. Macromol. 128, 761–767.

Zhong, R., Li, J.Q., Wu, S.W., He, X.M., Xuan, J.C., Long, H., Liu, H.Q., 2021. Transcriptome analysis reveals possible antitumor mechanism of *Chlorella* exopolysaccharide. Gene. 779, 145494.

Index

For Product Safety Concerns and Information please contact our EU
representative GPSR@taylorandfrancis.com
Taylor & Francis Verlag GmbH, Kaufingerstraße 24, 80331 München, Germany